Springer Undergraduate Mathematics Series

Editor-in-Chief
Endre Süli, Oxford, UK

Series Editors
Mark A. J. Chaplain, St. Andrews, UK
Angus Macintyre, Edinburgh, UK
Shahn Majid, London, UK
Nicole Snashall, Leicester, UK
Michael R. Tehranchi, Cambridge, UK

The Springer Undergraduate Mathematics Series (SUMS) is a series designed for undergraduates in mathematics and the sciences worldwide. From core foundational material to final year topics, SUMS books take a fresh and modern approach. Textual explanations are supported by a wealth of examples, problems and fully-worked solutions, with particular attention paid to universal areas of difficulty. These practical and concise texts are designed for a one- or two-semester course but the self-study approach makes them ideal for independent use.

Jörg Liesen • Volker Mehrmann

Linear Algebra

Second Edition

Jörg Liesen 🆔
Institute of Mathematics
Technical University of Berlin
Berlin, Germany

Volker Mehrmann 🆔
Institute of Mathematics
Technical University of Berlin
Berlin, Germany

ISSN 1615-2085　　　　　　ISSN 2197-4144　(electronic)
Springer Undergraduate Mathematics Series
ISBN 978-3-031-93259-5　　　ISBN 978-3-031-93260-1　(eBook)
https://doi.org/10.1007/978-3-031-93260-1

Mathematics Subject Classification: 15-01

First edition book titled "Linear Algebra" published by Springer in 2015

© The Editor(s) (if applicable) and The Author(s), under exclusive license to Springer Nature Switzerland AG 2015, 2025

This work is subject to copyright. All rights are solely and exclusively licensed by the Publisher, whether the whole or part of the material is concerned, specifically the rights of reprinting, reuse of illustrations, recitation, broadcasting, reproduction on microfilms or in any other physical way, and transmission or information storage and retrieval, electronic adaptation, computer software, or by similar or dissimilar methodology now known or hereafter developed.
The use of general descriptive names, registered names, trademarks, service marks, etc. in this publication does not imply, even in the absence of a specific statement, that such names are exempt from the relevant protective laws and regulations and therefore free for general use.
The publisher, the authors and the editors are safe to assume that the advice and information in this book are believed to be true and accurate at the date of publication. Neither the publisher nor the authors or the editors give a warranty, expressed or implied, with respect to the material contained herein or for any errors or omissions that may have been made. The publisher remains neutral with regard to jurisdictional claims in published maps and institutional affiliations.

This Springer imprint is published by the registered company Springer Nature Switzerland AG
The registered company address is: Gewerbestrasse 11, 6330 Cham, Switzerland

If disposing of this product, please recycle the paper.

Preface to the Second Edition

Compared with the first edition, which was published 10 years ago, the content has been significantly extended. Notable additions include the discussion of bases in infinite dimensional vector spaces in Sect. 9.2, a more detailed description of projections in Sect. 12.2, and the derivation of the Frobenius canonical form of endomorphisms in Sect. 16.4. Mathematical terms and concepts that are new in comparison with the first edition include bidual spaces, Boolean rings, derogatory matrices, the Euclidean algorithm, Fibonacci numbers, the Fréchet-Riesz isomorphism, greatest common divisors and least common multiples of polynomials, invariant factors, isometries, mathematical induction, and maximal vectors. We have also added about 70 new exercises throughout the book, 16 further historical personalities in the Index, a table with the matrix decompositions derived in this book (Appendix B), and a table with the Greek alphabet (Appendix C).

We have kept the established presentation of the book with its 20 chapters. Due to the many additions, the material has become so voluminous that it is hard to cover it completely in a two semester course. From a didactical point of view a detailed treatment of the most important topics is definitely preferable to a superficial treatment of all topics. The following topics may be skipped without losing the thread or disturbing the logical structure of the book: The eigenvalues of stochastic matrices (Sect. 8.3), the existence of bases in general K-vector spaces (Zorn's Lemma and Theorem 9.26), the vector product in $\mathbb{R}^{3,1}$ (Sect. 12.3), the proof of the Fundamental Theorem of Algebra (Theorem 15.20 and the associated lemmas), minimal polynomials of vectors and the Frobenius canonical form (Lemma 16.20 to Theorem 16.22), and the computation of the Jordan canonical form (Sect. 16.4). Furthermore, we consider the topics of matrix functions and systems of linear differential equations (Chap. 17), the singular value decomposition (Chap. 19) as well as the Kronecker product and linear matrix equations (Chap. 20) as extensions of the classical content of a Linear Algebra course. In a two semester course we usually cover only a selection of these topics.

Many people helped us with the fourth German edition, on which this book is based, or with proofreading the current text. In particular, we would like to thank

Marine Froideveaux, Riccardo Morandin, Justus Ramme, and Jan Zur. We thank Klaus Bongartz for helpful comments on our Lemma 16.21. Thanks also to the staff of Springer-Verlag, Heidelberg, for their support and assistance with editorial aspects of this book.

Berlin, Germany
March 2025

Jörg Liesen
Volker Mehrmann

Preface to the First Edition

This is a translation of the (slightly revised) second German edition of our book "Lineare Algebra", published by Springer Spektrum in 2015. Our general view of the field of Linear Algebra and the approach to it that we have chosen in this book were already described in our Preface to the first German edition, published by Vieweg+Teubner in 2012. In a nutshell, our exposition is matrix-oriented, and we aim at presenting a rather complete theory (including all details and proofs), while keeping an eye on the applicability of the results. Many of them, though appearing very theoretical at first sight, are of an immediate practical relevance. In our experience, the matrix-oriented approach to Linear Algebra leads to a better intuition and a deeper understanding of the abstract concepts, and therefore simplifies their use in real world applications.

Starting from basic mathematical concepts and algebraic structures we develop the classical theory of matrices, vectors spaces and linear maps, culminating in the proof of the Jordan canonical form. In addition to the characterization of important special classes of matrices or endomorphisms, the last chapters of the book are devoted to special topics: Matrix functions and systems of differential equations, the singular value decomposition, the Kronecker product, and linear matrix equations. These chapters can be used as starting points of more advanced courses or seminars in Applied Linear Algebra.

Many people helped us with the first two German editions and this English edition of the book. In addition to those mentioned in the Preface to the first German edition, we would like to particularly thank Olivier Sète, who carefully worked through the entire draft of the second edition and gave numerous comments, as well as Leonhard Batzke, Carl De Boor, Sadegh Jokar, Robert Luce, Christian Mehl, Helia Niroomand Rad, Jan Peter Schäfermeier, Daniel Wachsmuth and Gisbert Wüstholz. Thanks also to the staff of Springer Spektrum, Heidelberg, and Springer-Verlag, London, for their support and assistance with editorial aspects of this English edition.

Berlin, Germany
July 2015

Jörg Liesen
Volker Mehrmann

Preface to the First German Edition

Mathematics is the instrument that links theory and practice, thinking and observing; it establishes the connecting bridge and builds it stronger and stronger. This is why our entire culture these days, as long as it is concerned with understanding and harnessing nature, has Mathematics as its foundation.[1]

This assessment of the famous mathematician David Hilbert (1862–1943) is even more true today. Mathematics is found not only throughout the classical natural sciences, Biology, Chemistry and Physics, its methods have become indispensable in Engineering, Economics, Medicine and many other areas of life. This continuing mathematization of the world is possible because of the *transversal strength* of Mathematics. The abstract objects and operations developed in Mathematics can be used for the description and solution of problems in numerous different situations.

While the high level of abstraction of modern Mathematics continuously increases its potential for applications, it represents a challenge for students. This is particularly true in the first years, when they have to become familiar with a lot of new and complicated terminology. In order to get students excited about mathematics and capture their imagination, it is important for us teachers of basic courses such as Linear Algebra to present Mathematics as a living science in its global context. The short historical notes in the text and the list of some historical papers at the end of this book show that Linear Algebra is the result of a human endeavor.

An important guideline of the book is to demonstrate the *immediate practical relevance* of the developed theory. Right in the beginning we illustrate several concepts of Linear Algebra in every day life situations. We discuss mathematical basics of the search engine Google and of the premium rate calculations of car insurances. These and other applications will be investigated in later chapters using theoretical results. Here the goal is not to study the concrete examples or their

---

[1] "Das Instrument, welches die Vermittlung bewirkt zwischen Theorie und Praxis, zwischen Denken und Beobachten, ist die Mathematik; sie baut die verbindende Brücke und gestaltet sie immer tragfähiger. Daher kommt es, dass unsere ganze gegenwärtige Kultur, soweit sie auf der geistigen Durchdringung und Dienstbarmachung der Natur beruht, ihre Grundlage in der Mathematik findet."

solutions, but the presentation of the transversal strength of mathematical methods in the Linear Algebra context.

The central object for our approach to Linear Algebra is the *matrix*, which we introduce early on, immediately after discussing some of the basic mathematical foundations. Several chapters deal with some of their most important properties, before we finally make the big step to abstract vector spaces and homomorphisms. In our experience the matrix-oriented approach to Linear Algebra leads to a better intuition and a deeper understanding of the abstract concepts.

The same goal should be reached by the MATLAB-Minutes[2] that are scattered throughout the text and that allow readers to comprehend the concepts and results via computer experiments. The required basics for these short exercises are introduced in the Appendix. Besides the MATLAB-Minutes there is a large number of classical exercises, which just require a pencil and paper.

Another advantage of the matrix-oriented approach to Linear Algebra is given by the simplifications when transferring theoretical results into practical algorithms. Matrices show up wherever data are systematically ordered and processed, which happens in almost all future job areas of bachelor students in the mathematical sciences. This has also motivated the topics in the last chapters of this book: matrix functions, the singular value decomposition, and the Kronecker product.

Despite many comments on algorithmic and numerical aspects, the focus in this book is on the theory of Linear Algebra. The German physicist Gustav Robert Kirchhoff (1824–1887) is attributed to have said:

A good theory is the most practical thing there is.[3]

This is exactly how we view our approach to the field.

This book is based on our lectures at TU Chemnitz and TU Berlin. We would like to thank all students, co-workers and colleagues that helped in preparing and proof-reading the manuscript, in the formulation of exercises and with the content of lectures. Our special thanks goes to André Gaul, Florian Goßler, Daniel Kreßner, Robert Luce, Christian Mehl, Matthias Pester, Robert Polzin, Timo Reis, Olivier Sète, Tatjana Stykel, Elif Topcu, Wolfgang Wülling, and Andreas Zeiser.

We also thank the staff of the Vieweg+Teubner Verlag and, in particular, Ulrike Schmickler-Hirzebruch, who strongly supported this endeavor.

Berlin, Germany Jörg Liesen
July 2011 Volker Mehrmann

---

[2] MATLAB® is a registered trademark of The MathWorks, Inc.
[3] "Eine gute Theorie ist das Praktischste, was es gibt."

Declarations

Competing Interests The authors have no competing interests to declare that are relevant to the content of this manuscript.

Contents

1	**Linear Algebra in Every Day Life**	1
	1.1 The PageRank Algorithm	1
	1.2 No Claim Discounting in Car Insurances	3
	1.3 Production Planning in a Plant	5
	1.4 Predicting Future Profits	6
	1.5 Circuit Simulation	7
2	**Basic Mathematical Concepts**	9
	2.1 Sets and Mathematical Logic	9
	2.2 Maps	16
	2.3 Relations	20
	2.4 Mathematical Induction	23
	Exercises	24
3	**Algebraic Structures**	27
	3.1 Groups	27
	3.2 Rings	30
	3.3 Fields	35
	Exercises	38
4	**Matrices**	43
	4.1 Basic Definitions and Operations	43
	4.2 Matrix Groups and Rings	51
	Exercises	57
5	**The Echelon Form and the Rank of Matrices**	61
	5.1 Elementary Matrices	61
	5.2 The Echelon Form and Gaussian Elimination	63
	5.3 Rank and Equivalence of Matrices	73
	Exercises	79
6	**Linear Systems of Equations**	81
	Exercises	87
7	**Determinants of Matrices**	89
	7.1 Definition of the Determinant	89
	7.2 Properties of the Determinant	93

	7.3	Minors and the Laplace Expansion	100
		Exercises	107
8	**The Characteristic Polynomial and Eigenvalues of Matrices**		111
	8.1	The Characteristic Polynomial and the Cayley-Hamilton Theorem	111
	8.2	Eigenvalues and Eigenvectors	116
	8.3	Eigenvectors of Stochastic Matrices	120
		Exercises	123
9	**Vector Spaces**		125
	9.1	Basic Definitions and Properties	125
	9.2	Bases and Dimension of Vector Spaces	128
	9.3	Coordinates and Changes of the Basis	137
	9.4	Relations Between Vector Spaces and Their Dimensions	142
		Exercises	146
10	**Linear Maps**		149
	10.1	Basic Definitions and Properties	149
	10.2	Linear Maps and Matrices	159
		Exercises	168
11	**Linear Forms and Bilinear Forms**		173
	11.1	Linear Forms and Dual Spaces	173
	11.2	Bilinear Forms	180
	11.3	Sesquilinear Forms	185
		Exercises	188
12	**Euclidean and Unitary Vector Spaces**		193
	12.1	Scalar Products and Norms	193
	12.2	Orthogonality	199
	12.3	The Vector Product in $\mathbb{R}^{3,1}$	213
		Exercises	215
13	**Adjoints of Linear Maps**		221
	13.1	Adjoints in Finite Dimensional K-vector Spaces	221
	13.2	Adjoints in Finite Dimensional Euclidean and Unitary Vector Spaces	224
		Exercises	232
14	**Eigenvalues of Endomorphisms**		235
	14.1	Basic Definitions and Properties	235
	14.2	Diagonalization	240
	14.3	Triangulation and Schur's Theorem	245
		Exercises	249
15	**Polynomials and the Fundamental Theorem of Algebra**		253
	15.1	Polynomials	253

	15.2 The Fundamental Theorem of Algebra	260
	Exercises	267
16	**The Jordan and the Frobenius Canonical Form**	**271**
	16.1 Cyclic f-invariant Subspaces and Duality	271
	16.2 The Jordan Canonical Form	278
	16.3 The Minimal Polynomial and the Frobenius Canonical Form	285
	16.4 Computation of the Jordan Canonical Form	295
	Exercises	301
17	**Matrix Functions and Systems of Differential Equations**	**305**
	17.1 Matrix Functions and the Matrix Exponential Function	305
	17.2 Systems of Linear Ordinary Differential Equations	315
	Exercises	323
18	**Special Classes of Endomorphisms**	**325**
	18.1 Normal Endomorphisms	325
	18.2 Orthogonal and Unitary Endomorphisms	331
	18.3 Selfadjoint Endomorphisms	336
	Exercises	347
19	**The Singular Value Decomposition**	**351**
	Exercises	359
20	**The Kronecker Product and Linear Matrix Equations**	**361**
	Exercises	369
A	**A Short Introduction to MATLAB**	**371**
B	**Matrix Decompositions**	**375**
C	**The Greek Alphabet**	**377**
Selected Historical Works on Linear Algebra		**379**
References		**381**
Index		**383**

Linear Algebra in Every Day Life

> *One has to familiarize the student with actual questions from applications, so that he learns to deal with real world problems. (Man muss den Lernenden mit konkreten Fragestellungen aus den Anwendungen vertraut machen, dass er lernt, konkrete Fragen zu behandeln.)*
>
> Lothar Collatz (1910–1990)

In this chapter we present some examples from everyday life in which Linear Algebra is used for the mathematical modeling and solution of problems. These examples include determining the importance of documents using the PageRank algorithm and no claims classes in a car insurance policy, production planning in a manufacturing company, predicting profits or losses using linear regression, and circuit simulation.

1.1 The PageRank Algorithm

The *PageRank algorithm* is a method to assess the "importance" of documents with mutual *links*, such as web pages, on the basis of the link structure. It was developed by Sergei Brin and Larry Page, the founders of Google Inc., at Stanford University in the late 1990s. The basic idea of the algorithm is the following:

Instead of counting links, PageRank essentially interprets a link of page A to page B as a vote of page A for page B. PageRank then assesses the importance of a page by the number of received votes. PageRank also considers the importance of the page that casts the vote, since votes of some pages have a higher value, and thus also assign a higher value to the page they point to. Important pages will be rated higher and thus lead to a higher position in the search results.[1]

Let us describe (model) this idea mathematically. Our presentation uses ideas from the article [1]. For a given set of web pages, every page k will be assigned an importance value $x_k \geq 0$. A page k is more important than a page j if $x_k > x_j$. If a page k has a link to a page j, we say that page j has a *backlink* from page k. In the above description these backlinks are the votes. As an example, consider the following link structure:

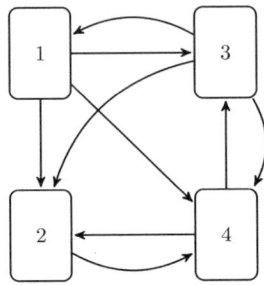

Here the page 1 has links to the pages 2, 3 and 4, and a backlink from page 3.

The easiest approach to define importance of web pages is to count its backlinks; the more votes are cast for a page, the more important the page is. In our example this gives the importance values

$$x_1 = 1, \quad x_2 = 3, \quad x_3 = 2, \quad x_4 = 3.$$

The pages 2 and 4 are thus the most important pages, and they are equally important.

However, the intuition and also the above description from Google suggests that backlinks from important pages are more important for the value of a page than those from less important pages. This idea can be modeled by defining x_k as the sum of all importance values of the backlinks of the page k. In our example this results in four equations that have to be satisfied simultaneously,

$$x_1 = x_3, \quad x_2 = x_1 + x_3 + x_4, \quad x_3 = x_1 + x_4, \quad x_4 = x_1 + x_2 + x_3.$$

A disadvantage of this approach is that it does not consider the number of links of the pages. Thus, it would be possible to (significantly) increase the importance of a page just by adding links to that page. In order to avoid this, the importance values of the backlinks in the PageRank algorithm are divided by the number of links of the corresponding page. This creates a kind of "internet democracy": Every page

---

[1] From a text found in 2010 on www.google.de/corporate/tech.html.

1.2 No Claim Discounting in Car Insurances

can vote for other pages, where in total it can cast one vote. In our example this gives the equations

$$x_1 = \frac{x_3}{3}, \quad x_2 = \frac{x_1}{3} + \frac{x_3}{3} + \frac{x_4}{2}, \quad x_3 = \frac{x_1}{3} + \frac{x_4}{2}, \quad x_4 = \frac{x_1}{3} + x_2 + \frac{x_3}{3}. \quad (1.1)$$

These are four equations for the four unknowns, and all equations are *linear*,[2] i.e., the unknowns occur only in first power. In Chap. 6 we will see how to write the equations in (1.1) in form of a *linear system of equations*. Analyzing and solving such systems is one of the most important tasks of Linear Algebra. The example of the PageRank algorithm shows that Linear Algebra presents a powerful modeling tool: We have turned the real world problem of assessing the importance of web pages into a problem of Linear Algebra. This problem will be examined further in Sect. 8.3.

For completeness, we mention that a solution for the four unknowns (computed with MATLAB and rounded to the second significant digit) is given by

$$x_1 = 0.14, \quad x_2 = 0.54, \quad x_3 = 0.41, \quad x_4 = 0.72.$$

Thus, page 4 is the most important one. It is possible to multiply the solution, i.e., the importance values x_k, by a positive constant. Such a multiplication or scaling is often advantageous for computational methods or for the visual display of the results. For example, the scaling could be used to give the most important page the value 1.00. A scaling is allowed, since it does not change the ranking of the pages, which is the essential information provided by the PageRank algorithm.

1.2 No Claim Discounting in Car Insurances

Insurance companies compute the premiums for their customers on the basis of the insured risk: the higher the risk, the higher the premium. It is therefore important to identify the factors that lead to higher risk. In the case of a car insurance these factors include the number of miles driven per year, the distance between home and work, the marital status, the engine power, or the age of the driver. Using such information, the company calculates the initial premium.

Usually the best indicator for future accidents, and hence future insurance claims, is the number of accidents of the individual customer in the past, i.e., the claims history. In order to incorporate this information into the premium rates, insurers establish a system of *risk classes*, which divide the customers into homogeneous risk groups with respect to their previous claims history. Customers with fewer accidents

---

[2] The term "linear" originates from the Latin word "linea", which means "(straight) line", and "linearis" means "consisting of (straight) lines".

in the past get a discount on their premium. This approach is called a *no claims discounting scheme*.

For a mathematical model of this scheme we need a set of risk classes and a *transition rule* for moving between the classes. At the end of a policy year, the customer may move to a different class depending on the claims made during the year. The discount is given in percent of the premium in the initial class. As a simple example we consider four risk classes,

	C_1	C_2	C_3	C_4
% discount	0	10	20	40

and the following transition rules:

- No accident: Step up one class (or stay in C_4).
- One accident: Step back one class (or stay in C_1).
- More than one accident: Step back to class C_1 (or stay in C_1).

Next, the insurance company has to estimate the probability that a customer who is in the class C_i in this year will move to the class C_j. This probability is denoted by p_{ij}. Let us assume, for simplicity, that the probability of exactly one accident for every customer is 0.1, i.e., 10%, and the probability of two or more accidents for every customer is 0.05, i.e., 5%. (Of course, in practice the insurance companies determine these probabilities in dependence of the classes.)

For example, a customer in the class C_1 will stay in C_1 in case of at least one accident. This happens with the probability 0.15, so that $p_{11} = 0.15$. A customer in C_1 has no accident with the probability 0.85, so that $p_{12} = 0.85$. There is no chance to move from C_1 to C_3 or C_4 in the next year, so that $p_{13} = p_{14} = 0.00$. In this way we obtain 16 values p_{ij}, $i, j = 1, 2, 3, 4$, which we can arrange in a (4×4)-*matrix* as follows:

$$\begin{bmatrix} p_{11} & p_{12} & p_{13} & p_{14} \\ p_{21} & p_{22} & p_{23} & p_{24} \\ p_{31} & p_{32} & p_{33} & p_{34} \\ p_{41} & p_{42} & p_{43} & p_{44} \end{bmatrix} = \begin{bmatrix} 0.15 & 0.85 & 0.00 & 0.00 \\ 0.15 & 0.00 & 0.85 & 0.00 \\ 0.05 & 0.10 & 0.00 & 0.85 \\ 0.05 & 0.00 & 0.10 & 0.85 \end{bmatrix}. \quad (1.2)$$

All entries of this matrix are nonnegative real numbers, and the sum of all entries in each row is equal to 1.00, i.e.,

$$p_{i1} + p_{i2} + p_{i3} + p_{i4} = 1.00 \quad \text{for each } i = 1, 2, 3, 4.$$

Such a matrix is called *row-stochastic*.

The analysis of matrix properties is a central topic of Linear Algebra that is developed throughout this book. As in the example with the PageRank algorithm, we have translated a practical problem into the language of Linear Algebra, and we can now study it using Linear Algebra techniques. This example of premium rates will be discussed further in Example 4.7.

1.3 Production Planning in a Plant

The production planning in a plant has to consider many different factors, in particular commodity prices, labor costs, and available capital, in order to determine a production plan. We consider a simple example:

A company produces the products P_1 and P_2. If x_i units of the product P_i are produced, where $i = 1, 2$, then the pair (x_1, x_2) is called a *production plan*. Suppose that the raw materials and labor for the production of one unit of the product P_i cost a_{1i} and a_{2i} Euros, respectively. If b_1 Euros are available for the purchase of raw materials and b_2 Euros for the payment of labor costs, then a production plan must satisfy the *constraint inequalities*

$$a_{11}x_1 + a_{12}x_2 \leq b_1 \quad \text{and} \quad a_{21}x_1 + a_{22}x_2 \leq b_2.$$

If a production plan satisfies these constraints, it is called *feasible*. Let p_i be the profit from selling one unit of product P_i. Then the goal is to determine a production plan that maximizes the *profit function*

$$\Phi(x_1, x_2) = p_1 x_1 + p_2 x_2.$$

How can we find this maximum?

The two linear equations

$$a_{11}x_1 + a_{12}x_2 = b_1 \quad \text{and} \quad a_{21}x_1 + a_{22}x_2 = b_2$$

describe straight lines in the coordinate system that has the variables x_1 and x_2 on its axes. These two lines form boundary lines of the feasible production plans, which are "below" the lines; see the figure below. Note that we also must have $x_i \geq 0$, since we cannot produce negative units of a product. For planned profits y_i, $i = 1, 2, 3, \ldots$, the equations $p_1 x_1 + p_2 x_2 = y_i$ describe parallel straight lines in the coordinate system; see the dashed lines in the figure. If x_1 and x_2 satisfy $p_1 x_1 + p_2 x_2 = y_i$, then $\Phi(x_1, x_2) = y_i$. The profit maximization problem can now be solved by moving the dashed lines until one of them reaches the corner with the maximal y:

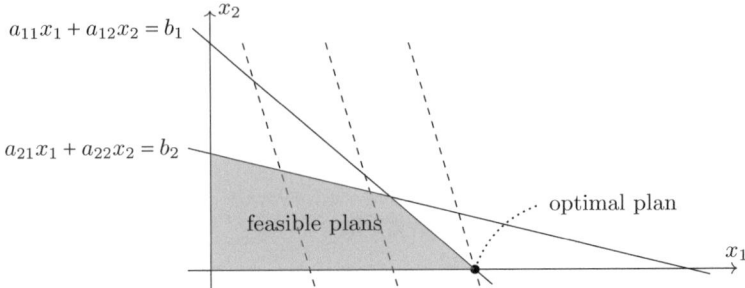

In case of more variables we cannot draw such a simple figure and obtain the solution "graphically". But the general idea of finding a corner with the maximum profit is still the same. This is an example of a *linear optimization problem*. As before, we have formulated a real world problem in the language of Linear Algebra, and we can use mathematical methods for its solution.

1.4 Predicting Future Profits

The prediction of profits or losses of a company is a central planning instrument of economics. Analogous problems arise in many areas of political decision making, for example in budget planning, tax estimates or the planning of new infrastructures. We consider a specific example:

In the four quarters of a year a company has profits of 10, 8, 9, 11 million Euros. The board now wants to predict the future profits development on the basis of these values. Evidence suggests, that the profits behave *linearly*. If this was true, then the profits would form a straight line $y(t) = \alpha t + \beta$ that connects the points (1, 10), (2, 8), (3, 9), (4, 11) in the coordinate system having "time" and "profit" as its axes. This, however, does neither hold in this example nor in practice. Therefore one tries to find a straight line that deviates "as little as possible" from the given points. One possible approach is to choose the parameters α and β in order to minimize the sum of the squared distances between the given points and the straight line. Once the parameters α and β have been determined, the resulting line $y(t)$ can be used for estimating or predicting the future profits, as illustrated in the following figure:

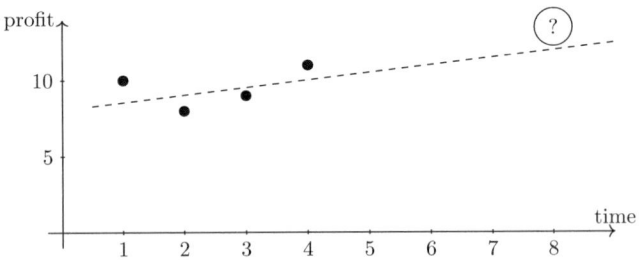

1.5 Circuit Simulation

The determination of the parameters α and β that minimize a sum of squares is called a *least squares problem*. We will solve least squares problems using methods of Linear Algebra in Example 12.17. The approach itself is sometimes called a *parameter identification*. In Statistics, the modeling of given data (here the company profits) using a linear predictor function (here $y(t) = \alpha t + \beta$) is known as *linear regression*.

1.5 Circuit Simulation

The current development of electronic devices is very rapid. In short intervals, nowadays often less than a year, new models of laptops or mobile phones have to be issued to the market. To achieve this, continuously new generations of computer chips have to be developed. These typically become smaller and more powerful, and naturally should use as little energy as possible. An important factor in this development is to plan and simulate the chips *virtually*, i.e., in the computer and without producing a physical prototype. This model-based planning and optimization of products is a central method in many high technology areas, and it is based on modern mathematics.

Usually, the switching behavior of a chip is modeled by a mathematical system consisting of differential and algebraic equations that describe the relation between currents and voltages. Without going into details, consider the following circuit:

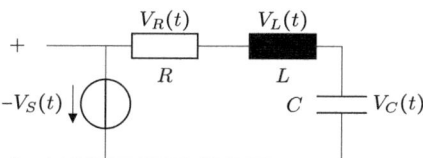

In this circuit description, $V_S(t)$ is the given input current at time t, and the characteristic values of the components are R for the resistor, L for the inductor, and C for the capacitor. The functions for the potential differences at the three components are denoted by $V_R(t)$, $V_L(t)$, and $V_C(t)$; $I(t)$ is the current.

Applying the Kirchhoff laws[3] of electrical engineering leads to the following system of linear equations and differential equations that model the dynamic behavior of the circuit:

$$L \frac{d}{dt} I = V_L,$$

$$C \frac{d}{dt} V_C = I,$$

---

[3] Gustav Robert Kirchhoff (1824–1887).

$$RI = V_R,$$
$$V_L + V_C + V_R = V_S.$$

In this example it is easy to solve the last two equations for V_L and V_R, and hence to obtain a system of differential equations

$$\frac{d}{dt}I = -\frac{R}{L}I - \frac{1}{L}V_C + \frac{1}{L}V_S,$$
$$\frac{d}{dt}V_C = \frac{1}{C}I,$$

for the functions I und V_C. We will discuss and solve this system in Example 17.15.

This simple example demonstrates that for the simulation of a circuit a system of linear differential equations and algebraic equations has to be solved. Modern computer chips in industrial practice require solving such systems with millions of *differential-algebraic equations*. Linear Algebra is one of the central tools for the theoretical analysis of such systems as well as the development of efficient solution methods.

Basic Mathematical Concepts 2

In this chapter we introduce the mathematical concepts that form the basis for the developments in the following chapters. We begin with sets and basic mathematical logic. Then we consider maps between sets and their most important properties. Finally we discuss relations and in particular equivalence relations on a set.

2.1 Sets and Mathematical Logic

We begin our development with the concept of a set and use the following definition of Cantor.[1]

Definition 2.1 A *set* is a collection M of *well determined* and *distinguishable* objects x of our perception or our thinking. The objects are called the *elements* of M.

The objects x in this definition are well determined, and therefore we can uniquely decide whether x belongs to a set M or not. We write $x \in M$ if x is an element of the set M, otherwise we write $x \notin M$. Furthermore, the elements are distinguishable, which means that all elements of M are (pairwise) distinct.

If two objects x and y are equal, then we write $x = y$, otherwise $x \neq y$. For mathematical objects we usually have to give a formal definition of *equality*. As an example consider the equality of sets; see Definition 2.2 below.

---

[1] Georg Cantor (1845–1918), one of the founders of set theory. Cantor published this definition in the journal "Mathematische Annalen" in 1895.

© The Author(s), under exclusive license to Springer Nature Switzerland AG 2025
J. Liesen, V. Mehrmann, *Linear Algebra*, Springer Undergraduate Mathematics Series, https://doi.org/10.1007/978-3-031-93260-1_2

We describe sets with curly brackets { } that contain either a list of the elements, for example

$$\{\text{red, yellow, green}\}, \quad \{1, 2, 3, 4\}, \quad \{2, 4, 6, \ldots\},$$

or a defining property, for example

$$\{x \mid x \text{ is a positive even number}\},$$
$$\{x \mid x \text{ is a person owning a bike}\}.$$

Some of the well known sets of numbers are denoted as follows:

$\mathbb{N} = \{1, 2, 3, \ldots\}$	the natural numbers
$\mathbb{N}_0 = \{0, 1, 2, \ldots\}$	the natural numbers including zero
$\mathbb{Z} = \{\ldots, -2, -1, 0, 1, 2, \ldots\}$	the integers
$\mathbb{Q} = \{x \mid x = a/b \text{ with } a \in \mathbb{Z} \text{ and } b \in \mathbb{N}\}$	the rational numbers
$\mathbb{R} = \{x \mid x \text{ is a real number}\}$	the real numbers

The construction and characterization of the real numbers $\mathbb{R}$ is usually done in an introductory course in Real Analysis.

To describe a set via its defining property we formally write $\{x \mid P(x)\}$. Here P is a *predicate* which may hold for an object x or not, and $P(x)$ is the *assertion* "P holds for x".

In general, an assertion is a statement that can be classified as either "true" or "false". For instance the statement "The set $\mathbb{N}$ has infinitely many elements" is true. The sentence "Tomorrow the weather will be good" is not an assertion, since the meaning of the term "good weather" is unclear and the weather prediction in general is uncertain. Instead of "assertion A is true" we also say "assertion A holds".

The *negation* of an assertion A is the assertion "not A", which we denote by $\neg A$. This assertion is true if and only if A is false, and false if and only if A is true. For instance, the negation of the true assertion "The set $\mathbb{N}$ has infinitely many elements" is given by "The set $\mathbb{N}$ does not have infinitely many elements" (or "The set $\mathbb{N}$ has finitely many elements"), which is false.

Two assertions A and B can be combined via logical compositions to a new assertion. The following is a list of the most common logical compositions, together with their mathematical short hand notation:

2.1 Sets and Mathematical Logic

Composition	Notation	Wording
Conjunction	$\wedge$	A and B
Disjunction	$\vee$	A or B
Implication	$\Rightarrow$	A implies B
		If A then B
		A is a sufficient condition for B
		B is a necessary condition for A
Equivalence	$\Leftrightarrow$	A and B are equivalent
		A is true if and only if B is true
		A is necessary and sufficient for B
		B is necessary and sufficient for A

For example, we can write the assertion "x is a real number and x is negative" as

$$x \in \mathbb{R} \wedge x < 0.$$

Whether an assertion that is composed of two assertions A and B is true or false, depends on the logical values of A and B. We have the following *table of logical values* ("t" and "f" denote true and false, respectively):

A	B	$A \wedge B$	$A \vee B$	$A \Rightarrow B$	$A \Leftrightarrow B$
t	t	t	t	t	t
t	f	f	t	f	f
f	t	f	t	t	f
f	f	f	f	t	t

For example, the assertion $A \wedge B$ is true only when A and B are both true. The assertion $A \Rightarrow B$ is false only when A is true and B is false. In particular, if A is false, then $A \Rightarrow B$ is true, regardless of the logical value of B. At first glance, this convention seems strange, since it can lead to meaningless statements that are formally true as for instance:

"If the Earth is a disk, then the Sun rotates around Mars."

A typical task in mathematics is to prove statements. When we have to prove that a certain implication $A \Rightarrow B$ is true, then it is sufficient to show that the truth of A implies the truth of B. In this context, the meaning of $A \Rightarrow B$ is that the truth of this statement implies that if A holds, then also B holds, and that "nothing special" follows from the falseness of A. If A is false, then we also say that the implication $A \Rightarrow B$ is *trivially true*, or *holds trivially*.

For example, consider the statement

$$\text{``For every } x \in \mathbb{Z} : |x| \geq 2 \Rightarrow x^2 \geq 4.\text{''}$$

The implication "$|x| \geq 2 \Rightarrow x^2 \geq 4$" is trivially true if the first part is false, i.e., if $|x| < 2$. In order to prove that the statment holds for every $x \in \mathbb{Z}$, it is therefore sufficient to show that $|x| \geq 2$ implies $x^2 \geq 4$, and this is certainly true.

Using truth tables the following equivalence can be shown:

$$(A \Rightarrow B) \quad \Leftrightarrow \quad (\neg B \Rightarrow \neg A).$$

(As an exercise create the table of logical values for $\neg B \Rightarrow \neg A$ and compare it with the table for $A \Rightarrow B$.) The truth of $A \Rightarrow B$ can therefore be proved by showing that the truth of $\neg B$ implies the truth of $\neg A$, i.e., that "B is false" implies "A is false". The assertion $\neg B \Rightarrow \neg A$ is called the *contraposition* of the assertion $A \Rightarrow B$ and the conclusion from $A \Rightarrow B$ to $\neg B \Rightarrow \neg A$ is called proof by contraposition.

Together with assertions we also often use so-called *quantifiers*:

Quantifier	Notation	Wording
Universal	$\forall$	For all
Existential	$\exists$	There exists

Now we return to set theory and introduce subsets and the equality of sets.

Definition 2.2 Let M, N be sets.

(1) M is called a *subset* of N, denoted by $M \subseteq N$, if every element of M is also an element of N. We write $M \not\subseteq N$, if this does not hold.
(2) M and N are called *equal*, denoted by $M = N$, if $M \subseteq N$ and $N \subseteq M$. We write $M \neq N$ is this does not hold.
(3) M is called a *proper subset* of N, denoted by $M \subset N$, if both $M \subseteq N$ and $M \neq N$ hold.

Instead of $M \subseteq N$ or $M \subset N$, we can also write $N \supseteq M$, respectively $N \supset M$, and then we call N a *(proper) superset* of M.

Using the notation of mathematical logic we can write this definition as follows:

(1) $M \subseteq N \quad \Leftrightarrow \quad (\forall x : x \in M \Rightarrow x \in N)$.
(2) $M = N \quad \Leftrightarrow \quad (M \subseteq N \wedge N \subseteq M)$.
(3) $M \subset N \quad \Leftrightarrow \quad (M \subseteq N \wedge M \neq N)$.

2.1 Sets and Mathematical Logic

The assertion on the right side of the equivalence in (1) reads as follows: For all objects x the truth of $x \in M$ implies the truth of $x \in N$. Or shorter: For all x, if $x \in M$ holds, then $x \in N$ holds.

A very special set is the set with no elements, which we define formally as follows.

Definition 2.3 The set $\emptyset := \{x \mid x \neq x\}$ is called the *empty set*.

The notation " $:=$ " means *is defined as*. We have introduced the empty set by a defining property: Every object x with $x \neq x$ is any element of $\emptyset$. This cannot hold for any object, and hence $\emptyset$ does not contain any element. A set that contains at least one element is called *nonempty*.

Theorem 2.4 *For every set M the following assertions hold:*

(1) $\emptyset \subseteq M$.
(2) $M \subseteq \emptyset \Rightarrow M = \emptyset$.

Proof

(1) We have to show that the assertion "$\forall x : x \in \emptyset \Rightarrow x \in M$" is true. Since there is no $x \in \emptyset$, the assertion "$x \in \emptyset$" is false, and therefore "$x \in \emptyset \Rightarrow x \in M$" is true for every x (cp. the remarks on the implication $A \Rightarrow B$).
(2) Let $M \subseteq \emptyset$. From (1) we know that $\emptyset \subseteq M$ and hence $M = \emptyset$ follows by (2) in Definition 2.2. □

Theorem 2.5 *Let M, N, L be sets. Then the following assertions hold for the subset relation "$\subseteq$":*

(1) $M \subseteq M$ *(reflexivity)*.
(2) If $M \subseteq N$ and $N \subseteq L$, then $M \subseteq L$ (transitivity).

Proof

(1) We have to show that the assertion "$\forall x : x \in M \Rightarrow x \in M$" is true. If "$x \in M$" is true, then "$x \in M \Rightarrow x \in M$" is an implication with two true assertions, and hence it is true.
(2) We have to show that the assertion "$\forall x : x \in M \Rightarrow x \in L$" is true. If "$x \in M$" is true, then also "$x \in N$" is true, since $M \subseteq N$. The truth of "$x \in N$" implies that "$x \in L$" is true, since $N \subseteq L$. Hence the assertion "$x \in M \Rightarrow x \in L$" is true. □

Definition 2.6 Let M, N be sets.

(1) The *union*[2] of M and N is $M \cup N := \{x \mid x \in M \vee x \in N\}$.
(2) The *intersection* of M and N is $M \cap N := \{x \mid x \in M \wedge x \in N\}$.
    If $M \cap N = \emptyset$, then the sets M and N are called *disjoint*.
(3) The *difference* of M and N is $M \setminus N := \{x \mid x \in M \wedge x \notin N\}$.

Example 2.7

(1) Let $-\mathbb{N} := \{-n \mid n \in \mathbb{N}\}$, then $\mathbb{N} \cup (-\mathbb{N}) = \mathbb{Z} \setminus \{0\}$ and $\mathbb{N} \cap (-\mathbb{N}) = \emptyset$.
(2) Important examples of sets are *intervals* of real numbers. For $a, b \in \mathbb{R}$ with $a \leq b$ we define:

$$[a, b] := \{x \mid x \in \mathbb{R} \text{ and } a \leq x \leq b\} \quad \text{(closed interval)},$$

$$]a, b] := \{x \mid x \in \mathbb{R} \text{ and } a < x \leq b\} \quad \text{(half open interval)},$$

$$[a, b[:= \{x \mid x \in \mathbb{R} \text{ and } a \leq x < b\} \quad \text{(half open interval)},$$

$$]a, b[:= \{x \mid x \in \mathbb{R} \text{ and } a < x < b\} \quad \text{(open interval)}.$$

For the intervals $M = [-2, 2]$ and $N =]-1, 3]$ we have

$$M \cup N = [-2, 3], \quad M \cap N =]-1, 2],$$
$$M \setminus N = [-2, -1], \quad N \setminus M =]2, 3].$$

The set operations union and intersection can be extended to more than two sets: If $I \neq \emptyset$ is a set and if for all $i \in I$ there is a set M_i, then

$$\bigcup_{i \in I} M_i := \{x \mid \exists i \in I \text{ with } x \in M_i\} \quad \text{and}$$

$$\bigcap_{i \in I} M_i := \{x \mid \forall i \in I \text{ we have } x \in M_i\}.$$

The set I is called an *index set*. For $I = \{1, 2, \ldots, n\} \subset \mathbb{N}$ we write the union and intersection of the sets $M_1, M_2, \ldots, M_n$ as

$$\bigcup_{i=1}^{n} M_i \quad \text{and} \quad \bigcap_{i=1}^{n} M_i.$$

---

[2] The notations $M \cup N$ and $M \cap N$ for union and intersection of sets M and N were introduced in 1888 by Giuseppe Peano (1858–1932), one of the founders of formal logic. The notation of the "smallest common multiple $\mathfrak{M}(M, N)$" and "largest common divisor $\mathfrak{D}(M, N)$" of the sets M and N suggested by Georg Cantor (1845–1918) did not catch on.

2.1 Sets and Mathematical Logic

Theorem 2.8 *Let $M \subseteq N$ for two sets M, N. Then the following are equivalent:*

(1) $M \subset N$.
(2) $N \setminus M \neq \emptyset$.

Proof We show that $(1) \Rightarrow (2)$ and $(2) \Rightarrow (1)$ hold.

$(1) \Rightarrow (2)$: Since $M \neq N$, there exists an $x \in N$ with $x \notin M$. Thus $x \in N \setminus M$, so that $N \setminus M \neq \emptyset$ holds.

$(2) \Rightarrow (1)$: There exists an $x \in N$ with $x \notin M$, and hence $N \neq M$. Since $M \subseteq N$ holds, we see that $M \subset N$ holds. □

Theorem 2.9 *Let M, N, L be sets. Then the following assertions hold:*

(1) $M \cap N \subseteq M$ and $M \subseteq M \cup N$.
(2) Commutativity: $M \cap N = N \cap M$ and $M \cup N = N \cup M$.
(3) Associativity: $M \cap (N \cap L) = (M \cap N) \cap L$ and $M \cup (N \cup L) = (M \cup N) \cup L$.
(4) Distributivity: $M \cup (N \cap L) = (M \cup N) \cap (M \cup L)$ and $M \cap (N \cup L) = (M \cap N) \cup (M \cap L)$.
(5) $M \setminus N \subseteq M$.
(6) $M \setminus (N \cap L) = (M \setminus N) \cup (M \setminus L)$ and $M \setminus (N \cup L) = (M \setminus N) \cap (M \setminus L)$.

Proof Exercise. □

Definition 2.10 Let M be a set.

(1) The *cardinality* of M, denoted by $|M|$, is the number of elements of M.
(2) The *power set* of M, denoted by $\mathcal{P}(M)$, is the set of all subsets of M, i.e., $\mathcal{P}(M) := \{N \mid N \subseteq M\}$.

The empty set $\emptyset$ has cardinality zero and $\mathcal{P}(\emptyset) = \{\emptyset\}$, thus $|\mathcal{P}(\emptyset)| = 1$. For $M = \{1, 3\}$ the cardinality is $|M| = 2$ and

$$\mathcal{P}(M) = \{\emptyset, \{1\}, \{3\}, M\},$$

and hence $|\mathcal{P}(M)| = 4 = 2^{|M|}$. One can show that for every set M with finitely many elements, i.e., finite cardinality, $|\mathcal{P}(M)| = 2^{|M|}$ holds.

2.2 Maps

In this section we discuss maps between sets.

Definition 2.11 Let X, Y be nonempty sets.

(1) A *map* f from X to Y is a rule that assigns to each $x \in X$ exactly one $y = f(x) \in Y$. We write this as

$$f : X \to Y, \qquad x \mapsto y = f(x).$$

Instead of $x \mapsto y = f(x)$ we also write $f(x) = y$. The sets X and Y are called *domain* and *codomain* of f.

(2) Two maps $f : X \to Y$ and $g : X \to Y$ are called *equal* when $f(x) = g(x)$ holds for all $x \in X$. We then write $f = g$.

In Definition 2.11 we have assumed that X and Y are nonempty, since otherwise there can be no rule that assigns an element of Y to each element of X. If one of these sets is empty, we can define an *empty map*. However, in the following we will always assume (but not always explicitly state) that the sets between which a given map acts are nonempty.

Example 2.12 Two maps from $X = \mathbb{R}$ to $Y = \mathbb{R}$ are given by

$$f : X \to Y, \quad f(x) = x^2, \tag{2.1}$$

$$g : X \to Y, \quad x \mapsto \begin{cases} 0, & x \leq 0, \\ 1, & x > 0. \end{cases} \tag{2.2}$$

To analyze the properties of maps we need some further terminology.

Definition 2.13 Let X, Y be nonempty sets.

(1) The map $\mathrm{Id}_X : X \to X$, $x \mapsto x$, is called the *identity on X*.
(2) Let $f : X \to Y$ be a map and let $M \subseteq X$ and $N \subseteq Y$. Then

$$f(M) := \{ f(x) \in Y \mid x \in M \} \subseteq Y \quad \text{is called the } \textit{image} \text{ of } M \text{ under } f,$$

$$f^{-1}(N) := \{ x \in X \mid f(x) \in N \} \quad \text{is called the } \textit{pre-image} \text{ of } N \text{ under } f.$$

(3) If $f : X \to Y$, $x \mapsto f(x)$ is a map and $\emptyset \neq M \subseteq X$, then $f|_M : M \to Y$, $x \mapsto f(x)$, is called the *restriction of f to M*.

2.2 Maps

One should note that in this definition $f^{-1}(N)$ is a set, and hence the symbol f^{-1} here does not mean the inverse map of f. (This map will be introduced below in Definition 2.22.)

Example 2.14 For the maps with domain $X = \mathbb{R}$ in (2.1) and (2.2) we have the following properties:

$$f(X) = \{x \in \mathbb{R} \mid x \geq 0\}, \quad f^{-1}(\mathbb{R}_-) = \{0\}, \quad f^{-1}(\{-1\}) = \emptyset,$$
$$g(X) = \{0, 1\}, \quad g^{-1}(\mathbb{R}_-) = g^{-1}(\{0\}) = \mathbb{R}_-,$$

where $\mathbb{R}_- := \{x \in \mathbb{R} \mid x \leq 0\}$.

Definition 2.15 Let X, Y be nonempty sets. A map $f : X \to Y$ is called

(1) *injective*, if for all $x_1, x_2 \in X$ the equality $f(x_1) = f(x_2)$ implies that $x_1 = x_2$,
(2) *surjective*, if $f(X) = Y$,
(3) *bijective*, if f is injective and surjective.

Since $x_1 \neq x_2 \Rightarrow f(x_1) \neq f(x_2)$ is the contraposition of the statement $f(x_1) = f(x_2) \Rightarrow x_1 = x_2$, we can also define injectivity as follows: For all $x_1, x_2 \in X$ it follows from $x_1 \neq x_2$ that $f(x_1) \neq f(x_2)$.

For every nonempty set X the simplest example of a bijective map from X to X is Id_X, the identity on X.

Example 2.16 Let $\mathbb{R}_+ := \{x \in \mathbb{R} \mid x \geq 0\}$, then
$f : \mathbb{R} \to \mathbb{R}$, $f(x) = x^2$, is neither injective nor surjective.
$f : \mathbb{R} \to \mathbb{R}_+$, $f(x) = x^2$, is surjective but not injective.
$f : \mathbb{R}_+ \to \mathbb{R}$, $f(x) = x^2$, is injective but not surjective.
$f : \mathbb{R}_+ \to \mathbb{R}_+$, $f(x) = x^2$, is bijective.
In these assertions we have used the continuity of the map $f(x) = x^2$ that is discussed in the basic courses on analysis. In particular, we have used the fact that continuous functions map real intervals to real intervals. The assertions also show why it is important to include the domain and codomain in the definition of a map.

Theorem 2.17 *A map $f : X \to Y$ is bijective if and only if for every $y \in Y$ there exists exactly one $x \in X$ with $f(x) = y$.*

Proof

$\Rightarrow$: Let f be bijective and let $y_1 \in Y$. Since f is surjective, there exists an $x_1 \in X$ with $f(x_1) = y_1$. If some $x_2 \in X$ also satisfies $f(x_2) = y_1$, then $x_1 = x_2$ follows from the injectivity of f. Therefore, there exists a unique $x_1 \in X$ with $f(x_1) = y_1$.

$\Leftarrow$: Since for all $y \in Y$ there exists a unique $x \in X$ with $f(x) = y$, it follows that $f(X) = Y$. Thus, f surjective. Let now $x_1, x_2 \in X$ with $f(x_1) = f(x_2) = y \in Y$. Then the assumption implies $x_1 = x_2$, so that f is also injective. □

One can show that between two sets X and Y of finite cardinality there exists a bijective map if and only if $|X| = |Y|$.

Lemma 2.18 *For sets X, Y with $|X| = |Y| = m \in \mathbb{N}$, there exist exactly $m! := 1 \cdot 2 \cdot \ldots \cdot m$ pairwise distinct bijective maps between X and Y.*

Proof Exercise. □

Definition 2.19 Let $f : X \to Y, x \mapsto f(x)$, and $g : Y \to Z, y \mapsto g(y)$ be maps. Then the *composition* of f and g is the map

$$g \circ f : X \to Z, \qquad x \mapsto g(f(x)).$$

The expression $g \circ f$ should be read "g after f", which stresses the order of the composition: First f is applied to x and then g to $f(x)$. We immediately see that $f \circ \mathrm{Id}_X = f = \mathrm{Id}_Y \circ f$ for every map $f : X \to Y$.

Theorem 2.20 *Let $f : W \to X, g : X \to Y, h : Y \to Z$ be maps. Then*

(1) $h \circ (g \circ f) = (h \circ g) \circ f$, i.e., the composition of maps is associative.
(2) If f and g are injective/surjective/bijective, then $g \circ f$ is injective/surjective/ bijective.
(3) If $g \circ f$ is injective, then f is injective.
(4) If $g \circ f$ is surjective, then g is surjective.

Proof Exercise. □

Theorem 2.21 *A map $f : X \to Y$ is bijective if and only if there exists a map $g : Y \to X$ with*

$$g \circ f = \mathrm{Id}_X \quad \text{and} \quad f \circ g = \mathrm{Id}_Y.$$

Proof

$\Rightarrow$: If f is bijective, then by Theorem 2.17 for every $y \in Y$ there exists an $x = x_y \in X$ with $f(x_y) = y$. We define the map g by

$$g : Y \to X, \quad g(y) = x_y.$$

2.2 Maps

Let $\tilde{y} \in Y$ be given, then

$$(f \circ g)(\tilde{y}) = f(g(\tilde{y})) = f(x_{\tilde{y}}) = \tilde{y}, \quad \text{hence} \quad f \circ g = \mathrm{Id}_Y.$$

If, on the other hand, $\tilde{x} \in X$ is given, then $\tilde{y} = f(\tilde{x}) \in Y$. By Theorem 2.17, there exists a unique $x_{\tilde{y}} \in X$ with $f(x_{\tilde{y}}) = \tilde{y}$ such that $\tilde{x} = x_{\tilde{y}}$. So with

$$(g \circ f)(\tilde{x}) = g(f(x_{\tilde{y}})) = g(\tilde{y}) = x_{\tilde{y}} = \tilde{x},$$

we have $g \circ f = \mathrm{Id}_X$.

$\Leftarrow$: By assumption $g \circ f = \mathrm{Id}_X$, thus $g \circ f$ is injective and thus also f is injective (see (3) in Theorem 2.20). Moreover, $f \circ g = \mathrm{Id}_Y$, thus $f \circ g$ is surjective and hence also f is surjective (see (4) in Theorem 2.20). Therefore, f is bijective. □

The map $g : Y \to X$ that was characterized in Theorem 2.21 is unique: If there were another map $h : Y \to X$ with $h \circ f = \mathrm{Id}_X$ and $f \circ h = \mathrm{Id}_Y$, then

$$h = \mathrm{Id}_X \circ h = (g \circ f) \circ h = g \circ (f \circ h) = g \circ \mathrm{Id}_Y = g.$$

This leads to the following definition.

Definition 2.22 If $f : X \to Y$ is a bijective map, then the unique map $g : Y \to X$ from Theorem 2.21 is called the *inverse* (or *inverse map*) of f. We denote the inverse of f by f^{-1}.

To show that a given map $g : Y \to X$ is the unique inverse of the bijective map $f : X \to Y$, it is sufficient to show one of the equations $g \circ f = \mathrm{Id}_X$ or $f \circ g = \mathrm{Id}_Y$. Indeed, if f is bijective and $g \circ f = \mathrm{Id}_X$, then

$$g = g \circ \mathrm{Id}_Y = g \circ (f \circ f^{-1}) = (g \circ f) \circ f^{-1} = \mathrm{Id}_X \circ f^{-1} = f^{-1}.$$

In the same way $g = f^{-1}$ follows from the assumption $f \circ g = \mathrm{Id}_Y$.

Theorem 2.23 *If $f : X \to Y$ and $g : Y \to Z$ are bijective maps, then the following assertions hold:*

(1) f^{-1} *is bijective with* $(f^{-1})^{-1} = f$.
(2) $g \circ f$ *is bijective with* $(g \circ f)^{-1} = f^{-1} \circ g^{-1}$.

Proof

(1) Exercise.
(2) We know from Theorem 2.20 that $g \circ f : X \to Z$ is bijective. Therefore, there exists a (unique) inverse of $g \circ f$. For the map $f^{-1} \circ g^{-1}$ we have

$$(f^{-1} \circ g^{-1}) \circ (g \circ f) = f^{-1} \circ (g^{-1} \circ (g \circ f)) = f^{-1} \circ ((g^{-1} \circ g) \circ f)$$
$$= f^{-1} \circ (\mathrm{Id}_Y \circ f) = f^{-1} \circ f = \mathrm{Id}_X.$$

Hence, $f^{-1} \circ g^{-1}$ is the inverse of $g \circ f$. $\square$

If the map $g \circ f$ is bijective, then this does not imply that the maps f or g are bijective. If, for instance,

$$f : \mathbb{R}_+ \to \mathbb{R}, \quad f(x) = x^2, \quad \text{and} \quad g : \mathbb{R} \to \mathbb{R}_+, \quad g(x) = x^2,$$

then $g \circ f : \mathbb{R}_+ \to \mathbb{R}_+$, $(g \circ f)(x) = x^4$, is bijective despite the fact that neither f nor g are bijective. On the other hand, here f is injective and g is surjective (cp. (3) and (4) in Theorem 2.20).

2.3 Relations

We first introduce the Cartesian product[3] of sets.

Definition 2.24 If M, N are sets, then the set

$$M \times N := \{(x, y) \mid x \in M \;\wedge\; y \in N\}$$

is the *Cartesian product* of M and N. The n-fold Cartesian product of sets $M_1, \ldots, M_n$ is

$$M_1 \times \cdots \times M_n := \{(x_1, \ldots, x_n) \mid x_i \in M_i \text{ for } i = 1, \ldots, n\}.$$

The n-fold Cartesian product of a single set M is

$$M^n := \underbrace{M \times \cdots \times M}_{n \text{ times}} = \{(x_1, \ldots, x_n) \mid x_i \in M \text{ for } i = 1, \ldots, n\}.$$

An element $(x, y) \in M \times N$ is called an *(ordered) pair*, and an element $(x_1, \ldots, x_n) \in M_1 \times \cdots \times M_n$ is called an *(ordered) n-tuple*.

If in these definitions at least one of the occurring sets is empty, then the resulting Cartesian product is the empty set as well.

---

[3] Named after René Descartes (1596–1650), one of the founders of Analytic Geometry. Georg Cantor (1845–1918) used in 1895 the name "connection set of M and N" ("Verbindungsmenge von M und N") and the notation $(M.N) = \{(m, n)\}$.

2.3 Relations

Definition 2.25 If M, N are sets, then a set $R \subseteq M \times N$ is called a *relation* between M and N. If $M = N$, then R is called a relation on M. Instead of $(x, y) \in R$ we also write $x \sim_R y$ or $x \sim y$, if it is clear which relation is considered.

If in this definition at least one of the sets M and N is empty, then every relation between M and N is also the empty set, since then $M \times N = \emptyset$.
If, for instance $M = \mathbb{N}$ and $N = \mathbb{Q}$, then

$$R = \{(x, y) \in M \times N \mid xy = 1\}$$

is a relation between M and N that can be expressed as

$$R = \{(1, 1), (2, 1/2), (3, 1/3), \ldots\} = \{(n, 1/n) \mid n \in \mathbb{N}\}.$$

Definition 2.26 A relation R on a set M is called

(1) *reflexive*, if $x \sim x$ holds for all $x \in M$,
(2) *symmetric*, if $(x \sim y) \Rightarrow (y \sim x)$ holds for all $x, y \in M$,
(3) *transitive*, if $(x \sim y \wedge y \sim z) \Rightarrow (x \sim z)$ holds for all $x, y, z \in M$.

If R is reflexive, transitive and symmetric, then it is called an *equivalence relation* on M.

Example 2.27

(1) Let $R = \{(x, y) \in \mathbb{Q}^2 \mid x = -y\}$. Then R is not reflexive, since $x = -x$ holds only for $x = 0$. If $x = -y$, then also $y = -x$, and hence R is symmetric. Finally, R is not transitive. For example, $(x, y) = (1, -1) \in R$ and $(y, z) = (-1, 1) \in R$, but $(x, z) = (1, 1) \notin R$.
(2) The relation $R = \{(x, y) \in \mathbb{Z}^2 \mid x \leq y\}$ is reflexive and transitive, but not symmetric.
(3) If $f : \mathbb{R} \to \mathbb{R}$ is a map, then $R = \{(x, y) \in \mathbb{R}^2 \mid f(x) = f(y)\}$ is an equivalence relation on $\mathbb{R}$, since the following properties hold:

- Reflexivity: $f(x) = f(x)$ for all $x \in \mathbb{R}$.
- Symmetry: $f(x) = f(y) \Rightarrow f(y) = f(x)$ for all $x, y \in \mathbb{R}$.
- Transitivity: $f(x) = f(y) \wedge f(y) = f(z) \Rightarrow f(x) = f(z)$ for all $x, y, z \in \mathbb{R}$.

Definition 2.28 Let R be an equivalence relation on the set M. Then, for $x \in M$ the set

$$[x]_R := \{y \in M \mid (x, y) \in R\} = \{y \in M \mid x \sim y\}$$

is called the *equivalence class* of x with respect to R. The set of equivalence classes

$$M/R := \{[x]_R \mid x \in M\}$$

is called the *quotient set* of M with respect to R.

The equivalence class $[x]_R$ of elements $x \in M$ is never the empty set, since always $x \sim x$ (reflexivity) and therefore $x \in [x]_R$. If it is clear which equivalence relation R is meant, we often write $[x]$ instead oft $[x]_R$ and also skip the additional "with respect to R".

Theorem 2.29 *If R is an equivalence relation on the set M and if $x, y \in M$, then the following are equivalent:*

(1) $[x] = [y]$.
(2) $[x] \cap [y] \neq \emptyset$.
(3) $x \sim y$.

Proof

(1) $\Rightarrow$ (2): Since $x \sim x$, it follows that $x \in [x]$. From $[x] = [y]$ it follows that $x \in [y]$ and thus $x \in [x] \cap [y]$.
(2) $\Rightarrow$ (3): Since $[x] \cap [y] \neq \emptyset$, there exists a $z \in [x] \cap [y]$. For this element z we have $x \sim z$ and $y \sim z$, and thus $x \sim z$ and $z \sim y$ (symmetry) and, therefore, $x \sim y$ (transitivity).
(3) $\Rightarrow$ (1): Let $x \sim y$ and $z \in [x]$, i.e., $x \sim z$. Using symmetry and transitivity, we obtain $y \sim z$, and hence $z \in [y]$. This means that $[x] \subseteq [y]$. In an analogous way one shows that $[y] \subseteq [x]$, and hence $[x] = [y]$ holds. $\square$

Theorem 2.29 shows that for two equivalence classes $[x]$ and $[y]$ we have either $[x] = [y]$ or $[x] \cap [y] = \emptyset$. Thus every $x \in M$ is contained in exactly one equivalence class (namely in $[x]$), so that an equivalence relation R yields a partitioning or decomposition of M into mutually disjoint subsets. Every element of $[x]$ is called a *representative* of the equivalence class $[x]$. A very useful and general approach that we will often use in this book is to partition a set of objects (e.g. sets of matrices) into equivalence classes, and to find in each such class a representative with a particularly simple structure. Such a representative is called a *canonical form* or *normal form* with respect to the given equivalence relation.

Example 2.30 For a given number $n \in \mathbb{N}$ the set

$$R_n := \{(a, b) \in \mathbb{Z}^2 \mid a - b \text{ is divisible by } n \text{ without remainder}\}$$

is an equivalence relation on $\mathbb{Z}$, since the following properties hold:

- Reflexivity: $a - a = 0$ is divisible by n without remainder.
- Symmetry: If $a - b$ is divisible by n without remainder, then also $b - a$.
- Transitivity: Let $a - b$ and $b - c$ be divisible by n without remainder and write $a - c = (a - b) + (b - c)$. Both summands on the right are divisible by n without remainder and hence this also holds for $a - c$.

For $a \in \mathbb{Z}$ the equivalence class $[a]$ is called *residue class of a modulo n*, and $[a] = a + n\mathbb{Z} := \{a + nz \mid z \in \mathbb{Z}\}$. The equivalence relation R_n yields a partitioning of $\mathbb{Z}$ into n mutually disjoint subsets. In particular, we have

$$[0] \cup [1] \cup \cdots \cup [n-1] = \bigcup_{a=0}^{n-1} [a] = \mathbb{Z}.$$

The set of all residue classes modulo n, i.e., the quotient set with respect to R_n, is often denoted by $\mathbb{Z}/n\mathbb{Z}$. Thus, $\mathbb{Z}/n\mathbb{Z} := \{[0], [1], \ldots, [n-1]\}$. This set plays an important role in the mathematical field of Number Theory.

2.4 Mathematical Induction

An important proof technique that we will frequently use is the *proof by induction*. In this technique, a statement $S(n)$ is shown for all natural numbers $n \in \mathbb{N}$ via the following three steps:

(1) In the *initial step* the statement $S(1)$ is proved.
(2) In the *induction hypothesis* one assumes that the statement $S(n)$ holds for some $n \geq 1$.
(3) In the *induction step* the statement "$S(n) \Rightarrow S(n+1)$" is shown.

The validity of this proof technique relies on the *induction axiom* of Peano:[4] If for a subset $M \subseteq \mathbb{N}$, it holds that $1 \in M$, and for every $n \in M$ also $n + 1 \in M$, then $M = \mathbb{N}$.

If one wants to show the truth of a statement $S(n)$ for all $n \geq n_1$ with $n_1 \in \mathbb{N}_0$, then in the initial step one has to prove the statement $S(n_1)$, and in the induction hypothesis one has to assume that the statement $S(n)$ holds for some

---

[4] Giuseppe Peano (1858–1932).

$n \geq n_1$. Later we will also use "inductive proofs" for statements about finite subsets $M = \{n_1, n_2, \ldots, n_k\} \subseteq \mathbb{N}$ (cp., e.g., the proofs of Theorems 4.13 and 5.2).

Example 2.31 A simple example for a proof by induction is that of the *Gauß summation formula*[5]

$$1 + 2 + 3 + \ldots + n = \frac{1}{2}n(n+1) \quad \text{for all } n \in \mathbb{N}.$$

(1) Initial step: For $n = 1$ the left side is equal to 1 and the right side is $\frac{1}{2} \cdot 1 \cdot 2 = 1$. Hence the formula holds for $n = 1$.
(2) Induction hypothesis: The formula holds for some $n \geq 1$.
(3) Induction step: Using the induction hypothesis we obtain

$$1 + 2 + \ldots + n + (n+1) = \frac{1}{2}n(n+1) + (n+1) = \frac{1}{2}(n+1)(n+2),$$

i.e., the formula holds for $n + 1$, and therefore for all $n \in \mathbb{N}$.

Exercises

2.1 Let A, B, C be assertions. Show that the following assertions are true:
   (a) For $\wedge$ and $\vee$ the associative laws

$$[(A \wedge B) \wedge C] \Leftrightarrow [A \wedge (B \wedge C)],$$
$$[(A \vee B) \vee C] \Leftrightarrow [A \vee (B \vee C)]$$

   hold.
   (b) For $\wedge$ and $\vee$ the commutative laws

$$(A \wedge B) \Leftrightarrow (B \wedge A), \quad (A \vee B) \Leftrightarrow (B \vee A)$$

   hold.
   (c) For $\wedge$ and $\vee$ the distributive laws

$$[(A \wedge B) \vee C] \Leftrightarrow [(A \vee C) \wedge (B \vee C)],$$
$$[(A \vee B) \wedge C] \Leftrightarrow [(A \wedge C) \vee (B \wedge C)]$$

   hold.

---

[5] Carl Friedrich Gauß (1777–1855).

2.2 Let A, B, C be assertions. Show that the following assertions are true:
   (a) $A \wedge B \Rightarrow A$.
   (b) $[A \Leftrightarrow B] \Leftrightarrow [(A \Rightarrow B) \wedge (B \Rightarrow A)]$.
   (c) $\neg(A \vee B) \Leftrightarrow [(\neg A) \wedge (\neg B)]$.
   (d) $\neg(A \wedge B) \Leftrightarrow [(\neg A) \vee (\neg B)]$.
   (e) $[(A \Rightarrow B) \wedge (B \Rightarrow C)] \Rightarrow [A \Rightarrow C]$.
   (f) $[A \Rightarrow (B \vee C)] \Leftrightarrow [(A \wedge \neg B) \Rightarrow C]$.
   (The assertions (c) and (d) are called the *De Morgan laws*[6] for $\wedge$ and $\vee$.)

2.3 Show that the following assertion for statements A, B is true:

$$(A \Rightarrow B) \Leftrightarrow \neg(A \wedge \neg B).$$

(This equivalence allows to prove the implication $A \Rightarrow B$ by *contradiction*. For this one assumes that A is true and B is false (i.e., $\neg B$ is true). Then one concludes that $A \wedge \neg B$ are false (and hence $\neg(A \wedge \neg B)$ true). This shows that $A \Rightarrow B$ is true.)

2.4 Prove Theorem 2.9.

2.5 Show that for two sets M, N the following holds:

$$N \subseteq M \quad \Leftrightarrow \quad M \cap N = N \quad \Leftrightarrow \quad M \cup N = M.$$

2.6 Let M be a set with $|M| = m \in \mathbb{N}$. How many elements does the set $\mathcal{P}(\mathcal{P}(M))$ have?

2.7 Let M, N be sets. Prove the following assertions:
   (a) $\mathcal{P}(M) \cap \mathcal{P}(N) = \mathcal{P}(M \cap N)$.
   (b) $\mathcal{P}(M) \cup \mathcal{P}(N) \subseteq \mathcal{P}(M \cup N)$.
   Construct sets M and N, for which there is no equality in (b).

2.8 Let X, Y be nonempty sets, $U, V \subseteq Y$ nonempty subsets and let $f : X \to Y$ be a map. Show that $f^{-1}(U \cap V) = f^{-1}(U) \cap f^{-1}(V)$. Let $U, V \subseteq X$ be nonempty. Check whether $f(U \cup V) = f(U) \cup f(V)$ holds.

2.9 Are the following maps injective, surjective, bijective?
   (a) $f_1 : \mathbb{R} \setminus \{0\} \to \mathbb{R}, \ x \mapsto \frac{1}{x}$.
   (b) $f_2 : [0, 1] \to [1, 2], \ x \mapsto 1 + \frac{x^2}{2}$.
   (c) $f_3 : \mathbb{R}^2 \to \mathbb{R}, \ (x, y) \mapsto x + y$.
   (d) $f_4 : \mathbb{R}^2 \to \mathbb{R}, \ (x, y) \mapsto x^2 + y^2 - 1$.
   (e) $f_5 : \mathbb{N} \to \mathbb{Z}, \ n \mapsto \begin{cases} \frac{n}{2}, & n \text{ even,} \\ -\frac{n-1}{2}, & n \text{ odd.} \end{cases}$

2.10 Let X be a set with finite cardinality, and let $f : X \to X$ be injective. Show that then f is also surjective, and hence bijective.

2.11 Let $a \in \mathbb{Z}$ be given. Show that the map $f_a : \mathbb{Z} \to \mathbb{Z}, \ f_a(x) = x + a$ is bijective and determine the inverse f_a^{-1}.

---

[6] Augustus De Morgan (1806–1871).

2.12 Prove Lemma 2.18.
2.13 Prove Theorem 2.20.
2.14 Prove Theorem 2.23 (1).
2.15 Let X, Y be nonempty sets, and let $f : X \to Y$ as well as $g, h : Y \to X$ be maps. Prove the following assertion: If $g \circ f = \mathrm{Id}_X$ and $f \circ h = \mathrm{Id}_Y$, then $g = h$.
2.16 Find two maps $f, g : \mathbb{N} \to \mathbb{N}$, so that simultaneously
  (a) f is not surjective,
  (b) g is not injective, and
  (c) $g \circ f$ is bijective.
2.17 Determine all equivalence relations on the set $\{1, 2\}$.
2.18 Determine a symmetric and transitive relation on the set $\{a, b, c\}$ that is not reflexive.
2.19 Let M be a nonempty set, let I be an index set, and let R_i be an equivalence relation on M for each $i \in I$. Show that the intersection $R := \bigcap_{i \in I} R_i$ is also an equivalence relation on M.
2.20 Show the following assertions by mathematical induction:
  (a) For every $n \in \mathbb{N}$ it holds that $1 + 3 + 5 + \ldots + (2n - 1) = n^2$.
  (b) For every $x \in \mathbb{R} \setminus \{1\}$ it holds that $1 + x + x^2 + \ldots + x^n = \frac{x^{n+1} - 1}{x - 1}$.
  (c) For every $n \in \mathbb{N}$ the number $5^n + 7$ is divisible by 4.
  (d) For every natural number $n \geq 10$ it holds that $2^n > n^3$.
  (e) For every natural number $n \geq 4$ it holds that $n! > 2^n$.

Algebraic Structures 3

An algebraic structure is a set with operations between its elements that follow certain rules. As an example of such a structure consider the integers and the operation '+'. What are the properties of this addition? Already in elementary school one learns that the sum $a + b$ of two integers a and b is another integer. Moreover, there is a number 0 such that $0 + a = a$ for every integer a, and for every integer a there exists an integer $-a$ such that $(-a) + a = 0$. The analysis of the properties of such concrete examples leads to definitions of abstract concepts that are built on a few simple axioms. For the integers and the operation addition, this leads to the algebraic structure of a group, which is the starting point in this chapter.

This principle of abstraction from concrete examples is one of the strengths and basic working principles of Mathematics. By "extracting and completely exposing the mathematical kernel" (David Hilbert) we also simplify our further work: Every proved assertion about an abstract concept automatically holds for all concrete examples. Moreover, by combining defined concepts we can move to further generalizations and in this way extend the mathematical theory step by step. Hermann Graßmann (1809–1877) described this procedure as follows:[1] "... the mathematical method moves forward from the simplest concepts to combinations of them and gains via such combinations new and more general concepts."

3.1 Groups

We begin with a set and an operation with specific properties.

Definition 3.1 A *group* is a set G with a map, called *operation*,

---

[1] "... die mathematische Methode hingegen schreitet von den einfachsten Begriffen zu den zusammengesetzteren fort, und gewinnt so durch Verknüpfung des Besonderen neue und allgemeinere Begriffe."

$$\oplus : G \times G \to G, \quad (a,b) \mapsto a \oplus b,$$

that satisfies the following:

(1) The operation $\oplus$ is associative, i.e., $(a \oplus b) \oplus c = a \oplus (b \oplus c)$ holds for all $a, b, c \in G$.
(2) There exists an element $e \in G$, called a *neutral element*, for which
   (a) $e \oplus a = a$ for all $a \in G$, and
   (b) for every $a \in G$ there exists an $\tilde{a} \in G$, called an *inverse element* of a, with $\tilde{a} \oplus a = e$.

If $a \oplus b = b \oplus a$ holds for all $a, b \in G$, then the group is called *commutative* or *Abelian*.[2]

As short hand notation for a group we use $(G, \oplus)$ or just G, if is clear which operation is used.

Theorem 3.2 *For every group $(G, \oplus)$ the following assertions hold:*

(1) If $e \in G$ is a neutral element and if $a, \tilde{a} \in G$ with $\tilde{a} \oplus a = e$, then also $a \oplus \tilde{a} = e$.
(2) If $e \in G$ is a neutral element and if $a \in G$, then also $a \oplus e = a$.
(3) G contains exactly one neutral element.
(4) For every $a \in G$ there exists a unique inverse element.

Proof

(1) Let $e \in G$ be a neutral element and let $a, \tilde{a} \in G$ satisfy $\tilde{a} \oplus a = e$. Then by Definition 3.1 there exists an element $a_1 \in G$ with $a_1 \oplus \tilde{a} = e$. Thus,

$$a \oplus \tilde{a} = e \oplus (a \oplus \tilde{a}) = (a_1 \oplus \tilde{a}) \oplus (a \oplus \tilde{a}) = a_1 \oplus ((\tilde{a} \oplus a) \oplus \tilde{a})$$
$$= a_1 \oplus (e \oplus \tilde{a}) = a_1 \oplus \tilde{a} = e.$$

(2) Let $e \in G$ be a neutral element and let $a \in G$. Then there exists $\tilde{a} \in G$ with $\tilde{a} \oplus a = e$. By (1) then also $a \oplus \tilde{a} = e$ and it follows that

$$a \oplus e = a \oplus (\tilde{a} \oplus a) = (a \oplus \tilde{a}) \oplus a = e \oplus a = a.$$

(3) Let $e, e_1 \in G$ be two neutral elements. Then $e_1 \oplus e = e$, since e_1 is a neutral element. Since e is also a neutral element, it follows that $e_1 = e \oplus e_1 = e_1 \oplus e$, where for the second identity we have used assertion (2). Hence, $e = e_1$.

---

[2] Named after Niels Henrik Abel (1802–1829), one of the founders of group theory.

3.1 Groups

(4) Let $\tilde{a}, a_1 \in G$ be two inverse elements of $a \in G$ and let $e \in G$ be the (unique) neutral element. Then with (1) and (2) it follows that

$$\tilde{a} = e \oplus \tilde{a} = (a_1 \oplus a) \oplus \tilde{a} = a_1 \oplus (a \oplus \tilde{a}) = a_1 \oplus e = a_1. \qquad \square$$

Example 3.3

(1) $(\mathbb{Z}, +)$, $(\mathbb{Q}, +)$ and $(\mathbb{R}, +)$ are commutative groups. In all these groups the neutral element is the number 0 (zero) and the inverse of a is the number $-a$. Instead of $a + (-b)$ we usually write $a - b$. Since the operation is the addition, these groups are also called *additive groups*.

The natural numbers $\mathbb{N}$ with the addition do not form a group, since there is no neutral element in $\mathbb{N}$. If we consider the set $\mathbb{N}_0$, which includes also the number 0 (zero), then $0 + a = a + 0 = a$ for all $a \in \mathbb{N}_0$, but only $a = 0$ has an inverse element in $\mathbb{N}$. Hence also $\mathbb{N}_0$ with the addition does not form a group.

(2) The sets $\mathbb{Q} \setminus \{0\}$ and $\mathbb{R} \setminus \{0\}$ with the usual multiplication form commutative groups. In these *multiplicative groups*, the neutral element is the number 1 (one) and the inverse element of a is the number $\frac{1}{a}$ (or a^{-1}). Instead of $a \cdot b^{-1}$ we also write $\frac{a}{b}$ or a/b.

The integers $\mathbb{Z}$ with the usual multiplication do not form a group. The set $\mathbb{Z}$ includes the number 1, for which $1 \cdot a = a \cdot 1 = a$ for all $a \in \mathbb{Z}$, but no $a \in \mathbb{Z} \setminus \{-1, 1\}$ has an inverse element in $\mathbb{Z}$.

Definition 3.4 Let $(G, \oplus)$ be a group and $H \subseteq G$. If $(H, \oplus)$ is a group, then it is called a *subgroup* of $(G, \oplus)$.

The next theorem gives an alternative characterization of a subgroup.

Theorem 3.5 *$(H, \oplus)$ is a subgroup of the group $(G, \oplus)$ if and only if the following properties hold:*

(1) $\emptyset \neq H \subseteq G$.
(2) $a \oplus b \in H$ for all $a, b \in H$.
(3) For every $a \in H$ also the inverse element satisfies $\tilde{a} \in H$.

Proof Exercise. $\qquad \square$

The following definition characterizes maps between two groups which are compatible with the respective group operations.

Definition 3.6 Let $(G_1, \oplus)$ and $(G_2, \circledast)$ be groups. A map

$$\varphi : G_1 \to G_2, \quad g \mapsto \varphi(g),$$

is called a *group homomorphism*, if

$$\varphi(a \oplus b) = \varphi(a) \circledast \varphi(b) \quad \text{for all } a, b \in G_1.$$

A bijective group homomorphism is called a *group isomorphism*.

3.2 Rings

We now want to extend the concept of a group and discuss mathematical structures that are characterized by two operations. As motivating example consider the integers with the addition, i.e., the group $(\mathbb{Z}, +)$. We can multiply the elements of $\mathbb{Z}$ and this multiplication is associative, i.e., $(a \cdot b) \cdot c = a \cdot (b \cdot c)$ for all $a, b, c \in \mathbb{Z}$. Furthermore the addition and multiplication satisfy the distributive laws $a \cdot (b + c) = a \cdot b + a \cdot c$ and $(a + b) \cdot c = a \cdot c + b \cdot c$ for all integers a, b, c. These properties make $\mathbb{Z}$ with addition and multiplication into a ring.

Definition 3.7 A *ring* is a set R with two operations

$$+ : R \times R \to R, \qquad (a, b) \mapsto a + b, \qquad \text{(addition)}$$
$$* : R \times R \to R, \qquad (a, b) \mapsto a * b, \qquad \text{(multiplication)}$$

that satisfy the following:

(1) $(R, +)$ is a commutative group.
    We call the neutral element in this group *zero*, and write 0. We denote the inverse element of $a \in R$ by $-a$, and write $a - b$ instead of $a + (-b)$.
(2) The multiplication is associative, i.e., $(a * b) * c = a * (b * c)$ for all $a, b, c \in R$.
(3) The distributive laws hold, i.e., for all $a, b, c \in R$ we have

$$a * (b + c) = a * b + a * c,$$
$$(a + b) * c = a * c + b * c.$$

A ring is called *commutative* if $a * b = b * a$ for all $a, b \in R$.
An element $1 \in R$ is called unit if $1 * a = a * 1 = a$ for all $a \in R$. In this case R is called a *ring with unit*.

On the right hand side of the two distributive laws we have omitted the parentheses, since multiplication is supposed to bind stronger than addition, i.e., $a + (b * c) = a + b * c$. If it is useful for illustration purposes we nevertheless use parentheses, e.g., we sometimes write $(a * b) + (c * d)$ instead of $a * b + c * d$.

Analogous to the notation for groups we denote a ring with $(R, +, *)$ or just with R, if the operations are clear from the context.

3.2 Rings

The standard example of a commutative ring with unit is given by $(\mathbb{Z}, +, *)$. Another example is given next.

Example 3.8 Let M be a nonempty set and let R be the set of maps $f : M \to \mathbb{R}$. Then $(R, +, *)$ with the operations

$$+ : R \times R \to R, \quad (f, g) \mapsto f + g, \quad (f + g)(x) := f(x) + g(x),$$
$$* : R \times R \to R, \quad (f, g) \mapsto f * g, \quad (f * g)(x) := f(x) \cdot g(x),$$

is a commutative ring with unit. Here $f(x) + g(x)$ and $f(x) \cdot g(x)$ are the sum and product of two real numbers, i.e., the addition and the multiplication are defined "pointwise". The zero in this ring is the map $0_R : M \to \mathbb{R}, x \mapsto 0$, and the unit is the map $1_R : M \to \mathbb{R}, x \mapsto 1$, where 0 and 1 are the real numbers zero and one.

Theorem 3.9 *If R is a ring with unit, then the following assertions hold:*

(1) The unit in R is uniquely determined.
(2) It holds that $1 = 0$ if and only if $R = \{0\}$.

Proof

(1) If $1, e \in R$ with $1 * a = a * 1 = a$ and $e * a = a * e = a$ for all $a \in R$, then $1 = e * 1 = e$.
(2) Exercise. $\square$

The ring $R = \{0\}$ is called the *zero ring*. When considering rings with unit we usually assume that $1 \neq 0$, and hence we exclude the (trivial) case of the zero ring.

For $a_1, a_2, \ldots, a_n \in R$ we use the following abbreviations for the *sum* and *product* of these elements:

$$\sum_{j=1}^{n} a_j := a_1 + a_2 + \ldots + a_n \quad \text{and} \quad \prod_{j=1}^{n} a_j := a_1 * a_2 * \ldots * a_n.$$

Moreover, $a^n := \prod_{j=1}^{n} a$ for all $a \in R$ and $n \in \mathbb{N}$. If $\ell > k$, then we define the *empty sum* as

$$\sum_{j=\ell}^{k} a_j := 0.$$

In a ring with unit we also define for $\ell > k$ the *empty product* as

$$\prod_{j=\ell}^{k} a_j := 1.$$

Theorem 3.10 *For every ring R the following assertions hold:*

(1) $0 * a = a * 0 = 0$ *for all* $a \in R$.
(2) $a * (-b) = -(a * b) = (-a) * b$ *and* $(-a) * (-b) = a * b$ *for all* $a, b \in R$.

Proof

(1) For every $a \in R$ we have $0 * a = (0 + 0) * a = (0 * a) + (0 * a)$. Adding $-(0 * a)$ on the left and right hand sides of this equality we obtain $0 = 0 * a$. In the same way we can show that $a * 0 = 0$ for all $a \in R$.

(2) Since $(a * b) + (a * (-b)) = a * (b + (-b)) = a * 0 = 0$, it follows that $a * (-b)$ is the (unique) additive inverse of $a * b$, i.e., $a * (-b) = -(a * b)$. In the same way we can show that $(-a) * b = -(a * b)$. Furthermore, we have

$$-(a * b) + (-a) * (-b) = a * (-b) + (-a) * (-b)$$
$$= (a + (-a)) * (-b) = 0 * (-b) = 0,$$

and thus $(-a) * (-b) = a * b$. □

Even though $a * 0 = 0 * a = 0$ holds in every ring R, $a * b = 0$ does not imply that $a = 0$ or $b = 0$. An element $a \in R$ is called a *zero divisor*,[3] if there exists an element $b \in R \setminus \{0\}$ with $a * b = 0$. The element $a = 0$ (i.e., the zero itself) is called the *trivial zero divisor*. Using the example of matrices we will later discuss rings with non-trivial zero divisors (see e.g. (4.3) in the proof of Theorem 4.9).

Analogous to the case of groups we can also identify subsets of rings that are again rings.

Definition 3.11 Let $(R, +, *)$ be a ring and $S \subseteq R$. If $(S, +, *)$ is a ring, then we call it a *subring* of $(R, +, *)$.

For example, $\mathbb{Z}$ is a subring of $\mathbb{Q}$, and $\mathbb{Q}$ is a subring of $\mathbb{R}$, where we use as operations the usual addition and multiplication in these sets of numbers.

In the definition of a ring only additive inverse elements occur. We will now formally define the concept of a multiplicative inverse.

---

[3] The concept of zero divisors was introduced in 1883 by Karl Weierstraß (1815–1897).

3.2 Rings

Definition 3.12 Let $(R, +, *)$ be a ring with unit. An element $b \in R$ is called an *inverse* of $a \in R$ (with respect to $*$), if $a * b = b * a = 1$. An element of R that has an inverse is called *invertible*.

It is clear from the definition that $b \in R$ is an inverse of $a \in R$ if and only if $a \in R$ is an inverse of $b \in R$. In general, however, not every element in a ring must be (or is) invertible. For example, let R be a ring with unit and $1 \neq 0$. Then $0 \in R$ is not invertible, since $0 * b = 0 \neq 1$ holds for all $b \in R$.

If an element $a \in R$ is invertible, then it has a unique inverse, as shown in the following theorem.

Theorem 3.13 Let $(R, +, *)$ be a ring with unit.

(1) If $a \in R$ is invertible, then the inverse is unique and we denote it by a^{-1}.
(2) If $a, b \in R$ are invertible then $a * b \in R$ is invertible and $(a*b)^{-1} = b^{-1} * a^{-1}$.

Proof

(1) If $b, \tilde{b} \in R$ are inverses of $a \in R$, then $b = b * 1 = b * (a * \tilde{b}) = (b * a) * \tilde{b} = 1 * \tilde{b} = \tilde{b}$.
(2) Since a and b are invertible, $b^{-1} * a^{-1} \in R$ is well defined and

$$(b^{-1} * a^{-1}) * (a*b) = ((b^{-1} * a^{-1}) * a) * b = (b^{-1} * (a^{-1} * a)) * b = b^{-1} * b = 1.$$

In the same way we can show that $(a*b) * (b^{-1} * a^{-1}) = 1$, and thus $(a*b)^{-1} = b^{-1} * a^{-1}$. □

The next theorem shows that for finite rings with unit already one of the conditions $a * b = 1$ or $b * a = 1$ is sufficient for the invertibility of a (and b).

Theorem 3.14 *If R is a ring with unit and $|R| \in \mathbb{N}$, then $a * b = 1$ for $a, b \in R$ implies that also $b * a = 1$.*

Proof Let $a, b \in R$ with $a*b = 1$. We define the map $f : R \to R$ with $f(c) = b*c$ for all $c \in R$. Let $c_1, c_2 \in R$ with $f(c_1) = f(c_2)$, and therefore $b * c_1 = b * c_2$. If we multiply this equation from the left with a, then we get $(a*b) * c_1 = (a*b) * c_2$, and $a * b = 1$ implies that $c_1 = c_2$, i.e., f is injective. Since $|R|$ is finite, f is also surjective, and hence bijective (cp. Exercise 2.10). Since f is bijective, there exists $c \in R$ with $1 = f(c) = b * c$. Multiplication of this equation from the left by a gives $a = c$, and hence $1 = b * a$. □

We now discuss the important example of the *ring of polynomials*.

Example 3.15 Let $(R, +, \cdot)$ be a commutative ring with unit. A *polynomial over* R and in the indeterminate or variable t is an expression of the form

$$p = \alpha_0 \cdot t^0 + \alpha_1 \cdot t^1 + \ldots + \alpha_n \cdot t^n,$$

where $\alpha_0, \alpha_1, \ldots, \alpha_n \in R$ are the *coefficients* of the polynomial. Instead of $\alpha_0 \cdot t^0$, t^1, $1 \cdot t^j$ and $\alpha_j \cdot t^j$ we often just write α_0, t, t^j and $\alpha_j t^j$. The set of all polynomials over R is denoted by $R[t]$.

Let

$$p = \alpha_0 + \alpha_1 \cdot t + \ldots + \alpha_n \cdot t^n, \quad q = \beta_0 + \beta_1 \cdot t + \ldots + \beta_m \cdot t^m$$

be two polynomials in $R[t]$ with $n \geq m$. If $n > m$, then we set $\beta_j = 0$ for $j = m+1, \ldots, n$ and call p and q *equal*, written $p = q$, if $\alpha_j = \beta_j$ for $j = 0, 1, \ldots, n$. In particular, we have

$$\alpha_0 + \alpha_1 \cdot t + \ldots + \alpha_n \cdot t^n = \alpha_n \cdot t^n + \ldots + \alpha_1 \cdot t + \alpha_0,$$
$$0 + 0 \cdot t + \ldots + 0 \cdot t^n = 0.$$

The *degree* of the polynomial $p = \alpha_0 + \alpha_1 \cdot t + \ldots + \alpha_n \cdot t^n$, denoted by $\deg(p)$, is defined as the largest index j, for which $\alpha_j \neq 0$. If no such index exists, then the polynomial is the *zero polynomial* $p = 0$ and we set $\deg(p) := -\infty$.

Let $p, q \in R[t]$ as above have degrees n, m, respectively, with $n \geq m$. If $n > m$, then we again set $\beta_j = 0$, $j = m+1, \ldots, n$. We define the following operations on $R[t]$:

$$p + q := (\alpha_0 + \beta_0) + (\alpha_1 + \beta_1) \cdot t + \ldots + (\alpha_n + \beta_n) \cdot t^n,$$
$$p * q := \gamma_0 + \gamma_1 \cdot t + \ldots + \gamma_{n+m} \cdot t^{n+m}, \quad \gamma_k := \sum_{i+j=k} \alpha_i \beta_j.$$

With these operations $(R[t], +, *)$ is a commutative ring with unit. The zero is given by the polynomial $p = 0$ and the unit is $p = 1 \cdot t^0 = 1$. But not every polynomial $p \in R[t] \setminus \{0\}$ is invertible. For example, for $p = t$ and any other polynomial $q = \beta_0 + \beta_1 t + \cdots + \beta_m t^m \in R[t]$ we have

$$p * q = \beta_0 t + \beta_1 t^2 + \cdots + \beta_m t^{m+1} \neq 1,$$

and hence p is not invertible.

In a polynomial we can "substitute" the variable t by some other object when the resulting expression can be evaluated algebraically. For example, we may substitute t by any $\lambda \in R$ and interpret the addition and multiplication as the corresponding operations in the ring R. This defines a map from R to R by

$$\lambda \mapsto p(\lambda) = \alpha_0 \cdot \lambda^0 + \alpha_1 \cdot \lambda^1 + \ldots + \alpha_n \cdot \lambda^n, \quad \lambda^k := \underbrace{\lambda \cdot \ldots \cdot \lambda}_{k \text{ times}}, \quad k = 0, 1, \ldots, n,$$

where $\lambda^0 = 1 \in R$ (this is an empty product). Here one should not confuse the ring element $p(\lambda)$ with the polynomial p itself, but rather think of $p(\lambda)$ as an *evaluation* of p at λ. We will study the properties of polynomials in more detail later on, and we will also evaluate polynomials at other objects such as matrices or endomorphisms.

3.3 Fields

From an algebraic point of view the difference between the integers on the one hand, and the rational or real numbers on the other, is that in the sets $\mathbb{Q}$ and $\mathbb{R}$ every element (except for the number zero) is invertible. This "additional structure" makes $\mathbb{Q}$ and $\mathbb{R}$ into fields.

Definition 3.16 A commutative ring R with unit is called a *field*, if $0 \neq 1$ and every $a \in R \setminus \{0\}$ is invertible.

By definition, every field is a commutative ring with unit, but the converse does not hold. One can also introduce the concept of a field based on the concept of a group (cp. Exercise 3.18).

Definition 3.17 A *field* is a set K with two operations[4]

$$+ : K \times K \to K, \quad (a, b) \mapsto a + b, \quad \text{(addition)}$$
$$* : K \times K \to K, \quad (a, b) \mapsto a * b, \quad \text{(multiplication)}$$

---

[4] The German term for a field is "Körper" (which literally means body in English). Because of this term we have been using the notation K since the first German edition of this book, and in order to have the notation consistent in all editions, we continue to use K here. Historically, the term "Körper" appeared much earlier than the term field. Richard Dedekind (1831–1916) introduced "Zahlkörper" in his lectures on Number Theory in the late 1850s, and "Körper" for certain infinite sets of real and complex numbers in 1871. Dedekind's motivation for the term, according to Felix Klein (1849–1925), was the German term "Köperschaft", which refers to an "organized entity". Another term Dedekind had used already in the late 1850s was "rationales Gebiet", which was the same as the "Rationalitätsbereich" of Leopold Kronecker (1823–1891) from 1881. The German terms "Gebiet" or "Bereich" can be translated as area, region, or field, which may have been the motivation for Eliakim Hastings Moore (1862–1932) to introduce the latter in the English mathematical literature in 1893. In the same year, Heinrich Weber (1842–1913) gave the first abstract definition of a general field as we define it today.

that satisfy the following:

(1) $(K, +)$ is a commutative group.
We call the neutral element in this group *zero*, and write 0. We denote the inverse element of $a \in K$ by $-a$, and write $a - b$ instead of $a + (-b)$.
(2) $(K \setminus \{0\}, *)$ is a commutative group.
We call the neutral element in this group *unit*, and write 1. We denote the inverse element of $a \in K \setminus \{0\}$ by a^{-1}.
(3) The distributive laws hold, i.e., for all $a, b, c \in K$ we have

$$a * (b + c) = a * b + a * c,$$
$$(a + b) * c = a * c + b * c.$$

We now show a few useful properties of fields.

Lemma 3.18 *For every field K the following assertions hold:*

(1) K has at least two elements.
(2) $0 * a = a * 0 = 0$ for all $a \in K$.
(3) $a * b = a * c$ and $a \neq 0$ imply that $b = c$ for all $a, b, c \in K$.
(4) $a * b = 0$ implies that $a = 0$ or $b = 0$ for all $a, b \in K$.

Proof

(1) This follows from the definition, since $0, 1 \in K$ with $0 \neq 1$.
(2) This has already been shown for rings (cp. Theorem 3.10).
(3) Since $a \neq 0$, we know that a^{-1} exists. Multiplying both sides of $a * b = a * c$ from the left with a^{-1} yields $b = c$.
(4) Suppose that $a * b = 0$. If $a = 0$, then we are finished. If $a \neq 0$, then a^{-1} exists and multiplying both sides of $a * b = 0$ from the left with a^{-1} yields $b = 0$. □

Property (4) in Lemma 3.18 means that fields contain only the trivial zero divisor. There are also rings in which property (4) holds, for instance the ring of integers $\mathbb{Z}$.

The following definition is analogous to the concepts of a subgroup (cp. Definition 3.4) and a subring (cp. Definition 3.11).

Definition 3.19 Let $(K, +, *)$ be a field and $L \subseteq K$. If $(L, +, *)$ is a field, then it is called a *subfield* of $(K, +, *)$.

We now discuss the important example of the *field of complex numbers*.

Example 3.20 The set of *complex numbers* is defined as

$$\mathbb{C} := \{(x, y) \mid x, y \in \mathbb{R}\} = \mathbb{R} \times \mathbb{R}.$$

3.3 Fields

On this set we define the following operations as addition and multiplication:

$$+ : \mathbb{C} \times \mathbb{C} \to \mathbb{C}, \quad (x_1, y_1) + (x_2, y_2) := (x_1 + x_2, y_1 + y_2),$$
$$\cdot : \mathbb{C} \times \mathbb{C} \to \mathbb{C}, \quad (x_1, y_1) \cdot (x_2, y_2) := (x_1 \cdot x_2 - y_1 \cdot y_2, x_1 \cdot y_2 + x_2 \cdot y_1).$$

On the right hand sides we here use the addition and the multiplication in the field $\mathbb{R}$. Then $(\mathbb{C}, +, \cdot)$ is a field with the neutral elements with respect to addition and multiplication given by

$$0_\mathbb{C} = (0, 0),$$
$$1_\mathbb{C} = (1, 0),$$

and the inverse elements with respect to addition and multiplication given by

$$-(x, y) = (-x, -y) \quad \text{for all} \quad (x, y) \in \mathbb{C},$$
$$(x, y)^{-1} = \left(\frac{x}{x^2 + y^2}, -\frac{y}{x^2 + y^2} \right) \quad \text{for all} \quad (x, y) \in \mathbb{C} \setminus \{(0, 0)\}.$$

In the multiplicative inverse element we have written $\frac{a}{b}$ instead of $a \cdot b^{-1}$, which is the common notation in $\mathbb{R}$.

Considering the subset $L := \{(x, 0) \mid x \in \mathbb{R}\} \subset \mathbb{C}$, we can identify every $x \in \mathbb{R}$ with an element of the set L via the (bijective) map $x \mapsto (x, 0)$. In particular, $0_\mathbb{R} \mapsto (0, 0) = 0_\mathbb{C}$ and $1_\mathbb{R} \mapsto (1, 0) = 1_\mathbb{C}$. Thus, we can interpret $\mathbb{R}$ as subfield of $\mathbb{C}$ (although $\mathbb{R}$ is not really a subset of $\mathbb{C}$), and we do not have to distinguish between the zero and unit elements in $\mathbb{R}$ and $\mathbb{C}$.

A special complex number is the *imaginary unit* $(0, 1)$, which satisfies

$$(0, 1) \cdot (0, 1) = (0 \cdot 0 - 1 \cdot 1, 0 \cdot 1 + 1 \cdot 0) = (-1, 0) = -1.$$

Here again we have identified the real number -1 with the complex number $(-1, 0)$. The imaginary unit is denoted by $\mathbf{i}$, i.e.,

$$\mathbf{i} := (0, 1),$$

and hence we can write $\mathbf{i}^2 = -1$. Using the identification of $x \in \mathbb{R}$ with $(x, 0) \in \mathbb{C}$ we can write $z = (x, y) \in \mathbb{C}$ as

$$(x, y) = (x, 0) + (0, y) = (x, 0) + (0, 1) \cdot (y, 0) = x + \mathbf{i}y = \text{Re}(z) + \mathbf{i}\,\text{Im}(z).$$

In the last expression $\text{Re}(z) = x$ and $\text{Im}(z) = y$ are the abbreviations for *real part* and *imaginary part* of the complex number $z = (x, y)$. Since $(0, 1) \cdot (y, 0) = (y, 0) \cdot (0, 1)$, i.e., $\mathbf{i}y = y\mathbf{i}$, it is justified to write the complex number $x + \mathbf{i}y$ as $x + y\mathbf{i}$.

For a given complex number $z = (x, y)$ or $z = x + \mathbf{i}y$ the number $\bar{z} := (x, -y)$, respectively $\bar{z} := x - \mathbf{i}y$, is called the associated *complex conjugate* number. Using the (real) square root, the *modulus* or *absolute value* of a complex number is defined as

$$|z| := (z\bar{z})^{1/2} = \left((x + \mathbf{i}y)(x - \mathbf{i}y)\right)^{1/2} = \left(x^2 - \mathbf{i}xy + \mathbf{i}yx - \mathbf{i}^2 y^2\right)^{1/2}$$
$$= (x^2 + y^2)^{1/2}.$$

(Again, for simplicity we have omitted the multiplication sign between two complex numbers.) This equation shows that the absolute value of a complex number is a nonnegative real number. Further properties of complex numbers are stated in Exercise 3.25.

Exercises

3.1 Determine for the following $(M, \oplus)$ whether they form a group:
   (a) $M = \{x \in \mathbb{R} \mid x > 0\}$ and $\oplus : M \times M \to M$, $(a, b) \mapsto a^b$.
   (b) $M = \mathbb{R} \setminus \{0\}$ and $\oplus : M \times M \to M$, $(a, b) \mapsto \frac{a}{b}$.

3.2 Let $a, b \in \mathbb{R}$, the map

$$f_{a,b} : \mathbb{R} \times \mathbb{R} \to \mathbb{R} \times \mathbb{R}, \quad (x, y) \mapsto (ax - by, ay),$$

and the set $G = \{f_{a,b} \mid a, b \in \mathbb{R}, a \neq 0\}$ be given. Show that $(G, \circ)$ is a commutative group, where the operation $\circ$ is the composition of maps (cp. Definition 2.19).

3.3 Let $X \neq \emptyset$ be a set and let $S(X) = \{f : X \to X \mid f \text{ is bijective}\}$. Show that $(S(X), \circ)$ is a group.

3.4 Let $(G, \oplus)$ be a group. For $a \in G$ denote by $-a \in G$ the (unique) inverse element. Show the following rules for elements of G:
   (a) $-(-a) = a$.
   (b) $-(a \oplus b) = (-b) \oplus (-a)$.
   (c) $a \oplus b_1 = a \oplus b_2 \Rightarrow b_1 = b_2$.
   (d) $a_1 \oplus b = a_2 \oplus b \Rightarrow a_1 = a_2$.

3.5 Prove Theorem 3.5.

3.6 Let $(G, \oplus)$ be a group and for a fixed $a \in G$ let $Z_G(a) = \{g \in G \mid a \oplus g = g \oplus a\}$. Show that $Z_G(a)$ is a subgroup of G.
   (This subgroup of all elements of G that commute with a is called *centralizer* of a.)

3.7 Let $\varphi : G \to H$ be a group homomorphism. Show the following assertions:
   (a) If $U \subseteq G$ is a subgroup, then also $\varphi(U) \subseteq H$ is a subgroup. If, furthermore, G is commutative, then also $\varphi(U)$ is commutative (even if H is not commutative).
   (b) If $V \subseteq H$ is a subgroup, then also $\varphi^{-1}(V) \subseteq G$ is a subgroup.

3.8 Let $\varphi : G \to H$ be a group homomorphism and let e_G and e_H be the neutral elements of the groups G and H, respectively.
   (a) Show that $\varphi(e_G) = e_H$.
   (b) Let $\ker(\varphi) := \{g \in G \mid \varphi(g) = e_H\}$. Show that φ is injective if and only if $\ker(\varphi) = \{e_G\}$.

3.9 Show the properties in Definition 3.7 for $(R, +, *)$ from Example 3.8 in order to prove that $(R, +, *)$ is a commutative ring with unit. Suppose that in Example 3.8 we replace the codomain $\mathbb{R}$ of the maps by a commutative ring with unit. Is $(R, +, *)$ then still a commutative ring with unit?

3.10 For sets M, N we define the *symmetric difference* $M \triangle N := (M \setminus N) \cup (N \setminus M)$. Show that for every nonempty set M the set $(\mathcal{P}(M), \triangle, \cap)$ is a commutative ring with unit.
(This is an example of a *Boolean ring*,[5] in which every element a is *idempotent*, i.e.. satisfies $a^2 = a$.)

3.11 Let R be a ring and $n \in \mathbb{N}$. Show the following assertions:
   (a) For all $a \in R$ we have $(-a)^n = \begin{cases} a^n, & \text{if } n \text{ is even,} \\ -a^n, & \text{if } n \text{ is odd.} \end{cases}$
   (b) If there exists a unit in R and if $a^n = 0$ for $a \in R$, then $1 - a$ is invertible.
(An element $a \in R$ with $a^n = 0$ for some $n \in \mathbb{N}$ is called *nilpotent*.)

3.12 Prove Theorem 3.9 (2).

3.13 Let $R := \mathbb{R} \cup \{-\infty\}$, and let for $a, b \in R$ the operations $\oplus$ and $\otimes$ be defined by

$$a \oplus b = \max\{a, b\} \quad \text{and} \quad a \otimes b = a + b,$$

where we assume that $-\infty \leq a$ and $-\infty + a = -\infty = a + (-\infty)$ for all $a \in R$. Examine which properties of a commutative ring with unit hold for $(R, \oplus, \otimes)$, and which do not hold.

3.14 Let $(R, +, *)$ be a ring with unit and let $R^\times$ denote the set of all invertible elements of R.
   (a) Show that $(R^\times, *)$ is a group (called the *group of units* of R).
   (b) Determine the sets $\mathbb{Z}^\times$, $K^\times$, and $K[t]^\times$, when K is a field.

3.15 For fixed $n \in \mathbb{N}$ let $n\mathbb{Z} = \{nk \mid k \in \mathbb{Z}\}$ and $\mathbb{Z}/n\mathbb{Z} = \{[0], [1], \ldots, [n-1]\}$ be as in Example 2.30.
   (a) Show that $n\mathbb{Z}$ is a subgroup of $\mathbb{Z}$.
   (b) Define by

$$\oplus : \mathbb{Z}/n\mathbb{Z} \times \mathbb{Z}/n\mathbb{Z} \to \mathbb{Z}/n\mathbb{Z}, \quad ([a], [b]) \mapsto [a] \oplus [b] = [a + b],$$

$$\odot : \mathbb{Z}/n\mathbb{Z} \times \mathbb{Z}/n\mathbb{Z} \to \mathbb{Z}/n\mathbb{Z}, \quad ([a], [b]) \mapsto [a] \odot [b] = [a \cdot b],$$

---

[5] George Boole (1815–1864).

an addition and multiplication in $\mathbb{Z}/n\mathbb{Z}$, (with $+$ and $\cdot$ being the addition and multiplication in $\mathbb{Z}$). Show the following assertions:

(i) $\oplus$ and $\odot$ are well defined.

(ii) $(\mathbb{Z}/n\mathbb{Z}, \oplus, \odot)$ is a commutative ring with unit.

(iii) $(\mathbb{Z}/n\mathbb{Z}, \oplus, \odot)$ is a field if and only if n is a prime number.

3.16 Let $(R, +, *)$ be a ring with $1 \neq 0$ that has no non-trivial zero divisors, i.e., $a * b = 0$ for $a, b \in R$ implies that $a = 0$ or $b = 0$. Show that then for all $a, b, c \in R, a \neq 0$, the *cancellation law*

$$ab = ac \quad \Leftrightarrow \quad b = c$$

holds. Can you find a ring with $1 \neq 0$ in which this cancellation law does not hold?

3.17 Let $(R, +, *)$ be a ring. Show that $(S, +, *)$ is a subring of $(R, +, *)$ (cp. Definition 3.11) if and only if the following properties hold:

(1) $S \subseteq R$.

(2) $0_R \in S$.

(3) For all $r, s \in S$ also $r + s \in S$ and $r * s \in S$.

(4) For all $r \in S$ also $-r \in S$.

3.18 Show that the Definitions 3.16 and 3.17 of a field describe the same mathematical structure.

3.19 Let $(K, +, *)$ be a field. Show that $(L, +, *)$ is a subfield of $(K, +, *)$ (cp. Definition 3.19), if and only if the following properties hold:

(1) $L \subseteq K$.

(2) $0_K, 1_K \in L$.

(3) For all $a, b \in L$ also $a + b \in L$ and $a * b \in L$.

(4) For all $a \in L$ also $-a \in L$.

(5) For all $a \in L \setminus \{0\}$ also $a^{-1} \in L$.

3.20 Show that in a field $1 + 1 = 0$ holds if and only if $1 + 1 + 1 + 1 = 0$.

3.21 Let $(R, +, *)$ be a commutative ring with $1 \neq 0$ that does not contain non-trivial zero divisors, i.e., if $a * b = 0$ for some $a, b \in R$, then $a = 0$ or $b = 0$. (Such a ring is called an *integral domain*.)

(a) Define on $M = R \times R \setminus \{0\}$ a relation by

$$(x, y) \sim (\widehat{x}, \widehat{y}) \quad \Leftrightarrow \quad x * \widehat{y} = y * \widehat{x}.$$

Show that this is an equivalence relation.

(b) Denote the equivalence class $[(x, y)]$ by $\frac{x}{y}$. Show that the following maps are well defined:

$$\oplus : (M/\sim) \times (M/\sim) \to (M/\sim) \quad \text{with} \quad \frac{x}{y} \oplus \frac{\widehat{x}}{\widehat{y}} := \frac{x * \widehat{y} + y * \widehat{x}}{y * \widehat{y}},$$

$$\odot : (M/\sim) \times (M/\sim) \to (M/\sim) \quad \text{with} \quad \frac{x}{y} \odot \frac{\widehat{x}}{\widehat{y}} := \frac{x * \widehat{x}}{y * \widehat{y}},$$

where $M/\sim$ denotes the quotient set with respect to $\sim$ (cp. Definition 2.28).
   (c) Show that $(M/\sim, \oplus, \odot)$ is a field. (This field is called the *quotient field* associated with R.)
   (d) Which field is $(M/\sim, \oplus, \odot)$ for $R = \mathbb{Z}$?
3.22 In Exercise 3.21 consider $R = K[t]$, the ring of polynomials over the field K, and construct in this way the field of *rational functions*.
3.23 Let $a = 2 + \mathbf{i} \in \mathbb{C}$ and $b = 1 - 3\mathbf{i} \in \mathbb{C}$. Determine $-a, -b, a+b, a-b$, $a^{-1}, b^{-1}, a^{-1}a, b^{-1}b, ab, ba$.
3.24 Show the following rules for complex numbers:
   (a) $\overline{z_1 + z_2} = \overline{z}_1 + \overline{z}_2$ and $\overline{z_1 z_2} = \overline{z}_1 \overline{z}_2$ for all $z_1, z_2 \in \mathbb{C}$.
   (b) $\overline{z^{-1}} = (\overline{z})^{-1}$ and $\text{Re}(z^{-1}) = \frac{1}{|z|^2}\text{Re}(z)$ for all $z \in \mathbb{C} \setminus \{0\}$.
3.25 Show that the absolute value of complex numbers satisfies the following properties:
   (a) $|z_1 z_2| = |z_1||z_2|$ for all $z_1, z_2 \in \mathbb{C}$.
   (b) $|z| \geq 0$ for all $z \in \mathbb{C}$ with equality if and only if $z = 0$.
   (c) $|z_1 + z_2| \leq |z_1| + |z_2|$ for all $z_1, z_2 \in \mathbb{C}$. (This inequality is called *triangle inequality*.)
   (Because of these three properties, the absolute value of complex numbers is a *norm*. We will discuss this concept in detail in Chap. 12.)

Matrices

4

In this chapter we define matrices with their most important operations and we study several groups and rings of matrices. James Joseph Sylvester (1814–1897) coined the term *matrix*[1] in 1850 and described matrices as "an oblong arrangement of terms". The matrix operations defined in this chapter were introduced by Arthur Cayley (1821–1895) in 1858. His article "A memoir on the theory of matrices" was the first to consider matrices as independent algebraic objects. In our book matrices form the central approach to the theory of Linear Algebra.

4.1 Basic Definitions and Operations

We begin with a formal definition of matrices.

Definition 4.1 Let R be a commutative ring with unit and let $n, m \in \mathbb{N}_0$. An array of the form

$$A = [a_{ij}] = \begin{bmatrix} a_{11} & a_{12} & \cdots & a_{1m} \\ a_{21} & a_{22} & \cdots & a_{2m} \\ \vdots & \vdots & & \vdots \\ a_{n1} & a_{n2} & \cdots & a_{nm} \end{bmatrix}$$

---

[1] The Latin word "matrix" means "womb". Sylvester considered matrices as objects "out of which we may form various systems of determinants" (cp. Chap. 5). Interestingly, the English writer Charles Lutwidge Dodgson (1832–1898), better known by his pen name Lewis Carroll, objected to Sylvester's term and wrote in 1867: "I am aware that the word 'Matrix' is already in use to express the very meaning for which I use the word 'Block'; but surely the former word means rather the mould, or form, into which algebraic quantities may be introduced, than an actual assemblage of such quantities". Dodgson also objected to the notation a_{ij} for the matrix entries: "... most of the space is occupied by a number of a's, which are wholly superfluous, while the only important part of the notation is reduced to minute subscripts, alike difficult to the writer and the reader."

with $a_{ij} \in R$, $i = 1, \ldots, n$, $j = 1, \ldots, m$, is called a *matrix* of size $n \times m$ over R. The a_{ij} are called the *entries* or *coefficients* of the matrix. The set of all such matrices is denoted by $R^{n,m}$.

In the following we usually assume (without explicitly mentioning it) that $1 \neq 0$ in R. This excludes the trivial case of the zero ring (cp. (2) in Theorem 3.9.)

Formally, in Definition 4.1 for $n = 0$ or $m = 0$ we obtain "empty matrices" of the size $0 \times m$, $n \times 0$ or 0×0. We denote such matrices by [ ]. They will be used for technical reasons in some of the proofs below. When we analyze algebraic properties of matrices, however, we always consider $n, m \geq 1$.

The *zero matrix* in $R^{n,m}$, denoted by $0_{n,m}$ or just 0, is the matrix that has all its entries equal to $0 \in R$.

A matrix of size $n \times n$ is called a *square matrix* or just *square*. The entries a_{ii} for $i = 1, \ldots, n$ are called the *diagonal entries* of A. The *identity matrix* in $R^{n,n}$ is the matrix $I_n := [\delta_{ij}]$, where

$$\delta_{ij} := \begin{cases} 1, & \text{if } i = j, \\ 0, & \text{if } i \neq j. \end{cases} \tag{4.1}$$

is the *Kronecker delta-function*.[2] If it is clear which n is considered, then we just write I instead of I_n. For $n = 0$ we set $I_0 := $ [ ].

The ith *row* of $A \in R^{n,m}$ is $[a_{i1}, a_{i2}, \ldots, a_{im}] \in R^{1,m}$, $i = 1, \ldots, n$, where we use commas for the optical separation of the entries. The jth *column* of A is

$$\begin{bmatrix} a_{1j} \\ a_{2j} \\ \vdots \\ a_{nj} \end{bmatrix} \in R^{n,1}, \quad j = 1, \ldots, m.$$

Thus, the rows and columns of a matrix are again matrices.

If $1 \times m$ matrices $a_i := [a_{i1}, a_{i2}, \ldots, a_{im}] \in R^{1,m}$, $i = 1, \ldots, n$, are given, then we can combine them to the matrix

$$A = \begin{bmatrix} a_1 \\ a_2 \\ \vdots \\ a_n \end{bmatrix} = \begin{bmatrix} a_{11} & a_{12} & \cdots & a_{1m} \\ a_{21} & a_{22} & \cdots & a_{2m} \\ \vdots & \vdots & & \vdots \\ a_{n1} & a_{n2} & \cdots & a_{nm} \end{bmatrix} \in R^{n,m}.$$

---

[2] Leopold Kronecker (1823–1891).

4.1 Basic Definitions and Operations

We then do not write square brackets around the rows of A. In the same way we can combine the $n \times 1$ matrices

$$a_j := \begin{bmatrix} a_{1j} \\ a_{2j} \\ \vdots \\ a_{nj} \end{bmatrix} \in R^{n,1}, \quad j = 1, \ldots, m,$$

to the matrix

$$A = [a_1, a_2, \ldots, a_m] = \begin{bmatrix} a_{11} & a_{12} & \cdots & a_{1m} \\ a_{21} & a_{22} & \cdots & a_{2m} \\ \vdots & \vdots & & \vdots \\ a_{n1} & a_{n2} & \cdots & a_{nm} \end{bmatrix} \in R^{n,m}.$$

If $n_1, n_2, m_1, m_2 \in \mathbb{N}_0$ and $A_{ij} \in R^{n_i, m_j}$, $i, j = 1, 2$, then we can combine these four matrices to the matrix

$$A = \begin{bmatrix} A_{11} & A_{12} \\ A_{21} & A_{22} \end{bmatrix} \in R^{n_1+n_2, m_1+m_2}.$$

The matrices A_{ij} are then called *blocks* of the *block matrix* A. In order to visualize the partitioning of matrices into blocks, we sometimes use vertical and horizontal lines, e.g.,

$$A = \left[\begin{array}{c|c} A_{11} & A_{12} \\ \hline A_{21} & A_{22} \end{array} \right].$$

We now introduce four operations for matrices and begin with the *addition*:

$$+ : R^{n,m} \times R^{n,m} \to R^{n,m}, \quad (A, B) \mapsto A + B := [a_{ij} + b_{ij}].$$

The addition in $R^{n,m}$ operates *entrywise*, based on the addition in R. Note that the addition is only defined for matrices of equal size.

The *multiplication* of two matrices is defined as follows:

$$* : R^{n,m} \times R^{m,s} \to R^{n,s}, \quad (A, B) \mapsto A * B = [c_{ij}], \quad c_{ij} := \sum_{k=1}^{m} a_{ik} b_{kj}.$$

Thus, the entry c_{ij} of the product $A * B$ is constructed by successive multiplication and summing up the entries in the ith row of A and the jth column of B. Clearly, in order to define the product $A * B$, the number of columns of A must be equal to the number of rows in B.

In the definition of the entries c_{ij} of the matrix $A * B$ we have not written the multiplication symbol for the elements in R. This follows the usual convention of omitting the multiplication sign when it is clear which multiplication is considered. Eventually we will also omit the multiplication sign between matrices.

We can illustrate the multiplication rule "c_{ij} equals ith row of A times jth column of B" as follows:

$$\begin{bmatrix} b_{11} & \cdots \\ \vdots & \\ b_{m1} & \cdots \end{bmatrix} \begin{bmatrix} b_{1j} \\ \vdots \\ b_{mj} \end{bmatrix} \begin{matrix} \cdots \\ \\ \cdots \end{matrix} \begin{bmatrix} b_{1s} \\ \vdots \\ b_{ms} \end{bmatrix}$$

$$\begin{bmatrix} a_{11} & \cdots & a_{1m} \\ \vdots & & \vdots \\ [\, a_{i1} & \cdots & a_{im} \,] \\ \vdots & & \vdots \\ a_{n1} & \cdots & a_{nm} \end{bmatrix} \quad \begin{bmatrix} & \downarrow & \\ \longrightarrow & c_{ij} & \\ & & \end{bmatrix}$$

It is important to note that the matrix multiplication in general is not commutative.

Example 4.2 For the matrices

$$A = \begin{bmatrix} 1 & 2 & 3 \\ 4 & 5 & 6 \end{bmatrix} \in \mathbb{Z}^{2,3}, \quad B = \begin{bmatrix} -1 & 1 \\ 0 & 0 \\ 1 & -1 \end{bmatrix} \in \mathbb{Z}^{3,2}$$

we have

$$A * B = \begin{bmatrix} 2 & -2 \\ 2 & -2 \end{bmatrix} \in \mathbb{Z}^{2,2}.$$

On the other hand, $B * A \in \mathbb{Z}^{3,3}$. Although $A * B$ and $B * A$ are both defined, we obviously have $A * B \neq B * A$. In this case one recognizes the non-commutativity of the matrix multiplication from the fact that $A * B$ and $B * A$ have different sizes. But even if $A * B$ and $B * A$ are both defined and have the same size, in general $A * B \neq B * A$. For example,

$$A = \begin{bmatrix} 1 & 2 \\ 0 & 3 \end{bmatrix} \in \mathbb{Z}^{2,2}, \quad B = \begin{bmatrix} 4 & 0 \\ 5 & 6 \end{bmatrix} \in \mathbb{Z}^{2,2}$$

4.1 Basic Definitions and Operations

yield the two products

$$A * B = \begin{bmatrix} 14 & 12 \\ 15 & 18 \end{bmatrix} \quad \text{and} \quad B * A = \begin{bmatrix} 4 & 8 \\ 5 & 28 \end{bmatrix}.$$

The matrix multiplication is, however, associative and distributive with respect to the matrix addition.

Lemma 4.3 *For $A, \widehat{A} \in R^{n,m}$, $B, \widehat{B} \in R^{m,\ell}$ and $C \in R^{\ell,k}$ the following assertions hold:*

(1) $A * (B * C) = (A * B) * C$.
(2) $(A + \widehat{A}) * B = A * B + \widehat{A} * B$.
(3) $A * (B + \widehat{B}) = A * B + A * \widehat{B}$.
(4) $I_n * A = A * I_m = A$.

Proof We only show property (1); the others are exercises. Let $A \in R^{n,m}$, $B \in R^{m,\ell}$, $C \in R^{\ell,k}$ as well as $(A * B) * C = [d_{ij}]$ and $A * (B * C) = [\widetilde{d}_{ij}]$. By the definition of the matrix multiplication and using the associative and distributive law in R, we get

$$d_{ij} = \sum_{s=1}^{\ell} \left(\sum_{t=1}^{m} a_{it} b_{ts} \right) c_{sj} = \sum_{s=1}^{\ell} \sum_{t=1}^{m} (a_{it} b_{ts}) c_{sj} = \sum_{s=1}^{\ell} \sum_{t=1}^{m} a_{it} (b_{ts} c_{sj})$$

$$= \sum_{t=1}^{m} a_{it} \left(\sum_{s=1}^{\ell} b_{ts} c_{sj} \right) = \widetilde{d}_{ij},$$

for $1 \leq i \leq n$ and $1 \leq j \leq k$, which implies that $(A * B) * C = A * (B * C)$. □

On the right hand sides of (2) and (3) in Lemma 4.3 we have not written parentheses, since we will use the common convention that the multiplication of matrices binds stronger than the addition.

For $A \in R^{n,n}$ we define

$$A^k := \underbrace{A * \ldots * A}_{k \text{ times}} \quad \text{for } k \in \mathbb{N},$$

$$A^0 := I_n.$$

Another multiplicative operation for matrices is the multiplication with a *scalar*,[3] which is defined as follows:

$$\cdot : R \times R^{n,m} \to R^{n,m}, \quad (\lambda, A) \mapsto \lambda \cdot A := [\lambda a_{ij}]. \tag{4.2}$$

We easily see that $0 \cdot A = 0_{n,m}$ and $1 \cdot A = A$ for all $A \in R^{n,m}$. For $A = [a_{ij}]$ we write $-A := (-1) \cdot A = [-a_{ij}]$. In addition, the scalar multiplication has the following properties.

Lemma 4.4 *For $A, B \in R^{n,m}$, $C \in R^{m,\ell}$ and $\lambda, \mu \in R$ the following assertions hold:*

(1) $(\lambda \mu) \cdot A = \lambda \cdot (\mu \cdot A)$.
(2) $(\lambda + \mu) \cdot A = \lambda \cdot A + \mu \cdot A$.
(3) $\lambda \cdot (A + B) = \lambda \cdot A + \lambda \cdot B$.
(4) $(\lambda \cdot A) * C = \lambda \cdot (A * C) = A * (\lambda \cdot C)$.

Proof Exercise. □

The fourth matrix operation that we introduce is the *transposition*:

$$T : R^{n,m} \to R^{m,n}, \quad A = [a_{ij}] \mapsto A^T = [b_{ij}], \quad b_{ij} := a_{ji}.$$

For example,

$$A = \begin{bmatrix} 1 & 2 & 3 \\ 4 & 5 & 6 \end{bmatrix} \in \mathbb{Z}^{2,3}, \quad A^T = \begin{bmatrix} 1 & 4 \\ 2 & 5 \\ 3 & 6 \end{bmatrix} \in \mathbb{Z}^{3,2}.$$

The matrix A^T is called the *transpose* of A.

Definition 4.5 If $A \in R^{n,n}$ satisfies $A = A^T$, then A is called *symmetric*. If $A = -A^T$, then A is called *skew-symmetric*.

For the transposition we have the following properties.

Lemma 4.6 *For $A, \widehat{A} \in R^{n,m}$, $B \in R^{m,\ell}$ and $\lambda \in R$ the following assertions hold:*

(1) $(A^T)^T = A$.
(2) $(A + \widehat{A})^T = A^T + \widehat{A}^T$.

---

[3] The term "scalar" was introduced in 1845 by Sir William Rowan Hamilton (1805–1865). It originates from the Latin word "scale" which means "ladder".

4.1 Basic Definitions and Operations

(3) $(\lambda \cdot A)^T = \lambda \cdot A^T$.
(4) $(A * B)^T = B^T * A^T$.

Proof Properties (1)–(3) are exercises. For the proof of (4) let $A * B = [c_{ij}]$ with $c_{ij} = \sum_{k=1}^{m} a_{ik} b_{kj}$, $A^T = [\widetilde{a}_{ij}]$, $B^T = [\widetilde{b}_{ij}]$ and $(A * B)^T = [\widetilde{c}_{ij}]$. Then

$$\widetilde{c}_{ij} = c_{ji} = \sum_{k=1}^{m} a_{jk} b_{ki} = \sum_{k=1}^{m} \widetilde{a}_{kj} \widetilde{b}_{ik} = \sum_{k=1}^{m} \widetilde{b}_{ik} \widetilde{a}_{kj},$$

from which we see that $(A * B)^T = B^T * A^T$. □

MATLAB-Minute 1

Carry out the following commands in order to get used to the matrix operations of this chapter in MATLAB notation:
A=ones(6,3), A+A, A-2*A, A', A*A, A'*A, A*A'.
B=ones(3,6), A*B, B*A.
C=ones(3,0), C', A*C, C'*B, C'*C, C*C'.
(In order to see MATLAB's output, do not put a semicolon at the end of the command.)

Example 4.7 Consider again the example of car insurance premiums from Chap. 1. Recall that p_{ij} denotes the probability that a customer in class C_i in this year will move to the class C_j. Our example consists of four such classes, and the 16 probabilities can be associated with a row-stochastic 4×4 matrix (cp. (1.2)), which we denote by P. Suppose that the insurance company has the following distribution of customers in the four classes: 40% in class C_1, 30% in class C_2, 20% in class C_3, and 10% in class C_4. Then the 1×4 matrix

$$p_0 := [0.4,\ 0.3,\ 0.2,\ 0.1]$$

describes the initial customer distribution. Using the matrix multiplication we now compute

$$p_1 := p_0 * P = [0.4,\ 0.3,\ 0.2,\ 0.1] * \begin{bmatrix} 0.15 & 0.85 & 0.00 & 0.00 \\ 0.15 & 0.00 & 0.85 & 0.00 \\ 0.05 & 0.10 & 0.00 & 0.85 \\ 0.05 & 0.00 & 0.10 & 0.85 \end{bmatrix}$$

$$= [0.12,\ 0.36,\ 0.265,\ 0.255].$$

Then p_1 contains the distribution of the customers in the next year. As an example, consider the entry of $p_0 * P$ in position $(1, 4)$, which is computed by

$$0.4 \cdot 0.00 + 0.3 \cdot 0.00 + 0.2 \cdot 0.85 + 0.1 \cdot 0.85 = 0.255.$$

A customer in the classes C_1 or C_2 in this year cannot move to the class C_4. Thus, the respective initial percentages are multiplied by the probabilities $p_{14} = 0.00$ and $p_{24} = 0.00$. A customer in the class C_3 or C_4 will be in the class C_4 with the probabilities $p_{34} = 0.85$ or $p_{44} = 0.85$, respectively. This yields the two products $0.2 \cdot 0.85$ and $0.1 \cdot 0.85$.

Continuing in the same way we obtain after k years the distribution

$$p_k := p_0 * P^k, \quad k = 0, 1, 2, \ldots.$$

(This formula also holds for $k = 0$, since $P^0 = I_4$.) The insurance company can use this formula to compute the revenue from the payments of premium rates in the coming years. Assume that the full premium rate (class C_1) is 500 Euros per year. Then the rates in classes C_2, C_3, and C_4 are 450, 400 and 300 Euros (10, 20 and 40% discount). If there are 1000 customers initially, then the revenue in the first year (in Euros) is

$$1000 \cdot \left(p_0 * [500, 450, 400, 300]^T \right) = 445{,}000.$$

If no customer cancels the contract, then this model yields the revenue in year $k \geq 0$ as

$$1000 \cdot \left(p_k * [500, 450, 400, 300]^T \right)$$
$$= 1000 \cdot \left(p_0 * (P^k * [500, 450, 400, 300]^T) \right).$$

For example, the revenue in the next 4 years is 404,500, 372,025, 347,340 and 341,819 (rounded to full Euros). These numbers decrease annually, but the rate of the decrease seems to slow down. Does there exists a "stationary state", i.e., a state when the revenue is not changing (significantly) any more? Which properties of the model guarantee the existence of such a state? These are important practical questions for the insurance company. Only the existence of a stationary state guarantees significant revenues in the long-time future. Since the formula depends essentially on the entries of the matrix P^k, we have reached an interesting problem of Linear Algebra: the analysis of the properties of row-stochastic matrices. We will analyze these properties in Sect. 8.3.

4.2 Matrix Groups and Rings

In this section we study algebraic structures that are formed by certain sets of matrices and the matrix operations introduced above. We begin with the addition in $R^{n,m}$.

Theorem 4.8 *($R^{n,m}, +$) is a commutative group. The neutral element is $0 \in R^{n,m}$ (the zero matrix) and for $A = [a_{ij}] \in R^{n,m}$ the inverse element is $-A = [-a_{ij}] \in R^{n,m}$. (We write $A - B$ instead of $A + (-B)$.)*

Proof Using the associativity of the addition in R, for arbitrary $A, B, C \in R^{n,m}$, we obtain

$$(A + B) + C = [a_{ij} + b_{ij}] + [c_{ij}] = [(a_{ij} + b_{ij}) + c_{ij}] = [a_{ij} + (b_{ij} + c_{ij})]$$
$$= [a_{ij}] + [b_{ij} + c_{ij}] = A + (B + C).$$

Thus, the addition in $R^{n,m}$ is associative.

The zero matrix $0 \in R^{n,m}$ satisfies $0 + A = [0] + [a_{ij}] = [0 + a_{ij}] = [a_{ij}] = A$. For a given $A = [a_{ij}] \in R^{n,m}$ and $-A := [-a_{ij}] \in R^{n,m}$ we have $-A + A = [-a_{ij}] + [a_{ij}] = [-a_{ij} + a_{ij}] = [0] = 0$.

Finally, the commutativity of the addition in R implies that $A + B = [a_{ij}] + [b_{ij}] = [a_{ij} + b_{ij}] = [b_{ij} + a_{ij}] = B + A$. □

Note that (2) in Lemma 4.6 implies that the transposition is a homomorphism (even an isomorphism) between the groups $(R^{n,m}, +)$ and $(R^{m,n}, +)$ (cp. Definition 3.6).

Theorem 4.9 *($R^{n,n}, +, *$) is a ring with unit given by the identity matrix I_n. This ring is commutative only for $n = 1$.*

Proof We have already shown that $(R^{n,n}, +)$ is a commutative group (cp. Theorem 4.8). The other properties of a ring (associativity, distributivity and the existence of a unit element) follow from Lemma 4.3. The commutativity for $n = 1$ holds because of the commutativity of the multiplication in the ring R. The example

$$\begin{bmatrix} 0 & 1 \\ 0 & 0 \end{bmatrix} * \begin{bmatrix} 1 & 0 \\ 0 & 0 \end{bmatrix} = \begin{bmatrix} 0 & 0 \\ 0 & 0 \end{bmatrix} \neq \begin{bmatrix} 0 & 1 \\ 0 & 0 \end{bmatrix} = \begin{bmatrix} 1 & 0 \\ 0 & 0 \end{bmatrix} * \begin{bmatrix} 0 & 1 \\ 0 & 0 \end{bmatrix} \quad (4.3)$$

shows that the ring $R^{n,n}$ is not commutative for $n \geq 2$. □

The example in (4.3) shows that for $n \geq 2$ the ring $R^{n,n}$ has non-trivial zero-divisors, i.e., there exist matrices $A, B \in R^{n,n} \setminus \{0\}$ with $A * B = 0$. These exist even when R is a field.

Let us now consider the *invertibility* of matrices in the ring $R^{n,n}$ (with respect to the matrix multiplication). For a given matrix $A \in R^{n,n}$, an inverse $\widetilde{A} \in R^{n,n}$ must satisfy the two equations $\widetilde{A} * A = I_n$ and $A * \widetilde{A} = I_n$ (cp. Definition 3.12). If an inverse of $A \in R^{n,n}$ exists, i.e., if A is *invertible*, then the inverse is unique and denoted by A^{-1} (cp. Theorem 3.13). An invertible matrix is sometimes called *non-singular*, while a non-invertible matrix is called *singular*. We will show in Corollary 7.19 that the existence of the inverse already is implied by one of the two equations $\widetilde{A} * A = I_n$ and $A * \widetilde{A} = I_n$, i.e., if one of them holds, then A is invertible and $A^{-1} = \widetilde{A}$. Until then, to be correct, we will have to check the validity of both equations.

Not all matrices $A \in R^{n,n}$ are invertible. Simple examples are the non-invertible matrices

$$A = [0] \in R^{1,1} \quad \text{and} \quad A = \begin{bmatrix} 1 & 0 \\ 0 & 0 \end{bmatrix} \in R^{2,2}.$$

Another non-invertible matrix is

$$A = \begin{bmatrix} 1 & 1 \\ 0 & 2 \end{bmatrix} \in \mathbb{Z}^{2,2}.$$

However, considered as an element of $\mathbb{Q}^{2,2}$, the (unique) inverse of A is given by

$$A^{-1} = \begin{bmatrix} 1 & -\frac{1}{2} \\ 0 & \frac{1}{2} \end{bmatrix} \in \mathbb{Q}^{2,2}.$$

Lemma 4.10 *If $A, B \in R^{n,n}$ are invertible, then the following assertions hold:*

(1) A^T is invertible with $(A^T)^{-1} = (A^{-1})^T$. (We also write this matrix as A^{-T}.)
(2) $A * B$ is invertible with $(A * B)^{-1} = B^{-1} * A^{-1}$.

Proof

(1) Using (4) in Lemma 4.6 we have

$$(A^{-1})^T * A^T = (A * A^{-1})^T = I_n^T = I_n = I_n^T = (A^{-1} * A)^T = A^T * (A^{-1})^T,$$

and thus $(A^{-1})^T$ is the inverse of A^T.
(2) This was already shown in Theorem 3.13 for general rings with unit and thus it holds, in particular, for the ring $(R^{n,n}, +, *)$. □

Our next result shows that the invertible matrices form a multiplicative group.

4.2 Matrix Groups and Rings

Theorem 4.11 *The set of invertible $n \times n$ matrices over R forms a group with respect to the matrix multiplication. We denote this group by $GL_n(R)$ ("GL" abbreviates "general linear (group)").*

Proof The associativity of the multiplication in $GL_n(R)$ is clear. As shown in (2) in Lemma 4.10, the product of two invertible matrices is an invertible matrix. The neutral element in $GL_n(R)$ is the identity matrix I_n, and since every $A \in GL_n(R)$ is assumed to be invertible, A^{-1} exists with $(A^{-1})^{-1} = A \in GL_n(R)$. □

We next introduce some important classes of matrices.

Definition 4.12 Let $A = [a_{ij}] \in R^{n,n}$.

(1) A is called *upper triangular*, if $a_{ij} = 0$ for all $i > j$.
    A is called *lower triangular*, if $a_{ij} = 0$ for all $j > i$ (i.e., A^T is upper triangular).
(2) A is called *diagonal*, if $a_{ij} = 0$ for all $i \neq j$ (i.e., A is upper and lower triangular). We write a diagonal matrix as $A = \text{diag}(a_{11}, \ldots, a_{nn})$.

We now investigate these sets of matrices with respect to their group properties, beginning with the invertible upper and lower triangular matrices.

Theorem 4.13 *The sets of the invertible upper triangular $n \times n$ matrices and of the invertible lower triangular $n \times n$ matrices over R form subgroups of $GL_n(R)$.*

Proof We will only show the result for the upper triangular matrices; the proof for the lower triangular matrices is analogous. In order to establish the subgroup property we will prove the three properties from Theorem 3.5.

Since I_n is an invertible upper triangular matrix, the set of the invertible upper triangular matrices is a nonempty subset of $GL_n(R)$.

Next we show that for two invertible upper triangular matrices $A, B \in R^{n,n}$ the product $C = A * B$ is again an invertible upper triangular matrix. The invertibility of $C = [c_{ij}]$ follows from (2) in Lemma 4.10. For $i > j$ we have

$$c_{ij} = \sum_{k=1}^{n} a_{ik} b_{kj} \quad \text{(here } b_{kj} = 0 \text{ for } k > j\text{)}$$

$$= \sum_{k=1}^{j} a_{ik} b_{kj} \quad \text{(here } a_{ik} = 0 \text{ for } k = 1, \ldots, j, \text{ since } i > j\text{)}$$

$$= 0.$$

Therefore, C is upper triangular.

It remains to prove that the inverse A^{-1} of an invertible upper triangular matrix A is an upper triangular matrix. For $n = 1$ the assertion holds trivially, so we assume that $n \geq 2$. Let $A^{-1} = [c_{ij}]$, then the equation $A * A^{-1} = I_n$ can be written as a system of n equations

$$\begin{bmatrix} a_{11} & \cdots & \cdots & a_{1n} \\ 0 & \ddots & & \vdots \\ \vdots & \ddots & \ddots & \vdots \\ 0 & \cdots & 0 & a_{nn} \end{bmatrix} * \begin{bmatrix} c_{1j} \\ \vdots \\ \vdots \\ c_{nj} \end{bmatrix} = \begin{bmatrix} \delta_{1j} \\ \vdots \\ \vdots \\ \delta_{nj} \end{bmatrix}, \quad j = 1, \ldots, n. \qquad (4.4)$$

Here, δ_{ij} is the Kronecker delta-function defined in (4.1).

We will now prove inductively for $i = n, n-1, \ldots, 1$ that the diagonal entry a_{ii} of A is invertible with $a_{ii}^{-1} = c_{ii}$, and that

$$c_{ij} = a_{ii}^{-1} \left(\delta_{ij} - \sum_{\ell=i+1}^{n} a_{i\ell} c_{\ell j} \right), \quad j = 1, \ldots, n. \qquad (4.5)$$

This formula implies, in particular, that $c_{ij} = 0$ for $i > j$.

For $i = n$ the last row of (4.4) is given by

$$a_{nn} c_{nj} = \delta_{nj}, \quad j = 1, \ldots, n.$$

For $j = n$ we have $a_{nn} c_{nn} = 1 = c_{nn} a_{nn}$, where in the second equation we use the commutativity of the multiplication in R. Therefore, a_{nn} is invertible with $a_{nn}^{-1} = c_{nn}$, and thus

$$c_{nj} = a_{nn}^{-1} \delta_{nj}, \quad j = 1, \ldots, n.$$

This is equivalent to (4.5) for $i = n$. (Note that for $i = n$ in (4.5) the sum is empty and thus equal to zero.) In particular, $c_{nj} = 0$ for $j = 1, \ldots, n-1$.

Now assume that our assertion holds for $i = n, \ldots, k+1$, where $1 \leq k \leq n-1$. Then, in particular, $c_{ij} = 0$ for $k+1 \leq i \leq n$ and $i > j$. In words, the rows $i = n, \ldots, k+1$ of A^{-1} are in "upper triangular from". In order to prove the assertion for $i = k$, we consider the kth row in (4.4), which is given by

$$a_{kk} c_{kj} + a_{k,k+1} c_{k+1,j} + \ldots + a_{kn} c_{nj} = \delta_{kj}, \quad j = 1, \ldots, n. \qquad (4.6)$$

For $j = k \, (< n)$ we obtain

$$a_{kk} c_{kk} + a_{k,k+1} c_{k+1,k} + \ldots + a_{kn} c_{nk} = 1.$$

4.2 Matrix Groups and Rings

By the induction hypothesis, we have $c_{k+1,k} = \cdots = c_{n,k} = 0$. This implies $a_{kk}c_{kk} = 1 = c_{kk}a_{kk}$, where we have used the commutativity of the multiplication in R. Hence a_{kk} is invertible with $a_{kk}^{-1} = c_{kk}$. From (4.6) we get

$$c_{kj} = a_{kk}^{-1}\left(\delta_{kj} - a_{k,k+1}c_{k+1,j} - \cdots - a_{kn}c_{nj}\right), \quad j = 1,\ldots,n,$$

and hence (4.5) holds for $i = k$. If $k > j$, then $\delta_{kj} = 0$ and $c_{k+1,j} = \cdots = c_{nj} = 0$, which gives $c_{kj} = 0$. □

We point out that (4.5) represents a *recursive formula* for computing the entries of the inverse of an invertible upper triangular matrix. Using this formula the entries are computed "from bottom to top" and "from right to left". This process is sometimes called *backward substitution*.

In the following we will frequently partition matrices into blocks and make use of the *block multiplication*: For every $k \in \{1,\ldots,n-1\}$, we can write $A \in R^{n,n}$ as

$$A = \begin{bmatrix} A_{11} & A_{12} \\ A_{21} & A_{22} \end{bmatrix} \quad \text{with } A_{11} \in R^{k,k} \text{ and } A_{22} \in R^{n-k,n-k}.$$

If $A, B \in R^{n,n}$ are both partitioned like this, then the product $A * B$ can be evaluated blockwise, i.e.,

$$\begin{bmatrix} A_{11} & A_{12} \\ A_{21} & A_{22} \end{bmatrix} * \begin{bmatrix} B_{11} & B_{12} \\ B_{21} & B_{22} \end{bmatrix} = \begin{bmatrix} A_{11}*B_{11} + A_{12}*B_{21} & A_{11}*B_{12} + A_{12}*B_{22} \\ A_{21}*B_{11} + A_{22}*B_{21} & A_{21}*B_{12} + A_{22}*B_{22} \end{bmatrix}. \tag{4.7}$$

In particular, if

$$A = \begin{bmatrix} A_{11} & A_{12} \\ 0 & A_{22} \end{bmatrix}$$

with $A_{11} \in GL_k(R)$ and $A_{22} \in GL_{n-k}(R)$, then $A \in GL_n(R)$ and a direct computation shows that

$$A^{-1} = \begin{bmatrix} A_{11}^{-1} & -A_{11}^{-1}*A_{12}*A_{22}^{-1} \\ 0 & A_{22}^{-1} \end{bmatrix}. \tag{4.8}$$

> **MATLAB-Minute 2**
> Create block matrices in MATLAB by carrying out the following commands:
> ```
> k=5;
> A11=gallery('tridiag',-ones(k-1,1),2*ones(k,1),-ones(k-1,1));
> A12=zeros(k,2); A12(1,1)=1; A12(2,2)=1;
> A22=-eye(2);
> A=full([A11 A12; A12' A22])
> B=full([A11 A12; zeros(2,k) -A22])
> ```
> Investigate the meaning of the command `full`. Compute the products A*B and B*A as well as the inverses `inv(A)` and `inv(B)`. Compute the inverse of B in MATLAB with the formula (4.8).

Corollary 4.14 *The set of the invertible diagonal $n \times n$ matrices over R forms a commutative subgroup (with respect to the matrix multiplication) of the invertible upper (or lower) triangular $n \times n$ matrices over R.*

Proof Since I_n is an invertible diagonal matrix, the invertible diagonal $n \times n$ matrices form a nonempty subset of the invertible upper (or lower) triangular $n \times n$ matrices. If $A = \text{diag}(a_{11}, \ldots, a_{nn})$ and $B = \text{diag}(b_{11}, \ldots, b_{nn})$ are invertible, then $A * B$ is invertible (cp. (2) in Lemma 4.10) and diagonal, since

$$A * B = \text{diag}(a_{11}, \ldots, a_{nn}) * \text{diag}(b_{11}, \ldots, b_{nn}) = \text{diag}(a_{11}b_{11}, \ldots, a_{nn}b_{nn}).$$

Moreover, if $A = \text{diag}(a_{11}, \ldots, a_{nn})$ is invertible, then $a_{ii} \in R$ is invertible for all $i = 1, \ldots, n$ (cp. the proof of Theorem 4.13). The inverse A^{-1} is given by the invertible diagonal matrix $\text{diag}(a_{11}^{-1}, \ldots, a_{nn}^{-1})$. Finally, the commutativity property $A * B = B * A$ follows directly from the commutativity in R. □

Definition 4.15 A matrix $P \in R^{n,n}$ is called a *permutation matrix*, if in every row and every column of P there is exactly one unit and all other entries are zero.

The term "permutation" means "exchange". If a matrix $A \in R^{n,n}$ is multiplied with a permutation matrix from the left or from the right, then its rows or columns, respectively, are exchanged (or permuted). For example, if

$$P = \begin{bmatrix} 0 & 0 & 1 \\ 0 & 1 & 0 \\ 1 & 0 & 0 \end{bmatrix}, \quad A = \begin{bmatrix} 1 & 2 & 3 \\ 4 & 5 & 6 \\ 7 & 8 & 9 \end{bmatrix} \in \mathbb{Z}^{3,3},$$

then

$$P * A = \begin{bmatrix} 7 & 8 & 9 \\ 4 & 5 & 6 \\ 1 & 2 & 3 \end{bmatrix} \quad \text{and} \quad A * P = \begin{bmatrix} 3 & 2 & 1 \\ 6 & 5 & 4 \\ 9 & 8 & 7 \end{bmatrix}.$$

Theorem 4.16 *The set of the $n \times n$ permutation matrices over R forms a subgroup of $GL_n(R)$. In particular, if $P \in R^{n,n}$ is a permutation matrix, then P is invertible with $P^{-1} = P^T$.*

Proof Exercise. □

From now on we will omit the multiplication sign in the matrix multiplication and write AB instead of $A * B$.

Exercises

(In the following exercises R is a commutative ring with unit.)

4.1 Consider the following matrices over $\mathbb{Z}$:

$$A = \begin{bmatrix} 1 & -2 & 4 \\ -2 & 3 & -5 \end{bmatrix}, \quad B = \begin{bmatrix} 2 & 4 \\ 3 & 6 \\ 1 & -2 \end{bmatrix} \quad C = \begin{bmatrix} -1 & 0 \\ 1 & 1 \end{bmatrix}.$$

Determine, if possible, the matrices CA, BC, $B^T A$, $A^T C$, $(-A)^T C$, $B^T A^T$, AC and CB.

4.2 Consider the matrices

$$A = [a_{ij}] \in R^{n,m}, \quad x = \begin{bmatrix} x_1 \\ \vdots \\ x_n \end{bmatrix} \in R^{n,1}, \quad y = [y_1, \ldots, y_m] \in R^{1,m}.$$

Which of the following expressions are well defined for $m \neq n$ or $m = n$?

(a) xy, (b) $x^T y$, (c) yx, (d) yx^T, (e) xAy, (f) $x^T Ay$,
(g) xAy^T, (h) $x^T Ay^T$, (i) xyA, (j) xyA^T, (k) Axy, (l) $A^T xy$.

4.3 Show the following computational rules:

$$\mu_1 x_1 + \mu_2 x_2 = [x_1, x_2]\begin{bmatrix} \mu_1 \\ \mu_2 \end{bmatrix} \quad \text{and} \quad A[x_1, x_2] = [Ax_1, Ax_2]$$

for $A \in R^{n,m}$, $x_1, x_2 \in R^{m,1}$ and $\mu_1, \mu_2 \in R$.

4.4 Prove Lemma 4.3 (2)–(4).
4.5 Prove Lemma 4.4.
4.6 Prove Lemma 4.6 (1)–(3).
4.7 Let $A = \begin{bmatrix} 0 & 1 & 1 \\ 0 & 0 & 1 \\ 0 & 0 & 0 \end{bmatrix} \in \mathbb{Z}^{3,3}$. Determine A^n for all $n \in \mathbb{N}_0$.
4.8 Let $A = \begin{bmatrix} i & 1 \\ 0 & -i \end{bmatrix} \in \mathbb{C}^{2,2}$. Determine A^n for all $n \in \mathbb{N}_0$.
4.9 Let $p = \alpha_n t^n + \ldots + \alpha_1 t + \alpha_0 t^0 \in R[t]$ be a polynomial (cp. Example 3.15) and $A \in R^{m,m}$. We define $p(A) \in R^{m,m}$ as $p(A) := \alpha_n A^n + \ldots + \alpha_1 A + \alpha_0 I_m$.
  (a) Determine $p(A)$ for $p = t^2 - 2t + 1 \in \mathbb{Z}[t]$ and $A = \begin{bmatrix} 1 & 0 \\ 3 & 1 \end{bmatrix} \in \mathbb{Z}^{2,2}$.
  (b) For a fixed matrix $A \in R^{m,m}$ consider the map $f_A : R[t] \to R^{m,m}$, $p \mapsto p(A)$. Show that $f_A(p+q) = f_A(p) + f_A(q)$ and $f_A(pq) = f_A(p)f_A(q)$ for all $p, q \in R[t]$.
  (The map f_A is a *ring homomorphism* between the rings $R[t]$ and $R^{m,m}$.)
  (c) Show that $f_A(R[t]) = \{p(A) \mid p \in R[t]\}$ is a commutative subring of $R^{m,m}$, i.e., that $f_A(R[t])$ is a subring of $R^{m,m}$ and that the multiplication in this subring is commutative.
  (d) Is the map f_A surjective?
4.10 Let R, S be commutative rings with unit, and let $\varphi : R \to S$ be a ring homomorphism of the rings R and S, i.e., $\varphi(a+b) = \varphi(a) + \varphi(b)$ and $\varphi(ab) = \varphi(a)\varphi(b)$ hold for all $a, b \in R$. Show that then
$$\widetilde{\varphi} : R^{n,n} \to S^{n,n}, \quad [a_{ij}] \mapsto [\varphi(a_{ij})],$$
is a ring homomorphism of the rings $R^{n,n}$ and $S^{n,n}$.
4.11 Let K be a field with $1 + 1 \neq 0$. Show that every matrix $A \in K^{n,n}$ can be written as $A = M + S$ with a symmetric matrix $M \in K^{n,n}$ (i.e., $M^T = M$) and a skew-symmetric matrix $S \in K^{n,n}$ (i.e., $S^T = -S$).
Does this also hold in a field with $1+1 = 0$? Give a proof or a counterexample.
4.12 Show the *binomial formula* for commuting matrices: If $A, B \in R^{n,n}$ with $AB = BA$, then $(A + B)^k = \sum_{j=0}^{k} \binom{k}{j} A^j B^{k-j}$, where $\binom{k}{j} := \frac{k!}{j!(k-j)!}$.
4.13 Let $A \in R^{n,n}$ be a matrix for which $I_n - A$ is invertible. Show that $(I_n - A)^{-1}(I_n - A^{m+1}) = \sum_{j=0}^{m} A^j$ holds for every $m \in \mathbb{N}$.
4.14 Let $A \in R^{n,n}$ be a matrix for which an $m \in \mathbb{N}$ with $A^m = I_n$ exists and let m be smallest natural number with this property.
  (a) Investigate whether A is invertible, and if so, give a particularly simple representation of the inverse.
  (b) Determine the cardinality of the set $\{A^k \mid k \in \mathbb{N}\}$.
4.15 Let $\mathcal{S}$ be a subring of $R^{n,n}$, and let $\mathcal{T} = \{A^T \in R^{n,n} \mid A \in \mathcal{S}\}$. Show that $\mathcal{T}$ is also a subring of $R^{n,n}$.

4.16 Let $\mathcal{A} = \{[a_{ij}] \in R^{n,n} \mid a_{nj} = 0 \text{ for } j = 1, \ldots, n\}$.
   (a) Show that $\mathcal{A}$ is a subring of $R^{n,n}$.
   (b) Show that $AM \in \mathcal{A}$ for all $M \in R^{n,n}$ and $A \in \mathcal{A}$.
   (A subring with this property is called a *right ideal* of $R^{n,n}$.)
   (c) Determine an analogous subring $\mathcal{B}$ of $R^{n,n}$, such that $MB \in \mathcal{B}$ for all $M \in R^{n,n}$ and $B \in \mathcal{B}$.
   (A subring with this property is called a *left ideal* of $R^{n,n}$.)

4.17 Examine whether $(G, *)$ with
$$G = \left\{ \begin{bmatrix} \cos(\alpha) & \sin(\alpha) \\ -\sin(\alpha) & \cos(\alpha) \end{bmatrix} \middle| \alpha \in \mathbb{R} \right\}$$
is a subgroup of $GL_2(\mathbb{R})$.

4.18 Let $A \in GL_n(R)$. Prove the following assertions:
   (a) $G := \{A^k \mid k \in \mathbb{Z}\}$ is a commutative subgroup of $GL_n(R)$.
   (b) The map $\varphi : (\mathbb{Z}, +) \to (G, *)$, $k \mapsto A^k$, is a surjective group homomorphism.

4.19 Generalize the block multiplication (4.7) to matrices $A \in R^{n,m}$ and $B \in R^{m,\ell}$.

4.20 Determine all invertible upper triangular matrices $A \in R^{n,n}$ with $A^{-1} = A^T$.

4.21 Let $A_{11} \in R^{n_1,n_1}$, $A_{12} \in R^{n_1,n_2}$, $A_{21} \in R^{n_2,n_1}$, $A_{22} \in R^{n_2,n_2}$ and
$$A = \begin{bmatrix} A_{11} & A_{12} \\ A_{21} & A_{22} \end{bmatrix} \in R^{n_1+n_2, n_1+n_2}.$$
   (a) Let $A_{11} \in GL_{n_1}(R)$. Show that A is invertible if and only if $A_{22} - A_{21}A_{11}^{-1}A_{12}$ is invertible and derive in this case a formula for A^{-1}.
   (b) Let $A_{22} \in GL_{n_2}(R)$. Show that A is invertible if and only if $A_{11} - A_{12}A_{22}^{-1}A_{21}$ is invertible and derive in this case a formula for A^{-1}.

4.22 Let $A \in GL_n(R)$, $U \in R^{n,m}$ and $V \in R^{m,n}$. Show the following assertions:
   (a) $A + UV \in GL_n(R)$ holds if and only if $I_m + VA^{-1}U \in GL_m(R)$.
   (b) If $I_m + VA^{-1}U \in GL_m(R)$, then
$$(A + UV)^{-1} = A^{-1} - A^{-1}U(I_m + VA^{-1}U)^{-1}VA^{-1}.$$
   (This last equation is called the *Sherman-Morrison-Woodbury formula*; named after Jack Sherman, Winifred J. Morrison and Max A. Woodbury.)

4.23 Show that the set of block upper triangular matrices with invertible 2×2 diagonal blocks, i.e., the set of matrices
$$\begin{bmatrix} A_{11} & A_{12} & \cdots & A_{1m} \\ 0 & A_{22} & \cdots & A_{2m} \\ \vdots & & \ddots & \vdots \\ 0 & \cdots & 0 & A_{mm} \end{bmatrix}, \quad A_{ii} \in GL_2(R), \quad i = 1, \ldots, m,$$
is a group with respect to the matrix multiplication.

4.24 Prove Theorem 4.16. Is the group of permutation matrices commutative?

4.25 Show that the following is an equivalence relation on $R^{n,n}$:

$$A \sim B \quad \Leftrightarrow \quad \text{There exists a permutation matrix } P \text{ with } A = P^T B P.$$

4.26 A company produces from four raw materials R_1, R_2, R_3, R_4 five intermediate products Z_1, Z_2, Z_3, Z_4, Z_5, and from these three final products E_1, E_2, E_3. The following tables show how many units of R_i and Z_j are required for producing one unit of Z_k and E_ℓ, respectively:

	Z_1	Z_2	Z_3	Z_4	Z_5
R_1	0	1	1	1	2
R_2	5	0	1	2	1
R_3	1	1	1	1	0
R_4	0	2	0	1	0

	E_1	E_2	E_3
Z_1	1	1	1
Z_2	1	2	0
Z_3	0	1	1
Z_4	4	1	1
Z_5	3	1	1

For instance, five units of R_2 and one unit of R_3 are required for producing one unit of Z_1.

(a) Determine, with the help of matrix operations, a corresponding table which shows how many units of R_i are required for producing one unit of E_ℓ.

(b) Determine how many units of the four raw materials are required for producing 100 units of E_1, 200 units of E_2 and 300 units of E_3.

The Echelon Form and the Rank of Matrices   5

In this chapter we develop a systematic method for transforming a matrix A with entries from a field into a special form which is called the echelon form of A. The transformation consists of a sequence of multiplications of A from the left by certain "elementary matrices". If A is invertible, then its echelon form is the identity matrix, and the inverse A^{-1} is the product of the inverses of the elementary matrices. For a non-invertible matrix its echelon form is, in some sense, the "closest possible" matrix to the identity matrix. This form motivates the concept of the rank of a matrix, which we introduce in this chapter and will use frequently later on.

5.1 Elementary Matrices

Let R be a commutative ring with unit, $n \in \mathbb{N}$ and $i, j \in \{1, \ldots, n\}$. Let $I_n \in R^{n,n}$ be the identity matrix and let e_i be its ith column, i.e., $I_n = [e_1, \ldots, e_n]$.

We define

$$E_{ij} := e_i e_j^T = [0, \ldots, 0, \underbrace{e_i}_{\text{column } j}, 0, \ldots, 0] \in R^{n,n},$$

i.e., the entry (i, j) of E_{ij} is 1, all other entries are 0.

For $n \geq 2$ and $i < j$ we define

$$P_{ij} := [e_1, \ldots, e_{i-1}, e_j, e_{i+1}, \ldots, e_{j-1}, e_i, e_{j+1}, \ldots, e_n] \in R^{n,n}. \tag{5.1}$$

Thus, P_{ij} is a permutation matrix (cp. Definition 4.15) obtained by exchanging the columns i and j of I_n. A multiplication of $A \in R^{n,m}$ from the left with P_{ij} means an exchange of the rows i and j of A. For example,

$$A = \begin{bmatrix} 1 & 2 & 3 \\ 4 & 5 & 6 \\ 7 & 8 & 9 \end{bmatrix}, \quad P_{13} = [e_3, e_2, e_1] = \begin{bmatrix} 0 & 0 & 1 \\ 0 & 1 & 0 \\ 1 & 0 & 0 \end{bmatrix}, \quad P_{13}A = \begin{bmatrix} 7 & 8 & 9 \\ 4 & 5 & 6 \\ 1 & 2 & 3 \end{bmatrix}.$$

For $\lambda \in R$ we define

$$M_i(\lambda) := [e_1, \ldots, e_{i-1}, \lambda e_i, e_{i+1}, \ldots, e_n] \in R^{n,n}. \tag{5.2}$$

Thus, $M_i(\lambda)$ is a diagonal matrix obtained by replacing the ith column of I_n by λe_i. A multiplication of $A \in R^{n,m}$ from the left with $M_i(\lambda)$ means a multiplication of the ith row of A by λ. For example,

$$A = \begin{bmatrix} 1 & 2 & 3 \\ 4 & 5 & 6 \\ 7 & 8 & 9 \end{bmatrix}, \quad M_2(-1) = [e_1, -e_2, e_3] = \begin{bmatrix} 1 & 0 & 0 \\ 0 & -1 & 0 \\ 0 & 0 & 1 \end{bmatrix},$$

$$M_2(-1)A = \begin{bmatrix} 1 & 2 & 3 \\ -4 & -5 & -6 \\ 7 & 8 & 9 \end{bmatrix}.$$

For $n \geq 2$, $i < j$ and $\lambda \in R$ we define

$$G_{ij}(\lambda) := I_n + \lambda E_{ji} = [e_1, \ldots, e_{i-1}, e_i + \lambda e_j, e_{i+1}, \ldots, e_n] \in R^{n,n}. \tag{5.3}$$

Thus, the lower triangular matrix $G_{ij}(\lambda)$ is obtained by replacing the ith column of I_n by $e_i + \lambda e_j$. A multiplication of $A \in R^{n,m}$ from the left with $G_{ij}(\lambda)$ means that λ times the ith row of A is added to the jth row of A. Similarly, a multiplication of $A \in R^{n,m}$ from the left by the upper triangular matrix $G_{ij}(\lambda)^T$ means that λ times the jth row of A is added to the ith row of A. For example,

$$A = \begin{bmatrix} 1 & 2 & 3 \\ 4 & 5 & 6 \\ 7 & 8 & 9 \end{bmatrix}, \quad G_{23}(-1) = [e_1, e_2 - e_3, e_3] = \begin{bmatrix} 1 & 0 & 0 \\ 0 & 1 & 0 \\ 0 & -1 & 1 \end{bmatrix},$$

$$G_{23}(-1)A = \begin{bmatrix} 1 & 2 & 3 \\ 4 & 5 & 6 \\ 3 & 3 & 3 \end{bmatrix}, \quad G_{23}(-1)^T A = \begin{bmatrix} 1 & 2 & 3 \\ -3 & -3 & -3 \\ 7 & 8 & 9 \end{bmatrix}.$$

Lemma 5.1 *The elementary matrices P_{ij}, $M_i(\lambda)$ for invertible $\lambda \in R$, and $G_{ij}(\lambda)$ defined in (5.1), (5.2), and (5.3), respectively, are invertible and have the following inverses:*

(1) $P_{ij}^{-1} = P_{ij}^T = P_{ij}$.
(2) $M_i(\lambda)^{-1} = M_i(\lambda^{-1})$.
(3) $G_{ij}(\lambda)^{-1} = G_{ij}(-\lambda)$.

Proof

(1) The invertibility of P_{ij} with $P_{ij}^{-1} = P_{ij}^T$ was already shown in Theorem 4.16; the symmetry of P_{ij} is easily seen.
(2) Since $\lambda \in R$ is invertible, the matrix $M_i(\lambda^{-1})$ is well defined. A straightforward computation now shows that $M_i(\lambda^{-1})M_i(\lambda) = M_i(\lambda)M_i(\lambda^{-1}) = I_n$.
(3) Since $e_j^T e_i = 0$ for $i < j$, we have $E_{ji}^2 = (e_i e_j^T)(e_i e_j^T) = 0$, and therefore

$$G_{ij}(\lambda)G_{ij}(-\lambda) = (I_n + \lambda E_{ji})(I_n + (-\lambda)E_{ji}) = I_n + \lambda E_{ji}$$
$$+ (-\lambda)E_{ji} + (-\lambda^2)E_{ji}^2 = I_n.$$

A similar computation shows that $G_{ij}(-\lambda)G_{ij}(\lambda) = I_n$. □

5.2 The Echelon Form and Gaussian Elimination

The constructive proof of the following theorem relies on the *Gaussian elimination algorithm*.[1] For a given matrix $A \in K^{n,m}$, where K is a field, this algorithm constructs a matrix $S \in GL_n(K)$ such that $SA = C$ is *quasi*-upper triangular. We obtain this special form by left-multiplication of A with elementary matrices P_{ij}, $M_i(\lambda)$ and $G_{ij}(\lambda)$. Each of these left-multiplications corresponds to the application of one of the so-called "elementary row operations" to the matrix A:

- P_{ij}: exchange two rows of A.
- $M_i(\lambda)$: multiply a row of A with an invertible scalar.
- $G_{ij}(\lambda)$: add a multiple of one row of A to another row of A.

---

[1] Named after Carl Friedrich Gauß (1777–1855). A similar method was already described in Chapter 8, "Rectangular Arrays", of the "Nine Chapters on the Mathematical Art". This text developed in ancient China over several decades BC stated problems of every day life and gave practical mathematical solution methods. A detailed commentary and analysis was written by Liu Hui (approx. 220–280 AD) around 260 AD.

We assume that the entries of A are in a field (rather than a ring) because in the proof of the theorem we require that nonzero entries of A are invertible. A generalization of the result which holds over certain rings (e.g. the integers $\mathbb{Z}$) is given by the *Hermite canonical form*,[2] which plays an important role in Number Theory.

Theorem 5.2 *Let K be a field and let $A \in K^{n,m}$. Then there exist invertible matrices $S_1, \ldots, S_t \in K^{n,n}$ (these are products of elementary matrices) such that $C := S_t \cdots S_1 A$ is in echelon form, i.e., either $C = 0$ or*

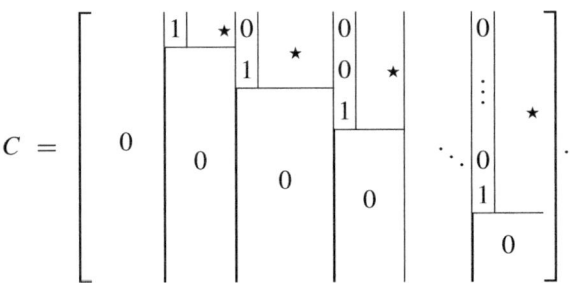

Here $\star$ denotes an arbitrary (zero or nonzero) entry of C.
More precisely, $C = [c_{ij}]$ is either the zero matrix, or there exists a sequence of natural numbers $j_1, \ldots, j_r$ (these are called the "steps" of the echelon form), where $1 \leq j_1 < \cdots < j_r \leq m$ and $1 \leq r \leq \min\{n, m\}$, such that

(1) $c_{ij} = 0$ for $1 \leq i \leq r$ and $1 \leq j < j_i$,
(2) $c_{ij} = 0$ for $r < i \leq n$ and $1 \leq j \leq m$,
(3) $c_{i,j_i} = 1$ for $1 \leq i \leq r$ and all other entries in column j_i are zero.

Proof If $A = 0$, then we set $t = 1$, $S_1 = I_n$, $C = 0$ and we are done.
    Now let $A \neq 0$ and let j_1 be the index of the first column of

$$A^{(1)} = \left[a_{ij}^{(1)} \right] := A$$

---

[2] Charles Hermite (1822–1901).

5.2 The Echelon Form and Gaussian Elimination

that does not consist of all zeros. Let $a^{(1)}_{i_1, j_1}$ be the first entry in this column that is nonzero, i.e., $A^{(1)}$ has the form

$$A^{(1)} = \left[\begin{array}{c|c|c} & 0 & \\ & \vdots & \\ & 0 & \\ 0 & a^{(1)}_{i_1, j_1} & \star \\ & \star & \\ & \vdots & \\ & \star & \\ & j_1 & \end{array} \right].$$

We then proceed as follows: First we *permute* the rows i_1 and 1 (if $i_1 > 1$). Then we *normalize* the new first row, i.e., we multiply it with $\left(a^{(1)}_{i_1, j_1} \right)^{-1}$. Finally we *eliminate* the nonzero entries below the first entry in column j_1. Permuting and normalizing leads to

$$\widetilde{A}^{(1)} = \left[\widetilde{a}^{(1)}_{ij} \right] := M_1 \left(\left(a^{(1)}_{i_1, j_1} \right)^{-1} \right) P_{1, i_1} A^{(1)} = \left[\begin{array}{c|c|c} & 1 & \\ & \widetilde{a}^{(1)}_{2, j_1} & \\ 0 & \vdots & \star \\ & \widetilde{a}^{(1)}_{n, j_1} & \\ & j_1 & \end{array} \right].$$

If $i_1 = 1$, then we set $P_{11} := I_n$. In order to eliminate below the 1 in column j_1, we multiply $\widetilde{A}^{(1)}$ from the left with the matrices

$$G_{1n} \left(-\widetilde{a}^{(1)}_{n, j_1} \right), \ldots, G_{12} \left(-\widetilde{a}^{(1)}_{2, j_1} \right).$$

Then we have

$$S_1 A^{(1)} = \left[\begin{array}{c|c|c} 0 & 1 & \star \\ \hline & 0 & \\ 0 & \vdots & A^{(2)} \\ & 0 & \\ & j_1 & \end{array} \right],$$

where

$$S_1 := G_{1n}\left(-\tilde{a}_{n,j_1}^{(1)}\right)\cdots G_{12}\left(-\tilde{a}_{2,j_1}^{(1)}\right) M_1 \left(\left(a_{i_1,j_1}^{(1)}\right)^{-1}\right) P_{1,i_1}$$

and $A^{(2)} = [a_{ij}^{(2)}]$ with $i = 2, \ldots, n$, $j = j_1 + 1, \ldots, m$, i.e., we keep the indices of the larger matrix $A^{(1)}$ in the smaller matrix $A^{(2)}$.

If $A^{(2)} = [\ ]$ or $A^{(2)} = 0$, then we are finished, since then $C := S_1 A^{(1)}$ is in echelon form. In this case $r = 1$.

If at least one of the entries of $A^{(2)}$ is nonzero, then we apply the steps described above to the matrix $A^{(2)}$. For $k = 2, 3, \ldots$ we define the matrices S_k recursively as

$$S_k = \begin{bmatrix} I_{k-1} & 0 \\ 0 & \tilde{S}_k \end{bmatrix}, \quad \text{where} \quad \tilde{S}_k A^{(k)} = \begin{bmatrix} 0 & 1 & \star \\ & 0 & \\ 0 & \vdots & A^{(k+1)} \\ & 0 & \\ & j_k & \end{bmatrix}.$$

Each matrix $\tilde{S}_k$ is constructed analogously to S_1: First we identify the first column j_k of $A^{(k)}$ that is not completely zero, as well as the first nonzero entry $a_{i_k,j_k}^{(k)}$ in that column. Then permuting and normalizing yields the matrix

$$\widetilde{A}^{(k)} = [\widetilde{a}_{ij}^{(k)}] := M_k \left(\left(a_{i_k,j_k}^{(k)}\right)^{-1}\right) P_{k,i_k} A^{(k)}.$$

If $k = i_k$, then we set $P_{kk} := I_{n-k+1}$. Now

$$\tilde{S}_k = G_{kn}\left(-\tilde{a}_{n,j_k}^{(k)}\right)\cdots G_{k,k+1}\left(-\tilde{a}_{k+1,j_k}^{(k)}\right) M_k \left(\left(a_{i_k,j_k}^{(k)}\right)^{-1}\right) P_{k,i_k},$$

so that S_k is indeed a product of elementary matrices of the form

$$\begin{bmatrix} I_{k-1} & 0 \\ 0 & T \end{bmatrix},$$

where T is an elementary matrix of size $(n - k + 1) \times (n - k + 1)$.

If we continue this procedure inductively, it will end after $r \leq \min\{n, m\}$ steps with either $A^{(r+1)} = 0$ or $A^{(r+1)} = [\ ]$.

5.2 The Echelon Form and Gaussian Elimination

After r steps we have

$$S_r \cdots S_1 A^{(1)} = \begin{bmatrix} 1 & \star & \star & & \star & & & \star & \\ & & 1 & \star & \star & \star & & & \\ 0 & & & & 1 & & & \vdots & \star \\ & 0 & & & & \ddots & & \star & \\ & & & 0 & & & & 1 & \\ & & & & & 0 & & & 0 \end{bmatrix} \quad (5.4)$$

By construction, the entries 1 in (5.4) are in the positions

$$(1, j_1), (2, j_2), \ldots, (r, j_r).$$

If $r = 1$, then $S_1 A^{(1)}$ is in echelon form (see the discussion at the beginning of the proof). If $r > 1$, then we still have to eliminate the nonzero entries above the 1 in columns $j_2, \ldots, j_r$. To do this, we denote the matrix in (5.4) by $U^{(1)} = [u_{ij}^{(1)}]$ and form for $k = 2, \ldots, r$ recursively

$$U^{(k)} = [u_{ij}^{(k)}] := S_{r+k-1} U^{(k-1)},$$

where

$$S_{r+k-1} := G_{1k} \left(-r_{1,j_k}^{(k-1)}\right)^T \cdots G_{k-1,k} \left(-r_{k-1,j_k}^{(k-1)}\right)^T.$$

For $t := 2r - 1$ we have $C := S_t S_{t-1} \cdots S_1 A$ in echelon form. $\square$

In the literature, the echelon form is sometimes called *reduced row* echelon form.

Corollary 5.3 *Let $A \in K^{n,n}$ and let $C = S_t \cdots S_1 A$ be in echelon form as in Theorem 5.2. The matrix A is invertible if and only if $C = I_n$. In this case $A^{-1} = S_t \cdots S_1$.*

Proof Let $A \in K^{n,n}$ and let $C = S_t S_{t-1} \cdots S_1 A$ be in echelon form. If A is invertible, then C is a product of invertible matrices and thus invertible. An invertible matrix cannot have a row containing only zeros, so that $r = n$ and hence $C = I_n$. If, on the other hand, $C = I_n$, then the invertibility of the elementary matrices implies that $S_1^{-1} \cdots S_t^{-1} = A$. As a product of invertible matrices, A is invertible and $A^{-1} = S_t \cdots S_1$. $\square$

If $A \in GL_n(K)$, then Corollary 5.3 shows that $A = S_1^{-1} \cdots S_t^{-1}$ for certain products of elementary matrices $S_1, \ldots, S_t \in GL_n(K)$. The inverses of elementary matrices are again elementary matrices (cp. Lemma 5.1), and thus every matrix $A \in GL_n(K)$ is a product of elementary matrices.

Example 5.4 Transformation of a matrix from $\mathbb{Q}^{3,5}$ to echelon form via left multiplication with elementary matrices:

$$\begin{bmatrix} 0 & 2 & 1 & 3 & 3 \\ 0 & 2 & 0 & 1 & 1 \\ 0 & 2 & 0 & 1 & 1 \end{bmatrix}$$

$$\begin{array}{c} j_1 = 2, i_1 = 1 \\ \longrightarrow \\ M_1\left(\frac{1}{2}\right) \end{array} \begin{bmatrix} 0 & 1 & \frac{1}{2} & \frac{3}{2} & \frac{3}{2} \\ 0 & 2 & 0 & 1 & 1 \\ 0 & 2 & 0 & 1 & 1 \end{bmatrix} \quad \xrightarrow{G_{13}(-2)} \quad \begin{bmatrix} 0 & 1 & \frac{1}{2} & \frac{3}{2} & \frac{3}{2} \\ 0 & 2 & 0 & 1 & 1 \\ 0 & 0 & -1 & -2 & -2 \end{bmatrix}$$

$$\xrightarrow{G_{12}(-2)} \begin{bmatrix} 0 & 1 & \frac{1}{2} & \frac{3}{2} & \frac{3}{2} \\ 0 & 0 & -1 & -2 & -2 \\ 0 & 0 & -1 & -2 & -2 \end{bmatrix} \quad \begin{array}{c} j_2 = 3, i_2 - 2 \\ \longrightarrow \\ M_2(-1) \end{array} \begin{bmatrix} 0 & 1 & \frac{1}{2} & \frac{3}{2} & \frac{3}{2} \\ 0 & 0 & 1 & 2 & 2 \\ 0 & 0 & -1 & -2 & -2 \end{bmatrix}$$

$$\xrightarrow{G_{23}(1)} \begin{bmatrix} 0 & 1 & \frac{1}{2} & \frac{3}{2} & \frac{3}{2} \\ 0 & 0 & 1 & 2 & 2 \\ 0 & 0 & 0 & 0 & 0 \end{bmatrix}$$

$$\xrightarrow{G_{12}\left(-\frac{1}{2}\right)^T} \begin{bmatrix} 0 & 1 & 0 & \frac{1}{2} & \frac{1}{2} \\ 0 & 0 & 1 & 2 & 2 \\ 0 & 0 & 0 & 0 & 0 \end{bmatrix}.$$

> **MATLAB-Minute 3**
> The echelon form is computed in MATLAB with the command rref ("reduced row echelon form"). Apply rref to the matrix [A eye(n+1)] in order to compute the inverse of the matrix A=full(gallery('tridiag',-ones(n,1),2*ones(n+1,1),-ones(n,1))) for n=1,2,3,4,5 (cp. Exercise 5.6).
> Formulate a conjecture about the general form of A^{-1}. (Can you prove your conjecture?)

5.2 The Echelon Form and Gaussian Elimination

The proof of Theorem 5.2 leads to the so-called *LU-decomposition* of a square matrix.

Theorem 5.5 *For every matrix $A \in K^{n,n}$, there exists a permutation matrix $P \in K^{n,n}$, a lower triangular matrix $L \in GL_n(K)$ with ones on the diagonal and an upper triangular matrix $U \in K^{n,n}$, such that $A = PLU$. The matrix U is invertible if and only if A is invertible.*

Proof For $A \in K^{n,n}$ the Eq. (5.4) has the form $S_n \cdots S_1 A = \widetilde{U}$, where $\widetilde{U}$ is upper triangular. If $r < n$, then we set $S_n = S_{n-1} = \cdots = S_{r+1} = I_n$. Since the matrices $S_1, \ldots, S_n$ are invertible, it follows that $\widetilde{U}$ is invertible if and only if A is invertible. For $i = 1, \ldots, n$ every matrix S_i has the form

$$S_i = \begin{bmatrix} 1 & & & & & & \\ & \ddots & & & & & \\ & & 1 & & & & \\ & & & s_{ii} & & & \\ & & & s_{i+1,i} & 1 & & \\ & & & \vdots & & \ddots & \\ & & & s_{ni} & & & 1 \end{bmatrix} P_{i,j_i},$$

where $j_i \geq i$ for $i = 1, \ldots, n$ and $P_{ii} := I_n$ (if $j_i = i$, then no permutation was necessary). Therefore,

$$S_n \cdots S_1 = \begin{bmatrix} 1 & & & \\ & \ddots & & \\ & & 1 & \\ & & & 1 \\ & & & s_{nn} \end{bmatrix} \begin{bmatrix} 1 & & & \\ & \ddots & & \\ & & 1 & \\ & & s_{n-1,n-1} & \\ & & s_{n,n-1} & 1 \end{bmatrix} P_{n-1,j_{n-1}}$$

$$\begin{bmatrix} 1 & & & & \\ & \ddots & & & \\ & & 1 & & \\ & & s_{n-2,n-2} & & \\ & & s_{n-1,n-2} & 1 & \\ & & s_{n,n-2} & 0 & 1 \end{bmatrix} P_{n-2,j_{n-2}} \cdots \begin{bmatrix} 1 & & & \\ & s_{22} & & \\ & s_{32} & 1 & \\ & \vdots & & \ddots \\ & s_{n2} & & & 1 \end{bmatrix} P_{2,j_2} \begin{bmatrix} s_{11} & & & \\ s_{21} & 1 & & \\ s_{31} & & 1 & \\ \vdots & & & \ddots \\ s_{n1} & & & & 1 \end{bmatrix} P_{1,j_1}.$$

The form of the permutation matrices for $k = 2, \ldots, n-1$ and $\ell = 1, \ldots, k-1$ implies that

$$P_{k,j_k} \begin{bmatrix} 1 & & & & & \\ & \ddots & & & & \\ & & 1 & & & \\ & & & s_{\ell\ell} & & \\ & & & s_{\ell+1,\ell} & 1 & \\ & & & \vdots & & \ddots \\ & & & s_{n\ell} & & & 1 \end{bmatrix} = \begin{bmatrix} 1 & & & & & \\ & \ddots & & & & \\ & & 1 & & & \\ & & & s_{\ell\ell} & & \\ & & & \widetilde{s}_{\ell+1,\ell} & 1 & \\ & & & \vdots & & \ddots \\ & & & \widetilde{s}_{n\ell} & & & 1 \end{bmatrix} P_{k,j_k}$$

holds for certain $\widetilde{s}_{j\ell} \in K$, $j = \ell+1, \ldots, n$. Hence,

$$S_n \cdots S_1 = \begin{bmatrix} 1 & & & & \\ & \ddots & & & \\ & & 1 & & \\ & & & s_{n-1,n-1} & \\ & & & s_{nn}s_{n,n-1} & s_{nn} \end{bmatrix} \begin{bmatrix} 1 & & & & \\ & \ddots & & & \\ & & 1 & & \\ & & & \widetilde{s}_{n-2,n-2} & & \\ & & & \widetilde{s}_{n-1,n-2} & 1 & \\ & & & \widetilde{s}_{n,n-2} & & 1 \end{bmatrix} \cdots$$

$$\begin{bmatrix} 1 & & & \\ & s_{22} & & \\ & \widetilde{s}_{32} & 1 & \\ & \vdots & & \ddots \\ & \widetilde{s}_{n2} & & & 1 \end{bmatrix} \begin{bmatrix} s_{11} & & & \\ \widetilde{s}_{21} & 1 & & \\ \widetilde{s}_{31} & & 1 & \\ \vdots & & & \ddots \\ \widetilde{s}_{n1} & & & & 1 \end{bmatrix} P_{n-1,j_{n-1}} \cdots P_{1,j_1}.$$

The invertible lower triangular matrices and the permutation matrices form groups with respect to the matrix multiplication (cp. Theorems 4.13 and 4.16). Thus, $S_n \cdots S_1 = \widetilde{L}\widetilde{P}$, where $\widetilde{L}$ is invertible and lower triangular, and $\widetilde{P}$ is a permutation matrix. Since $\widetilde{L} = [\widetilde{l}_{ij}]$ is invertible, also $D := \text{diag}(\widetilde{l}_{11}, \ldots, \widetilde{l}_{nn})$ is invertible, and we obtain $A = PLU$ with $P := \widetilde{P}^{-1} = \widetilde{P}^T$, $L := \widetilde{L}^{-1}D$ and $U := D^{-1}\widetilde{U}$. By construction, all diagonal entries of L are equal to one. □

5.2 The Echelon Form and Gaussian Elimination

Example 5.6 Computation of an LU-decomposition of a matrix from $\mathbb{Q}^{3,3}$:

$$\begin{bmatrix} 2 & 2 & 4 \\ 2 & 2 & 1 \\ 2 & 0 & 1 \end{bmatrix}$$

$j_1 = 1, i_1 = 1$
$\xrightarrow{M_1\left(\frac{1}{2}\right)}$
$\begin{bmatrix} 1 & 1 & 2 \\ 2 & 2 & 1 \\ 2 & 0 & 1 \end{bmatrix}$
$\xrightarrow{G_{13}(-2)}$
$\begin{bmatrix} 1 & 1 & 2 \\ 2 & 2 & 1 \\ 0 & -2 & -3 \end{bmatrix}$

$\xrightarrow{G_{12}(-2)}$
$\begin{bmatrix} 1 & 1 & 2 \\ 0 & 0 & -3 \\ 0 & -2 & -3 \end{bmatrix}$
$\xrightarrow{P_{23}}$
$\begin{bmatrix} 1 & 1 & 2 \\ 0 & -2 & -3 \\ 0 & 0 & -3 \end{bmatrix} = \tilde{U}.$

Hence, $\tilde{P} = P_{23}$,

$$\tilde{L} = G_{13}(-2) G_{12}(-2) M_1\left(\frac{1}{2}\right) = \begin{bmatrix} \frac{1}{2} & 0 & 0 \\ -1 & 1 & 0 \\ -1 & 0 & 1 \end{bmatrix}, \quad D = \operatorname{diag}\left(\frac{1}{2}, 1, 1\right),$$

and thus, $P = \tilde{P}^T = P_{23}^T = P_{23}$,

$$L = \tilde{L}^{-1} D = \begin{bmatrix} 1 & 0 & 0 \\ 1 & 1 & 0 \\ 1 & 0 & 1 \end{bmatrix}, \quad U = D^{-1}\tilde{U} = \begin{bmatrix} 2 & 2 & 4 \\ 0 & -2 & -3 \\ 0 & 0 & -3 \end{bmatrix}.$$

If $A \in GL_n(K)$, then the LU-decomposition yields $A^{-1} = U^{-1} L^{-1} P^T$. Hence after computing the LU-decomposition, we obtain the inverse of A essentially by inverting the two triangular matrices. Since this can be achieved by the efficient recursive formula (4.5), the LU-decomposition is a popular method in scientific computing applications that require the inversion of matrices or the solution of linear systems of equations (cp. Chap. 6). In this context, however, alternative strategies for the choice of the permutation matrices are used. For example, instead of the first nonzero entry in a column one chooses an entry with large (or largest) absolute value for the row exchange and the subsequent elimination. By this strategy the influence of rounding errors in the computation is reduced.

MATLAB-Minute 4

The *Hilbert matrix*[a] $A = [a_{ij}] \in \mathbb{Q}^{n,n}$ has the entries $a_{ij} = 1/(i+j-1)$ for $i, j = 1, \ldots, n$. It can be generated in MATLAB with the command hilb(n). Carry out the command [L,U,P]=lu(hilb(4)) in order to compute an LU-decomposition of the matrix hilb(4). How do the matrices P, L and U look like?

Compute also the LU-decomposition of the matrix
full(gallery('tridiag',-ones(3,1),2*ones(4,1),-ones(3,1)))
and study the corresponding matrices P, L and U.

[a]David Hilbert (1862–1943).

We will now show that, for a given matrix A, the matrix C in Theorem 5.2 is uniquely determined in a certain sense. For this we need the following definition.

Definition 5.7 If $C \in K^{n,m}$ is in echelon form (as in Theorem 5.2), then the positions of $(1, j_1), \ldots, (r, j_r)$ are called the *pivot positions* of C.

We also need the following results.

Lemma 5.8 *If* $Z \in GL_n(K)$ *and* $x \in K^{n,1}$, *then* $Zx = 0$ *if and only if* $x = 0$.

Proof Exercise. □

Theorem 5.9 *Let* $A, B \in K^{n,m}$ *be in echelon form. If* $A = ZB$ *for a matrix* $Z \in GL_n(K)$, *then* $A = B$.

Proof If B is the zero matrix, then $A = ZB = 0$, and hence $A = B$.

Let now $B \neq 0$ and let A, B have the respective columns $a_i, b_i, 1 \leq i \leq m$. Furthermore, let $(1, j_1), \ldots, (r, j_r)$ be the $r \geq 1$ pivot positions of B. We will show that every matrix $Z \in GL_n(K)$ with $A = ZB$ has the form

$$Z = \left[\begin{array}{c|c} I_r & \star \\ \hline 0 & Z_{n-r} \end{array}\right],$$

where $Z_{n-r} \in GL_{n-r}(K)$. Since B is in echelon form and all entries of B below its row r are zero, it then follows that $B = ZB = A$.

Since $(1, j_1)$ is the first pivot position of B, we have $b_i = 0 \in K^{n,1}$ for $1 \leq i \leq j_1 - 1$ and $b_{j_1} = e_1$ (the first column of I_n). Then $A = ZB$ implies $a_i = 0 \in K^{n,1}$ for $1 \leq i \leq j_1 - 1$ and $a_{j_1} = Zb_{j_1} = Ze_1$. Since Z is invertible, Lemma 5.8 implies that $a_{j_1} \neq 0 \in K^{n,1}$. Since A is in echelon form, $a_{j1} = e_1 = b_{j_1}$. Furthermore,

$$Z = Z_n := \begin{bmatrix} 1 & \star \\ \hline 0 & Z_{n-1} \end{bmatrix},$$

where $Z_{n-1} \in GL_{n-1}(K)$ (cp. Exercise 5.4). If $r = 1$, then we are done.

If $r > 1$, then we proceed with the other pivot positions in an analogous way: Since B is in echelon form, the kth pivot position gives $b_{j_k} = e_k$. From $a_{j_k} = Zb_{j_k}$ and the invertibility of Z_{n-k+1} we obtain $a_{j_k} = b_{j_k}$ and

$$Z = \begin{bmatrix} I_{k-1} & 0 & \star \\ 0 & 1 & \star \\ 0 & 0 & Z_{n-k} \end{bmatrix},$$

where $Z_{n-k} \in GL_{n-k}(K)$. $\square$

This result yields the uniqueness of the echelon form of a matrix and its invariance under left-multiplication with invertible matrices.

Corollary 5.10 *For $A \in K^{n,m}$ the following assertions hold:*

(1) There is a unique matrix $C \in K^{n,m}$ in echelon form to which A can be transformed by elementary row operations, i.e., by left-multiplication with elementary matrices. This matrix C is called the echelon form of A.
(2) If $M \in GL_n(K)$, then the matrix C in (1) is also the echelon form of MA, i.e., the echelon form of a matrix is invariant under left-multiplication with invertible matrices.

Proof

(1) If $S_1 A = C_1$ and $S_2 A = C_2$, where C_1, C_2 are in echelon form and S_1, S_2 are invertible, then $C_1 = \left(S_1 S_2^{-1}\right) C_2$. Theorem 5.9 now gives $C_1 = C_2$.
(2) If $M \in GL_n(K)$ and $S_3(MA) = C_3$ is in echelon form, then with $S_1 A = C_1$ from (1) we get $C_3 = \left(S_3 M S_1^{-1}\right) C_1$. Theorem 5.9 now gives $C_3 = C_1$. $\square$

5.3 Rank and Equivalence of Matrices

As we have seen in Corollary 5.10, the echelon form of $A \in K^{n,m}$ is unique. In particular, for every matrix $A \in K^{n,m}$, there exists a unique number of pivot positions (cp. Definition 5.7) in its echelon form. This justifies the following definition.

Definition 5.11 The number r of pivot positions in the echelon form of $A \in K^{n,m}$ is called the *rank*[3] of A and denoted by $\text{rank}(A)$.

We see immediately that for $A \in K^{n,m}$ always $0 \leq \text{rank}(A) \leq \min\{n, m\}$, where $\text{rank}(A) = 0$ if and only if $A = 0$. Moreover, Theorem 5.2 shows that $A \in K^{n,n}$ is invertible if and only if $\text{rank}(A) = n$. Further properties of the rank are summarized in the following theorem.

Theorem 5.12 *For $A \in K^{n,m}$ the following assertions hold:*

(1) There exist matrices $Q \in GL_n(K)$ and $Z \in GL_m(K)$ with

$$QAZ = \begin{bmatrix} I_r & 0_{r,m-r} \\ 0_{n-r,r} & 0_{n-r,m-r} \end{bmatrix}$$

if and only if $\text{rank}(A) = r$.
(2) If $Q \in GL_n(K)$ and $Z \in GL_m(K)$, then $\text{rank}(A) = \text{rank}(QAZ)$.
(3) If $A = BC$ with $B \in K^{n,\ell}$ and $C \in K^{\ell,m}$, then

$\quad\quad\quad$ (a) $\quad \text{rank}(A) \leq \text{rank}(B)$,

$\quad\quad\quad$ (b) $\quad \text{rank}(A) \leq \text{rank}(C)$.

(4) There exist matrices $B \in K^{n,\ell}$ and $C \in K^{\ell,m}$ with $A = BC$ if and only if $\text{rank}(A) \leq \ell$.
(5) $\text{rank}(A) = \text{rank}(A^T)$.

Proof

(3a) Let $Q \in GL_n(K)$ be such that QB is in echelon form. Then $QA = QBC$. In the matrix QBC at most the first $\text{rank}(B)$ rows contain nonzero entries. By Corollary 5.10, the echelon form of QA is equal to the echelon form of A. Thus, in the echelon form of A also at most the first $\text{rank}(B)$ rows will be nonzero, which implies $\text{rank}(A) \leq \text{rank}(B)$.

(2) Let $Q \in GL_n(K)$ and $Z \in GL_m(K)$. Using (3a) twice yields $\text{rank}(A) = \text{rank}(AZZ^{-1}) \leq \text{rank}(AZ) \leq \text{rank}(A)$, and hence $\text{rank}(A) = \text{rank}(AZ)$. Using (2) in Corollary 5.10 now gives $\text{rank}(A) = \text{rank}(AZ) = \text{rank}(QAZ)$.

(1) $\Rightarrow$: This follows directly from (3a).

$\quad\Leftarrow$: If $\text{rank}(A) = r = 0$, then $A = 0$ and the assertion holds for arbitrary matrices $Q \in GL_n(K)$ and $Z \in GL_m(K)$.

---

[3] The concept of the rank was introduced (in the context of bilinear forms) first in 1879 by Ferdinand Georg Frobenius (1849–1917).

5.3 Rank and Equivalence of Matrices

Now let $r \geq 1$, and let $Q \in GL_n(K)$ be such that QA is in echelon form with r pivot positions. Then there exists a permutation matrix $P \in K^{m,m}$, that is a product of elementary permutation matrices P_{ij}, with

$$PA^T Q^T = \begin{bmatrix} I_r & 0_{r,n-r} \\ V & 0_{m-r,n-r} \end{bmatrix}$$

for some matrix $V \in K^{m-r,r}$. If $r = m$, then $V = [\ ]$. In the following, for simplicity, we omit the sizes of the zero matrices. The matrix

$$Y := \begin{bmatrix} I_r & 0 \\ -V & I_{m-r} \end{bmatrix} \in K^{m,m}$$

is invertible with

$$Y^{-1} = \begin{bmatrix} I_r & 0 \\ V & I_{m-r} \end{bmatrix} \in K^{m,m}.$$

Thus,

$$YPA^T Q^T = \begin{bmatrix} I_r & 0 \\ 0 & 0 \end{bmatrix},$$

and with $Z := P^T Y^T \in GL_m(K)$ we obtain

$$QAZ = \begin{bmatrix} I_r & 0 \\ 0 & 0 \end{bmatrix}. \tag{5.5}$$

(4) Let $A = BC$ with $B \in K^{n,\ell}$ and $C \in K^{\ell,m}$. Then by (3a),

$$\text{rank}(A) = \text{rank}(BC) \leq \text{rank}(B) \leq \ell.$$

Let, on the other hand, $\text{rank}(A) = r \leq \ell$. Then there exist matrices $Q \in GL_n(K)$ and $Z \in GL_m(K)$ with $QAZ = \begin{bmatrix} I_r & 0 \\ 0 & 0 \end{bmatrix}$. Thus, we obtain

$$A = \left(Q^{-1} \begin{bmatrix} I_r & 0_{r,\ell-r} \\ 0_{n-r,r} & 0_{n-r,\ell-r} \end{bmatrix} \right) \left(\begin{bmatrix} I_r & 0_{r,m-r} \\ 0_{\ell-r,r} & 0_{\ell-r,m-r} \end{bmatrix} Z^{-1} \right) =: BC,$$

where $B \in K^{n,\ell}$ and $C \in K^{\ell,m}$.

(5) If rank$(A) = r$, then by (1) there exist matrices $Q \in GL_n(K)$ and $Z \in GL_m(K)$ with $QAZ = \begin{bmatrix} I_r & 0 \\ 0 & 0 \end{bmatrix}$. Therefore, using (2),

$$\text{rank}(A) = \text{rank}(QAZ) = \text{rank}\left(\begin{bmatrix} I_r & 0 \\ 0 & 0 \end{bmatrix}\right) = \text{rank}\left(\begin{bmatrix} I_r & 0 \\ 0 & 0 \end{bmatrix}^T\right) = \text{rank}((QAZ)^T)$$

$$= \text{rank}(Z^T A^T Q^T) = \text{rank}(A^T).$$

(3b) If $A = BC$, then

$$\text{rank}(A) = \text{rank}(A^T) = \text{rank}(C^T B^T) \le \text{rank}(C^T) = \text{rank}(C)$$

where we have used (3a) and (4). □

Let $A \in K^{n,m}$ with rank$(A) = r$. A factorization of the form $A = BC$ with $B \in K^{n,r}$ and $C \in K^{r,m}$ as in part (4) of Theorem 5.12 is called a *rank factorization* of A. If r is much smaller than $\min\{n, m\}$, then the matrices B and C have much fewer entries than A. In the special case $r = 1$ we often write the factorization as $A = bc^T$ with $b \in K^{n,1}$ and $c \in K^{m,1}$. The $n \cdot m$ entries of A are then completely represented by the $n + m$ entries of b and c.

A topic of great practical relevance in the field of Numerical Linear Algebra is the approximation of large matrices by a *low-rank factorization*. For a given matrix $A \in K^{n,m}$ we look for matrices $B \in K^{n,\ell}$ and $C \in K^{\ell,m}$ with $A \approx BC$ in an appropriate sense and with $\ell \le r$ as small as possible. We will address this question using the singular value decomposition in Chap. 19.

Example 5.13 The matrix

$$A = \begin{bmatrix} 0 & 2 & 1 & 3 & 3 \\ 0 & 2 & 0 & 1 & 1 \\ 0 & 2 & 0 & 1 & 1 \end{bmatrix} \in \mathbb{Q}^{3,5}$$

from Example 5.4 has the echelon form

$$\begin{bmatrix} 0 & 1 & 0 & \frac{1}{2} & \frac{1}{2} \\ 0 & 0 & 1 & 2 & 2 \\ 0 & 0 & 0 & 0 & 0 \end{bmatrix}.$$

5.3 Rank and Equivalence of Matrices

Since there are two pivot positions, we have $\text{rank}(A) = 2$. Multiplying A from the right by

$$B = \begin{bmatrix} 1 & 0 & 0 & 0 & 0 \\ 0 & 0 & 0 & 0 & 0 \\ 0 & 0 & 0 & 0 & 0 \\ 0 & 0 & 0 & -1 & -1 \\ 0 & 0 & 0 & 1 & 1 \end{bmatrix} \in \mathbb{Q}^{5,5},$$

yields $AB = 0 \in \mathbb{Q}^{3,5}$, and hence $\text{rank}(AB) = 0 < \text{rank}(A)$.

Assertion (1) in Theorem 5.12 motivates the following definition.

Definition 5.14 Two matrices $A, B \in K^{n,m}$ are called *equivalent*, if there exist matrices $Q \in GL_n(K)$ and $Z \in GL_m(K)$ with $A = QBZ$.

As the name suggests, this defines an equivalence relation on the set $K^{n,m}$, since the following properties hold:

- Reflexivity: $A = QAZ$ with $Q = I_n$ and $Z = I_m$.
- Symmetry: If $A = QBZ$, then $B = Q^{-1}AZ^{-1}$.
- Transitivity: If $A = Q_1 B Z_1$ and $B = Q_2 C Z_2$, then $A = (Q_1 Q_2) C (Z_2 Z_1)$.

The equivalence class of $A \in K^{n,m}$ is given by

$$[A] = \{QAZ \mid Q \in GL_n(K) \text{ and } Z \in GL_m(K)\}.$$

If $\text{rank}(A) = r$, then by (1) in Theorem 5.12 we have

$$\begin{bmatrix} I_r & 0_{r,m-r} \\ 0_{n-r,r} & 0_{n-r,m-r} \end{bmatrix} = \begin{bmatrix} I_r & 0 \\ 0 & 0 \end{bmatrix} \in [A]$$

and, therefore,

$$\left[\begin{bmatrix} I_r & 0 \\ 0 & 0 \end{bmatrix} \right] = [A].$$

Consequently, the rank of A fully determines the equivalence class $[A]$. The matrix

$$\begin{bmatrix} I_r & 0 \\ 0 & 0 \end{bmatrix} \in K^{n,m}$$

is called the *equivalence normal form* of A. We obtain

$$K^{n,m} = \bigcup_{r=0}^{\min\{n,m\}} \left[\begin{bmatrix} I_r & 0 \\ 0 & 0 \end{bmatrix}\right], \quad \text{where}$$

$$\left[\begin{bmatrix} I_r & 0 \\ 0 & 0 \end{bmatrix}\right] \cap \left[\begin{bmatrix} I_\ell & 0 \\ 0 & 0 \end{bmatrix}\right] = \emptyset, \quad \text{if } r \neq \ell.$$

Hence there are $1 + \min\{n, m\}$ pairwise distinct equivalence classes, and

$$\left\{ \begin{bmatrix} I_r & 0 \\ 0 & 0 \end{bmatrix} \in K^{n,m} \,\middle|\, r = 0, 1, \ldots, \min\{n, m\} \right\}$$

is a complete set of representatives.

From the proof of Theorem 4.9 we know that $(K^{n,n}, +, *)$ for $n \geq 2$ is a non-commutative ring with unit that contains non-trivial zero divisors. Using the equivalence normal form these can be characterized as follows:

- If $A \in K^{n,n}$ is invertible, then A cannot be a zero divisor, since then $AB = 0$ implies that $B = 0$.
- If $A \in K^{n,n} \setminus \{0\}$ is a zero divisor, then A cannot be invertible, and hence $1 \leq \text{rank}(A) = r < n$, so that the equivalence normal form of A is not the identity matrix I_n. Let $Q, Z \in GL_n(K)$ be given with

$$QAZ = \begin{bmatrix} I_r & 0 \\ 0 & 0 \end{bmatrix}.$$

Then for every matrix

$$V := \begin{bmatrix} 0_{r,r} & 0_{r,n-r} \\ V_{21} & V_{22} \end{bmatrix} \in K^{n,n}$$

and $B := ZV$ we have

$$AB = Q^{-1} \begin{bmatrix} I_r & 0 \\ 0 & 0 \end{bmatrix} \begin{bmatrix} 0_{r,r} & 0_{r,n-r} \\ V_{21} & V_{22} \end{bmatrix} = 0.$$

If $V \neq 0$, then $B \neq 0$, since Z is invertible.

Exercises

(In the following exercises K is an arbitrary field.)

5.1 Compute the echelon forms of the matrices

$$A = \begin{bmatrix} 1 & 2 & 3 \\ 2 & 4 & 48 \end{bmatrix} \in \mathbb{Q}^{2,3}, \quad B = \begin{bmatrix} 1 & i \\ i & 1 \end{bmatrix} \in \mathbb{C}^{2,2}, \quad C = \begin{bmatrix} 1 & i & -i & 0 \\ 0 & 0 & 0 & 1 \\ 5 & 0 & -6i & 0 \\ 0 & 1 & 0 & 0 \end{bmatrix} \in \mathbb{C}^{4,4},$$

$$D = \begin{bmatrix} 1 & 0 \\ 1 & 1 \\ 0 & 1 \end{bmatrix} \in (\mathbb{Z}/2\mathbb{Z})^{3,2}, \quad E = \begin{bmatrix} 1 & 0 & 2 & 0 \\ 2 & 0 & 1 & 1 \\ 1 & 2 & 0 & 2 \end{bmatrix} \in (\mathbb{Z}/3\mathbb{Z})^{3,4}.$$

(Here, for simplicity, the elements of $\mathbb{Z}/n\mathbb{Z}$ are denoted by k instead of $[k]$.) State the elementary matrices that carry out the transformations. If one of the matrices is invertible, then compute its inverse as a product of the elementary matrices.

5.2 Let $A = \begin{bmatrix} \alpha & \beta \\ \gamma & \delta \end{bmatrix} \in K^{2,2}$ with $\alpha\delta \neq \beta\gamma$. Determine the echelon form of A and a formula for A^{-1}.

5.3 Determine the rank of the following matrices over $\mathbb{R}$:

$$A = \begin{bmatrix} 0 & 1 & 1 \\ 2 & 1 & 0 \end{bmatrix}, \quad B = \begin{bmatrix} 2 & -2 \\ -1 & 4 \\ 1 & 2 \end{bmatrix}, \quad C = \begin{bmatrix} -1 & 2 & -3 \\ 2 & -1 & 1 \\ 1 & 1 & -2 \end{bmatrix}.$$

5.4 Let $A = \begin{bmatrix} 1 & A_{12} \\ 0 & B \end{bmatrix} \in K^{n,n}$ with $A_{12} \in K^{1,n-1}$ and $B \in K^{n-1,n-1}$. Show that $A \in GL_n(K)$ if and only if $B \in GL_{n-1}(K)$.

5.5 Consider the matrix

$$A = \begin{bmatrix} \frac{t+1}{t-1} & \frac{t-1}{t^2} \\ \frac{t^2}{t+1} & \frac{t-1}{t+1} \end{bmatrix} \in (K(t))^{2,2},$$

where $K(t)$ is the field of rational functions (cp. Exercise 3.22). Examine whether A is invertible and determine, if possible, A^{-1}. Verify your result by computing $A^{-1}A$ and AA^{-1}.

5.6 Show that if $A \in GL_n(K)$, then the echelon form of $[A, I_n] \in K^{n,2n}$ is given by $[I_n, A^{-1}]$.

(The inverse of an invertible matrix A can thus be computed via the transformation of $[A, I_n]$ to its echelon form.)

5.7 Prove or disprove (using a counterexample) the following assertions for $A, B \in \mathbb{R}^{2,2}$:
   (a) If $A \neq 0$, then A is invertible.
   (b) If A is not invertible, then A has a zero row.
   (c) If $A \neq 0$ and $B \neq 0$, then $AB \neq 0$.
   (d) If A and B are invertible, then $AB \neq 0$.

5.8 Find examples for matrices $A, B \in K^{n,n}$ with $AB = 0$ and $BA \neq 0$.

5.9 Two matrices $A, B \in K^{n,m}$ are called *left equivalent*, if there exists a matrix $Q \in GL_n(K)$ with $A = QB$. Show that this defines an equivalence relation on $K^{n,m}$ and determine a most simple representative for each equivalence class.

5.10 Prove Lemma 5.8.

5.11 Determine LU-decompositions (cp. Theorem 5.5) of the matrices

$$A = \begin{bmatrix} 1 & 2 & 3 & 0 \\ 4 & 0 & 0 & 1 \\ 5 & 0 & 6 & 0 \\ 0 & 1 & 0 & 0 \end{bmatrix}, \quad B = \begin{bmatrix} 2 & 0 & -2 & 0 \\ -4 & 0 & 4 & -1 \\ 0 & -1 & -1 & -2 \\ 0 & 0 & 1 & 1 \end{bmatrix} \in \mathbb{R}^{4,4}.$$

If one of these matrices is invertible, then determine its inverse using its LU-decomposition.

5.12 Let A be the 4×4 *Hilbert matrix* (cp. MATLAB-Minute 4 above Definition 5.7). Determine rank(A). Does A have an LU-decomposition as in Theorem 5.5 with $P = I_4$?

5.13 Determine the rank of the matrix

$$A = \begin{bmatrix} 0 & \alpha & \beta \\ -\alpha & 0 & \gamma \\ -\beta & -\gamma & 0 \end{bmatrix} \in \mathbb{R}^{3,3}$$

in dependence of $\alpha, \beta, \gamma \in \mathbb{R}$.

5.14 Let $A, B \in K^{n,n}$ be given. Show that

$$\text{rank}(A) + \text{rank}(B) \leq \text{rank}\left(\begin{bmatrix} A & C \\ 0 & B \end{bmatrix}\right)$$

for all $C \in K^{n,n}$. Examine when this inequality is strict.

5.15 Show that $A \in K^{n,n}$ has rank 1 if and only if there exist $a, b \in K^{n,1} \setminus \{0\}$ with $A = ab^T$.

5.16 Let $a, b, c \in \mathbb{R}^{n,1}$ and $M(a, b) := ba^T - ab^T$. Show the following assertions:
   (a) $M(a, b) = -M(b, a)$ and $M(a, b)c + M(b, c)a + M(c, a)b = 0$.
   (b) $M(\lambda a + \mu b, c) = \lambda M(a, c) + \mu M(b, c)$ for $\lambda, \mu \in \mathbb{R}$.
   (c) rank($M(a, b)$) = 0 holds if and only if there exist $\lambda, \mu \in \mathbb{R}$ with $\lambda \neq 0$ or $\mu \neq 0$ and $\lambda a + \mu b = 0$.
   (d) rank($M(a, b)$) $\in \{0, 2\}$.

Linear Systems of Equations

Solving linear systems of equations is a central problem of Linear Algebra that we discuss in an introductory way in this chapter. Such systems arise in numerous applications from engineering to the natural and social sciences. Major sources of linear systems of equations are the discretization of differential equations and the linearization of nonlinear equations. In this chapter we analyze the solution sets of linear systems of equations and we characterize the number of solutions using the echelon form from Chap. 5. We also develop an algorithm for the computation of the solutions.

Definition 6.1 A *linear system (of equations)* over a field K with n equations in m unknowns $x_1, \ldots, x_m$ has the form

$$
\begin{aligned}
a_{11}x_1 + \ldots + a_{1m}x_m &= b_1, \\
a_{21}x_1 + \ldots + a_{2m}x_m &= b_2, \\
&\vdots \\
a_{n1}x_1 + \ldots + a_{nm}x_m &= b_n
\end{aligned}
\tag{6.1}
$$

or

$$Ax = b, \tag{6.2}$$

where the *coefficient matrix* $A = [a_{ij}] \in K^{n,m}$ and the *right hand side* $b = [b_i] \in K^{n,1}$ are given. If $b = 0$, then the linear system is called *homogeneous*, otherwise *non-homogeneous*. Every $\widehat{x} \in K^{m,1}$ with $A\widehat{x} = b$ is called a *solution* of the linear system. All these $\widehat{x}$ form the *solution set* of the linear system, which we denote by $\mathscr{L}(A, b)$.

© The Author(s), under exclusive license to Springer Nature Switzerland AG 2025
J. Liesen, V. Mehrmann, *Linear Algebra*, Springer Undergraduate Mathematics Series, https://doi.org/10.1007/978-3-031-93260-1_6

The next result characterizes the solution set $\mathscr{L}(A, b)$ of the linear system $Ax = b$ using the solution set $\mathscr{L}(A, 0)$ of the associated homogeneous linear system $Ax = 0$.

Lemma 6.2 Let $A \in K^{n,m}$ and $b \in K^{n,1}$ with $\mathscr{L}(A, b) \neq \emptyset$ be given. If $\widehat{x} \in \mathscr{L}(A, b)$, then

$$\mathscr{L}(A, b) = \widehat{x} + \mathscr{L}(A, 0) := \{\widehat{x} + \widehat{z} \mid \widehat{z} \in \mathscr{L}(A, 0)\}.$$

Proof If $\widehat{z} \in \mathscr{L}(A, 0)$, and thus $\widehat{x} + \widehat{z} \in \widehat{x} + \mathscr{L}(A, 0)$, then

$$A(\widehat{x} + \widehat{z}) = A\widehat{x} + A\widehat{z} = b + 0 = b.$$

Hence $\widehat{x} + \widehat{z} \in \mathscr{L}(A, b)$, which shows that $\widehat{x} + \mathscr{L}(A, 0) \subseteq \mathscr{L}(A, b)$.

Let now $\widehat{x}_1 \in \mathscr{L}(A, b)$ and let $\widehat{z} := \widehat{x}_1 - \widehat{x}$. Then

$$A\widehat{z} = A\widehat{x}_1 - A\widehat{x} = b - b = 0,$$

i.e., $\widehat{z} \subset \mathscr{L}(A, 0)$. Hence $\widehat{x}_1 = \widehat{x} + \widehat{z} \in \widehat{x} + \mathscr{L}(A, 0)$, which shows that $\mathscr{L}(A, b) \subseteq \widehat{x} + \mathscr{L}(A, 0)$. □

We will have a closer look at the set $\mathscr{L}(A, 0)$: Clearly, $0 \in \mathscr{L}(A, 0) \neq \emptyset$. If $\widehat{z}_1, \widehat{z}_2 \in \mathscr{L}(A, 0)$ and $\alpha, \beta \in K$, then

$$A(\alpha \widehat{z}_1 + \beta \widehat{z}_2) = \alpha A\widehat{z}_1 + \beta A\widehat{z}_2 = 0 + 0 = 0. \tag{6.3}$$

Thus, $\mathscr{L}(A, 0)$ is a nonempty subset of $K^{m,1}$ that is closed under scalar multiplication and addition.

Lemma 6.3 If $A \in K^{n,m}$, $b \in K^{n,1}$ and $S \in K^{n,n}$, then $\mathscr{L}(A, b) \subseteq \mathscr{L}(SA, Sb)$. Moreover, if S is invertible, then $\mathscr{L}(A, b) = \mathscr{L}(SA, Sb)$.

Proof If $\widehat{x} \in \mathscr{L}(A, b)$, then also $SA\widehat{x} = Sb$, and thus $\widehat{x} \in \mathscr{L}(SA, Sb)$, which shows that $\mathscr{L}(A, b) \subseteq \mathscr{L}(SA, Sb)$. If S is invertible and $\widehat{y} \in \mathscr{L}(SA, Sb)$, then $SA\widehat{y} = Sb$. Multiplying from the left with S^{-1} yields $A\widehat{y} = b$. Since $\widehat{y} \in \mathscr{L}(A, b)$, we have $\mathscr{L}(SA, Sb) \subseteq \mathscr{L}(A, b)$. □

Consider the linear system of equations $Ax = b$. By Theorem 5.2 we can find a matrix $S \in GL_n(K)$ such that SA is in echelon form. Let $\widetilde{b} = [\widetilde{b}_i] := Sb$, then

$\mathscr{L}(A,b) = \mathscr{L}(SA,\widetilde{b})$ by Lemma 6.3, and the linear system $SAx = \widetilde{b}$ takes the form

$$\begin{bmatrix} 1 & \star & 0 & & 0 & & & 0 & \\ & & 1 & \star & 0 & \star & & & \\ & & & & 1 & & & & \\ 0 & & & & & \ddots & 0 & \star & \\ & 0 & & & & & & 1 & \\ & & & 0 & & & & & 0 \end{bmatrix} x = \begin{bmatrix} \widetilde{b}_1 \\ \vdots \\ \widetilde{b}_r \\ \widetilde{b}_{r+1} \\ \vdots \\ \widetilde{b}_n \end{bmatrix}. \tag{6.4}$$

Suppose that $\text{rank}(A) = r$, and let $j_1, j_2, \ldots, j_r$ be the pivot columns. If in the linear system (6.4) at least one of the entries $\widetilde{b}_{r+1}, \ldots, \widetilde{b}_n$ on the right hand side is nonzero, then no solution exists. If all these entries are zero, then we can reorder the linear system according to the "pivot variables" $x_{j_1}, \ldots, x_{j_r}$. The values of these variables can then be determined by choosing values for the other variables. In the following we show how this procedure works.

Consider the permutation matrix

$$P^T := [e_{j_1}, \ldots, e_{j_r}, e_1, \ldots, e_{j_1-1}, e_{j_1+1}, \ldots, e_{j_2-1}, e_{j_2+1}, \ldots, e_{j_r-1}, e_{j_r+1}, \ldots, e_m] \in K^{m,m}.$$

Then

$$\widetilde{A} := SAP^T = \begin{bmatrix} I_r & \widetilde{A}_{12} \\ 0_{n-r,r} & 0_{n-r,m-r} \end{bmatrix},$$

where $\widetilde{A}_{12} \in K^{r,m-r}$. (If $r = m$, then $\widetilde{A}_{12} = [\ ]$.) Thus, multiplying SA from the right with P^T permutes the r pivot columns of SA "upfront". This permutation leads to a simplification of the following presentation, but it is usually omitted in practical computations.

Since $P^T P = I_m$, we can write $SAx = \widetilde{b}$ as $\widetilde{A}Px = \widetilde{b}$. With $y := Px$ this system has the form

$$\begin{bmatrix} I_r & \widetilde{A}_{12} \\ \hline 0_{n-r,r} & 0_{n-r,m-r} \end{bmatrix} \begin{bmatrix} y_1 \\ \vdots \\ y_r \\ y_{r+1} \\ \vdots \\ y_m \end{bmatrix} = \begin{bmatrix} \widetilde{b}_1 \\ \vdots \\ \widetilde{b}_r \\ \widetilde{b}_{r+1} \\ \vdots \\ \widetilde{b}_n \end{bmatrix}. \tag{6.5}$$

The left-multiplication of x with P just means a different ordering of the unknowns $x_1, \ldots, x_m$.

The matrix

$$[\widetilde{A}, \widetilde{b}] \in K^{n,m+1},$$

is called the *extended coefficient matrix* of the linear system (6.5). This matrix satisfies

$$\text{rank}([\widetilde{A}, \widetilde{b}]) = \text{rank}([SAP^T, Sb]) = \text{rank}\left(S[A, b]\begin{bmatrix} P^T & 0 \\ 0 & 1 \end{bmatrix}\right) = \text{rank}([A, b]).$$

Obviously, $\text{rank}(A) = \text{rank}(\widetilde{A})$, and therefore

$$\text{rank}(A) = \text{rank}(\widetilde{A}) \leq \text{rank}([\widetilde{A}, \widetilde{b}]) = \text{rank}([A, b]),$$

with equality if and only if $\widetilde{b}_{r+1} = \cdots = \widetilde{b}_n = 0$.

If $\text{rank}(\widetilde{A}) < \text{rank}([\widetilde{A}, \widetilde{b}])$, then at least one of $\widetilde{b}_{r+1}, \ldots, \widetilde{b}_n$ is nonzero, and (6.5) cannot have a solution, i.e., $\mathscr{L}(A, b) = \emptyset$.

If, on the other hand, $\text{rank}(\widetilde{A}) = \text{rank}([\widetilde{A}, \widetilde{b}])$, then $\widetilde{b}_{r+1} = \cdots = \widetilde{b}_n = 0$ and (6.5) can be written as

$$\begin{bmatrix} y_1 \\ \vdots \\ y_r \end{bmatrix} = \begin{bmatrix} \widetilde{b}_1 \\ \vdots \\ \widetilde{b}_r \end{bmatrix} - \widetilde{A}_{12} \begin{bmatrix} y_{r+1} \\ \vdots \\ y_m \end{bmatrix}. \tag{6.6}$$

This representation yields, for the choice $y_{r+1} = \cdots = y_m = 0$, the particular solution

$$\widehat{y} := [\widetilde{b}_1, \ldots, \widetilde{b}_r, \underbrace{0, \ldots, 0}_{m-r}]^T \in \mathscr{L}(\widetilde{A}, \widetilde{b}) \neq \emptyset. \tag{6.7}$$

From Lemma 6.2 we know that $\mathscr{L}(\widetilde{A}, \widetilde{b}) = \widehat{y} + \mathscr{L}(\widetilde{A}, 0)$. In order to determine $\mathscr{L}(\widetilde{A}, 0)$ we set $\widetilde{b}_1 = \cdots = \widetilde{b}_r = 0$ in (6.6), which yields

$$\mathscr{L}(\widetilde{A}, 0) = \{ [y_1, \ldots, y_m]^T \mid y_{r+1}, \ldots, y_m \text{ arbitrary and} \tag{6.8}$$

$$[y_1, \ldots, y_r]^T = 0 - \widetilde{A}_{12}[y_{r+1}, \ldots, y_m]^T \}.$$

If $r = m$, then $\widetilde{A}_{12} = [\ ]$, $\mathscr{L}(\widetilde{A}, 0) = \{0\}$, and $\mathscr{L}(\widetilde{A}, \widetilde{b}) = \{\widehat{y}\}$. In this case the solution of $\widetilde{A}y = \widetilde{b}$ is uniquely determined. If $r < m$, then this system in general has "many" solutions.

Example 6.4 For the extended coefficient matrix

$$[\widetilde{A}, \widetilde{b}] = \begin{bmatrix} 1 & 0 & 3 & \widetilde{b}_1 \\ 0 & 1 & 4 & \widetilde{b}_2 \\ 0 & 0 & 0 & \widetilde{b}_3 \end{bmatrix} \in \mathbb{Q}^{3,4}$$

we have $\text{rank}(\widetilde{A}) = \text{rank}([\widetilde{A}, \widetilde{b}])$ if and only if $\widetilde{b}_3 = 0$. If $\widetilde{b}_3 = 0$, then $\widetilde{A}y = \widetilde{b}$ can be written as

$$\begin{bmatrix} y_1 \\ y_2 \end{bmatrix} = \begin{bmatrix} \widetilde{b}_1 \\ \widetilde{b}_2 \end{bmatrix} - \begin{bmatrix} 3 \\ 4 \end{bmatrix}[y_3].$$

Hence, $\widehat{y} = [\widetilde{b}_1, \widetilde{b}_2, 0]^T \in \mathscr{L}(\widetilde{A}, \widetilde{b})$ and

$$\mathscr{L}(\widetilde{A}, 0) = \left\{ [y_1, y_2, y_3]^T \mid y_3 \in \mathbb{Q} \text{ arbitrary and } [y_1, y_2]^T = -[3, 4]^T[y_3] \right\}$$
$$= \left\{ [-3y_3, -4y_3, y_3]^T \mid y_3 \in \mathbb{Q} \right\}.$$

On the other hand, if $\widetilde{b}_3 \neq 0$, then $\mathscr{L}(\widetilde{A}, \widetilde{b}) = \emptyset$.

We have $\widehat{y} \in \mathscr{L}(\widetilde{A}, \widetilde{b})$ if and only if $\widehat{x} := P^T\widehat{y} \in \mathscr{L}(SA, \widetilde{b}) = \mathscr{L}(A, b)$. We therefore have shown the following result about the solvability of the linear system $Ax = b$.

Theorem 6.5 Let $A \in K^{n,m}$ and $b \in K^{n,1}$ be given, and let $\widetilde{A} = SAP^T \in K^{n,m}$ and $\widetilde{b} = Sb \in K^{n,1}$ be as in (6.5). Then the following assertions hold:

(1) If $\text{rank}(A) < \text{rank}([A, b])$, then $\mathscr{L}(A, b) = \emptyset$.
(2) If $\text{rank}(A) = \text{rank}([A, b]) = m$, then $\mathscr{L}(A, b) = \{P^T\widehat{y}\}$ with $\widehat{y}$ in (6.7).
(3) If $\text{rank}(A) = \text{rank}([A, b]) < m$, then $\mathscr{L}(A, b) = \{P^T y \mid y \in \widehat{y} + \mathscr{L}(\widetilde{A}, 0)\}$ with $\widehat{y}$ in (6.7) and $\mathscr{L}(\widetilde{A}, 0)$ in (6.8).

In particular, it follows that the homogeneous linear system $Ax = 0$ with $A \in K^{n,m}$ has the uniquely determined solution $\widehat{x} = 0$ if and only if $\text{rank}(A) = m$.

If $A \in K^{n,m}$ and $\text{rank}(A) = \text{rank}([A, b]) < m$, then the solution of $Ax = b$ is not uniquely determined. If the field K has infinitely many elements (e.g. for $K = \mathbb{Q}$, $K = \mathbb{R}$ or $K = \mathbb{C}$), then infinitely many pairwise distinct solutions exist. We will discuss the different cases in Theorem 6.5 in more detail later (see, in particular, Example 10.8.)

Summarizing our considerations we have the following algorithm for solving a linear system of equations.

Algorithm 6.6 Given $A \in K^{n,m}$ and $b \in K^{n,1}$.

(1) Apply Gaussian elimination to the extended coefficient matrix $[A, b]$ to compute $S \in GL_n(K)$ such that SA is in echelon form. The last column of the resulting matrix is then $\tilde{b} = Sb$. Determine $r = \text{rank}(A)$.
(2a) If at least one of $\tilde{b}_{r+1}, \ldots, \tilde{b}_n$ is nonzero, then $\mathscr{L}(A, b) = \emptyset$.
(2b) If $\tilde{b}_{r+1} = \cdots = \tilde{b}_n = 0$ and $r = m$, then $\mathscr{L}(A, b) = \{P^T \hat{y}\}$ with $\hat{y}$ in (6.7).
(2c) If $\tilde{b}_{r+1} = \cdots = \tilde{b}_n = 0$ and $r < m$, then $\mathscr{L}(A, b) = \{P^T y \mid y \in \hat{y} + \mathscr{L}(\hat{A}, 0)\}$ with $\hat{y}$ in (6.7) and $\mathscr{L}(\hat{A}, 0)$ in (6.8).

Example 6.7 Let $K = \mathbb{Q}$ and consider the linear system of equations $Ax = b$ with

$$A = \begin{bmatrix} 1 & 2 & 2 & 1 \\ 0 & 1 & 0 & 3 \\ 1 & 0 & 3 & 0 \\ 2 & 3 & 5 & 4 \\ 1 & 1 & 3 & 3 \end{bmatrix}, \quad b = \begin{bmatrix} 1 \\ 0 \\ 2 \\ 3 \\ 2 \end{bmatrix}.$$

We form $[A, b]$ and apply the Gaussian elimination algorithm in order to transform A into echelon form:

$$[A, b] \rightsquigarrow \begin{bmatrix} 1 & 2 & 2 & 1 & | & 1 \\ 0 & 1 & 0 & 3 & | & 0 \\ 0 & -2 & 1 & -1 & | & 1 \\ 0 & -1 & 1 & 2 & | & 1 \\ 0 & -1 & 1 & 2 & | & 1 \end{bmatrix} \rightsquigarrow \begin{bmatrix} 1 & 2 & 2 & 1 & | & 1 \\ 0 & 1 & 0 & 3 & | & 0 \\ 0 & 0 & 1 & 5 & | & 1 \\ 0 & 0 & 1 & 5 & | & 1 \\ 0 & 0 & 1 & 5 & | & 1 \end{bmatrix} \rightsquigarrow \begin{bmatrix} 1 & 2 & 2 & 1 & | & 1 \\ 0 & 1 & 0 & 3 & | & 0 \\ 0 & 0 & 1 & 5 & | & 1 \\ 0 & 0 & 0 & 0 & | & 0 \\ 0 & 0 & 0 & 0 & | & 0 \end{bmatrix}$$

$$\rightsquigarrow \begin{bmatrix} 1 & 0 & 2 & -5 & | & 1 \\ 0 & 1 & 0 & 3 & | & 0 \\ 0 & 0 & 1 & 5 & | & 1 \\ 0 & 0 & 0 & 0 & | & 0 \\ 0 & 0 & 0 & 0 & | & 0 \end{bmatrix} \rightsquigarrow \begin{bmatrix} 1 & 0 & 0 & -15 & | & -1 \\ 0 & 1 & 0 & 3 & | & 0 \\ 0 & 0 & 1 & 5 & | & 1 \\ 0 & 0 & 0 & 0 & | & 0 \\ 0 & 0 & 0 & 0 & | & 0 \end{bmatrix} = [SA | \tilde{b}].$$

Here $\text{rank}(A) = 3$ and $\tilde{b}_4 = \tilde{b}_5 = 0$, and hence solutions exist. The pivot columns are $j_i = i$ for $i = 1, 2, 3$, so that $P = P^T = I_4$ and $\tilde{A} = SA$. Now $SAx = \tilde{b}$ can be written as

$$\begin{bmatrix} x_1 \\ x_2 \\ x_3 \end{bmatrix} = \begin{bmatrix} -1 \\ 0 \\ 1 \end{bmatrix} - \begin{bmatrix} -15 \\ 3 \\ 5 \end{bmatrix} [x_4].$$

Consequently, $\widehat{x} = [-1, 0, 1, 0]^T \in \mathscr{L}(A, b)$ and $\mathscr{L}(A, b) = \widehat{x} + \mathscr{L}(A, 0)$, where

$$\mathscr{L}(A, 0) = \left\{ [x_1, x_2, x_3, x_4]^T \mid x_4 \in \mathbb{Q} \text{ arbitrary and } [x_1, x_2, x_3]^T \right.$$
$$= -[-15, 3, 5]^T [x_4] \right\}$$
$$= \left\{ [15x_4, -3x_4, -5x_4, x_4]^T \mid x_4 \in \mathbb{Q} \right\}.$$

Exercises

6.1 Find a field K and matrices $A \in K^{n,m}$, $S \in K^{n,n}$ and $b \in K^{n,1}$ with $\mathscr{L}(A, b) \neq \mathscr{L}(SA, Sb)$.

6.2 Determine $\mathscr{L}(A, b)$ for the following A and b:

$$A = \begin{bmatrix} 1 & 1 & 1 \\ 1 & 2 & -1 \\ 1 & -1 & 6 \end{bmatrix} \in \mathbb{R}^{3,3}, \quad b = \begin{bmatrix} 1 \\ -2 \\ 3 \end{bmatrix} \in \mathbb{R}^{3,1},$$

$$A = \begin{bmatrix} 1 & 1 & 1 & 0 \\ 1 & 2 & -1 & -1 \\ 1 & -1 & 6 & 2 \end{bmatrix} \in \mathbb{R}^{3,4}, \quad b = \begin{bmatrix} 1 \\ -2 \\ 3 \end{bmatrix} \in \mathbb{R}^{3,1},$$

$$A = \begin{bmatrix} 1 & 1 & 1 \\ 1 & 2 & -1 \\ 1 & -1 & 6 \\ 1 & 1 & 1 \end{bmatrix} \in \mathbb{R}^{4,3}, \quad b = \begin{bmatrix} 1 \\ -2 \\ 3 \\ 1 \end{bmatrix} \in \mathbb{R}^{4,1},$$

$$A = \begin{bmatrix} 1 & 1 & 1 \\ 1 & 2 & -1 \\ 1 & -1 & 6 \\ 1 & 1 & 1 \end{bmatrix} \in \mathbb{R}^{4,3}, \quad b = \begin{bmatrix} 1 \\ -2 \\ 3 \\ 0 \end{bmatrix} \in \mathbb{R}^{4,1},$$

$$A = \begin{bmatrix} 1 & 2\mathbf{i} & 1 \\ \mathbf{i} & -1 & 1 \\ 1+\mathbf{i} & 0 & 2 \end{bmatrix} \in \mathbb{C}^{3,3} \quad b = \begin{bmatrix} \mathbf{i} \\ -\mathbf{i} \\ 0 \end{bmatrix} \in \mathbb{C}^{3,1},$$

$$A = \begin{bmatrix} 1 & 2 & 0 & 1 \\ 0 & 1 & 1 & 2 \\ 2 & 2 & 1 & 1 \end{bmatrix} \in (\mathbb{Z}/3\mathbb{Z})^{3,4}, \quad b = \begin{bmatrix} 1 \\ 0 \\ 2 \end{bmatrix} \in (\mathbb{Z}/3\mathbb{Z})^{3,1}.$$

(For simplicity, the elements of $\mathbb{Z}/n\mathbb{Z}$ are denoted by k instead of $[k]$.)

6.3 Let $\alpha \in \mathbb{Q}$,

$$A = \begin{bmatrix} 3 & 2 & 1 \\ 1 & 1 & 1 \\ 2 & 1 & 0 \end{bmatrix} \in \mathbb{Q}^{3,3}, \quad b_\alpha = \begin{bmatrix} 6 \\ 3 \\ \alpha \end{bmatrix} \in \mathbb{Q}^{3,1}.$$

Determine $\mathscr{L}(A, 0)$ and $\mathscr{L}(A, b_\alpha)$ in dependence of α.

6.4 Let K be a field, $A \in K^{n,m}$ and $B \in K^{n,s}$. For $i = 1, \ldots, s$ denote by b_i the ith column of B. Show that the linear system of equations $AX = B$ has at least one solution $\widehat{X} \in K^{m,s}$ if and only if

$$\operatorname{rank}(A) = \operatorname{rank}([A, b_1]) = \operatorname{rank}([A, b_2]) = \cdots = \operatorname{rank}([A, b_s]).$$

Find conditions under which this solution is unique.

6.5 Let K be a field and $A \in K^{n,m}$. Show the following assertions:
 (a) A matrix $B \in K^{m,n}$ with $AB = I_n$ exists if and only if $\operatorname{rank}(A) = n$.
 (b) If $m \neq n$, then there exist either no or at least two distinct $B \in K^{m,n}$ with $AB = I_n$.

6.6 Let

$$A = \begin{bmatrix} 0 & \beta_1 & & \\ \alpha_2 & 0 & \ddots & \\ & \ddots & \ddots & \beta_n \\ & & \alpha_n & 0 \end{bmatrix} \in K^{n,n}, \quad b = \begin{bmatrix} b_1 \\ \vdots \\ b_n \end{bmatrix} \in K^{n,1}$$

be given with $\beta_i, \alpha_i \neq 0$ for all i. Determine a recursive formula for the entries of the solution of the linear system $Ax = b$.

Determinants of Matrices 7

The determinant is a map that assigns to every square matrix $A \in R^{n,n}$, where R is a commutative ring with unit, an element of R. This map has very interesting and important properties. For instance it yields a necessary and sufficient condition for the invertibility of $A \in R^{n,n}$. Moreover, it forms the basis for the definition of the characteristic polynomial of a matrix in Chap. 8. In the field of Analytic Geometry the determinant is used to calculate volumes of (polyhedral) sets, and in Real Analysis it plays an important role in transformation formulas for integrals of functions of several variables.

7.1 Definition of the Determinant

There are several different approaches to define the determinant of a matrix. We use the constructive approach via permutations.

Definition 7.1 Let $n \in \mathbb{N}$ be given. A bijective map

$$\sigma : \{1, 2, \ldots, n\} \to \{1, 2, \ldots, n\}, \quad j \mapsto \sigma(j),$$

is called a *permutation* of the numbers $\{1, 2, \ldots, n\}$. We denote the set of all these maps by S_n.

A permutation $\sigma \in S_n$ can be written in the form

$$[\sigma(1) \; \sigma(2) \; \ldots \; \sigma(n)]$$

For example, $S_1 = \{[1]\}$, $S_2 = \{[1\,2], [2\,1]\}$, and

$$S_3 = \{[1\,2\,3], [1\,3\,2], [2\,1\,3], [2\,3\,1], [3\,1\,2], [3\,2\,1]\}.$$

© The Author(s), under exclusive license to Springer Nature Switzerland AG 2025
J. Liesen, V. Mehrmann, *Linear Algebra*, Springer Undergraduate
Mathematics Series, https://doi.org/10.1007/978-3-031-93260-1_7

From Lemma 2.18 we know that $|S_n| = n! = 1 \cdot 2 \cdot \ldots \cdot n$.

The set S_n with the composition of maps "∘" forms a group (cp. Exercise 3.3), which is sometimes called the *symmetric group*. The neutral element in this group is the identity, i.e., the permutation $[1\,2\,\ldots\,n]$.

While S_1 and S_2 are commutative groups, the group S_n for $n \geq 3$ is non-commutative. As an example consider $n = 3$ and the permutations $\sigma_1 = [2\,3\,1]$, $\sigma_2 = [1\,3\,2]$. Then

$$\sigma_1 \circ \sigma_2 = [\sigma_1(\sigma_2(1))\ \sigma_1(\sigma_2(2))\ \sigma_1(\sigma_2(3))] = [\sigma_1(1)\ \sigma_1(3)\ \sigma_1(2)] = [2\,1\,3],$$

$$\sigma_2 \circ \sigma_1 = [\sigma_2(\sigma_1(1))\ \sigma_2(\sigma_1(2))\ \sigma_2(\sigma_1(3))] = [\sigma_2(2)\ \sigma_2(3)\ \sigma_2(1)] = [3\,2\,1].$$

Definition 7.2 Let $n \geq 2$ and $\sigma \in S_n$. A pair $(\sigma(i), \sigma(j))$ with $1 \leq i < j \leq n$ and $\sigma(i) > \sigma(j)$ is called an *inversion* of σ. If k is the number of inversions of σ, then $\mathrm{sgn}(\sigma) := (-1)^k$ is called the *sign* of σ. For $n = 1$ we define $\mathrm{sgn}([1]) := 1 = (-1)^0$.

In short, an inversion of a permutation σ is a pair that is "out of order". The term inversion should not be confused with the inverse map σ^{-1} (which exists, since σ is bijective). The sign of a permutation is sometimes also called the *signature*.

Example 7.3 The permutation $[2\,3\,1\,4] \in S_4$ has the inversions $(2, 1)$ and $(3, 1)$, so that $\mathrm{sgn}([2\,3\,1\,4]) = 1$. The permutation $[4\,1\,2\,3] \in S_4$ has the inversions $(4, 1)$, $(4, 2)$, $(4, 3)$, so that $\mathrm{sgn}([4\,1\,2\,3]) = -1$.

We can now define the determinant map.

Definition 7.4 Let R be a commutative ring with unit and let $n \in \mathbb{N}$. The map

$$\det : R^{n,n} \to R, \quad A = [a_{ij}] \mapsto \det(A) := \sum_{\sigma \in S_n} \mathrm{sgn}(\sigma) \prod_{i=1}^n a_{i,\sigma(i)}, \tag{7.1}$$

is called the *determinant*, and the ring element $\det(A)$ is called the *determinant of A*.

The formula (7.1) for $\det(A)$ is called the *signature formula of Leibniz*.[1] The term $\mathrm{sgn}(\sigma)$ in this definition is to be interpreted as an element of the ring R, i.e., either $\mathrm{sgn}(\sigma) = 1 \in R$ or $\mathrm{sgn}(\sigma) = -1 \in R$, where $-1 \in R$ is the unique additive inverse of the unit $1 \in R$.

---

[1] Gottfried Wilhelm Leibniz (1646–1716).

7.1 Definition of the Determinant

Example 7.5 For $n = 1$ we have $A = [a_{11}]$ and thus $\det(A) = \text{sgn}([1])a_{11} = a_{11}$. For $n = 2$ we get

$$\det(A) = \det\left(\begin{bmatrix} a_{11} & a_{12} \\ a_{21} & a_{22} \end{bmatrix}\right) = \text{sgn}([1\ 2])a_{11}a_{22} + \text{sgn}([2\ 1])a_{12}a_{21}$$

$$= a_{11}a_{22} - a_{12}a_{21}.$$

For $n = 3$ we have the *Sarrus rule*:[2]

$$\det(A) = a_{11}a_{22}a_{33} + a_{12}a_{23}a_{31} + a_{13}a_{21}a_{32}$$
$$- a_{11}a_{23}a_{32} - a_{12}a_{21}a_{33} - a_{13}a_{22}a_{31}.$$

In order to compute $\det(A)$ using the signature formula of Leibniz we have to form $n!$ products with n factors each. For large n this is too costly even on modern computers. As we will see after Theorem 7.15, there are more efficient ways for computing $\det(A)$. The signature formula is mostly of theoretical relevance, since it represents the determinant of A explicitly in terms of the entries of A. Considering the n^2 entries as variables, we can interpret $\det(A)$ as a polynomial in these variables. If $R = \mathbb{R}$ or $R = \mathbb{C}$, then standard techniques of Analysis show that $\det(A)$ is a continuous function of the entries of A.

We will now study the group of permutations in more detail. The permutation $\sigma = [3\ 2\ 1] \in S_3$ has the inversions $(3, 2)$, $(3, 1)$ and $(2, 1)$, so that $\text{sgn}(\sigma) = -1$. Moreover,

$$\prod_{1 \le i < j \le 3} \frac{\sigma(j) - \sigma(i)}{j - i} = \frac{\sigma(2) - \sigma(1)}{2 - 1} \cdot \frac{\sigma(3) - \sigma(1)}{3 - 1} \cdot \frac{\sigma(3) - \sigma(2)}{3 - 2}$$

$$= \frac{2 - 3}{2 - 1} \cdot \frac{1 - 3}{3 - 1} \cdot \frac{1 - 2}{3 - 2} = (-1)^3 = -1 = \text{sgn}(\sigma).$$

This observation can be generalized as follows.

Lemma 7.6 *For each $\sigma \in S_n$ we have*

$$\text{sgn}(\sigma) = \prod_{1 \le i < j \le n} \frac{\sigma(j) - \sigma(i)}{j - i}. \tag{7.2}$$

Proof If $n = 1$, then the left hand side of (7.2) is an empty product, which is defined to be 1 (cp. Sect. 3.2), so that (7.2) holds for $n = 1$.

---

[2] Pierre Frédéric Sarrus (1798–1861).

Let $n > 1$ and $\sigma \in S_n$ with $\mathrm{sgn}(\sigma) = (-1)^k$, i.e., k is the number of pairs $(\sigma(i), \sigma(j))$ with $i < j$ but $\sigma(i) > \sigma(j)$. Then

$$\prod_{1 \leq i < j \leq n} (\sigma(j) - \sigma(i)) = (-1)^k \prod_{1 \leq i < j \leq n} |\sigma(j) - \sigma(i)| = (-1)^k \prod_{1 \leq i < j \leq n} (j - i).$$

In the last equation we have used the fact that the two products have the same factors (except possibly for their order). □

Theorem 7.7 *For all $\sigma_1, \sigma_2 \in S_n$ we have $\mathrm{sgn}(\sigma_1 \circ \sigma_2) = \mathrm{sgn}(\sigma_1)\,\mathrm{sgn}(\sigma_2)$. In particular, $\mathrm{sgn}(\sigma^{-1}) = \mathrm{sgn}(\sigma)$ for all $\sigma \in S_n$.*

Proof By Lemma 7.6 we have

$$\mathrm{sgn}(\sigma_1 \circ \sigma_2) = \prod_{1 \leq i < j \leq n} \frac{\sigma_1(\sigma_2(j)) - \sigma_1(\sigma_2(i))}{j - i}$$

$$= \left(\prod_{1 \leq i < j \leq n} \frac{\sigma_1(\sigma_2(j)) - \sigma_1(\sigma_2(i))}{\sigma_2(j) - \sigma_2(i)} \right) \left(\prod_{1 \leq i < j \leq n} \frac{\sigma_2(j) - \sigma_2(i)}{j - i} \right)$$

$$= \left(\prod_{1 \leq \sigma_2(i) < \sigma_2(j) \leq n} \frac{\sigma_1(\sigma_2(j)) - \sigma_1(\sigma_2(i))}{\sigma_2(j) - \sigma_2(i)} \right) \mathrm{sgn}(\sigma_2)$$

$$= \left(\prod_{1 \leq i < j \leq n} \frac{\sigma_1(j) - \sigma_1(i)}{j - i} \right) \mathrm{sgn}(\sigma_2)$$

$$= \mathrm{sgn}(\sigma_1)\,\mathrm{sgn}(\sigma_2).$$

For each $\sigma \in S_n$ we have $1 = \mathrm{sgn}([1\,2\,\ldots\,n]) = \mathrm{sgn}(\sigma \circ \sigma^{-1}) = \mathrm{sgn}(\sigma)\,\mathrm{sgn}(\sigma^{-1})$, so that $\mathrm{sgn}(\sigma) = \mathrm{sgn}(\sigma^{-1})$. □

Theorem 7.7 shows that the map sgn is a homomorphism between the groups $(S_n, \circ)$ and $(\{1, -1\}, \cdot)$, where the operation in the second group is the standard multiplication of the integers 1 and -1 (cp. Exercise 7.3).

Definition 7.8 A *transposition* is a permutation $\tau \in S_n$, $n \geq 2$, that exchanges exactly two distinct elements $k, \ell \in \{1, 2, \ldots, n\}$, i.e., $\tau(k) = \ell$, $\tau(\ell) = k$ and $\tau(j) = j$ for all $j \in \{1, 2, \ldots, n\} \setminus \{k, \ell\}$.

Obviously $\tau^{-1} = \tau$ for every transposition $\tau \in S_n$.

Lemma 7.9 *Let $\tau \in S_n$ be the transposition that exchanges k and ℓ for some $1 \leq k < \ell \leq n$. Then τ has exactly $2(\ell - k) - 1$ inversions and, hence, $\mathrm{sgn}(\tau) = -1$.*

Proof We have $\ell = k + j$ for a $j \geq 1$ and thus τ is given by

$$\tau = [1, \ldots, k-1, \ k+j, \ k+1, \ldots, k+(j-1), \ k, \ \ell+1, \ldots, n],$$

where the points denote values of τ in increasing and thus "correct" order. A simple counting argument shows that τ has exactly $2j - 1 = 2(\ell - k) - 1$ inversions. These are given by

$$\underbrace{(k, k+1), \ldots, (k, k+j)}_{j \text{ inversions}}, \ \underbrace{(k+1, k+j), \ldots, (k+j-1, k+j)}_{j-1 \text{ inversions}}.$$

(For $j = 1$ the second list is empty.) In particular, we obtain $\text{sgn}(\tau) = (-1)^{2j-1} = -1$. □

7.2 Properties of the Determinant

In this section we prove important properties of the determinant map.

Lemma 7.10 *For $A \in R^{n,n}$ the following assertions hold:*

(1) For $\lambda \in R$,

$$\det\left(\left[\begin{array}{c|c}\lambda & \star \\ \hline 0_{n,1} & A\end{array}\right]\right) = \det\left(\left[\begin{array}{c|c}\lambda & 0_{1,n} \\ \hline \star & A\end{array}\right]\right) = \lambda \det(A).$$

(2) If $A = [a_{ij}]$ is upper or lower triangular, then $\det(A) = \prod_{i=1}^{n} a_{ii}$.
(3) If A has a zero row or column, then $\det(A) = 0$.
(4) If $n \geq 2$ and A has two equal rows or two equal columns, then $\det(A) = 0$.
(5) $\det(A) = \det(A^T)$.

Proof

(1) Exercise.
(2) This follows by repeated application of (1) to the upper (or lower) triangular matrix A.
(3) If A has a zero row or column, then for every $\sigma \in S_n$ at least one factor in the product $\prod_{i=1}^{n} a_{i,\sigma(i)}$ is equal to zero and thus $\det(A) = 0$.
(4) Let the rows k and ℓ, with $k < \ell$, of $A = [a_{ij}]$ be equal, i.e., $a_{kj} = a_{\ell j}$ for $j = 1, \ldots, n$. Let $\tau \in S_n$ be the transposition that exchanges the elements k and ℓ, and let

$$T_n := \{\sigma \in S_n \mid \sigma(k) < \sigma(\ell)\}.$$

Since the set T_n contains all permutations $\sigma \in S_n$ for which $\sigma(k) < \sigma(\ell)$, we have $|T_n| = |S_n|/2$ and
$$S_n \setminus T_n = \{\sigma \circ \tau \mid \sigma \in T_n\}.$$

Moreover,
$$a_{i,(\sigma \circ \tau)(i)} = \begin{cases} a_{i,\sigma(i)}, & i \neq k, \ell, \\ a_{k,\sigma(\ell)}, & i = k, \\ a_{\ell,\sigma(k)}, & i = \ell. \end{cases}$$

We have $a_{k,\sigma(\ell)} = a_{\ell,\sigma(\ell)}$ and $a_{\ell,\sigma(k)} = a_{k,\sigma(k)}$. Thus, using Theorem 7.7 and Lemma 7.9, we obtain

$$\sum_{\sigma \in S_n \setminus T_n} \operatorname{sgn}(\sigma) \prod_{i=1}^n a_{i,\sigma(i)} = \sum_{\sigma \in T_n} \operatorname{sgn}(\sigma \circ \tau) \prod_{i=1}^n a_{i,(\sigma \circ \tau)(i)}$$
$$= \sum_{\sigma \in T_n} (-\operatorname{sgn}(\sigma)) \prod_{i=1}^n a_{i,(\sigma \circ \tau)(i)}$$
$$= -\sum_{\sigma \in T_n} \operatorname{sgn}(\sigma) \prod_{i=1}^n a_{i,\sigma(i)}.$$

This implies
$$\det(A) = \sum_{\sigma \in S_n} \operatorname{sgn}(\sigma) \prod_{i=1}^n a_{i,\sigma(i)} = \sum_{\sigma \in T_n} \operatorname{sgn}(\sigma) \prod_{i=1}^n a_{i,\sigma(i)}$$
$$+ \sum_{\sigma \in S_n \setminus T_n} \operatorname{sgn}(\sigma) \prod_{i=1}^n a_{i,\sigma(i)} = 0.$$

The proof for the case of two equal columns is analogous.

(5) We observe first that
$$\{(\sigma(i), i) \mid 1 \leq i \leq n\} = \{(i, \sigma^{-1}(i)) \mid 1 \leq i \leq n\}$$

for every $\sigma \in S_n$. To see this, let i with $1 \leq i \leq n$ be fixed. Then $\sigma(i) = j$ if and only if $i = \sigma^{-1}(j)$. Thus, $(\sigma(i), i) = (j, i)$ is an element of the first set if and only if $(j, \sigma^{-1}(j)) = (j, i)$ is an element of the second set. Since σ is bijective, the two sets are equal.

7.2 Properties of the Determinant

Let $A = [a_{ij}]$ and $A^T = [b_{ij}]$ with $b_{ij} = a_{ji}$. Then

$$\det(A^T) = \sum_{\sigma \in S_n} \text{sgn}(\sigma) \prod_{i=1}^n b_{i,\sigma(i)} = \sum_{\sigma \in S_n} \text{sgn}(\sigma) \prod_{i=1}^n a_{\sigma(i),i}$$

$$= \sum_{\sigma \in S_n} \text{sgn}(\sigma^{-1}) \prod_{i=1}^n a_{\sigma(i),i} = \sum_{\sigma \in S_n} \text{sgn}(\sigma^{-1}) \prod_{i=1}^n a_{i,\sigma^{-1}(i)}$$

$$= \sum_{\sigma \in S_n} \text{sgn}(\sigma) \prod_{i=1}^n a_{i,\sigma(i)} = \det(A).$$

Here we have used that $\text{sgn}(\sigma) = \text{sgn}(\sigma^{-1})$ (cp. Theorem 7.7) and the fact that the two products $\prod_{i=1}^n a_{\sigma(i),i}$ and $\prod_{i=1}^n a_{i,\sigma^{-1}(i)}$ have the same factors. □

Example 7.11 For the matrices

$$A = \begin{bmatrix} 1 & 2 & 3 \\ 0 & 4 & 5 \\ 0 & 0 & 6 \end{bmatrix}, \quad B = \begin{bmatrix} 1 & 2 & 0 \\ 1 & 3 & 0 \\ 1 & 4 & 0 \end{bmatrix}, \quad C = \begin{bmatrix} 1 & 1 & 2 \\ 1 & 1 & 3 \\ 1 & 1 & 4 \end{bmatrix}$$

from $\mathbb{Z}^{3,3}$ we obtain $\det(A) = 1 \cdot 4 \cdot 6 = 24$ by (2) in Lemma 7.10, and $\det(B) = \det(C) = 0$ by (3) and (4) in Lemma 7.10. We may also compute these determinants using the Sarrus rule from Example 7.5.

Item (2) in Lemma 7.10 shows in particular that $\det(I_n) = 1$ for the identity matrix $I_n = [e_1, e_2, \ldots, e_n] \in R^{n,n}$. For this reason the determinant map is called *normalized*.

For $\sigma \in S_n$ the matrix

$$P_\sigma := [e_{\sigma(1)}, e_{\sigma(2)}, \ldots, e_{\sigma(n)}]$$

is called the permutation matrix associated with σ. This map from the group S_n to the group of permutation matrices in $R^{n,n}$ is bijective. The inverse of a permutation matrix is its transpose (cp. Theorem 4.16) and we can easily check that

$$P_\sigma^{-1} = P_\sigma^T = P_{\sigma^{-1}}.$$

If $A = [a_1, a_2, \ldots, a_n] \in R^{n,n}$, i.e., $a_j \in R^{n,1}$ is the jth column of A, then

$$AP_\sigma = [a_{\sigma(1)}, a_{\sigma(2)}, \ldots, a_{\sigma(n)}],$$

i.e., the right-multiplication of A with P_σ exchanges the columns of A according to the permutation σ. If, on the other hand, $a_i \in R^{1,n}$ is the ith row of A, then

$$P_\sigma^T A = \begin{bmatrix} a_{\sigma(1)} \\ a_{\sigma(2)} \\ \vdots \\ a_{\sigma(n)} \end{bmatrix},$$

i.e., the left-multiplication of A by P_σ^T exchanges the rows of A according to the permutation σ.

We next study the determinants of the elementary matrices.

Lemma 7.12

(1) For $\sigma \in S_n$ and the associated permutation matrix $P_\sigma \in R^{n,n}$ we have $\text{sgn}(\sigma) = \det(P_\sigma)$. If $n \geq 2$ and P_{ij} is defined as in (5.1), then $\det(P_{ij}) = -1$.
(2) If $M_i(\lambda)$ and $G_{ij}(\lambda)$ are defined as in (5.2) and (5.3), respectively, then $\det(M_i(\lambda)) = \lambda$ and $\det(G_{ij}(\lambda)) = 1$.

Proof

(1) If $\widetilde{\sigma} \in S_n$ and $P_{\widetilde{\sigma}} = [a_{ij}] \in R^{n,n}$, then $a_{\widetilde{\sigma}(j),j} = 1$ for $j = 1, 2, \ldots, n$, and all other entries of $P_{\widetilde{\sigma}}$ are zero. Hence

$$\det(P_{\widetilde{\sigma}}) = \det(P_{\widetilde{\sigma}}^T) = \sum_{\sigma \in S_n} \text{sgn}(\sigma) \underbrace{\prod_{j=1}^n a_{\sigma(j),j}}_{=0 \text{ for } \sigma \neq \widetilde{\sigma}} = \text{sgn}(\widetilde{\sigma}) \underbrace{\prod_{j=1}^n a_{\widetilde{\sigma}(j),j}}_{=1} = \text{sgn}(\widetilde{\sigma}).$$

The permutation matrix P_{ij} is associated with the transposition that exchanges i and j. Hence, $\det(P_{ij}) = -1$ follows from Lemma 7.9.
(2) Since $M_i(\lambda)$ and $G_{ij}(\lambda)$ are lower triangular matrices, the assertion follows from (2) in Lemma 7.10. □

These results lead to some important computational rules for determinants.

Lemma 7.13 For $A \in R^{n,n}$, $n \geq 2$, and $\lambda \in R$ the following assertions hold:

(1) The multiplication of a row of A by λ leads to the multiplication of $\det(A)$ by λ: $\det(M_i(\lambda)A) = \lambda \det(A) = \det(M_i(\lambda))\det(A)$.
(2) The addition of the λ-multiple of a row of A to another row of A does not change $\det(A)$:
$\det(G_{ij}(\lambda)A) = \det(A) = \det(G_{ij}(\lambda))\det(A)$, and
$\det(G_{ij}(\lambda)^T A) = \det(A) = \det(G_{ij}(\lambda)^T)\det(A)$.

7.2 Properties of the Determinant

(3) Exchanging two rows of A changes the sign of $\det(A)$*:*
$\det(P_{ij} A) = -\det(A) = \det(P_{ij})\det(A)$.

The analogous assertions hold for the columns of A.

Proof

(1) If $A = [a_{mk}]$ and $\widetilde{A} = M_i(\lambda)A = [\widetilde{a}_{mk}]$, then

$$\widetilde{a}_{mk} = \begin{cases} a_{mk}, & m \neq i, \\ \lambda a_{mk}, & m = i, \end{cases}$$

and hence

$$\det(\widetilde{A}) = \sum_{\sigma \in S_n} \text{sgn}(\sigma) \prod_{m=1}^{n} \widetilde{a}_{m,\sigma(m)} = \sum_{\sigma \in S_n} \text{sgn}(\sigma) \underbrace{\widetilde{a}_{i,\sigma(i)}}_{= \lambda a_{i,\sigma(i)}} \prod_{\substack{m=1 \\ m \neq i}}^{n} \underbrace{\widetilde{a}_{m,\sigma(m)}}_{= a_{m,\sigma(m)}}$$

$$= \lambda \det(A).$$

(2) If $A = [a_{mk}]$ and $\widetilde{A} = G_{ij}(\lambda)A = [\widetilde{a}_{mk}]$, then

$$\widetilde{a}_{mk} = \begin{cases} a_{mk}, & m \neq j, \\ a_{jk} + \lambda a_{ik}, & m = j, \end{cases}$$

and hence

$$\det(\widetilde{A}) = \sum_{\sigma \in S_n} \text{sgn}(\sigma) (a_{j,\sigma(j)} + \lambda a_{i,\sigma(j)}) \prod_{\substack{m=1 \\ m \neq j}}^{n} a_{m,\sigma(m)}$$

$$= \sum_{\sigma \in S_n} \text{sgn}(\sigma) \prod_{m=1}^{n} a_{m,\sigma(m)} + \lambda \sum_{\sigma \in S_n} \text{sgn}(\sigma) a_{i,\sigma(j)} \prod_{\substack{m=1 \\ m \neq j}}^{n} a_{m,\sigma(m)}.$$

The first term is equal to $\det(A)$, and the second is equal to the determinant of a matrix with two equal columns, and thus equal to zero. The proof for the matrix $G_{ij}(\lambda)^T A$ is analogous.

(3) The permutation matrix P_{ij} exchanges rows i and j of A, where $i < j$. This exchange can be expressed by the following four elementary row operations: Multiply row j by -1; add row i to row j; add the (-1)-multiple of row j to row i; add row i to row j. Therefore,

$$P_{ij} = G_{ij}(1)(G_{ij}(-1))^T G_{ij}(1) M_j(-1).$$

(One may verify this also by carrying out the matrix multiplications.) Using (1) and (2) we obtain

$$\det(P_{ij}A) = \det\left(G_{ij}(1)(G_{ij}(-1))^T G_{ij}(1)M_j(-1)A\right)$$
$$= \det(G_{ij}(1))\det((G_{ij}(-1))^T)\det(G_{ij}(1))\det(M_j(-1))\det(A)$$
$$= (-1)\det(A).$$

The analogous assertions hold for the columns of A, since $\det(A) = \det(A^T)$ (cp. (5) in Lemma 7.10). □

Example 7.14 Consider the matrices

$$A = \begin{bmatrix} 1 & 3 & 0 \\ 1 & 2 & 0 \\ 1 & 2 & 4 \end{bmatrix}, \quad B = \begin{bmatrix} 3 & 1 & 0 \\ 2 & 1 & 0 \\ 2 & 1 & 4 \end{bmatrix} \in \mathbb{Z}^{3,3}.$$

A simple calculation shows that $\det(A) = -4$. Since B is obtained from A by exchanging the first two columns we have $\det(B) = -\det(A) = 4$.

The determinant map can be interpreted as a map of $(R^{n,1})^n$ to R, i.e., as a map of the n columns of the matrix $A \in R^{n,n}$ to the ring R. If $a_i, a_j \in R^{n,1}$ are two columns of A,

$$A = [\ldots a_i \ldots a_j \ldots],$$

then

$$\det(A) = -\det([\ldots a_j \ldots a_i \ldots])$$

by (3) in Lemma 7.13. Due to this property the determinant map is called an *alternating* map of the columns of A. Analogously, the determinant map is an alternating map of the rows of A.

If the kth row of A has the form $\lambda a^{(1)} + \mu a^{(2)}$ for some $\lambda, \mu \in R$ and $a^{(j)} = \left[a_{k1}^{(j)}, \ldots, a_{kn}^{(j)}\right] \in R^{1,n}$, $j = 1, 2$, then

$$\det(A) = \det\left(\begin{bmatrix} \vdots \\ \lambda a^{(1)} + \mu a^{(2)} \\ \vdots \end{bmatrix}\right)$$

$$= \sum_{\sigma \in S_n} \text{sgn}(\sigma)\left(\lambda a_{k,\sigma(k)}^{(1)} + \mu a_{k,\sigma(k)}^{(2)}\right) \prod_{\substack{i=1 \\ i \neq k}}^{n} a_{i,\sigma(i)}$$

7.2 Properties of the Determinant

$$= \lambda \sum_{\sigma \in S_n} \text{sgn}(\sigma)\, a^{(1)}_{k,\sigma(k)} \prod_{\substack{i=1 \\ i \neq k}}^{n} a_{i,\sigma(i)} + \mu \sum_{\sigma \in S_n} \text{sgn}(\sigma)\, a^{(2)}_{k,\sigma(k)} \prod_{\substack{i=1 \\ i \neq k}}^{n} a_{i,\sigma(i)}$$

$$= \lambda \det\left(\begin{bmatrix} \vdots \\ a^{(1)} \\ \vdots \end{bmatrix}\right) + \mu \det\left(\begin{bmatrix} \vdots \\ a^{(2)} \\ \vdots \end{bmatrix}\right).$$

This property is called the *linearity* of the determinant map with respect to the rows of A. Analogously, we have the linearity with respect to the columns of A.

If the kth column of A has the form $\lambda a^{(1)} + \mu a^{(2)}$, then we obtain analogously

$$\det(A) = \det([\ldots, \lambda a^{(1)} + \mu a^{(2)}, \ldots])$$
$$= \lambda \det([\ldots, a^{(1)}, \ldots]) + \mu \det([\ldots, a^{(2)}, \ldots]), \tag{7.3}$$

i.e., the determinant is linear with respect to the columns of A. Linear maps will play a central role beginning from Chap. 9.

The next result is called the *multiplication theorem for determinants*.

Theorem 7.15 *If K is a field and $A, B \in K^{n,n}$, then $\det(AB) = \det(A)\det(B)$.*

Proof By Theorem 5.2 we know that for $A \in K^{n,n}$ there exist invertible elementary matrices $S_1, \ldots, S_t$ such that $\widetilde{A} = S_t \ldots S_1 A$ is in echelon form. By Lemma 7.13 we have

$$\det(A) = \det(S_1^{-1}) \cdots \det(S_t^{-1}) \det(\widetilde{A}),$$

as well as

$$\det(AB) = \det\left(S_1^{-1} \cdots S_t^{-1} \widetilde{A} B\right)$$
$$= \det(S_1^{-1}) \cdots \det(S_t^{-1}) \det(\widetilde{A} B).$$

There are two cases: If A is not invertible, then $\widetilde{A}$ and thus also $\widetilde{A}B$ have a zero row. Then $\det(\widetilde{A}) = \det(\widetilde{A}B) = 0$, which implies that $\det(A) = 0$, and hence $\det(AB) = 0 = \det(A)\det(B)$. On the other hand, if A is invertible, then $\widetilde{A} = I_n$, since $\widetilde{A}$ is in echelon form. Now $\det(I_n) = 1$ again gives $\det(AB) = \det(A)\det(B)$. □

Since our proof of Theorem 7.15 relies on Theorem 5.2, which is valid for matrices over a field K, we have formulated Theorem 7.15 for $A, B \in K^{n,n}$. However, the multiplication theorem for determinants also holds for matrices over a commutative ring R with unit. A direct proof based on the signature formula of

Leibniz can be found, for example, in the book "Advanced Linear Algebra" by Loehr [11, Section 5.13]. That book also contains a proof of the *Cauchy-Binet formula*[3] for $\det(AB)$ with $A \in R^{n,m}$ and $B \in R^{m,n}$ for $n \leq m$. Below we will sometimes use that $\det(AB) = \det(A)\det(B)$ holds for all $A, B \in R^{n,n}$, although we have shown the result in Theorem 7.15 only for $A, B \in K^{n,n}$.

The proof of Theorem 7.15 suggests that $\det(A)$ can be easily computed while transforming $A \in K^{n,n}$ into its echelon form using elementary row operations: If $\widetilde{A} = S_t \ldots S_1 A$ is in echelon form, then $\widetilde{A}$ has a zero row and hence $\det(A) = 0$, or $\widetilde{A} = I_n$ and hence $\det(A) = (\det(S_1))^{-1} \cdots (\det(S_t))^{-1}$.

As shown in Theorem 5.5, every matrix $A \in K^{n,n}$ can be factorized as $A = PLU$, and hence $\det(A) = \det(P)\det(L)\det(U)$. The determinants of the matrices on the right hand side are easily computed, since these are permutation and triangular matrices. An LU-decomposition of a matrix A therefore yields an efficient way to compute $\det(A)$.

MATLAB-Minute 5
Look at the matrices `wilkinson(n)` for n=2,3,...,10 in MATLAB. Can you find a general formula for their entries? For n=2,3,...,10 compute
`A=wilkinson(n)`
`[L,U,P]=lu(A)` (LU-decomposition; cp. MATLAB-Minute 4 above Definition 5.7)
`det(L), det(U), det(P), det(P)*det(L)*det(U), det(A)`
Which permutation is associated with the computed matrix P? Why is `det(A)` an integer for odd n?

7.3 Minors and the Laplace Expansion

We now show that the determinant can be used for deriving formulas for the inverse of an invertible matrix and for the solution of linear systems of equations. These formulas are, however, more of theoretical than practical relevance.

Definition 7.16 Let R be a commutative ring with unit and let $A \in R^{n,n}$, $n \geq 2$. Then the matrix $A(j,i) \in R^{n-1,n-1}$ that is obtained by deleting the jth row and ith column of A is called a *minor*[4] of A. The matrix

$$\mathrm{adj}(A) = [b_{ij}] \in R^{n,n} \quad \text{with} \quad b_{ij} := (-1)^{i+j} \det(A(j,i)),$$

is called the *adjunct* of A.

---

[3] Augustin Louis Cauchy (1789–1857) and Jacques Philippe Marie Binet (1786–1856).
[4] This term was introduced in 1850 by James Joseph Sylvester (1814–1897).

7.3 Minors and the Laplace Expansion

The adjunct is also called *adjungate* or *classical adjoint* of A.

Theorem 7.17 *For $A \in R^{n,n}$, $n \geq 2$, we have*

$$A \operatorname{adj}(A) = \operatorname{adj}(A) A = \det(A) I_n.$$

In particular A is invertible if and only if $\det(A) \in R$ is invertible. In this case $(\det(A))^{-1} = \det(A^{-1})$ and $A^{-1} = (\det(A))^{-1} \operatorname{adj}(A)$.

Proof Let $B = [b_{ij}]$ have the entries $b_{ij} = (-1)^{i+j} \det(A(j, i))$. Then $C = [c_{ij}] = \operatorname{adj}(A) A$ satisfies

$$c_{ij} = \sum_{k=1}^{n} b_{ik} a_{kj} = \sum_{k=1}^{n} (-1)^{i+k} \det(A(k, i)) a_{kj}.$$

Let a_ℓ be the ℓth column of A and let

$$\widetilde{A}(k, i) := [a_1, \ldots, a_{i-1}, e_k, a_{i+1}, \ldots, a_n] \in R^{n,n},$$

where e_k is the kth column of the identity matrix I_n. Then there exist permutation matrices P and Q that perform $k-1$ row and $i-1$ column exchanges, respectively, such that

$$P \widetilde{A}(k, i) Q = \left[\begin{array}{c|c} 1 & \star \\ \hline 0 & A(k, i) \end{array}\right].$$

Using (1) in Lemma 7.10 we obtain

$$\det(A(k, i)) = \det\left(\left[\begin{array}{c|c} 1 & \star \\ \hline 0 & A(k, i) \end{array}\right]\right) = \det(P \widetilde{A}(k, i) Q)$$

$$= \det(P) \det(\widetilde{A}(k, i)) \det(Q)$$

$$= (-1)^{(k-1)+(i-1)} \det(\widetilde{A}(k, i))$$

$$= (-1)^{k+i} \det(\widetilde{A}(k, i)).$$

The linearity of the determinant with respect to the columns now gives

$$c_{ij} = \sum_{k=1}^{n} (-1)^{i+k} (-1)^{k+i} a_{kj} \det(\widetilde{A}(k, i))$$

$$= \det([a_1, \ldots, a_{i-1}, a_j, a_{i+1}, \ldots, a_n])$$

$$= \begin{cases} 0, & i \neq j \\ \det(A), & i = j \end{cases}$$
$$= \delta_{ij} \det(A),$$

and thus $\operatorname{adj}(A)A = \det(A)I_n$. Analogously we can show that $A\operatorname{adj}(A) = \det(A)I_n$.

If $\det(A) \in R$ is invertible, then

$$I_n = (\det(A))^{-1} \operatorname{adj}(A) A = A (\det(A))^{-1} \operatorname{adj}(A),$$

i.e., A is invertible with $A^{-1} = (\det(A))^{-1} \operatorname{adj}(A)$. If, on the other hand, A is invertible, then

$$1 = \det(I_n) = \det(AA^{-1}) = \det(A)\det(A^{-1}) = \det(A^{-1})\det(A),$$

where we have used the multiplication theorem for determinants over R (cp. our comment following the proof of Theorem 7.15). Thus, $\det(A)$ is invertible with $(\det(A))^{-1} = \det(A^{-1})$, and again $A^{-1} = (\det(A))^{-1} \operatorname{adj}(A)$. □

Example 7.18

(1) For

$$A = \begin{bmatrix} 4 & 1 \\ 2 & 1 \end{bmatrix} \in \mathbb{Z}^{2,2}$$

we have $\det(A) = 2$ and thus A is not invertible. But A is invertible when considered as an element of $\mathbb{Q}^{2,2}$, since in this case $\det(A^{-1}) = (\det(A))^{-1} = \frac{1}{2}$.

(2) For

$$A = \begin{bmatrix} t-1 & t-2 \\ t & t-1 \end{bmatrix} \in (\mathbb{Z}[t])^{2,2}$$

we have $\det(A) = 1$. The matrix A is invertible, since $1 \in \mathbb{Z}[t]$ is invertible.

Note that if $A \in R^{n,n}$ is invertible, then Theorem 7.17 shows that A^{-1} can be obtained by inverting only one ring element, $\det(A)$.

We now use Theorem 7.17 and the multiplication theorem for matrices over a commutative ring with unit to prove a result already announced in Sect. 4.2: In order to show that $\widetilde{A} \in R^{n,n}$ is the (unique) inverse of $A \in R^{n,n}$, only one of the two equations $\widetilde{A}A = I_n$ or $A\widetilde{A} = I_n$ needs to be checked.

7.3 Minors and the Laplace Expansion

Corollary 7.19 *Let $A \in R^{n,n}$. If a matrix $\widetilde{A} \in R^{n,n}$ exists with $\widetilde{A}A = I_n$ or $A\widetilde{A} = I_n$, then A is invertible and $\widetilde{A} = A^{-1}$.*

Proof If $\widetilde{A}A = I_n$, then the multiplication theorem for determinants yields

$$1 = \det(I_n) = \det(\widetilde{A}A) = \det(\widetilde{A})\det(A) = \det(A)\det(\widetilde{A}),$$

i.e., $\det(A) \in R$ is invertible with $(\det(A))^{-1} = \det(\widetilde{A})$. Thus also A is invertible and has a unique inverse A^{-1}. For $n = 1$ this is obvious and for $n \geq 2$ it was shown in Theorem 7.17. If we multiply the equation $\widetilde{A}A = I_n$ from the right with A^{-1} we get $\widetilde{A} = A^{-1}$.

The proof starting from $A\widetilde{A} = I_n$ is analogous. □

Let us summarize the invertibility criteria for a square matrix over a field that we have shown so far:

$A \in GL_n(K) \overset{\text{Theorem 5.2}}{\Longleftrightarrow}$ The echelon form of A is the identity matrix I_n

$\overset{\text{Definition 5.10}}{\Longleftrightarrow} \operatorname{rank}(A) = n$

$\overset{\text{clear}}{\Longleftrightarrow} \operatorname{rank}(A) = \operatorname{rank}([A, b]) = n \text{ for all } b \in K^{n,1}$

$\overset{\text{Algorithm 6.6}}{\Longleftrightarrow} |\mathscr{L}(A, b)| = 1 \text{ for all } b \in K^{n,1}$

$\overset{\text{Theorem 7.18}}{\Longleftrightarrow} \det(A) \neq 0. \qquad (7.4)$

Alternatively we obtain:

$A \notin GL_n(K) \overset{\text{Theorem 5.2}}{\Longleftrightarrow}$ The echelon form of A has at least one zero row

$\overset{\text{Definition 5.10}}{\Longleftrightarrow} \operatorname{rank}(A) < n$

$\overset{\text{clear}}{\Longleftrightarrow} \operatorname{rank}([A, 0]) < n$

$\overset{\text{Algorithm 6.6}}{\Longleftrightarrow} \mathscr{L}(A, 0) \neq \{0\}$

$\overset{\text{Theorem 7.18}}{\Longleftrightarrow} \det(A) = 0. \qquad (7.5)$

In the fields $\mathbb{Q}$, $\mathbb{R}$ and $\mathbb{C}$ we have the (usual) absolute value $|\cdot|$ of numbers and can formulate the following useful invertibility criterion for matrices.

Theorem 7.20 *If $A \in K^{n,n}$ with $K \in \{\mathbb{Q}, \mathbb{R}, \mathbb{C}\}$ is diagonally dominant, i.e., if*

$$|a_{ii}| > \sum_{\substack{j=1 \\ j \neq i}}^{n} |a_{ij}| \quad \text{for all } i = 1, \ldots, n,$$

then $\det(A) \neq 0$.

Proof We prove the assertion by contraposition, i.e., by showing that $\det(A) = 0$ implies that A is not diagonally dominant.

If $\det(A) = 0$, then $\mathscr{L}(A, 0) \neq \{0\}$, i.e., the homogeneous linear system of equations $Ax = 0$ has at least one solution $\widehat{x} = [\widehat{x}_1, \ldots, \widehat{x}_n]^T \neq 0$. Let $\widehat{x}_m$ be an entry of $\widehat{x}$ with maximal absolute value, i.e., $|\widehat{x}_m| \geq |\widehat{x}_j|$ for all $j = 1, \ldots, n$. In particular, we then have $|\widehat{x}_m| > 0$. The mth row of $A\widehat{x} = 0$ is given by

$$a_{m1}\widehat{x}_1 + a_{m2}\widehat{x}_2 + \ldots + a_{mn}\widehat{x}_n = 0 \quad \Leftrightarrow \quad a_{mm}\widehat{x}_m = -\sum_{\substack{j=1 \\ j \neq m}}^{n} a_{mj}\widehat{x}_j.$$

We now take absolute values on both sides and use the triangle inequality (cp. Exercise 3.25), which yields

$$|a_{mm}||\widehat{x}_m| \leq \sum_{\substack{j=1 \\ j \neq m}}^{n} |a_{mj}||\widehat{x}_j| \leq \sum_{\substack{j=1 \\ j \neq m}}^{n} |a_{mj}||\widehat{x}_m|, \quad \text{hence} \quad |a_{mm}| \leq \sum_{\substack{j=1 \\ j \neq m}}^{n} |a_{mj}|,$$

so that A not diagonally dominant. $\square$

The converse of this theorem does not hold: For example, the matrix

$$A = \begin{bmatrix} 1 & 2 \\ 1 & 0 \end{bmatrix} \in \mathbb{Q}^{2,2}$$

has $\det(A) = -2 \neq 0$, but A is not diagonally dominant.

From Theorem 7.17 we obtain the *Laplace expansion*[5] of the determinant, which is particularly useful when A contains many zero entries (cp. Example 7.23 below).

---

[5] Pierre-Simon Laplace (1749–1827) published this expansion in 1772.

7.3 Minors and the Laplace Expansion

Corollary 7.21 *For $A \in R^{n,n}$, $n \geq 2$, the following assertions hold:*

(1) For each $i = 1, 2, \ldots, n$ we have

$$\det(A) = \sum_{j=1}^{n} (-1)^{i+j} a_{ij} \det(A(i, j)).$$

(Laplace expansion of $\det(A)$ with respect to the ith row A.)

(2) For each $j = 1, 2, \ldots, n$ we have

$$\det(A) = \sum_{i=1}^{n} (-1)^{i+j} a_{ij} \det(A(i, j)).$$

(Laplace expansion of $\det(A)$ with respect to the jth column of A.)

Proof The two expansions for $\det(A)$ follow immediately by comparison of the diagonal entries in the matrix equations $\det(A) I_n = A \operatorname{adj}(A)$ and $\det(A) I_n = \operatorname{adj}(A) A$. □

The Laplace expansion allows a recursive definition of the determinant: For $A \in R^{n,n}$ with $n \geq 2$, let $\det(A)$ be defined as in (1) or (2) in Corollary 7.21. We can choose an arbitrary row or column of A. The formula for $\det(A)$ then contains only matrices of size $(n-1) \times (n-1)$. For each of these we can use the Laplace expansion again, now expressing each determinant in terms of determinants of $(n-2) \times (n-2)$ matrices. We can do this recursively until only 1×1 matrices remain. For $A = [a_{11}] \in R^{1,1}$ we define $\det(A) := a_{11}$.

Finally we state *Cramer's rule*,[6] which gives an explicit formula for the solution of a linear system in form of determinants. This rule is only of theoretical value, because in order to compute the n components of the solution it requires the evaluation of $n + 1$ determinants of $n \times n$ matrices.

Corollary 7.22 *Let K be a field, $A = [a_1, \ldots, a_n] \in GL_n(K)$ and $b \in K^{n,1}$. Then the unique solution of the linear system of equations $Ax = b$ is given by*

$$\widehat{x} = [\widehat{x}_1, \ldots, \widehat{x}_n]^T = A^{-1}b = (\det(A))^{-1} \operatorname{adj}(A) b,$$

with

$$\widehat{x}_i = \frac{\det[a_1, \ldots, a_{i-1}, b, a_{i+1}, \ldots, a_n]}{\det(A)}, \quad i = 1, \ldots, n.$$

---

[6] Gabriel Cramer (1704–1752).

Proof If $\widehat{x} = A^{-1}b$ is the unique solution of $Ax = b$, then

$$b = A\widehat{x} = \sum_{j=1}^{n} \widehat{x}_j a_j.$$

For every fixed $i = 1, \ldots, n$, with (1) and (2) in Lemma 7.13 (applied to the ith column), we obtain

$$\det([a_1, \ldots, a_{i-1}, b, a_{i+1}, \ldots, a_n]) = \det\left([a_1, \ldots, a_{i-1}, \sum_{j=1}^{n} \widehat{x}_j a_j, a_{i+1}, \ldots, a_n]\right)$$

$$= \det([a_1, \ldots, a_{i-1}, \widehat{x}_i a_i, a_{i+1}, \ldots, a_n])$$

$$= \widehat{x}_i \det([a_1, \ldots, a_{i-1}, a_i, a_{i+1}, \ldots, a_n])$$

$$= \widehat{x}_i \det(A),$$

which yields the assertion. □

Example 7.23 Consider

$$A = \begin{bmatrix} 1 & 3 & 0 & 0 \\ 1 & 2 & 0 & 0 \\ 1 & 2 & 1 & 0 \\ 1 & 2 & 3 & 1 \end{bmatrix} \in \mathbb{Q}^{4,4}, \quad b = \begin{bmatrix} 1 \\ 2 \\ 1 \\ 0 \end{bmatrix} \in \mathbb{Q}^{4,1}.$$

The Laplace expansion with respect to the last column yields

$$\det(A) = 1 \cdot \det\left(\begin{bmatrix} 1 & 3 & 0 \\ 1 & 2 & 0 \\ 1 & 2 & 1 \end{bmatrix}\right) = 1 \cdot 1 \cdot \det\left(\begin{bmatrix} 1 & 3 \\ 1 & 2 \end{bmatrix}\right) = 1 \cdot 1 \cdot (-1) = -1.$$

Thus, A is invertible and $Ax = b$ has the unique solution $\widehat{x} = A^{-1}b \in \mathbb{Q}^{4,1}$, which by Cramer's rule has the following entries:

$$\widehat{x}_1 = \det\left(\begin{bmatrix} 1 & 3 & 0 & 0 \\ 2 & 2 & 0 & 0 \\ 1 & 2 & 1 & 0 \\ 0 & 2 & 3 & 1 \end{bmatrix}\right) / \det(A) = -4/(-1) = 4,$$

$$\widehat{x}_2 = \det\left(\begin{bmatrix} 1 & 1 & 0 & 0 \\ 1 & 2 & 0 & 0 \\ 1 & 1 & 1 & 0 \\ 1 & 0 & 3 & 1 \end{bmatrix}\right) / \det(A) = 1/(-1) = -1,$$

$$\widehat{x}_3 = \det\left(\begin{bmatrix} 1 & 3 & 1 & 0 \\ 1 & 2 & 2 & 0 \\ 1 & 2 & 1 & 0 \\ 1 & 2 & 0 & 1 \end{bmatrix}\right) / \det(A) = 1/(-1) = -1,$$

$$\widehat{x}_4 = \det\left(\begin{bmatrix} 1 & 3 & 0 & 1 \\ 1 & 2 & 0 & 2 \\ 1 & 2 & 1 & 1 \\ 1 & 2 & 3 & 0 \end{bmatrix}\right) / \det(A) = -1/(-1) = 1.$$

Exercises

7.1 A permutation $\sigma \in S_n$ is called an *r-cycle* if there exists a subset $\{i_1, \ldots, i_r\} \subseteq \{1, 2, \ldots, n\}$ with $r \geq 1$ elements and

$$\sigma(i_k) = i_{k+1} \text{ for } k = 1, 2, \ldots, r-1, \quad \sigma(i_r) = i_1,$$
$$\sigma(i) = i \text{ for } i \notin \{i_1, \ldots, i_r\}.$$

We write an r-cycle as $\sigma = (i_1, i_2, \ldots, i_r)$. In particular, a transposition $\tau \in S_n$ is a 2-cycle.

(a) Let $n = 4$ and the 2-cycles $\tau_{1,2} = (1, 2)$, $\tau_{2,3} = (2, 3)$ and $\tau_{3,4} = (3, 4)$ be given. Compute $\tau_{1,2} \circ \tau_{2,3}$, $\tau_{1,2} \circ \tau_{2,3} \circ \tau_{1,2}^{-1}$, and $\tau_{1,2} \circ \tau_{2,3} \circ \tau_{3,4}$.
(b) Let $n \geq 4$ and $\sigma = (1, 2, 3, 4)$. Determine σ^j for $j = 2, 3, 4, 5$.
(c) Show that the inverse of the cycle $(i_1, \ldots, i_r)$ is given by $(i_r, \ldots, i_1)$.
(d) Show that two cycles with disjoint elements, i.e., $(i_1, \ldots, i_r)$ and $(j_1, \ldots, j_s)$ with $\{i_1, \ldots, i_r\} \cap \{j_1, \ldots, j_s\} = \emptyset$, commute.
(e) Show that every permutation $\sigma \in S_n$ can be written as product of disjoint cycles that are, except for the order, uniquely determined by σ.

7.2 Prove Lemma 7.10 (1) using (7.1).

7.3 Show that the group homomorphism sgn : $(S_n, \circ) \to (\{1, -1\}, \cdot)$ satisfies the following assertions:
(a) The set $A_n = \{\sigma \in S_n \mid \text{sgn}(\sigma) = 1\}$ is a subgroup of S_n (cp. Exercise 3.8).
(b) For all $\sigma \in A_n$ and $\pi \in S_n$ we have $\pi \circ \sigma \circ \pi^{-1} \in A_n$.

7.4 Compute the determinants of the following matrices:
(a) $[e_n, e_{n-1}, \ldots, e_1] \in \mathbb{Z}^{n,n}$, where e_i is the ith column of the identity matrix.

(b)
$$A = \begin{bmatrix} 1 & 1 & 1 & 1 & \cdots & 1 \\ 1 & 2 & 2 & 2 & \cdots & 2 \\ 1 & 2 & 3 & 3 & \cdots & 3 \\ 1 & 2 & 3 & 4 & \cdots & 4 \\ \vdots & \vdots & \vdots & \vdots & \ddots & \vdots \\ 1 & 2 & 3 & 4 & \cdots & n \end{bmatrix} \in \mathbb{Z}^{n,n},$$

$$B = [b_{ij}] \in \mathbb{Z}^{n,n} \text{ with } b_{ij} = \begin{cases} 2 & \text{for } |i-j| = 0, \\ -1 & \text{for } |i-j| = 1, \\ 0 & \text{for } |i-j| \geq 2, \end{cases}$$

$$C = \begin{bmatrix} 1 & 0 & 1 & 0 & 0 & 0 & 0 \\ e & 0 & e^\pi & 4 & 5 & 1 & \sqrt{\pi} \\ e^2 & 1 & \frac{17}{31} & \sqrt{6} & \sqrt{7} & \sqrt{8} & \sqrt{10} \\ e^3 & 0 & -e & \pi & e & 0 & \pi^e \\ e^4 & 0 & 10001 & 0 & \pi^{-1} & 0 & e^2\pi \\ e^6 & 0 & \sqrt{2} & 0 & 0 & 0 & -1 \\ 0 & 0 & 1 & 0 & 0 & 0 & 0 \end{bmatrix} \in \mathbb{R}^{7,7}.$$

(c) The 4×4 *Wilkinson matrix*[7] (cp. MATLAB-Minute 5 at the end of Sect. 7.2).

7.5 Construct matrices $A, B \in \mathbb{R}^{n,n}$ for some $n \geq 2$ and with $\det(A + B) \neq \det(A) + \det(B)$.

7.6 Let R be a commutative ring with unit, $n \geq 2$ and $A \in R^{n,n}$. Show that the following assertions hold:
(a) $\text{adj}(I_n) = I_n$.
(b) $\text{adj}(AB) = \text{adj}(B)\text{adj}(A)$, if A and $B \in R^{n,n}$ are invertible.
(c) $\text{adj}(\lambda A) = \lambda^{n-1}\text{adj}(A)$ for all $\lambda \in R$.
(d) $\text{adj}(A^T) = \text{adj}(A)^T$.
(e) $\det(\text{adj}(A)) = (\det(A))^{n-1}$, if A is invertible.
(f) $\text{adj}(\text{adj}(A)) = \det(A)^{n-2}A$.
(g) $\text{adj}(A^{-1}) = \text{adj}(A)^{-1}$, if A is invertible.
Can we drop the requirement of invertibility in (b) or (e)?

7.7 Let $n \geq 2$ and $A = [a_{ij}] \in \mathbb{R}^{n,n}$ with $a_{ij} = \frac{1}{x_i + y_j}$ for some $x_1, \ldots, x_n$, $y_1, \ldots, y_n \in \mathbb{R}$. Hence, in particular, $x_i + y_j \neq 0$ for all i, j. (Such a matrix A is called a *Cauchy matrix*.[8])

---

[7] James Hardy Wilkinson (1919–1986).
[8] Augustin Louis Cauchy (1789–1857).

(a) Show that
$$\det(A) = \frac{\prod_{1 \leq i < j \leq n}(x_j - x_i)(y_j - y_i)}{\prod_{i,j=1}^{n}(x_i + y_j)}.$$

(b) Use (a) to derive a formula for the determinant of the $n \times n$ *Hilbert matrix* (cp. MATLAB-Minute 4 above Definition 5.7).

7.8 Let R be a commutative ring with unit. If $\alpha_1, \ldots, \alpha_n \in R$, $n \geq 2$, then

$$V_n := \left[\alpha_i^{j-1}\right] = \begin{bmatrix} 1 & \alpha_1 & \cdots & \alpha_1^{n-1} \\ 1 & \alpha_2 & \cdots & \alpha_2^{n-1} \\ \vdots & \vdots & & \vdots \\ 1 & \alpha_n & \cdots & \alpha_n^{n-1} \end{bmatrix} \in R^{n,n}$$

is called a *Vandermonde matrix*.[9]

(a) Show that
$$\det(V_n) = \prod_{1 \leq i < j \leq n}(\alpha_j - \alpha_i).$$

(b) Let K be a field and let $K[t]_{\leq n-1}$ be the set of polynomials in the variable t of degree at most $n-1$. Show that two polynomials $p, q \in K[t]_{\leq n-1}$ are equal if there exist pairwise distinct $\beta_1, \ldots, \beta_n \in K$ with $p(\beta_j) = q(\beta_j)$.

7.9 Show the following assertions:
(a) Let K be a field with $1 + 1 \neq 0$ and let $A \in K^{n,n}$ with $A^T = -A$. Then there exists a $\lambda \in K$ with $\det(A) = \lambda^2$, and if n is odd, then $\lambda = 0$.
(*Hint:* Show the assertion separately for even and odd n, and use mathematical induction for $n = 2m$.)

(b) If $A \in GL_n(\mathbb{R})$ with $A^T = A^{-1}$, then $\det(A) \in \{1, -1\}$.

7.10 Let K be a field and
$$A = \begin{bmatrix} A_{11} & A_{12} \\ A_{21} & A_{22} \end{bmatrix}$$

for some $A_{11} \in K^{n_1,n_1}$, $A_{12} \in K^{n_1,n_2}$, $A_{21} \in K^{n_2,n_1}$, $A_{22} \in K^{n_2,n_2}$. Show the following assertions:

(a) If $A_{11} \in GL_{n_1}(K)$, then $\det(A) = \det(A_{11})\det\left(A_{22} - A_{21}A_{11}^{-1}A_{12}\right)$.

(b) If $A_{22} \in GL_{n_2}(K)$, then $\det(A) = \det(A_{22})\det\left(A_{11} - A_{12}A_{22}^{-1}A_{21}\right)$.

(c) If $A_{21} = 0$, then $\det(A) = \det(A_{11})\det(A_{22})$.

Can you show this also when the matrices are defined over a commutative ring with unit?

---

[9] Alexandre-Théophile Vandermonde (1735–1796).

7.11 Construct matrices $A_{11}, A_{12}, A_{21}, A_{22} \in \mathbb{R}^{n,n}$ for $n \geq 2$ with

$$\det\left(\begin{bmatrix} A_{11} & A_{12} \\ A_{21} & A_{22} \end{bmatrix}\right) \neq \det(A_{11})\det(A_{22}) - \det(A_{12})\det(A_{21}).$$

7.12 Let $A = [a_{ij}] \in GL_n(\mathbb{R})$ with $a_{ij} \in \mathbb{Z}$ for $i, j = 1, \ldots, n$. Show that the following assertions hold:
(a) $A^{-1} \in \mathbb{Q}^{n,n}$.
(b) $A^{-1} \in \mathbb{Z}^{n,n}$ if and only if $\det(A) \in \{-1, 1\}$.
(c) The linear system of equations $Ax = b$ has a unique solution $\hat{x} \in \mathbb{Z}^{n,1}$ for every $b \in \mathbb{Z}^{n,1}$ if and only if $\det(A) \in \{-1, 1\}$.

7.13 Show that $G = \{A \in \mathbb{Z}^{n,n} \mid \det(A) \in \{-1, 1\}\}$ is a subgroup of $GL_n(\mathbb{Q})$. (The elements of G are called *unimodular matrices*.)

7.14 Show that $(G, *)$ with $G = \{A \in \mathbb{Z}^{n,n} \mid \det(A) \in \{-1, 1\}\}$ is a subgroup of $GL_n(\mathbb{Q})$.

7.15 Let K be a field and $A \in K^{n,n}$. Show that the following statements are equivalent:
(1) $A \in GL_n(K)$.
(2) There exists $B \in K^{n,n}$ with $BA = I_n$ (i.e., A has a *left inverse*).
(3) There exists $C \in K^{n,n}$ with $AC = I_n$ (i.e., A has a *right inverse*).

The Characteristic Polynomial and Eigenvalues of Matrices

8

We have already characterized matrices using their rank and their determinant. In this chapter we use the determinant map in order to assign to every square matrix a unique polynomial that is called the characteristic polynomial of the matrix. This polynomial contains important information about the matrix. For example, one can read off the determinant and thus see whether the matrix is invertible. Even more important are the roots of the characteristic polynomial, which are called the eigenvalues of the matrix.

8.1 The Characteristic Polynomial and the Cayley-Hamilton Theorem

Let R be a commutative ring with unit and let $R[t]$ be the corresponding ring of polynomials (cp. Example 3.15). For $A = [a_{ij}] \in R^{n,n}$ we set

$$tI_n - A := \begin{bmatrix} t - a_{11} & -a_{12} & \cdots & -a_{1n} \\ -a_{21} & t - a_{22} & \ddots & \vdots \\ \vdots & \ddots & \ddots & -a_{n-1,n} \\ -a_{n1} & \cdots & -a_{n,n-1} & t - a_{nn} \end{bmatrix} \in (R[t])^{n,n}.$$

The entries of the matrix $tI_n - A$ are elements of the commutative ring with unit $R[t]$, where the diagonal entries are polynomials of degree 1, and the other entries are constant polynomials. Using Definition 7.4 we can form the determinant of the matrix $tI_n - A$, which is an element of $R[t]$.

Definition 8.1 Let R be a commutative ring with unit and $A \in R^{n,n}$. Then

$$P_A := \det(tI_n - A) \in R[t]$$

is called the *characteristic polynomial* of A.

Example 8.2 If $n = 1$ and $A = [a_{11}]$, then

$$P_A = \det(tI_1 - A) = \det([t - a_{11}]) = t - a_{11}.$$

For $n = 2$ and

$$A = \begin{bmatrix} a_{11} & a_{12} \\ a_{21} & a_{22} \end{bmatrix}$$

we obtain

$$P_A = \det\left(\begin{bmatrix} t - a_{11} & -a_{12} \\ -a_{21} & t - a_{22} \end{bmatrix}\right) = t^2 - (a_{11} + a_{22})t + (a_{11}a_{22} - a_{12}a_{21}).$$

Using Definition 7.4 we see that the general form of P_A for a matrix $A \in R^{n,n}$ is given by

$$P_A = \sum_{\sigma \in S_n} \text{sgn}(\sigma) \prod_{i=1}^{n} \left(\delta_{i,\sigma(i)}t - a_{i,\sigma(i)}\right). \tag{8.1}$$

The following lemma presents basic properties of the characteristic polynomial.

Lemma 8.3 *For $A \in R^{n,n}$ we have $P_A = P_{A^T}$ and*

$$P_A = t^n - \alpha_{n-1}t^{n-1} + \ldots + (-1)^{n-1}\alpha_1 t + (-1)^n \alpha_0$$

with $\alpha_{n-1} = \sum_{i=1}^{n} a_{ii}$ and $\alpha_0 = \det(A)$.

Proof Using (5) in Lemma 7.10 we obtain

$$P_A = \det(tI_n - A) = \det((tI_n - A)^T) = \det(tI_n - A^T) = P_{A^T}.$$

Using P_A as in (8.1) we see that

$$P_A = \prod_{i=1}^{n}(t - a_{ii}) + \sum_{\substack{\sigma \in S_n \\ \sigma \neq [1,2,\ldots,n]}} \text{sgn}(\sigma) \prod_{i=1}^{n} \left(\delta_{i,\sigma(i)}t - a_{i,\sigma(i)}\right).$$

The first term on the right hand side is of the form

$$t^n - \left(\sum_{i=1}^{n} a_{ii}\right) t^{n-1} + \text{(polynomial of degree } \leq n - 2\text{)},$$

8.1 Characteristic Polynomial and the Cayley-Hamilton Theorem

and the second term is a polynomial of degree $\leq n-2$. Thus, $\alpha_{n-1} = \sum_{i=1}^{n} a_{ii}$ as claimed. Moreover, Definition 8.1 yields

$$P_A(0) = \det(-A) = (-1)^n \det(A),$$

so that $\alpha_0 = \det(A)$. □

This lemma shows that the characteristic polynomial of $A \in R^{n,n}$ always is of degree n. The coefficient of t^n is $1 \in R$. Such a polynomial is called *monic*. The coefficient of t^{n-1} is given by the sum of the diagonal entries of A. This quantity is called the *trace* of A, i.e.,

$$\text{trace}(A) := \sum_{i=1}^{n} a_{ii}.$$

The following lemma shows that for every monic polynomial $p \in R[t]$ of degree $n \geq 1$ there exists a matrix $A \in R^{n,n}$ with $P_A = p$.

Lemma 8.4 *If $n \in \mathbb{N}$ and $p = t^n + \beta_{n-1}t^{n-1} + \ldots + \beta_0 \in R[t]$, then p is the characteristic polynomial of the matrix*

$$A = \begin{bmatrix} 0 & & & -\beta_0 \\ 1 & \ddots & & \vdots \\ & \ddots & 0 & -\beta_{n-2} \\ & & 1 & -\beta_{n-1} \end{bmatrix} \in R^{n,n}.$$

(For $n = 1$ we have $A = [-\beta_0]$.) The matrix A is called the companion matrix of p.

Proof We prove the assertion by induction on n.

For $n = 1$ we have $p = t + \beta_0$, $A = [-\beta_0]$ and $P_A = \det([t + \beta_0]) = p$.

Let the assertion hold for some $n \geq 1$. We consider $p = t^{n+1} + \beta_n t^n + \ldots + \beta_0$ and

$$A = \begin{bmatrix} 0 & & & -\beta_0 \\ 1 & \ddots & & \vdots \\ & \ddots & 0 & -\beta_{n-1} \\ & & 1 & -\beta_n \end{bmatrix} \in R^{n+1,n+1}.$$

Using the Laplace expansion with respect to the first row (cp. Corollary 7.21) and the induction hypothesis we get

$$P_A = \det(t I_{n+1} - A)$$

$$= \det\left(\begin{bmatrix} t & & & \beta_0 \\ -1 & \ddots & & \vdots \\ & \ddots & t & \beta_{n-1} \\ & & -1 & t+\beta_n \end{bmatrix}\right)$$

$$= t \cdot \det\left(\begin{bmatrix} t & & & \beta_1 \\ -1 & \ddots & & \vdots \\ & \ddots & t & \beta_{n-1} \\ & & -1 & t+\beta_n \end{bmatrix}\right) + (-1)^{n+2} \cdot \beta_0 \cdot \det\left(\begin{bmatrix} -1 & t & & \\ & \ddots & \ddots & \\ & & \ddots & t \\ & & & -1 \end{bmatrix}\right)$$

$$= t \cdot (t^n + \beta_n t^{n-1} + \ldots + \beta_1) + (-1)^{2n+2} \beta_0$$

$$= t^{n+1} + \beta_n t^n + \ldots + \beta_1 t + \beta_0$$

$$= p. \qquad \square$$

Example 8.5 The polynomial $p = (t-1)^3 = t^3 - 3t^2 + 3t - 1 \in \mathbb{Z}[t]$ has the companion matrix

$$A = \begin{bmatrix} 0 & 0 & 1 \\ 1 & 0 & -3 \\ 0 & 1 & 3 \end{bmatrix} \in \mathbb{Z}^{3,3}.$$

The identity matrix I_3 has the characteristic polynomial

$$P_{I_3} = \det(t I_3 - I_3) = (t-1)^3 = P_A.$$

Thus, different matrices may have the same characteristic polynomial.

In Example 3.15 we have seen how to evaluate a polynomial $p \in R[t]$ at a scalar $\lambda \in R$. Analogously, we can evaluate p at a matrix $M \in R^{m,m}$ (cp. Exercise 4.9). For

$$p = \beta_n t^n + \beta_{n-1} t^{n-1} + \ldots + \beta_0 \in R[t]$$

8.1 Characteristic Polynomial and the Cayley-Hamilton Theorem

we define

$$p(M) := \beta_n M^n + \beta_{n-1} M^{n-1} + \ldots + \beta_0 I_m \in R^{m,m},$$

where the multiplication on the right hand side is the scalar multiplication of $\beta_j \in R$ and $M^j \in R^{m,m}$, $j = 0, 1, \ldots, n$. (Recall that $M^0 = I_m$.) Evaluating a given polynomial at matrices $M \in R^{m,m}$ therefore defines a map from $R^{m,m}$ to $R^{m,m}$.

In particular, using (8.1), the characteristic polynomial P_A of $A \in R^{n,n}$ satisfies

$$P_A(M) = \sum_{\sigma \in S_n} \mathrm{sgn}(\sigma) \prod_{i=1}^{n} \left(\delta_{i,\sigma(i)} M - a_{i,\sigma(i)} I_m \right) \quad \text{for all } M \in R^{m,m}.$$

Note that for $M \in R^{n,n}$ and $P_A = \det(t I_n - A)$ the "obvious" equation $P_A(M) = \det(M - A)$ is *wrong*. By definition, $P_A(M) \in R^{n,n}$ and $\det(M - A) \in R$, so that the two expressions cannot be the same, even for $n = 1$.

The following result is called the *Cayley-Hamilton theorem*.[1]

Theorem 8.6 *For every matrix $A \in R^{n,n}$ and its characteristic polynomial $P_A \in R[t]$ we have $P_A(A) = 0 \in R^{n,n}$.*

Proof For $n = 1$ we have $A = [a_{11}]$ and $P_A = t - a_{11}$, so that $P_A(A) = [a_{11}] - [a_{11}] = [0]$.

Let now $n \geq 2$ and let e_i be the ith column of the identity matrix $I_n \in R^{n,n}$. Then

$$A e_i = a_{1i} e_1 + a_{2i} e_2 + \ldots + a_{ni} e_n, \quad i = 1, \ldots, n,$$

which is equivalent to

$$(A - a_{ii} I_n) e_i + \sum_{\substack{j=1 \\ j \neq i}}^{n} (-a_{ji} I_n) e_j = 0, \quad i = 1, \ldots, n.$$

---

[1] Arthur Cayley (1821–1895) showed this theorem in 1858 for $n = 2$ and claimed that he had verified it for $n = 3$. He did not feel it necessary to give a proof for general n. Sir William Rowan Hamilton (1805–1865) proved the theorem for the case $n = 4$ in 1853 in the context of his investigations of quaternions. One of the first proofs for general n was given by Ferdinand Georg Frobenius (1849–1917) in 1878. James Joseph Sylvester (1814–1897) coined the name of the theorem in 1884 by calling it the "no-little-marvelous Hamilton-Cayley theorem".

The last n equations can be written as

$$\begin{bmatrix} A - a_{11}I_n & -a_{21}I_n & \cdots & -a_{n1}I_n \\ -a_{12}I_n & A - a_{22}I_n & \cdots & -a_{n2}I_n \\ \vdots & \vdots & & \vdots \\ -a_{1n}I_n & -a_{2n}I_n & \cdots & A - a_{nn}I_n \end{bmatrix} \begin{bmatrix} e_1 \\ e_2 \\ \vdots \\ e_n \end{bmatrix} = \begin{bmatrix} 0 \\ 0 \\ \vdots \\ 0 \end{bmatrix}, \quad \text{or} \quad B\widehat{e} = \widehat{0}.$$

Hence $B \in (R[A])^{n,n}$ with $R[A] := \{p(A) \mid p \in R[t]\} \subset R^{n,n}$. The set $R[A]$ forms a commutative ring with unit given by the identity matrix I_n (cp. Exercise 4.9). Using Theorem 7.17 we obtain

$$\mathrm{adj}(B)B = \det(B)\widehat{I_n},$$

where $\det(B) \in R[A]$ and $\widehat{I_n}$ is the identity matrix in $(R[A])^{n,n}$. The matrix $\widehat{I_n}$ has n times the identity matrix I_n on its diagonal. Multiplying this equation from the right by $\widehat{e}$ yields

$$\mathrm{adj}(B)B\widehat{e} = \det(B)\widehat{I_n}\widehat{e},$$

which implies that $\det(B) = 0 \in R^{n,n}$. Finally, using Lemma 8.3 gives

$$0 = \det(B) = \sum_{\sigma \in S_n} \mathrm{sgn}(\sigma) \prod_{i=1}^{n} (\delta_{i,\sigma(i)}A - a_{\sigma(i),i}I_n)$$

$$= \sum_{\sigma \in S_n} \mathrm{sgn}(\sigma) \prod_{i=1}^{n} (\delta_{\sigma(i),i}A - a_{\sigma(i),i}I_n)$$

$$= P_{A^T}(A)$$

$$= P_A(A),$$

which completes the proof. □

8.2 Eigenvalues and Eigenvectors

In this section we present an introduction to the topic of eigenvalues and eigenvectors of square matrices over a field K. These concepts will be studied in more detail in later chapters.

Definition 8.7 Let $A \in K^{n,n}$. If $\lambda \in K$ and $v \in K^{n,1} \setminus \{0\}$ satisfy $Av = \lambda v$, then λ is called an *eigenvalue* of A and v is called an *eigenvector* of A corresponding to λ.

8.2 Eigenvalues and Eigenvectors

While by definition $v = 0$ can never be an eigenvector of a matrix, $\lambda = 0$ may be an eigenvalue. For example,

$$\begin{bmatrix} 1 & -1 \\ -1 & 1 \end{bmatrix} \begin{bmatrix} 1 \\ 1 \end{bmatrix} = 0 \begin{bmatrix} 1 \\ 1 \end{bmatrix}.$$

If $\lambda \in K$ is an eigenvalue of A, then every associated eigenvector v satisfies the equation $Av = \lambda v$, i.e.,

$$(\lambda I_n - A)v = 0.$$

For an eigenvalue λ of A, therefore, the associated eigenvectors are the nonzero solutions of the homogeneous linear system $(\lambda I_n - A)x = 0$.

If v_1, v_2 are eigenvectors associated with the eigenvalue λ of A, and if $\alpha, \beta \in K$ are arbitrary with $\alpha v_1 + \beta v_2 \neq 0$, then also $\alpha v_1 + \beta v_2$ is an eigenvector of A associated with the eigenvalue λ, since

$$A(\alpha v_1 + \beta v_2) = \alpha A v_1 + \beta A v_2 = \alpha \lambda v_1 + \beta \lambda v_2 = \lambda(\alpha v_1 + \beta v_2).$$

We already have discussed this property of the solution set of homogeneous linear systems after Lemma 6.2.

The following theorem describes the important relationship between the eigenvalues of $A \in K^{n,n}$ and the roots of the characteristic polynomial P_A of A.

Theorem 8.8 *For $A \in K^{n,n}$ the following assertions hold:*

(1) λ is an eigenvalue of A if and only if λ is a root of the characteristic polynomial of A, i.e., $P_A(\lambda) = 0 \in K$.
(2) $\lambda = 0$ is an eigenvalue of A if and only if $\det(A) = 0$.
(3) λ is an eigenvalue of A if and only if λ is an eigenvalue of A^T.

Proof

(1) The equation $P_A(\lambda) = \det(\lambda I_n - A) = 0$ holds if and only if the matrix $\lambda I_n - A$ is not invertible (cp. (7.5)), and this is equivalent to $\mathscr{L}(\lambda I_n - A, 0) \neq \{0\}$. This, however, means that there exists a vector $\widehat{x} \neq 0$ with $(\lambda I_n - A)\widehat{x} = 0$, or $A\widehat{x} = \lambda \widehat{x}$.
(2) By (1), $\lambda = 0$ is an eigenvalue of A if and only if $P_A(0) = 0$. The assertion now follows from $P_A(0) = (-1)^n \det(A)$ (cp. Lemma 8.3).
(3) This follows from (1) and $P_A = P_{A^T}$ (cp. Lemma 8.3). □

The characteristic polynomial $P_A \in K[t]$ of $A \in K^{n,n}$ has degree n. Such a polynomial has at most n roots, as we will show in Corollary 15.5. In $\mathbb{Q}[t]$ and $\mathbb{R}[t]$

there exist non-constant polynomials that do not have roots, and hence there exist matrices over $\mathbb{Q}$ and over $\mathbb{R}$ that do not have eigenvalues. Moreover, item (3) in Theorem 8.8 shows that A and A^T have the same eigenvalues. An eigenvector of A, however, may not be an eigenvector of A^T. These observations are illustrated in the following example.

Example 8.9

(1) The matrix

$$A = \begin{bmatrix} 0 & 1 \\ 2 & 0 \end{bmatrix} \in \mathbb{Q}^{2,2}$$

has the characteristic polynomial $P_A = t^2 - 2 \in \mathbb{Q}[t]$. This polynomial has no roots because there are no rational solutions of the equation $t^2 - 2 = 0$. If we consider A as element of $\mathbb{R}^{2,2}$, then $P_A \in \mathbb{R}[t]$ has the real roots $\sqrt{2}$ and $-\sqrt{2}$, and these two real numbers are the eigenvalues of A.

(2) The matrix

$$A = \begin{bmatrix} 0 & 1 \\ -1 & 0 \end{bmatrix} \in \mathbb{R}^{2,2}$$

has the characteristic polynomial $P_A = t^2 + 1 \in \mathbb{R}[t]$. This polynomial has no roots because there are no real solutions of the equation $t^2 + 1 = 0$. If we consider A as an element of $\mathbb{C}^{2,2}$, then $P_A \in \mathbb{C}[t]$ has the roots $\mathbf{i}$ and $-\mathbf{i}$, and these two complex numbers are the eigenvalues of A.

(3) The matrix

$$A = \begin{bmatrix} 3 & 3 \\ 1 & 1 \end{bmatrix} \in \mathbb{R}^{2,2}$$

has the characteristic polynomial $P_A = t^2 - 4t = t \cdot (t - 4)$, and hence its eigenvalues are 0 and 4. We have

$$A \begin{bmatrix} 1 \\ -1 \end{bmatrix} = 0 \begin{bmatrix} 1 \\ -1 \end{bmatrix} \quad \text{and} \quad A^T \begin{bmatrix} 1 \\ -1 \end{bmatrix} = \begin{bmatrix} 2 \\ 2 \end{bmatrix} \neq \lambda \begin{bmatrix} 1 \\ -1 \end{bmatrix}$$

for all $\lambda \in \mathbb{R}$. Thus, $[1, -1]^T$ is an eigenvector of A corresponding to the eigenvalue 0, but it is not an eigenvector of A^T. On the other hand,

$$A^T \begin{bmatrix} 1 \\ -3 \end{bmatrix} = 0 \begin{bmatrix} 1 \\ -3 \end{bmatrix} \quad \text{and} \quad A \begin{bmatrix} 1 \\ -3 \end{bmatrix} = \begin{bmatrix} -6 \\ -2 \end{bmatrix} \neq \lambda \begin{bmatrix} 1 \\ -3 \end{bmatrix}$$

8.2 Eigenvalues and Eigenvectors

for all $\lambda \in \mathbb{R}$. Thus, $[1, -3]^T$ is an eigenvector of A^T corresponding to the eigenvalue 0, but it is not an eigenvector of A.

This example demonstrates that the existence of eigenvalues depends on the field over which a given matrix is considered. In Chap. 15 we will show that every non-constant polynomial over $\mathbb{C}$ has a least one root. Therefore, matrices $A \in \mathbb{C}^{n,n}$ always have eigenvalues.

Theorem 8.8 implies further criteria for the invertibility of $A \in K^{n,n}$ (cp. (7.4)):

$$A \in GL_n(K) \Leftrightarrow 0 \text{ is not an eigenvalue of } A$$
$$\Leftrightarrow 0 \text{ is not a root of } P_A.$$

Definition 8.10 Two matrices $A, B \in K^{n,n}$ are called *similar*, if there exists a matrix $Z \in GL_n(K)$ with $A = ZBZ^{-1}$.

One can easily show that this defines an equivalence relation on the set $K^{n,n}$ (cp. the proof following Definition 5.14).

Theorem 8.11 *If two matrices $A, B \in K^{n,n}$ are similar, then $P_A = P_B$.*

Proof If $A = ZBZ^{-1}$, then the multiplication theorem for determinants yields

$$P_A = \det(tI_n - A) = \det(tI_n - ZBZ^{-1}) = \det(Z(tI_n - B)Z^{-1})$$
$$= \det(Z)\det(tI_n - B)\det(Z^{-1}) = \det(tI_n - B)\det(ZZ^{-1})$$
$$= P_B$$

(cp. the remarks below Theorem 7.15). □

Theorem 8.11 and (1) in Theorem 8.8 show that two similar matrices have the same eigenvalues. The condition that A and B are similar is sufficient, but not necessary for $P_A = P_B$.

Example 8.12 Let

$$A = \begin{bmatrix} 1 & 1 \\ 0 & 1 \end{bmatrix}, \quad B = \begin{bmatrix} 1 & 0 \\ 0 & 1 \end{bmatrix} = I_2.$$

Then $P_A = (t-1)^2 = P_B$, but for every matrix $Z \in GL_n(K)$ we have $ZBZ^{-1} = I_2 \neq A$. Thus, we have $P_A = P_B$ although A and B are not similar (cp. also Example 8.5).

MATLAB-Minute 6
The roots of a polynomial $p = \alpha_n t^n + \alpha_{n-1} t^{n-1} + \ldots + \alpha_0$ can be computed (or approximated) in MATLAB using the command `roots(p)`, where p is a $1 \times (n+1)$ matrix with the entries $\texttt{p(i)} = \alpha_{n+1-i}$ for $i = 1, \ldots, n+1$. Compute `roots(p)` for the monic polynomial $p = t^3 - 3t^2 + 3t - 1 \in \mathbb{R}[t]$ and display the output using `format long`. What are the exact roots of p and how large is the numerical error in the computation of the roots using `roots(p)`?

Form the matrix $\texttt{A} = \texttt{compan(p)}$ and compare its structure with the one of the companion matrix from Lemma 8.4. Can you transfer the proof of Lemma 8.4 to the structure of the matrix A?

Compute the eigenvalues of A with the command `eig(A)` and compare the output with the one of `roots(p)`. What do you observe?

8.3 Eigenvectors of Stochastic Matrices

We now consider the eigenvalue problem presented in Sect. 1.1 in the context of the PageRank algorithm. The mathematical modeling leads to the Eq. (1.1), which can be written in the form $Ax = x$. Here $A = [a_{ij}] \in \mathbb{R}^{n,n}$ (n is the number of documents) satisfies

$$a_{ij} \geq 0 \quad \text{and} \quad \sum_{i=1}^{n} a_{ij} = 1 \quad \text{for} \quad j = 1, \ldots, n.$$

Such a matrix A is called *column-stochastic*. Note that A is column-stochastic if and only if A^T is row-stochastic. Such matrices also occurred in the car insurance application considered in Sects. 1.2 and Example 4.7. We want to determine $x = [x_1, \ldots, x_n]^T \in \mathbb{R}^{n,1} \setminus \{0\}$ with $Ax = x$, where the entry x_i describes the importance of document i. The importance values should be nonnegative, i.e., $x_i \geq 0$ for $i = 1, \ldots, n$. Thus, we want to determine an entrywise nonnegative eigenvector of A corresponding to the eigenvalue $\lambda = 1$.

We first check whether this problem has a solution, and then study whether the solution is unique. Our presentation is based on the article [1].

Lemma 8.13 *A column-stochastic matrix $A \in \mathbb{R}^{n,n}$ has an eigenvector corresponding to the eigenvalue 1.*

Proof Since A is column-stochastic, we have $A^T[1, \ldots, 1]^T = [1, \ldots, 1]^T$, so that 1 is an eigenvalue of A^T. Now (3) in Theorem 8.8 shows that also A has the eigenvalue 1, and hence there exists a corresponding eigenvector. $\square$

8.3 Eigenvectors of Stochastic Matrices

A matrix with real entries is called *positive*, if all its entries are positive.

Lemma 8.14 *If $A \in \mathbb{R}^{n,n}$ is positive and column-stochastic and if $x \in \mathbb{R}^{n,1}$ is an eigenvector of A corresponding to the eigenvalue 1, then either x or $-x$ is positive.*

Proof If $x = [x_1, \ldots, x_n]^T$ is an eigenvector of $A = [a_{ij}]$ corresponding to the eigenvalue 1, then

$$x_i = \sum_{j=1}^{n} a_{ij} x_j, \quad i = 1, \ldots, n.$$

Suppose that not all entries of x are positive or not all entries of x are negative. Then there exists at least one index k with

$$|x_k| = \left| \sum_{j=1}^{n} a_{kj} x_j \right| < \sum_{j=1}^{n} a_{kj} |x_j|,$$

which implies

$$\sum_{i=1}^{n} |x_i| < \sum_{i=1}^{n} \sum_{j=1}^{n} a_{ij} |x_j| = \sum_{j=1}^{n} \sum_{i=1}^{n} a_{ij} |x_j| = \sum_{j=1}^{n} \left(|x_j| \cdot \underbrace{\sum_{i=1}^{n} a_{ij}}_{=1} \right) = \sum_{j=1}^{n} |x_j|.$$

This is impossible, so that indeed x or $-x$ must be positive. □

We can now prove the following uniqueness result.

Theorem 8.15 *If $A \in \mathbb{R}^{n,n}$ is positive and column-stochastic, then there exists a unique positive $x = [x_1, \ldots, x_n]^T \in \mathbb{R}^{n,1}$ with $\sum_{i=1}^{n} x_i = 1$ and $Ax = x$.*

Proof By Lemma 8.14, A has a least one positive eigenvector corresponding to the eigenvalue 1. Suppose that $x^{(1)} = [x_1^{(1)}, \ldots, x_n^{(1)}]^T$ and $x^{(2)} = [x_1^{(2)}, \ldots, x_n^{(2)}]^T$ are two such eigenvectors. Suppose that these are normalized by $\sum_{i=1}^{n} x_i^{(j)} = 1$, $j = 1, 2$. This assumption can be made without loss of generality, since every nonzero multiple of an eigenvector is still an eigenvector.

We will show that $x^{(1)} = x^{(2)}$. For $\alpha \in \mathbb{R}$ we define $x(\alpha) := x^{(1)} + \alpha x^{(2)} \in \mathbb{R}^{n,1}$, then

$$Ax(\alpha) = Ax^{(1)} + \alpha Ax^{(2)} = x^{(1)} + \alpha x^{(2)} = x(\alpha).$$

If $\widetilde{\alpha} := -x_1^{(1)}/x_1^{(2)}$, then the first entry of $x(\widetilde{\alpha})$ is equal to zero and thus, by Lemma 8.14, $x(\widetilde{\alpha})$ cannot be an eigenvector of A corresponding to the eigenvalue 1.

Now $Ax(\widetilde{\alpha}) = x(\widetilde{\alpha})$ implies that $x(\widetilde{\alpha}) = 0$, and hence

$$x_i^{(1)} + \widetilde{\alpha} x_i^{(2)} = 0, \quad i = 1, \ldots, n. \tag{8.2}$$

Summing up these n equations yields

$$\underbrace{\sum_{i=1}^n x_i^{(1)}}_{=1} + \widetilde{\alpha} \underbrace{\sum_{i=1}^n x_i^{(2)}}_{=1} = 0,$$

so that $\widetilde{\alpha} = -1$. From (8.2) we get $x_i^{(1)} = x_i^{(2)}$ for $i = 1, \ldots, n$, and therefore $x^{(1)} = x^{(2)}$. □

The unique positive eigenvector x in Theorem 8.15 is called the *Perron eigenvector*[2] of the positive matrix A. The theory of eigenvalues and eigenvectors of positive (or more general nonnegative) matrices is an important area of Matrix Theory, since these matrices arise in many applications.

By construction, the matrix $A \in \mathbb{R}^{n,n}$ in the PageRank algorithm is column-stochastic but not positive, since there are (usually many) entries $a_{ij} = 0$. In order to obtain a uniquely solvable problem we can use the following trick:

Let $S = [s_{ij}] \in \mathbb{R}^{n,n}$ with $s_{ij} = 1/n$. Obviously, S is positive and column-stochastic. For a real number $\alpha \in (0, 1]$ we define the matrix

$$\widehat{A}(\alpha) := (1 - \alpha) A + \alpha S.$$

This matrix is positive and column-stochastic, and hence it has a unique positive eigenvector $\widehat{u}$ corresponding to the eigenvalue 1. We thus have

$$\widehat{u} = \widehat{A}(\alpha) \widehat{u} = (1 - \alpha) A \widehat{u} + \alpha S \widehat{u} = (1 - \alpha) A \widehat{u} + \frac{\alpha}{n} [1, \ldots, 1]^T.$$

For a very large number of documents (e.g. the entire internet) the number α/n is very small, so that $(1 - \alpha) A \widehat{u} \approx \widehat{u}$. Therefore a solution of the eigenvalue problem $\widehat{A}(\alpha) \widehat{u} = \widehat{u}$ for small α potentially gives a good approximation of a $u \in \mathbb{R}^{n,1}$ that satisfies $Au = u$. The practical solution of the eigenvalue problem with the matrix $\widehat{A}(\alpha)$ is a topic of the field of Numerical Linear Algebra.

The matrix S represents a link structure where all document are mutually linked and thus all documents are equally important. The matrix $\widehat{A}(\alpha) = (1 - \alpha) A + \alpha S$ therefore models the following internet "surfing behavior": A user follows a proposed link with the probability $1 - \alpha$ and an arbitrary link with the probability α. Originally, Google Inc. used the value $\alpha = 0.15$.

---

[2] Oskar Perron (1880–1975).

Exercises

(In the following exercises K is an arbitrary field.)

8.1 Determine the characteristic polynomials of the following matrices over $\mathbb{Q}$:

$$A = \begin{bmatrix} 2 & 0 \\ 0 & 2 \end{bmatrix}, \quad B = \begin{bmatrix} 4 & 4 \\ -1 & 0 \end{bmatrix}, \quad C = \begin{bmatrix} 2 & 1 \\ 0 & 2 \end{bmatrix}, \quad D = \begin{bmatrix} 2 & 0 & -1 \\ 0 & 2 & 0 \\ -4 & 0 & 2 \end{bmatrix}.$$

Verify the Cayley-Hamilton theorem in each case by direct computation. Are two of the matrices A, B, C similar?

8.2 Let R be a commutative ring with unit and $n \geq 2$.
  (a) Show that for every $A \in GL_n(R)$ there exists a polynomial $p \in R[t]$ of degree at most $n-1$ with $\mathrm{adj}(A) = p(A)$. Conclude that $A^{-1} = q(A)$ holds for a polynomial $q \in R[t]$ of degree at most $n-1$.
  (b) Let $A \in R^{n,n}$. Apply Theorem 7.17 to the matrix $tI_n - A \in (R[t])^{n,n}$ and derive an alternative proof of the Cayley-Hamilton theorem from the formula $\det(tI_n - A)\, I_n = (tI_n - A)\, \mathrm{adj}(tI_n - A)$.

8.3 Let $p \in K[t]$, $A \in K^{n,n}$, and let $v \in K^{n,1} \setminus \{0\}$ be an eigenvector of A corresponding to the eigenvalue $\lambda \in K$. Show that then v is an eigenvector of $p(A)$ corresponding to the eigenvalue $p(\lambda)$.

8.4 Let $A \in K^{n,n}$ be a matrix with $A^k = 0$ for some $k \in \mathbb{N}$. (Such a matrix is called *nilpotent*.)
  (a) Show that $\lambda = 0$ is the only eigenvalue of A.
  (b) Determine P_A and show that $A^n = 0$.
  (*Hint:* You may assume that P_A has the form $\prod_{i=1}^{n}(t - \lambda_i)$ for some $\lambda_1, \ldots, \lambda_n \in K$.)
  (c) Show that $\mu I_n - A$ is invertible if and only if $\mu \in K \setminus \{0\}$.
  (d) Show that $(I_n - A)^{-1} = I_n + A + A^2 + \ldots + A^{n-1}$.

8.5 Determine the eigenvalues and corresponding eigenvectors of the following matrices over $\mathbb{R}$:

$$A = \begin{bmatrix} 1 & 1 & 1 \\ 0 & 1 & 1 \\ 0 & 0 & 1 \end{bmatrix}, \quad B = \begin{bmatrix} 3 & 8 & 16 \\ 0 & 7 & 8 \\ 0 & -4 & -5 \end{bmatrix}, \quad C = \begin{bmatrix} 0 & -1 & 0 & 0 \\ 1 & 0 & 0 & 0 \\ 0 & 0 & -2 & 1 \\ 0 & 0 & 0 & -2 \end{bmatrix}.$$

Is there any difference when you consider A, B, C as matrices over $\mathbb{C}$?

8.6 Let $n \geq 3$ and $\varepsilon \in \mathbb{R}$. Consider the matrix

$$A(\varepsilon) = \begin{bmatrix} 1 & 1 & & \\ & \ddots & \ddots & \\ & & \ddots & 1 \\ \varepsilon & & & 1 \end{bmatrix}$$

as an element of $\mathbb{C}^{n,n}$ and determine all eigenvalues in dependence of ε. How many pairwise distinct eigenvalues does $A(\varepsilon)$ have?

8.7 Determine the eigenvalues and corresponding eigenvectors of

$$A = \begin{bmatrix} 2 & 2-a & 2-a \\ 0 & 4-a & 2-a \\ 0 & -4+2a & -2+2a \end{bmatrix} \in \mathbb{R}^{3,3}, \quad B = \begin{bmatrix} 1 & 1 & 0 \\ 1 & 0 & 1 \\ 0 & 1 & 1 \end{bmatrix} \in (\mathbb{Z}/2\mathbb{Z})^{3,3}.$$

(For simplicity, the elements of $\mathbb{Z}/2\mathbb{Z}$ are here denoted by k instead of $[k]$.)

8.8 Let $A \in K^{n,n}$, $B \in K^{m,m}$, $n \geq m$, and $C \in K^{n,m}$ with $\text{rank}(C) = m$ and $AC = CB$. Show that then every eigenvalue of B is an eigenvalue of A.

8.9 Show the following assertions:
(a) $\text{trace}(\lambda A + \mu B) = \lambda \, \text{trace}(A) + \mu \, \text{trace}(B)$ holds for all $\lambda, \mu \in K$ and $A, B \in K^{n,n}$.
(b) $\text{trace}(AB) = \text{trace}(BA)$ holds for all $A, B \in K^{n,n}$.
(c) If $A, B \in K^{n,n}$ are similar, then $\text{trace}(A) = \text{trace}(B)$.

8.10 Prove or disprove the following statements:
(a) There exist matrices $A, B \in K^{n,n}$ with $\text{trace}(AB) \neq \text{trace}(A) \, \text{trace}(B)$.
(b) There exist matrices $A, B \in K^{n,n}$ with $AB - BA = I_n$.

8.11 Suppose that the matrix $A = [a_{ij}] \in \mathbb{C}^{n,n}$ has only real entries a_{ij}. Show that if $\lambda \in \mathbb{C} \setminus \mathbb{R}$ is an eigenvalue of A with corresponding eigenvector $v = [v_1, \ldots, v_n]^T \in \mathbb{C}^{n,1}$, then also $\overline{\lambda}$ is an eigenvalue of A with corresponding eigenvector $\overline{v} := [\overline{v}_1, \ldots, \overline{v}_n]^T$.

Vector Spaces

9

In the previous chapters we have focussed on matrices and their properties. We have defined algebraic operations with matrices and derived important concepts associated with them, including their rank, determinant, characteristic polynomial, and eigenvalues. In this chapter we place these concepts in a more abstract framework by introducing the idea of a vector space. Matrices form one of the most important examples of vector spaces, and properties of certain (namely, finite dimensional) vector spaces can be studied in a transparent way using matrices. In the next chapter we will study (linear) maps between vector spaces, and there the connection with matrices will play a central role as well.

9.1 Basic Definitions and Properties

We begin with the definition of a vector space over a field K.

Definition 9.1 Let K be a field. A *vector space over K*, or shortly *K-vector space*, is a set $\mathcal{V}$ with two operations,

$$+ : \mathcal{V} \times \mathcal{V} \to \mathcal{V}, \quad (v, w) \mapsto v + w, \quad \text{(addition)}$$
$$\cdot : K \times \mathcal{V} \to \mathcal{V}, \quad (\lambda, v) \mapsto \lambda \cdot v, \quad \text{(scalar multiplication)}$$

that satisfy the following:

(1) $(\mathcal{V}, +)$ is a commutative group.
(2) For all $v, w \in \mathcal{V}$ and $\lambda, \mu \in K$ the following assertions hold:
   (a) $\lambda \cdot (\mu \cdot v) = (\lambda \mu) \cdot v$.
   (b) $1 \cdot v = v$.
   (c) $\lambda \cdot (v + w) = \lambda \cdot v + \lambda \cdot w$.
   (d) $(\lambda + \mu) \cdot v = \lambda \cdot v + \mu \cdot v$.

An element $v \in \mathcal{V}$ is called a *vector*,[1] an element $\lambda \in K$ is called a *scalar*.

Again, we usually omit the sign of the scalar multiplication, i.e., we usually write λv instead of $\lambda \cdot v$. If it is clear from the context (or not important) which field we are using, we often omit the explicit reference to K and simply write vector space instead of K-vector space.

Example 9.2

(1) The set $K^{n,m}$ with the matrix addition and the scalar multiplication forms a K-vector space. For obvious reasons, the elements of $K^{n,1}$ and $K^{1,m}$ are sometimes called *column* and *row* vectors, respectively.
(2) The set $K[t]$ forms a K-vector space, if the addition is defined as in Example 3.15 (usual addition of polynomials) and the scalar multiplication for $p = \alpha_0 + \alpha_1 t + \ldots + \alpha_n t^n \in K[t]$ is defined by

$$\lambda \cdot p := (\lambda \alpha_0) + (\lambda \alpha_1)t + \ldots + (\lambda \alpha_n)t^n.$$

(3) The continuous and real valued functions defined on the real interval $[0, 1]$ with the pointwise addition and scalar multiplication, i.e.,

$$(f + g)(x) := f(x) + g(x) \quad \text{and} \quad (\lambda \cdot f)(x) := \lambda f(x),$$

form an $\mathbb{R}$-vector space. This can be shown by using that the addition of two continuous functions as well as the multiplication of a continuous function by a real number yield again a continuous function.

Since, by definition, $(\mathcal{V}, +)$ is a commutative group, we already know some vector space properties from the theory of groups (cp. Chap. 3). In particular, every vector space contains a unique neutral element (with respect to addition) $0_\mathcal{V}$, which is called the *null vector*. Every vector $v \in \mathcal{V}$ has a unique (additive) inverse $-v \in \mathcal{V}$ with $v + (-v) = v - v = 0_\mathcal{V}$. As usual, we will write $v - w$ instead of $v + (-w)$.

Lemma 9.3 *Let $\mathcal{V}$ be a K-vector space. If 0_K and $0_\mathcal{V}$ are the neutral (null) elements of K and $\mathcal{V}$, respectively, then the following assertions hold:*

(1) $0_K \cdot v = 0_\mathcal{V}$ for all $v \in \mathcal{V}$.
(2) $\lambda \cdot 0_\mathcal{V} = 0_\mathcal{V}$ for all $\lambda \in K$.
(3) $-(\lambda \cdot v) = (-\lambda) \cdot v = \lambda \cdot (-v)$ for all $v \in \mathcal{V}$ and $\lambda \in K$.

---

[1] This term was introduced in 1845 by Sir William Rowan Hamilton (1805–1865) in the context of his *quaternions*. It is motivated by the Latin verb "vehi" ("vehor", "vectus sum") which means to ride or drive. Also the term "scalar" was introduced by Hamilton; see the footnote on the scalar multiplication (4.2).

9.1 Basic Definitions and Properties

Proof

(1) For all $v \in \mathcal{V}$ we have $0_K \cdot v = (0_K + 0_K) \cdot v = 0_K \cdot v + 0_K \cdot v$. Adding $-(0_K \cdot v)$ on both sides of this identity gives $0_\mathcal{V} = 0_K \cdot v$.
(2) For all $\lambda \in K$ we have $\lambda \cdot 0_\mathcal{V} = \lambda \cdot (0_\mathcal{V} + 0_\mathcal{V}) = \lambda \cdot 0_\mathcal{V} + \lambda \cdot 0_\mathcal{V}$. Adding $-(\lambda \cdot 0_\mathcal{V})$ on both sides of this identity gives $0_\mathcal{V} = \lambda \cdot 0_\mathcal{V}$.
(3) For all $\lambda \in K$ and $v \in \mathcal{V}$ we have $\lambda \cdot v + (-\lambda) \cdot v = (\lambda - \lambda) \cdot v = 0_K \cdot v = 0_\mathcal{V}$, as well as $\lambda \cdot v + \lambda \cdot (-v) = \lambda \cdot (v - v) = \lambda \cdot 0_\mathcal{V} = 0_\mathcal{V}$. $\square$

In the following we will write 0 instead of 0_K and $0_\mathcal{V}$ when it is clear which null element is meant.

As in groups, rings and fields we can identify substructures in vector spaces that are again vector spaces.

Definition 9.4 Let $(\mathcal{V}, +, \cdot)$ be a K-vector space and let $\mathcal{U} \subseteq \mathcal{V}$. If $(\mathcal{U}, +, \cdot)$ is a K-vector space, then it is called a *subspace* of $(\mathcal{V}, +, \cdot)$.

A substructure must be closed with respect to the given operations, which here are addition and scalar multiplication.

Lemma 9.5 $(\mathcal{U}, +, \cdot)$ *is a subspace of the K-vector space* $(\mathcal{V}, +, \cdot)$ *if and only if* $\emptyset \neq \mathcal{U} \subseteq \mathcal{V}$ *and the following assertions hold:*

(1) $v + w \in \mathcal{U}$ *for all* $v, w \in \mathcal{U}$,
(2) $\lambda v \in \mathcal{U}$ *for all* $\lambda \in K$ *and* $v \in \mathcal{U}$.

Proof Exercise. $\square$

Example 9.6

(1) Every vector space $\mathcal{V}$ has the trivial subspaces $\mathcal{U} = \mathcal{V}$ and $\mathcal{U} = \{0\}$.
(2) Let $A \in K^{n,m}$ and $\mathcal{U} = \mathscr{L}(A, 0) \subseteq K^{m,1}$, i.e., $\mathcal{U}$ is the solution set of the homogeneous linear system $Ax = 0$. We have $0 \in \mathcal{U}$, so $\mathcal{U}$ is not empty. If $v, w \in \mathcal{U}$, then

$$A(v + w) = Av + Aw = 0 + 0 = 0,$$

i.e., $v + w \in \mathcal{U}$. Furthermore, for all $\lambda \in K$,

$$A(\lambda v) = \lambda (Av) = \lambda 0 = 0,$$

i.e., $\lambda v \in \mathcal{U}$. Hence, $\mathcal{U}$ is a subspace of $K^{m,1}$ (cp. Eq. (6.3)).
(3) For every $n \in \mathbb{N}_0$ the set $K[t]_{\leq n} := \{p \in K[t] \mid \deg(p) \leq n\}$ is a subspace of $K[t]$.

Definition 9.7 Let $\mathcal{V}$ be a K-vector space, $n \in \mathbb{N}$, and $v_1, \ldots, v_n \in \mathcal{V}$. A vector of the form

$$\lambda_1 v_1 + \ldots + \lambda_n v_n = \sum_{i=1}^{n} \lambda_i v_i \in \mathcal{V}$$

is called a *linear combination* of $v_1, \ldots, v_n$ with the *coefficients* $\lambda_1, \ldots, \lambda_n \in K$. The *(linear) span* of $v_1, \ldots, v_n$ is the set

$$\operatorname{span}\{v_1, \ldots, v_n\} := \Big\{ \sum_{i=1}^{n} \lambda_i v_i \mid \lambda_1, \ldots, \lambda_n \in K \Big\}.$$

Let M be a nonempty set and suppose that for every $m \in M$ we have a vector $v_m \in \mathcal{V}$. Let the set of all these vectors, called the *system* of these vectors, be denoted by $\{v_m\}_{m \in M}$. Then the *(linear) span* of the system $\{v_m\}_{m \in M}$, denoted by $\operatorname{span}\{v_m\}_{m \in M}$, is defined as the set of all vectors $v \in \mathcal{V}$ that are linear combinations of finitely many vectors of the system.

This definition can be consistently extended to the case $n = 0$. In this case $v_1, \ldots, v_n$ is a list of length zero, or an *empty list*. If we define the empty sum of vectors as $0 \in \mathcal{V}$, then we obtain $\operatorname{span}\{v_1, \ldots, v_n\} = \operatorname{span} \emptyset = \{0\}$.

If in the following we consider a list of vectors $v_1, \ldots, v_n$ or a set of vectors $\{v_1, \ldots, v_n\}$, we usually mean that $n \geq 1$. The case of empty list and the associated *zero vector space* $\mathcal{V} = \{0\}$ will sometimes be discussed separately.

Example 9.8 The vector space $K^{1,3} = \{[\alpha_1, \alpha_2, \alpha_3] \mid \alpha_1, \alpha_2, \alpha_3 \in K\}$ is spanned by the vectors $[1, 0, 0], [0, 1, 0], [0, 0, 1]$. The set $\{[\alpha_1, \alpha_2, 0] \mid \alpha_1, \alpha_2 \in K\}$ forms a subspace of $K^{1,3}$ that is spanned by the vectors $[1, 0, 0], [0, 1, 0]$.

Lemma 9.9 *If $\mathcal{V}$ is a vector space and $v_1, \ldots, v_n \in \mathcal{V}$, then $\operatorname{span}\{v_1, \ldots, v_n\}$ is a subspace of $\mathcal{V}$.*

Proof It is clear that $\emptyset \neq \operatorname{span}\{v_1, \ldots, v_n\} \subseteq \mathcal{V}$. Furthermore, $\operatorname{span}\{v_1, \ldots, v_n\}$ is by definition closed with respect to addition and scalar multiplication, so that (1) and (2) in Lemma 9.5 are satisfied. □

9.2 Bases and Dimension of Vector Spaces

We will now discuss the central theory of bases and dimension of vector spaces, and start with the concept of linear independence.

9.2 Bases and Dimension of Vector Spaces

Definition 9.10 Let $\mathcal{V}$ be a K-vector space.

(1) The vectors $v_1, \ldots, v_n \in \mathcal{V}$ are called *linearly independent* if the equation

$$\sum_{i=1}^{n} \lambda_i v_i = 0 \quad \text{with} \quad \lambda_1, \ldots, \lambda_n \in K$$

always implies that $\lambda_1 = \cdots = \lambda_n = 0$. Otherwise, i.e., when $\sum_{i=1}^{n} \lambda_i v_i = 0$ holds for some scalars $\lambda_1, \ldots, \lambda_n \in K$ that are not all equal to zero, then the vectors $v_1, \ldots, v_n$ are called *linearly dependent*.
(2) The empty list is linearly independent.
(3) If M is a set and for every $m \in M$ we have a vector $v_m \in \mathcal{V}$, the corresponding system $\{v_m\}_{m \in M}$ is called *linearly independent* when finitely many vectors of the system are always linearly independent in the sense of (1). Otherwise the system is called *linearly dependent*.

The vectors $v_1, \ldots, v_n$ are linearly independent if and only if the zero vector can be linearly combined only in the trivial way $0 = 0 \cdot v_1 + \ldots + 0 \cdot v_n$.

Lemma 9.11 *If $\mathcal{V}$ is a K-vector space, then the following assertions hold:*

(1) A single vector $v \in \mathcal{V}$ is linearly independent if and only if $v \neq 0$.
(2) If $v_1, \ldots, v_n \in \mathcal{V}$ are linearly independent and $\{u_1, \ldots, u_m\} \subseteq \{v_1, \ldots, v_n\}$, then $u_1, \ldots, u_m$ are linearly independent.
(3) If $v_1, \ldots, v_n \in \mathcal{V}$ are linearly dependent and $u_1, \ldots, u_m \in V$, then $v_1, \ldots, v_n, u_1, \ldots, u_m$ are also linearly dependent.

Proof Exercise. □

The following result gives a useful characterization of the linear independence of finitely many (but at least two) given vectors.

Lemma 9.12 *The vectors $v_1, \ldots, v_n$, $n \geq 2$, are linearly independent if and only if no vector v_i, $i = 1, \ldots, n$, can be written as a linear combination of the others.*

Proof We prove the assertion by contraposition. The vectors $v_1, \ldots, v_n$ are linearly dependent if and only if

$$\sum_{i=1}^{n} \lambda_i v_i = 0$$

with at least one scalar $\lambda_j \neq 0$. Equivalently,

$$v_j = -\sum_{\substack{i=1 \\ i \neq j}}^{n} (\lambda_j^{-1} \lambda_i) v_i,$$

so that v_j is a linear combination of the other vectors. □

Using the concept of linear independence we can now define the concept of the basis of a vector space.

Definition 9.13 Let $\mathcal{V}$ be a vector space.

(1) A set $\{v_1, \ldots, v_n\} \subseteq \mathcal{V}$ is called a *basis* of $\mathcal{V}$, when $v_1, \ldots, v_n$ are linearly independent and $\mathrm{span}\{v_1, \ldots, v_n\} = \mathcal{V}$.
(2) The set $\emptyset$ is the basis of the zero vector space $\mathcal{V} = \{0\}$.
(3) Let M be a set and suppose that for every $m \in M$ we have a vector $v_m \in \mathcal{V}$. The set $\{v_m \mid m \in M\}$ is called a *basis* of $\mathcal{V}$ if the corresponding system $\{v_m\}_{m \in M}$ is linearly independent and $\mathrm{span}\{v_m\}_{m \in M} = \mathcal{V}$.

In short, a basis is a *linearly independent spanning set* of a vector space.

Example 9.14

(1) Let $E_{ij} \in K^{n,m}$ be the matrix with entry 1 in position (i, j) and all other entries 0 (cp. Sect. 5.1). Then the set

$$\{E_{ij} \mid 1 \leq i \leq n \text{ and } 1 \leq j \leq m\} \tag{9.1}$$

is a basis of the vector space $K^{n,m}$ (cp. (1) in Example 9.2): The matrices $E_{ij} \in K^{n,m}$, $1 \leq i \leq n$ and $1 \leq j \leq m$, are linearly independent, since

$$0 = \sum_{i=1}^{n} \sum_{j=1}^{m} \lambda_{ij} E_{ij} = [\lambda_{ij}]$$

implies that $\lambda_{ij} = 0$ for $i = 1, \ldots, n$ and $j = 1, \ldots, m$. For any $A = [a_{ij}] \in K^{n,m}$ we have

$$A = \sum_{i=1}^{n} \sum_{j=1}^{m} a_{ij} E_{ij},$$

and hence

$$\mathrm{span}\{E_{ij} \mid 1 \leq i \leq n \text{ and } 1 \leq j \leq m\} = K^{n,m}.$$

9.2 Bases and Dimension of Vector Spaces

The basis (9.1) is called the *canonical* or *standard* basis of the vector space $K^{n,m}$. For $m = 1$ we denote the canonical basis vectors of $K^{n,1}$ by

$$e_1 := \begin{bmatrix} 1 \\ 0 \\ 0 \\ \vdots \\ 0 \end{bmatrix}, \quad e_2 := \begin{bmatrix} 0 \\ 1 \\ 0 \\ \vdots \\ 0 \end{bmatrix}, \quad \ldots, \quad e_n := \begin{bmatrix} 0 \\ \vdots \\ 0 \\ 0 \\ 1 \end{bmatrix}.$$

These vectors are also called *unit vectors*; they are the n columns of the identity matrix I_n.

(2) A basis of the vector space $K[t]$ (cp. (2) in Example 9.2) is given by the set $\{t^m \mid m \in \mathbb{N}_0\}$, since the corresponding system $\{t^m\}_{m \in \mathbb{N}_0}$ is linearly independent, and every polynomial $p \in K[t]$ is a linear combination of finitely many vectors of the system.

The next result is called the *basis extension theorem*.

Theorem 9.15 *Let V be a vector space and let $v_1, \ldots, v_r, w_1, \ldots, w_\ell \in V$, where $r, \ell \in \mathbb{N}_0$. If $v_1, \ldots, v_r$ are linearly independent and $\text{span}\{v_1, \ldots, v_r, w_1, \ldots, w_\ell\} = V$, then the set $\{v_1, \ldots, v_r\}$ can be extended to a basis of V using vectors from the set $\{w_1, \ldots, w_\ell\}$.*

Proof Note that for $r = 0$ the list $v_1, \ldots, v_r$ is empty and hence linearly independent due to (2) in Definition 9.10.

We prove the assertion by induction on ℓ. If $\ell = 0$, then $\text{span}\{v_1, \ldots, v_r\} = V$, and the linear independence of $\{v_1, \ldots, v_r\}$ shows that this set is a basis of V.

Let the assertion hold for some $\ell \geq 0$. Suppose that $v_1, \ldots, v_r, w_1, \ldots, w_{\ell+1} \in V$ are given, where $v_1, \ldots, v_r$ are linearly independent and $\text{span}\{v_1, \ldots, v_r, w_1, \ldots, w_{\ell+1}\} = V$. If $\{v_1, \ldots, v_r\}$ already is a basis of V, then we are done. Suppose, therefore, that $\text{span}\{v_1, \ldots, v_r\} \subset V$. Then there exists at least one j, $1 \leq j \leq \ell+1$, such that $w_j \notin \text{span}\{v_1, \ldots, v_r\}$. In particular, we have $w_j \neq 0$. Then

$$\lambda w_j + \sum_{i=1}^{r} \lambda_i v_i = 0$$

implies that $\lambda = 0$ (otherwise we would have $w_j \in \text{span}\{v_1, \ldots, v_r\}$) and, therefore, $\lambda_1 = \cdots = \lambda_r = 0$ due to the linear independence of $v_1, \ldots, v_r$. Thus, $v_1, \ldots, v_r, w_j$ are linearly independent. By the induction hypothesis we can extend the set $\{v_1, \ldots, v_r, w_j\}$ to a basis of V using vectors from the set $\{w_1, \ldots, w_{\ell+1}\} \setminus \{w_j\}$, which contains ℓ elements. □

Example 9.16 Let $\mathcal{V} = K[t]_{\leq 5}$ (cp. (3) in Example 9.6) and consider the sets

$$V = \{v_m = t^m \mid m = 1, 2, 3, 4, 5\} \quad \text{and} \quad W = \{w_1 = t^2 + 1, w_2 = t^5 - t^3\}.$$

Then $\mathcal{V} = \text{span}\{v_1, \ldots, v_5, w_1, w_2\}$, i.e., the vector space $\mathcal{V}$ is spanned by the elements of V and W. The elements of V are linearly independent, but this set does not form a basis of $\mathcal{V}$, since in particular the polynomial $t^0 = 1 \in K[t]_{\leq 5}$ is not a linear combination of the elements of V. Extending V by the vector w_1 yields the set $\{v_1, v_2, v_3, v_4, v_5, w_1\}$, which forms a basis of $K[t]_{\leq 5}$.

As an immediate consequence of the basis extension theorem we have the following result.

Corollary 9.17 *If* $\mathcal{V} = \text{span}\{v_1, \ldots, v_n\}$ *is a K-vector space, then there exists a subset of* $\{v_1, \ldots, v_n\}$ *that forms a basis of* $\mathcal{V}$.

Proof The assertion is clear for $\mathcal{V} = \{0\}$. If $\mathcal{V} = \text{span}\{v_1, \ldots, v_n\} \neq \{0\}$, then at least one of the vectors $v_1, \ldots, v_n$ is nonzero, and this vector is therefore linearly independent. By Theorem 9.15 we can extend this vector with appropriate elements from the set $\{v_1, \ldots, v_n\}$ to construct a basis of $\mathcal{V}$. $\square$

Corollary 9.17 shows that every vector space that is spanned by finitely many vectors has a basis with finitely many elements.

A central result of the theory of vector spaces is that every such basis has the same number of elements. In order to show this result we first prove the following *exchange lemma*.

Lemma 9.18 *Let $\mathcal{V}$ be a vector space, let* $v_1, \ldots, v_m \in \mathcal{V}$ *and let* $w = \sum_{i=1}^m \lambda_i v_i \in \mathcal{V}$ *with* $\lambda_1 \neq 0$. *Then* $\text{span}\{w, v_2, \ldots, v_m\} = \text{span}\{v_1, v_2, \ldots, v_m\}$.

Proof By assumption we have

$$v_1 = \lambda_1^{-1} w - \sum_{i=2}^m \left(\lambda_1^{-1} \lambda_i\right) v_i.$$

If $y \in \text{span}\{v_1, \ldots, v_m\}$, say $y = \sum_{i=1}^m \gamma_i v_i$, then

$$y = \gamma_1 \left(\lambda_1^{-1} w - \sum_{i=2}^m \left(\lambda_1^{-1} \lambda_i\right) v_i\right) + \sum_{i=2}^m \gamma_i v_i$$

$$= \left(\gamma_1 \lambda_1^{-1}\right) w + \sum_{i=2}^m \left(\gamma_i - \gamma_1 \lambda_1^{-1} \lambda_i\right) v_i \in \text{span}\{w, v_2, \ldots, v_m\}.$$

9.2 Bases and Dimension of Vector Spaces

If, on the other hand, $y = \alpha_1 w + \sum_{i=2}^{m} \alpha_i v_i \in \mathrm{span}\{w, v_2, \ldots, v_m\}$, then

$$y = \alpha_1 \left(\sum_{i=1}^{m} \lambda_i v_i \right) + \sum_{i=2}^{m} \alpha_i v_i$$

$$= \alpha_1 \lambda_1 v_1 + \sum_{i=2}^{m} (\alpha_1 \lambda_i + \alpha_i) v_i \in \mathrm{span}\{v_1, \ldots, v_m\},$$

and thus $\mathrm{span}\{w, v_2, \ldots, v_m\} = \mathrm{span}\{v_1, v_2, \ldots, v_m\}$. $\square$

Using this lemma we now prove the *exchange theorem*.[2]

Theorem 9.19 *Let $W = \{w_1, \ldots, w_n\}$ and $U = \{u_1, \ldots, u_m\}$ be finite subsets of a vector space, and let $w_1, \ldots, w_n$ be linearly independent. If $W \subseteq \mathrm{span}\{u_1, \ldots, u_m\}$, then $n \leq m$ and furthermore, n elements of U, if numbered appropriately the elements $u_1, \ldots, u_n$, can be exchanged against n elements of W in such a way that*

$$\mathrm{span}\{w_1, \ldots, w_n, u_{n+1}, \ldots, u_m\} = \mathrm{span}\{u_1, \ldots, u_n, u_{n+1}, \ldots, u_m\}.$$

Proof By assumption we have $w_1 = \sum_{i=1}^{m} \lambda_i u_i$ for some scalars $\lambda_1, \ldots, \lambda_m$ that are not all zero (otherwise $w_1 = 0$, which contradicts the linear independence of $w_1, \ldots, w_n$). After an appropriate renumbering we have $\lambda_1 \neq 0$, and Lemma 9.18 yields

$$\mathrm{span}\{w_1, u_2, \ldots, u_m\} = \mathrm{span}\{u_1, u_2, \ldots, u_m\}.$$

Suppose that for some r, $1 \leq r \leq n-1$, we have exchanged the vectors $u_1, \ldots, u_r$ against $w_1, \ldots, w_r$ so that

$$\mathrm{span}\{w_1, \ldots, w_r, u_{r+1}, \ldots, u_m\} = \mathrm{span}\{u_1, \ldots, u_r, u_{r+1}, \ldots, u_m\}.$$

It is then clear that $r \leq m$.

By assumption we have $w_{r+1} \in \mathrm{span}\{u_1, \ldots, u_m\}$, and thus

$$w_{r+1} = \sum_{i=1}^{r} \lambda_i w_i + \sum_{i=r+1}^{m} \lambda_i u_i$$

for some scalars $\lambda_1, \ldots, \lambda_m$. One of the scalars $\lambda_{r+1}, \ldots, \lambda_m$ must be nonzero (otherwise $w_{r+1} \in \mathrm{span}\{w_1, \ldots, w_r\}$, which contradicts the linear independence

---

[2] In the literature, his theorem is sometimes called the *Steinitz exchange theorem* after Ernst Steinitz (1871–1928). The result was first proved in 1862 by Hermann Graßmann (1809–1877).

of $w_1, \ldots, w_m$). After an appropriate renumbering we have $\lambda_{r+1} \neq 0$, and Lemma 9.18 yields

$$\mathrm{span}\{w_1, \ldots, w_{r+1}, u_{r+2}, \ldots, u_m\} = \mathrm{span}\{w_1, \ldots, w_r, u_{r+1}, \ldots, u_m\}.$$

If we continue this construction until $r = n - 1$, then we obtain

$$\mathrm{span}\{w_1, \ldots, w_n, u_{n+1}, \ldots, u_m\} = \mathrm{span}\{u_1, \ldots, u_n, u_{n+1}, \ldots u_m\},$$

where in particular $n \leq m$. $\square$

Using this fundamental theorem, the following result about the unique number of basis elements is a simple corollary.

Corollary 9.20 *If a vector space V is spanned by finitely many vectors, then V has a basis consisting of finitely many elements, and any two bases of V have the same number of elements.*

Proof We know from Corollary 9.17 that V has a basis with finitely many elements. Let $U := \{u_1, \ldots, u_\ell\}$ and $W := \{w_1, \ldots, w_k\}$ be two such bases. Then

$$W \subseteq V = \mathrm{span}\{u_1, \ldots, u_\ell\} \stackrel{\text{Theorem 9.18}}{\Longrightarrow} k \leq \ell,$$

$$U \subseteq V = \mathrm{span}\{w_1, \ldots, w_k\} \stackrel{\text{Theorem 9.18}}{\Longrightarrow} \ell \leq k,$$

and thus $\ell = k$. $\square$

We can now define the dimension of a vector space.

Definition 9.21

(1) If there exists a basis of a K-vector space V that consists of finitely many elements, then V is called *finite dimensional*, and the unique number of basis elements is called the *dimension* of V. We denote the dimension by $\dim_K(V)$ or $\dim(V)$, if it is clear which field is meant.

(2) If V is a K-vector space that is not spanned by finitely many vectors, then V is called *infinite dimensional*, and we write $\dim_K(V) = \infty$ or $\dim(V) = \infty$.

Note that the zero vector space $V = \{0\}$ has the basis $\emptyset$ and thus it has dimension zero (cp. (2) in Definition 9.13).

Lemma 9.22 *Let V be a K-vector space with $n = \dim(V) \in \mathbb{N}$ and let $v_1, \ldots, v_n \in V$. Then the following statements are equivalent:*

(1) $v_1, \ldots, v_n$ are linearly independent.

9.2 Bases and Dimension of Vector Spaces

(2) span$\{v_1, \ldots, v_n\} = \mathcal{V}$.
(3) $\{v_1, \ldots, v_n\}$ is a basis of $\mathcal{V}$.

Proof Exercise. □

In particular, if $\dim(\mathcal{V}) \in \mathbb{N}$, then vectors $v_1, \ldots, v_m \in \mathcal{V}$ with $m > \dim(\mathcal{V})$ are always linearly dependent.

Example 9.23

(1) The basis (9.1) of the vector space $K^{n,m}$ has $n \cdot m$ elements, and hence $\dim(K^{n,m}) = n \cdot m$. Matrices $A_1, \ldots, A_k \in K^{n,m}$ with $k > n \cdot m$ are linearly dependent.
(2) The vector space $K[t]$ is not spanned by finitely many vectors (cp. (2) in Example 9.14), and hence is infinite dimensional.
(3) Let $\mathcal{V}$ be the vector space of continuous and real valued functions on the real interval $[0, 1]$ (cp. (3) in Example 9.2). Define for $n = 1, 2, \ldots$ the function $f_n \in \mathcal{V}$ by

$$f_n(x) = \begin{cases} 0, & x < \frac{1}{n+1}, \\ 0, & \frac{1}{n} < x, \\ 2n(n+1)x - 2n, & \frac{1}{n+1} \leq x \leq \frac{1}{2}\left(\frac{1}{n} + \frac{1}{n+1}\right), \\ -2n(n+1)x + 2n + 2, & \frac{1}{2}\left(\frac{1}{n} + \frac{1}{n+1}\right) < x \leq \frac{1}{n}. \end{cases}$$

The following picture shows the function f_n in the interval $[0, 1]$:

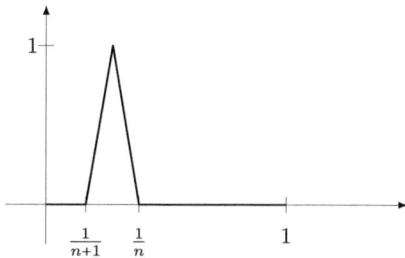

Every linear combination $\sum_{n=1}^{k} \lambda_n f_n$ is a continuous function that has the value λ_n at $\frac{1}{2}\left(\frac{1}{n} + \frac{1}{n+1}\right)$. Thus, the equation $\sum_{n=1}^{k} \lambda_n f_n = 0 \in \mathcal{V}$ implies that all λ_n must be zero, so that $f_1, \ldots, f_k \in \mathcal{V}$ are linearly independent for all $k \in \mathbb{N}$. Consequently, $\dim(\mathcal{V}) = \infty$.

Finally, we would like to show that every vector space has a basis. For this we need some notation from set theory.

Let M be a set and let for all $m \in M$ a set S_m be given. We define $\mathcal{S} := \{S_m \mid m \in M\}$. The subset relation $\subseteq$ is reflexive, transitive and *anti-symmetric*, i.e., for all $S, \widetilde{S} \in \mathcal{S}$ with $S \subseteq \widetilde{S}$ and $\widetilde{S} \subseteq S$ we have $S = \widetilde{S}$. A relation with these properties is called a *partial order*, and the set $\mathcal{S}$ is called *partially ordered*. A set $\widehat{S} \in \mathcal{S}$ is called a *maximal element* of $\mathcal{S}$ (with respect to $\subseteq$), if $S \in \mathcal{S}$ and $\widehat{S} \subseteq S$ imply $\widehat{S} = S$. A nonempty subset $\mathcal{K} \subseteq \mathcal{S}$ is called a *chain* (with respect to $\subseteq$), if for all $K_1, K_2 \in \mathcal{K}$ we have that $K_1 \subseteq K_2$ or $K_2 \subseteq K_1$. An element $\widetilde{K} \in \mathcal{S}$ is called an *upper bound* of the chain $\mathcal{K}$ (with respect to $\subseteq$), if $K \subseteq \widetilde{K}$ holds for every $K \in \mathcal{K}$.

Example 9.24 If $\mathcal{S} = \mathcal{P}(\{1, 2, \ldots, 10\})$ then

$$\mathcal{K} = \{\emptyset, \{1\}, \{1, 3, 4\}, \{1, 3, 4, 7\}, \{1, 3, 4, 7, 8, 10\}\} \subseteq \mathcal{S}$$

is a chain. The set $\{1, 3, 4, 7, 8, 10\}$ is an upper bound of the chain $\mathcal{K}$, and the set $\{1, 2, \ldots, 10\}$ is a maximal element of $\mathcal{S}$. The set $\{\{1, 2\}, \{2, 3\}\}$ does not form a chain.

The following *Lemma of Zorn*[3] is a fundamental result that we will use without giving a proof; see [10] for a simple proof.

Lemma 9.25 *A partially ordered set in which each chain has an upper bound contains at least one maximal element.*

Let $\mathcal{V}$ be a K-vector space and let $V \subseteq \mathcal{V}$ be a set with linearly independent elements or a linearly independent system (cp. (3) in Definition 9.10). Such a set is called *maximal linearly independent*, if the elements of each set W with $V \subset W$ are linearly dependent. A subset $B \subseteq \mathcal{V}$ is a basis of $\mathcal{V}$, if B is a maximal linearly independent subset of $\mathcal{V}$ (cp. Exercise 9.22). With this observation and the Lemma of Zorn we can prove the following theorem.

Theorem 9.26 *Let $\mathcal{V}$ be a K-vector space and let $V \subseteq \mathcal{V}$ be a set with linearly independent elements or a linearly independent system. Then there exists a basis B of $\mathcal{V}$ with $V \subseteq B$.*

Proof Let $\mathcal{S} \subseteq \mathcal{P}(\mathcal{V})$ be the set of all subsets S of $\mathcal{V}$ that consist of linearly independent elements, and with $V \subseteq S$. Since $V \in \mathcal{S}$, the set $\mathcal{S}$ is nonempty. Furthermore, $\mathcal{S}$ is partially ordered.

Let $\mathcal{K}$ be a chain of subsets of $\mathcal{S}$ and let $L := \bigcup_{K \in \mathcal{K}} K$. Then the elements of L are linearly independent. We can show this by contradiction: If the elements of L are linearly dependent, then there exists a finite subset of L consisting of linearly

---
[3] Max Zorn (1906–1993).

dependent elements $\ell_1, \ldots, \ell_m$. Each of these vectors ℓ_j is contained in (at least) one subset K_j. Since $\mathcal{K}$ is a chain, one of the subsets $K_1, \ldots, K_m$ must contain all the others. This set then contains the linearly dependent vectors $\ell_1, \ldots, \ell_m$, in contradiction to the linear independence of the elements of each subset of $\mathcal{S}$.

Since the elements of L are linearly independent, we have $L \in \mathcal{S}$. Thus, by construction, $S \subseteq L$ for every $S \in \mathcal{K}$, i.e., L is an upper bound of $\mathcal{K}$. The Lemma of Zorn now implies that $\mathcal{S}$ contains a maximal element B. By construction, this is a maximal linearly independent subset of $\mathcal{V}$ and, therefore, forms the desired basis of $\mathcal{V}$, since from $B \in \mathcal{S}$ we also have $\mathcal{V} \subseteq B$. □

The application of Theorem 9.26 to the set $V = \emptyset$ shows that every K-vector space has a basis. This is, however, only an existence result. From the theorem and its proof we cannot obtain a method for the construction or calculation of a basis.

A basis of an infinite dimensional vector space, for which every vector is a finite linear combination of basis elements (as assumed in Definition 9.13), is called a *Hamel basis*.[4] In the field of Functional Analysis one also studies *Schauder bases*,[5] in which every vector is represented by a unique linear combination, but not necessarily of finitely many vectors.

9.3 Coordinates and Changes of the Basis

We will now study the linear combinations of basis vectors of a finite dimensional vector space. In particular, we will study what happens with a linear combination if we change to another basis of the vector space.

Lemma 9.27 *Let $\mathcal{V}$ be a K-vector space. A set $\{v_1, \ldots, v_n\} \subseteq \mathcal{V}$ is a basis $\mathcal{V}$, if and only if for every $v \in \mathcal{V}$ there exist uniquely determined scalars $\lambda_1, \ldots, \lambda_n \in K$ with $v = \lambda_1 v_1 + \ldots + \lambda_n v_n$. These scalars are called the coordinates of v with respect to the basis $\{v_1, \ldots, v_n\}$.*

Proof If $\{v_1, \ldots, v_n\}$ is a basis of $\mathcal{V}$, then $v \in \text{span}\{v_1, \ldots, v_n\}$ holds for every $v \in \mathcal{V}$. Let $v = \sum_{i=1}^{n} \lambda_i v_i = \sum_{i=1}^{n} \mu_i v_i$ for some scalars $\lambda_i, \mu_i \in K, i = 1, \ldots, n$, then

$$0 = v - v = \sum_{i=1}^{n} (\lambda_i - \mu_i) v_i.$$

The linear independence of $v_1, \ldots, v_n$ implies that $\lambda_i = \mu_i$ for $i = 1, \ldots, n$.

---
[4] Georg Hamel (1877–1954).
[5] Juliusz Schauder (1899–1943).

On the other hand, suppose that every $v \in \mathcal{V}$ is a linear combination of vectors from $\{v_1, \ldots, v_n\}$ with uniquely determined scalars. Then the zero vector can only be combined trivially, i.e., with all coefficients equal to zero, and hence $v_1, \ldots, v_n$ are linearly independent. Moreover, $\text{span}\{v_1, \ldots, v_n\} = \mathcal{V}$, and thus $\{v_1, \ldots, v_n\}$ is a basis of $\mathcal{V}$. □

By definition, the coordinates of a vector depend on the given basis. In particular, they depend on the ordering (or numbering) of the basis vectors. Because of this, some authors distinguish between the basis as "set", i.e., a collection of elements without a particular ordering, and an "ordered basis". In this book we will keep the set notation for a basis $\{v_1, \ldots, v_n\}$, where the indices indicate the ordering of the basis vectors.

Let $\mathcal{V}$ be a K-vector space, $v_1, \ldots, v_n \in \mathcal{V}$ (they need not be linearly independent) and

$$v = \lambda_1 v_1 + \ldots + \lambda_n v_n$$

for some coefficients $\lambda_1, \ldots, \lambda_n \in K$. Let us write

$$(v_1, \ldots, v_n) \begin{bmatrix} \lambda_1 \\ \vdots \\ \lambda_n \end{bmatrix} := \lambda_1 v_1 + \ldots + \lambda_n v_n. \tag{9.2}$$

Here $(v_1, \ldots, v_n)$ is an n-tuple over $\mathcal{V}$, i.e.,

$$(v_1, \ldots, v_n) \in \mathcal{V}^n = \underbrace{\mathcal{V} \times \cdots \times \mathcal{V}}_{n \text{ times}}.$$

For $n = 1$ we have $\mathcal{V}^1 = \mathcal{V}$. We then skip the parentheses and write v instead of (v) for a 1-tuple. The notation (9.2) formally defines a "multiplication" as map from $\mathcal{V}^n \times K^{n,1}$ to $\mathcal{V}$.

For all $\alpha \in K$ we have

$$\alpha \cdot v = (\alpha \lambda_1) v_1 + \ldots + (\alpha \lambda_n) v_n = (v_1, \ldots, v_n) \begin{bmatrix} \alpha \lambda_1 \\ \vdots \\ \alpha \lambda_n \end{bmatrix}.$$

If $\mu_1, \ldots, \mu_n \in K$ and

$$u = \mu_1 v_1 + \ldots + \mu_n v_n = (v_1, \ldots, v_n) \begin{bmatrix} \mu_1 \\ \vdots \\ \mu_n \end{bmatrix},$$

then

$$v + u = (\lambda_1 + \mu_1)v_1 + \ldots + (\lambda_n + \mu_n)v_n = (v_1, \ldots, v_n)\begin{bmatrix} \lambda_1 + \mu_1 \\ \vdots \\ \lambda_n + \mu_n \end{bmatrix}.$$

This shows that if vectors are given by linear combinations, then the operations scalar multiplication and addition correspond to operations with the coefficients of the vectors with respect to the linear combinations.

We can further extend this notation. Let $A = [a_{ij}] \in K^{n,m}$ and let

$$u_j = (v_1, \ldots, v_n)\begin{bmatrix} a_{1j} \\ \vdots \\ a_{nj} \end{bmatrix}, \quad j = 1, \ldots, m.$$

Then we write the m linear combinations for $u_1, \ldots, u_m$ as the system

$$(u_1, \ldots, u_m) =: (v_1, \ldots, v_n)A. \tag{9.3}$$

On both sides of this equation we have elements of $\mathcal{V}^m$. The right-multiplication of an arbitrary n-tuple $(v_1, \ldots, v_n) \in \mathcal{V}^n$ with a matrix $A \in K^{n,m}$ thus corresponds to forming m linear combinations of the vectors $v_1, \ldots, v_n$, with the corresponding coefficients given by the entries of A. Formally, this defines a "multiplication" as a map from $\mathcal{V}^n \times K^{n,m}$ to $\mathcal{V}^m$.

Now consider also a matrix $B \in K^{m,\ell}$. Using (9.3) we obtain

$$(u_1, \ldots, u_m)B = ((v_1, \ldots, v_n)A)B.$$

The multiplication on the right hand side is associative in the following sense.

Lemma 9.28 *Let $\mathcal{V}$ be a K-vector space and $v_1, \ldots, v_n \in \mathcal{V}$. Then*

$$((v_1, \ldots, v_n)A)B = (v_1, \ldots, v_n)(AB)$$

for all $A \in K^{n,m}$ and $B \in K^{m,\ell}$.

Proof Exercise. □

Lemma 9.28 is used in the proof of the following result.

Lemma 9.29 *Let $\mathcal{V}$ be a K-vector space, let $v_1, \ldots, v_n \in \mathcal{V}$ be linearly independent, let $A \in K^{n,m}$, and let $(u_1, \ldots, u_m) = (v_1, \ldots, v_n)A$. Then the vectors $u_1, \ldots, u_m$ are linearly independent if and only if $\mathrm{rank}(A) = m$.*

Proof Let $\lambda_1, \ldots, \lambda_m \in K$ be arbitrary, and consider the equation

$$0 = \lambda_1 u_1 + \ldots + \lambda_m u_m = (u_1, \ldots, u_m) \begin{bmatrix} \lambda_1 \\ \vdots \\ \lambda_m \end{bmatrix} = ((v_1, \ldots, v_n)A) \begin{bmatrix} \lambda_1 \\ \vdots \\ \lambda_m \end{bmatrix}$$

$$= (v_1, \ldots, v_n) \left(A \begin{bmatrix} \lambda_1 \\ \vdots \\ \lambda_m \end{bmatrix} \right).$$

Since $v_1, \ldots, v_n$ are linearly independent, this equation holds if and only if $A[\lambda_1, \ldots, \lambda_m]^T = 0$, i.e., $[\lambda_1, \ldots, \lambda_m]^T \in \mathscr{L}(A, 0)$. The assertion now follows from Theorem 6.5, which shows that $\mathscr{L}(A, 0) = \{0\}$ holds if and only if $\text{rank}(A) = m$. $\square$

If $n = m$ in Lemma 9.29, then the vectors $u_1, \ldots, u_n$ are linearly independent if and only if $A \in GL_n(K)$. If $m > n$, then $\text{rank}(A) \leq \min\{n, m\} \leq n < m$, and the vectors $u_1, \ldots, u_m$ must be linearly dependent. Thus, from n linearly independent vectors $v_1, \ldots, v_n$ we can linearly combine at most n further linearly independent vectors.

Let $\{v_1, \ldots, v_n\}$ and $\{w_1, \ldots, w_n\}$ be bases of $\mathcal{V}$ and let $v \in \mathcal{V}$. By Lemma 9.27 there exist (unique) coordinates $\lambda_1, \ldots, \lambda_n$ and $\mu_1, \ldots, \mu_n$, respectively, with

$$v = (v_1, \ldots, v_n) \begin{bmatrix} \lambda_1 \\ \vdots \\ \lambda_n \end{bmatrix} = (w_1, \ldots, w_n) \begin{bmatrix} \mu_1 \\ \vdots \\ \mu_n \end{bmatrix}.$$

We will now describe a method for transforming the coordinates $\lambda_1, \ldots, \lambda_n$ with respect to the basis $\{v_1, \ldots, v_n\}$ into the coordinates $\mu_1, \ldots, \mu_n$ with respect to the basis $\{w_1, \ldots, w_n\}$, and vice versa.

For every basis vector v_j, $j = 1, \ldots, n$, there exist (unique) coordinates p_{ij}, $i = 1, \ldots, n$, such that

$$v_j = (w_1, \ldots, w_n) \begin{bmatrix} p_{1j} \\ \vdots \\ p_{nj} \end{bmatrix}, \quad j = 1, \ldots, n.$$

Defining $P = [p_{ij}] \in K^{n,n}$ we can write these n equations for the vectors v_j analogously to (9.3) as

$$(v_1, \ldots, v_n) = (w_1, \ldots, w_n) P. \qquad (9.4)$$

9.3 Coordinates and Changes of the Basis

In the same way, for every basis vector w_j, $j = 1, \ldots, n$, there exist (unique) coordinates q_{ij}, $i = 1, \ldots, n$, such that

$$w_j = (v_1, \ldots, v_n) \begin{bmatrix} q_{1j} \\ \vdots \\ q_{nj} \end{bmatrix}, \quad j = 1, \ldots, n.$$

If we set $Q = [q_{ij}] \in K^{n,n}$, then analogously to (9.4) we get

$$(w_1, \ldots, w_n) = (v_1, \ldots, v_n) Q.$$

Thus,

$$(w_1, \ldots, w_n) = (v_1, \ldots, v_n) Q = ((w_1, \ldots, w_n) P) Q = (w_1, \ldots, w_n)(PQ).$$

Consequently, $PQ = I_n$, since $\{w_1, \ldots, w_n\}$ is a basis and the coordinates with respect to a basis are uniquely determined. Analogously we obtain the equation $QP = I_n$. Therefore the matrix $P \in K^{n,n}$ is invertible with $P^{-1} = Q$.

Furthermore, we have

$$v = (v_1, \ldots, v_n) \begin{bmatrix} \lambda_1 \\ \vdots \\ \lambda_n \end{bmatrix} = ((w_1, \ldots, w_n) P) \begin{bmatrix} \lambda_1 \\ \vdots \\ \lambda_n \end{bmatrix} = (w_1, \ldots, w_n) \left(P \begin{bmatrix} \lambda_1 \\ \vdots \\ \lambda_n \end{bmatrix} \right).$$

Due to the uniqueness of the coordinates of v with respect to the basis $\{w_1, \ldots, w_n\}$ we obtain

$$\begin{bmatrix} \mu_1 \\ \vdots \\ \mu_n \end{bmatrix} = P \begin{bmatrix} \lambda_1 \\ \vdots \\ \lambda_n \end{bmatrix}, \quad \text{or} \quad \begin{bmatrix} \lambda_1 \\ \vdots \\ \lambda_n \end{bmatrix} = P^{-1} \begin{bmatrix} \mu_1 \\ \vdots \\ \mu_n \end{bmatrix}.$$

Hence a multiplication with the matrix P transforms the coordinates of v with respect to the basis $\{v_1, \ldots, v_n\}$ into those with respect to the basis $\{w_1, \ldots, w_n\}$; a multiplication with P^{-1} yields the inverse transformation. Therefore, P and P^{-1} are called *coordinate transformation matrices*.

We can summarize the results obtained above as follows.

Theorem 9.30 Let $\{v_1, \ldots, v_n\}$ and $\{w_1, \ldots, w_n\}$ be bases of a K-vector space $\mathcal{V}$. Then the uniquely determined matrix $P \in K^{n,n}$ in (9.4) is invertible and yields the coordinate transformation from $\{v_1, \ldots, v_n\}$ to $\{w_1, \ldots, w_n\}$: If

$$v = (v_1, \ldots, v_n) \begin{bmatrix} \lambda_1 \\ \vdots \\ \lambda_n \end{bmatrix} = (v_1, \ldots, v_n) \begin{bmatrix} \mu_1 \\ \vdots \\ \mu_n \end{bmatrix},$$

then

$$\begin{bmatrix} \mu_1 \\ \vdots \\ \mu_n \end{bmatrix} = P \begin{bmatrix} \lambda_1 \\ \vdots \\ \lambda_n \end{bmatrix}.$$

Example 9.31 Consider the vector space $\mathcal{V} = \mathbb{R}^2 = \{(\alpha_1, \alpha_2) \mid \alpha_1, \alpha_2 \in \mathbb{R}\}$ with the entrywise addition and scalar multiplication. A basis of $\mathcal{V}$ is given by the set $\{e_1 = (1, 0),\ e_2 = (0, 1)\}$, and we have $(\alpha_1, \alpha_2) = \alpha_1 e_1 + \alpha_2 e_2$ for all $(\alpha_1, \alpha_2) \in \mathcal{V}$. Another basis of $\mathcal{V}$ is the set $\{v_1 = (1, 1),\ v_2 = (1, 2)\}$. The corresponding coordinate transformation matrices can be obtained from the defining equations $(v_1, v_2) = (e_1, e_2)P$ and $(e_1, e_2) = (v_1, v_2)Q$ as

$$P = \begin{bmatrix} 1 & 1 \\ 1 & 2 \end{bmatrix}, \quad Q = P^{-1} = \begin{bmatrix} 2 & -1 \\ -1 & 1 \end{bmatrix}.$$

9.4 Relations Between Vector Spaces and Their Dimensions

Our first result describes the relation between a vector space and a subspace.

Lemma 9.32 If $\mathcal{V}$ is a finite dimensional vector space and $\mathcal{U} \subseteq \mathcal{V}$ is a subspace, then $\dim(\mathcal{U}) \leq \dim(\mathcal{V})$ with equality if and only if $\mathcal{U} = \mathcal{V}$.

Proof Let $\mathcal{U} \subseteq \mathcal{V}$ and let $\{u_1, \ldots, u_m\}$ be a basis of $\mathcal{U}$, where $\{u_1, \ldots, u_m\} = \emptyset$ for $\mathcal{U} = \{0\}$. Using Theorem 9.15 we can extend this set to a basis of $\mathcal{V}$. If $\mathcal{U}$ is a proper subset of $\mathcal{V}$, then at least one basis vector needs to be added and hence $\dim(\mathcal{U}) < \dim(\mathcal{V})$. If $\mathcal{U} = \mathcal{V}$, then every basis of $\mathcal{V}$ is also a basis of $\mathcal{U}$, and thus $\dim(\mathcal{U}) = \dim(\mathcal{V})$. □

If $\mathcal{U}_1$ and $\mathcal{U}_2$ are subspaces of a vector space $\mathcal{V}$, then their *intersection* is given by

$$\mathcal{U}_1 \cap \mathcal{U}_2 = \{u \in \mathcal{V} \mid u \in \mathcal{U}_1 \ \wedge\ u \in \mathcal{U}_2\}$$

9.4 Relations Between Vector Spaces and Their Dimensions

(cp. Definition 2.6). The *sum* of the two subspaces is defined as

$$\mathcal{U}_1 + \mathcal{U}_2 := \{u_1 + u_2 \in \mathcal{V} \mid u_1 \in \mathcal{U}_1 \wedge u_2 \in \mathcal{U}_2\}.$$

Lemma 9.33 *If $\mathcal{U}_1, \mathcal{U}_2, \mathcal{U}_3$ are subspaces of a vector space $\mathcal{V}$, then the following assertions hold:*

(1) $\mathcal{U}_1 \cap \mathcal{U}_2$ *and* $\mathcal{U}_1 + \mathcal{U}_2$ *are subspaces of* $\mathcal{V}$.
(2) $\mathcal{U}_1 + (\mathcal{U}_2 + \mathcal{U}_3) = (\mathcal{U}_1 + \mathcal{U}_2) + \mathcal{U}_3$ *and* $\mathcal{U}_1 + \mathcal{U}_2 = \mathcal{U}_2 + \mathcal{U}_1$.
(3) $\mathcal{U}_1 + \{0\} = \mathcal{U}_1$ *and* $\mathcal{U}_1 + \mathcal{U}_1 = \mathcal{U}_1$.
(4) $\mathcal{U}_1 \subseteq \mathcal{U}_1 + \mathcal{U}_2$, *with equality if and only if* $\mathcal{U}_2 \subseteq \mathcal{U}_1$.

Proof Exercise. □

The associativity of the addition of subspaces of a vector space $\mathcal{V}$, and the existence of the (neutral) element $\{0\}$ with $\mathcal{U} + \{0\} = \mathcal{U}$ for every subspace $\mathcal{U} \subseteq \mathcal{V}$ makes the algebraic structure

$$(\{\mathcal{U} \mid \mathcal{U} \subseteq \mathcal{V} \text{ is a subspace}\}, +)$$

a *monoid*. It is not a group, since for $\mathcal{U} \neq \{0\}$ there does not exist a subspace $\widetilde{\mathcal{U}}$ with $\mathcal{U} + \widetilde{\mathcal{U}} = \{0\}$, i.e., with respect to the addition of subspaces there do not exist inverse elements. Further examples of monoids are $(\mathbb{N}_0, +)$ and $(K^{n,n}, *)$, for which the neutral elements are the number 0 and the identity matrix I_n, respectively.

An important result is the following *dimension formula for subspaces*.

Theorem 9.34 *If $\mathcal{U}_1$ and $\mathcal{U}_2$ are finite dimensional subspaces of a vector space $\mathcal{V}$, then*

$$\dim(\mathcal{U}_1 \cap \mathcal{U}_2) + \dim(\mathcal{U}_1 + \mathcal{U}_2) = \dim(\mathcal{U}_1) + \dim(\mathcal{U}_2).$$

Proof Let $\{v_1, \ldots, v_r\}$ be a basis of $\mathcal{U}_1 \cap \mathcal{U}_2$. We extend this set to a basis $\{v_1, \ldots, v_r, w_1, \ldots, w_\ell\}$ of $\mathcal{U}_1$ and to a basis $\{v_1, \ldots, v_r, x_1, \ldots, x_k\}$ of $\mathcal{U}_2$, where we assume that $r, \ell, k \geq 1$. (If one of the lists is empty, then the following argument is easily modified.)

If suffices to show that $\{v_1, \ldots, v_r, w_1, \ldots, w_\ell, x_1, \ldots, x_k\}$ is a basis of $\mathcal{U}_1 + \mathcal{U}_2$. Obviously,

$$\text{span}\{v_1, \ldots, v_r, w_1, \ldots, w_\ell, x_1, \ldots, x_k\} = \mathcal{U}_1 + \mathcal{U}_2,$$

and hence it suffices to show that $v_1, \ldots, v_r, w_1, \ldots, w_\ell, x_1, \ldots, x_k$ are linearly independent. Let

$$\sum_{i=1}^{r} \lambda_i v_i + \sum_{i=1}^{\ell} \mu_i w_i + \sum_{i=1}^{k} \gamma_i x_i = 0,$$

then

$$\sum_{i=1}^{k} \gamma_i x_i = -\left(\sum_{i=1}^{r} \lambda_i v_i + \sum_{i=1}^{\ell} \mu_i w_i\right).$$

On the left hand side of this equation we have, by definition, a vector in $\mathcal{U}_2$; on the right hand side a vector in $\mathcal{U}_1$. Therefore, $\sum_{i=1}^{k} \gamma_i x_i \in \mathcal{U}_1 \cap \mathcal{U}_2$. By construction, however, $\{v_1, \ldots, v_r\}$ is a basis of $\mathcal{U}_1 \cap \mathcal{U}_2$ and the vectors $v_1, \ldots, v_r, w_1, \ldots, w_\ell$ are linearly independent. Therefore, $\sum_{i=1}^{\ell} \mu_i w_i = 0$ implies that $\mu_1 = \cdots = \mu_\ell = 0$. But then also

$$\sum_{i=1}^{r} \lambda_i v_i + \sum_{i=1}^{k} \gamma_i x_i = 0,$$

and hence $\lambda_1 = \cdots = \lambda_r = \gamma_1 = \cdots = \gamma_k = 0$ due to the linear independence of $v_1, \ldots, v_r, x_1, \ldots, x_k$. □

If at least one of the subspaces in Theorem 9.34 is infinite dimensional, then the assertion is still formally correct, since in this case $\dim(\mathcal{U}_1 + \mathcal{U}_2) = \infty$ and $\dim(\mathcal{U}_1) + \dim(\mathcal{U}_2) = \infty$.

Example 9.35 For the subspaces

$$\mathcal{U}_1 = \{[\alpha_1, \alpha_2, 0] \mid \alpha_1, \alpha_2 \in K\} \quad \text{and} \quad \mathcal{U}_2 = \{[0, \alpha_2, \alpha_3] \mid \alpha_2, \alpha_3 \in K\}$$

of $K^{1,3}$ we have $\dim(\mathcal{U}_1) = \dim(\mathcal{U}_2) = 2$,

$$\mathcal{U}_1 \cap \mathcal{U}_2 = \{[0, \alpha_2, 0] \mid \alpha_2 \in K\}, \quad \dim(\mathcal{U}_1 \cap \mathcal{U}_2) = 1,$$
$$\mathcal{U}_1 + \mathcal{U}_2 = K^{1,3}, \qquad\qquad\qquad \dim(\mathcal{U}_1 + \mathcal{U}_2) = 3.$$

The above definition of the sum can be extended to an arbitrary (but finite) number of subspaces: If $\mathcal{U}_1, \ldots, \mathcal{U}_k$, $k \geq 2$, are subspaces of the vector space $\mathcal{V}$, then we define

$$\mathcal{U}_1 + \ldots + \mathcal{U}_k = \sum_{j=1}^{k} \mathcal{U}_j := \left\{\sum_{j=1}^{k} u_j \mid u_j \in \mathcal{U}_j, \ j = 1, \ldots, k\right\}.$$

9.4 Relations Between Vector Spaces and Their Dimensions

This sum is called *direct*, if

$$\mathcal{U}_i \cap \sum_{\substack{j=1\\j\neq i}}^{k} \mathcal{U}_j = \{0\} \quad \text{for } i = 1, \ldots, k,$$

and in this case we write the (direct) sum as

$$\mathcal{U}_1 \oplus \ldots \oplus \mathcal{U}_k = \bigoplus_{j=1}^{k} \mathcal{U}_j.$$

In particular, a sum $\mathcal{U}_1 + \mathcal{U}_2$ of two subspaces $\mathcal{U}_1, \mathcal{U}_2 \subseteq \mathcal{V}$ is direct if $\mathcal{U}_1 \cap \mathcal{U}_2 = \{0\}$.

The following theorem presents two equivalent characterizations of the direct sum of subspaces.

Theorem 9.36 *If $\mathcal{U} = \mathcal{U}_1 + \ldots + \mathcal{U}_k$ is a sum of $k \geq 2$ subspaces of a vector space $\mathcal{V}$, then the following statements are equivalent:*

(1) *The sum $\mathcal{U}$ is direct, i.e., $\mathcal{U}_i \cap \sum_{j \neq i} \mathcal{U}_j = \{0\}$ for $i = 1, \ldots, k$.*
(2) *Every vector $u \in \mathcal{U}$ has a representation of the form $u = \sum_{j=1}^{k} u_j$ with uniquely determined $u_j \in \mathcal{U}_j$ for $j = 1, \ldots, k$.*
(3) $\sum_{j=1}^{k} u_j = 0$ *with $u_j \in \mathcal{U}_j$ for $j = 1, \ldots, k$ implies that $u_j = 0$ for $j = 1, \ldots, k$.*

Proof

(1) $\Rightarrow$ (2): Let $u = \sum_{j=1}^{k} u_j = \sum_{j=1}^{k} \widetilde{u}_j$ with $u_j, \widetilde{u}_j \in \mathcal{U}_j$, $j = 1, \ldots, k$. For every $i = 1, \ldots, k$ we then have

$$u_i - \widetilde{u}_i = -\sum_{j \neq i}(u_j - \widetilde{u}_j) \in \mathcal{U}_i \cap \sum_{j \neq i} \mathcal{U}_j.$$

Now $\mathcal{U}_i \cap \sum_{j \neq i} \mathcal{U}_j = \{0\}$ implies that $u_i - \widetilde{u}_i = 0$, and hence $u_i = \widetilde{u}_i$ for $i = 1, \ldots, k$.

(2) $\Rightarrow$ (3): This is obvious.

(3) $\Rightarrow$ (1): For a given i, let $u \in \mathcal{U}_i \cap \sum_{j \neq i} \mathcal{U}_j$. Then $u = \sum_{j \neq i} u_j$ for some $u_j \in \mathcal{U}_j$, $j \neq i$, and hence $-u + \sum_{j \neq i} u_j = 0$. In particular, this implies that $u = 0$, and thus $\mathcal{U}_i \cap \sum_{j \neq i} \mathcal{U}_j = \{0\}$. □

If $\mathcal{U}_1, \ldots, \mathcal{U}_k$, $k \geq 3$, are subspaces of a K-vector space $\mathcal{V}$, then in general the property $\mathcal{U}_i \cap \mathcal{U}_j = \{0\}$ for all $i \neq j$ does not imply that $\mathcal{U}_1 + \ldots + \mathcal{U}_k$ is a direct sum.

Example 9.37 For the subspaces

$$\mathcal{U}_1 = \operatorname{span}\left\{\begin{bmatrix}1\\0\\0\end{bmatrix}\right\}, \quad \mathcal{U}_2 = \operatorname{span}\left\{\begin{bmatrix}0\\1\\0\end{bmatrix}\right\}, \quad \mathcal{U}_3 = \operatorname{span}\left\{\begin{bmatrix}1\\1\\0\end{bmatrix}\right\}$$

of $K^{3,1}$ we have $\mathcal{U}_i \cap \mathcal{U}_j = \{0\}$ for all $i \neq j$, but the sum is not direct, since $\mathcal{U}_3 \cap (\mathcal{U}_1 + \mathcal{U}_2) = \mathcal{U}_3 \neq \{0\}$. Furthermore,

$$0 = \alpha \begin{bmatrix}1\\0\\0\end{bmatrix} + \alpha \begin{bmatrix}0\\1\\0\end{bmatrix} - \alpha \begin{bmatrix}1\\1\\0\end{bmatrix},$$

for every $\alpha \in K$, i.e., the zero vector does not have a representation of the form $0 = u_1 + u_2 + u_3$ with uniquely determined vectors $u_j \in \mathcal{U}_j$, $j = 1, 2, 3$.

If we replace $\mathcal{U}_3$ by $\mathcal{U}_4 = \operatorname{span}\{[1, 1, 1]^T\}$, then $\mathcal{U}_1 + \mathcal{U}_2 + \mathcal{U}_4$ is a direct sum.

Exercises

(In the following exercises K is an arbitrary field.)

9.1 Let $\mathcal{V}$ be a K-vector space and let $v_1, v_2, v_3 \in \mathcal{V}$ with $v_1 + v_2 + v_3 = 0$. Show that then $\operatorname{span}\{v_1, v_2\} = \operatorname{span}\{v_1, v_3\} = \operatorname{span}\{v_2, v_3\}$.

9.2 Show that $a_1, \ldots, a_n \in K^{n,1}$ are linearly independent if and only if $\det([a_1, \ldots, a_n]) \neq 0$.

9.3 Show that $(K^{n,m}, +, \cdot)$ is a K-vector space (cp. (1) in Example 9.2).

9.4 Let $A \in K^{n,n}$. Show that $\mathcal{U} := \{X \in K^{n,n} \mid AX = XA\}$ is a subspace of $(K^{n,n}, +, \cdot)$.

9.5 Let $A \in K^{n,n}$ and $B \in K^{m,m}$. Show that $\mathcal{U} := \{AX + XB \mid X \in K^{n,m}\}$ is a subspace of $(K^{n,m}, +, \cdot)$.

9.6 Show that $(K[t], +, \cdot)$ is a K-vector space (cp. (2) in Example 9.2). Show further that $K[t]_{\leq n}$ is a subspace of $K[t]$ (cp. (3) in Example 9.6) and determine $\dim(K[t]_{\leq n})$.

9.7 Which of the following sets (with the usual addition and scalar multiplication) are $\mathbb{R}$-vector spaces?

$$\left\{[\alpha_1, \alpha_2] \in \mathbb{R}^{1,2} \mid \alpha_1 = \alpha_2\right\}, \quad \left\{[\alpha_1, \alpha_2] \in \mathbb{R}^{1,2} \mid \alpha_1^2 + \alpha_2^2 = 1\right\},$$

$$\left\{[\alpha_1, \alpha_2] \in \mathbb{R}^{1,2} \mid \alpha_1 \geq \alpha_2\right\},$$

$$\left\{[\alpha_1, \alpha_2] \in \mathbb{R}^{1,2} \mid \alpha_1 - \alpha_2 = 0 \text{ and } 2\alpha_1 + \alpha_2 = 0\right\}.$$

Determine, if possible, a basis and the dimension.

Exercises

9.8 Consider the $\mathbb{Q}$-vector space $\mathbb{R}$, and show that $1, \sqrt{2}, \sqrt{3} \in \mathbb{R}$ are linearly independent in this vector space. Is $\sqrt{6}$ a linear combination of these three vectors?
(*Hint:* You can use that $\sqrt{2}, \sqrt{3}, \sqrt{6} \notin \mathbb{Q}$.)

9.9 Let $\mathcal{V}$ be a K-vector space, Ω a nonempty set and $\text{Map}(\Omega, \mathcal{V})$ the set of maps from Ω to $\mathcal{V}$. On $\text{Map}(\Omega, \mathcal{V})$ we define an addition and a scalar multiplication pointwise, i.e., for $f, g \in \text{Map}(\Omega, \mathcal{V})$ and $\lambda \in K$ we define

$$(f + g)(x) := f(x) + g(x) \quad \text{und} \quad (\lambda \cdot f)(x) := \lambda f(x)$$

for all $x \in \Omega$. Show that $(\text{Map}(\Omega, \mathcal{V}), +, \cdot)$ is a K-vector space.

9.10 Let $\mathcal{V}$ be a finite dimensional K-vector space and Ω a nonempty set. Show that $\dim(\text{Map}(\Omega, \mathcal{V})) = |\Omega| \cdot \dim(\mathcal{V})$.

9.11 Show that the functions sin and cos in $\text{Map}(\mathbb{R}, \mathbb{R})$ are linearly independent.

9.12 Let $\mathcal{V}$ be a $\mathbb{C}$-vector space. Show that $\mathcal{V}$ then also is an $\mathbb{R}$-vector space.

9.13 Determine a basis of the $\mathbb{R}$-vector space $\mathbb{C}$ and $\dim_{\mathbb{R}}(\mathbb{C})$. Determine a basis of the $\mathbb{C}$-vector space $\mathbb{C}$ and $\dim_{\mathbb{C}}(\mathbb{C})$.

9.14 Determine all $\alpha \in \mathbb{R}$, so that $\{[1 + \alpha, 2], [1, 2 + \alpha]\}$ is a basis of $\mathbb{R}^{1,2}$.

9.15 Show that the polynomials $p_1 = t^5 + t^4$, $p_2 = t^5 - 7t^3$, $p_3 = t^5 - 1$, $p_4 = t^5 + 3t$ are linearly independent in $\mathbb{Q}[t]_{\leq 5}$ and extend $\{p_1, p_2, p_3, p_4\}$ to a basis of $\mathbb{Q}[t]_{\leq 5}$.

9.16 Let $n \in \mathbb{N}$ and

$$K[t_1, t_2] := \left\{ \sum_{i,j=0}^{n} \alpha_{ij} t_1^i t_2^j \,\Big|\, \alpha_{ij} \in K \right\}.$$

An element of $K[t_1, t_2]$ is called *bivariate polynomial* over K in the unknowns t_1 and t_2. Define a scalar multiplication and an addition so that $K[t_1, t_2]$ becomes a vector space. Determine a basis of $K[t_1, t_2]$.

9.17 Prove Lemma 9.5.

9.18 Prove Lemma 9.11.

9.19 Let $A \in K^{n,m}$ and $b \in K^{n,1}$. Is the solution set $\mathscr{L}(A, b)$ of $Ax = b$ a subspace of $K^{m,1}$?

9.20 Let $A \in K^{n,n}$ and let $\lambda \in K$ be an eigenvalue of A. Show that the set $\{v \in K^{n,1} \mid Av = \lambda v\}$ is a subspace of $K^{n,1}$.

9.21 Let $A \in K^{n,n}$ and let $\lambda_1 \neq \lambda_2$ be two eigenvalues of A. Show that any two associated eigenvectors v_1 and v_2 are linearly independent.

9.22 Let $\mathcal{V}$ be a K-vector space and let $B \subseteq \mathcal{V}$. Show that the following statements are equivalent:
   (1) B is a basis of $\mathcal{V}$.
   (2) B is a *minimal generating system* of $\mathcal{V}$, i.e., for no proper subset $M \subset B$ we have $\text{span}(M) = \mathcal{V}$.
   (3) B is a maximal linearly independent subset of $\mathcal{V}$, i.e., if $B \subset W$, then the elements of W (or the system formed from them) are linearly dependent.

9.23 Show that $B = \{B_1, B_2, B_3, B_4\}$ and $C = \{C_1, C_2, C_3, C_4\}$ with

$$B_1 = \begin{bmatrix} 1 & 1 \\ 0 & 0 \end{bmatrix}, \quad B_2 = \begin{bmatrix} 1 & 0 \\ 0 & 0 \end{bmatrix}, \quad B_3 = \begin{bmatrix} 1 & 0 \\ 1 & 0 \end{bmatrix}, \quad B_4 = \begin{bmatrix} 1 & 1 \\ 0 & 1 \end{bmatrix}$$

and

$$C_1 = \begin{bmatrix} 1 & 0 \\ 0 & 1 \end{bmatrix}, \quad C_2 = \begin{bmatrix} 1 & 0 \\ 1 & 0 \end{bmatrix}, \quad C_3 = \begin{bmatrix} 1 & 0 \\ 0 & 0 \end{bmatrix}, \quad C_4 = \begin{bmatrix} 0 & 1 \\ 1 & 0 \end{bmatrix}$$

are bases of the vector space $K^{2,2}$, and determine corresponding coordinate transformation matrices.

9.24 Examine the elements of the following sets for linear independence in the vector space $K[t]_{\leq 3}$:

$$U_1 = \{t, t^2 + 2t, t^2 + 3t + 1, t^3\}, \quad U_2 = \{1, t, t + t^2, t^2 + t^3\},$$
$$U_3 = \{1, t^2 - t, t^2 + t, t^3\}.$$

Determine the dimensions of the subspaces spanned by the elements of U_1, U_2, U_3. Is one of these sets a basis of $K[t]_{\leq 3}$?

9.25 Show that the set of sequences $\{(\alpha_1, \alpha_2, \alpha_3, \ldots) \mid \alpha_i \in K, i \in \mathbb{N}\}$ with entrywise addition and scalar multiplication forms an infinite dimensional vector space, and determine a basis system.

9.26 Prove Lemma 9.22.

9.27 Prove Lemma 9.28.

9.28 Prove Lemma 9.33.

9.29 Let $\mathcal{U}_1, \mathcal{U}_2$ be finite dimensional subspaces of a vector space $\mathcal{V}$. Show that the sum $\mathcal{U}_1 + \mathcal{U}_2$ is direct if $\dim(\mathcal{U}_1 + \mathcal{U}_2) = \dim(\mathcal{U}_1) + \dim(\mathcal{U}_2)$.

9.30 Let $\mathcal{U}_1, \ldots, \mathcal{U}_k$, $k \geq 3$, be finite dimensional subspaces of a vector space $\mathcal{V}$. Suppose that $\mathcal{U}_i \cap \mathcal{U}_j = \{0\}$ for all $i \neq j$. Is the sum $\mathcal{U}_1 + \ldots + \mathcal{U}_k$ direct?

9.31 Let $\mathcal{U}$ be a subspace of a finite dimensional vector space $\mathcal{V}$. Show that there exists another subspace $\widetilde{\mathcal{U}}$ with $\mathcal{U} \oplus \widetilde{\mathcal{U}} = \mathcal{V}$. (The subspace $\widetilde{\mathcal{U}}$ is called a *complement* of $\mathcal{U}$.)

9.32 Determine three subspaces $\mathcal{U}_1, \mathcal{U}_2, \mathcal{U}_3$ of $\mathcal{V} = \mathbb{R}^{3,1}$ with $\mathcal{U}_2 \neq \mathcal{U}_3$ and $\mathcal{V} = \mathcal{U}_1 \oplus \mathcal{U}_2 = \mathcal{U}_1 \oplus \mathcal{U}_3$. Is there a subspace $\mathcal{U}_1$ of $\mathcal{V}$ with a uniquely determined complement?

9.33 Let $\mathcal{U}_1, \ldots, \mathcal{U}_k$, $k \geq 2$, be finite dimensional subspaces of a K-vector space $\mathcal{V}$. Show that $\dim(\mathcal{U}_1 + \ldots + \mathcal{U}_k) \leq \dim(\mathcal{U}_1) + \ldots + \dim(\mathcal{U}_k)$. When do we have equality?

Linear Maps

10

In this chapter we study maps between vector spaces that are compatible with the two vector space operations, addition and scalar multiplication. These maps are called linear maps or homomorphisms. We first investigate their most important properties and then show that in the case of finite dimensional vector spaces every linear map can be represented by a matrix, when bases in the respective spaces have been chosen. If the bases are chosen in a clever way, then we can read off important properties of a linear map from its matrix representation. This central idea will arise frequently in later chapters.

10.1 Basic Definitions and Properties

We start our investigations with the definition of linear maps between vector spaces.

Definition 10.1 Let $\mathcal{V}$ and $\mathcal{W}$ be K-vector spaces. A map $f : \mathcal{V} \to \mathcal{W}$ is called *linear*, if

(1) $f(\lambda v) = \lambda f(v)$, and
(2) $f(v + w) = f(v) + f(w)$,

hold for all $v, w \in \mathcal{V}$ and $\lambda \in K$. The set of all these maps is denoted by $\mathcal{L}(\mathcal{V}, \mathcal{W})$.

A linear map $f : \mathcal{V} \to \mathcal{W}$ is also called a *linear transformation* or (vector space) *homomorphism*. A bijective linear map is called an *isomorphism*. If there exists an isomorphism between $\mathcal{V}$ and $\mathcal{W}$, then the spaces $\mathcal{V}$ and $\mathcal{W}$ are called *isomorphic*, which we denote by

$$\mathcal{V} \cong \mathcal{W}.$$

A map $f \in \mathcal{L}(\mathcal{V}, \mathcal{V})$ is called an *endomorphism*, and a bijective endomorphism is called an *automorphism*.

It is an easy exercise to show that the conditions (1) and (2) in Definition 10.1 hold if and only if

$$f(\lambda v + \mu w) = \lambda f(v) + \mu f(w)$$

holds for all $\lambda, \mu \in K$ and $v, w \in \mathcal{V}$.

Example 10.2

(1) Every matrix $A \in K^{n,m}$ defines a map

$$A : K^{m,1} \to K^{n,1}, \quad x \mapsto Ax.$$

This map is linear, since

$$A(\lambda x) = \lambda Ax \quad \text{for all } x \in K^{m,1} \text{ and } \lambda \in K,$$
$$A(x + y) = Ax + Ay \quad \text{for all } x, y \in K^{m,1}$$

(cp. Lemmas 4.3 and 4.4).

(2) The map trace : $K^{n,n} \to K$, $A = [a_{ij}] \mapsto \text{trace}(A) := \sum_{i=1}^{n} a_{ii}$, is linear (cp. Exercise 8.9).

(3) The map

$$f : \mathbb{Q}[t]_{\leq 3} \to \mathbb{Q}[t]_{\leq 2}, \quad \alpha_3 t^3 + \alpha_2 t^2 + \alpha_1 t + \alpha_0 \mapsto 2\alpha_2 t^2 + 3\alpha_1 t + 4\alpha_0,$$

is linear. (Show this as an exercise). The map

$$g : \mathbb{Q}[t]_{\leq 3} \to \mathbb{Q}[t]_{\leq 2}, \quad \alpha_3 t^3 + \alpha_2 t^2 + \alpha_1 t + \alpha_0 \mapsto \alpha_2 t^2 + \alpha_1 t + \alpha_0^2,$$

is not linear. For example, if $p_1 = t + 2$ and $p_2 = t + 1$, then $g(p_1 + p_2) = 2t + 9 \neq 2t + 5 = g(p_1) + g(p_2)$.

(4) Let $n \in \mathbb{N}, k \in \{1, \ldots, n\}$ and $a_1, \ldots, a_{k-1}, a_{k+1}, \ldots, a_n \in K^{n,1}$. Then

$$f : K^{n,1} \to K, \quad v \mapsto \det([a_1, \ldots, a_{k-1}, v, a_{k+1}, \ldots, a_n]),$$

is a linear map (cp. Eq. (7.3)). This linear map is, in fact, a *linear form* on $K^{n,1}$ (cp. Definition 11.1). Since this holds for every $k \in \{1, \ldots, n\}$, the determinant map is also called a *multilinear form* on $K^{n,1}$.

10.1 Basic Definitions and Properties

(5) Let $\mathcal{V} = \{(\alpha_1, \alpha_2, \alpha_3, \ldots) \mid \alpha_i \in K, i \in \mathbb{N}\}$ be the infinite dimensional K-vector space of sequences (cp. Exercise 9.25). Then

$$f : \mathcal{V} \to \mathcal{V}, \quad (\alpha_1, \alpha_2, \alpha_3, \ldots) \mapsto (0, \alpha_1, \alpha_2, \alpha_3, \ldots),$$

$$g : \mathcal{V} \to \mathcal{V}, \quad (\alpha_1, \alpha_2, \alpha_3, \ldots) \mapsto (\alpha_2, \alpha_3, \alpha_4, \ldots),$$

are two linear maps called *right shift operator* and *left shift operator*, respectively. (As an exercise prove the linearity of the two maps.)

The set of linear maps between vector spaces forms a vector space itself.

Lemma 10.3 *Let $\mathcal{V}$ and $\mathcal{W}$ be K-vector spaces. For $f, g \in \mathcal{L}(\mathcal{V}, \mathcal{W})$ and $\lambda \in K$ define $f + g$ and $\lambda \cdot f$ by*

$$(f + g)(v) := f(v) + g(v),$$
$$(\lambda \cdot f)(v) := \lambda f(v),$$

for all $v \in \mathcal{V}$. Then $(\mathcal{L}(\mathcal{V}, \mathcal{W}), +, \cdot)$ is a K-vector space.

Proof Cp. Exercise 9.9. □

The next result deals with the existence and uniqueness of linear maps.

Theorem 10.4 *Let $\mathcal{V}$ and $\mathcal{W}$ be K-vector spaces, let $\{v_1, \ldots, v_m\}$ be a basis of $\mathcal{V}$, and let $w_1, \ldots, w_m \in \mathcal{W}$. Then there exists a unique linear map $f \in \mathcal{L}(\mathcal{V}, \mathcal{W})$ with $f(v_i) = w_i$ for $i = 1, \ldots, m$.*

Proof For every $v \in \mathcal{V}$ there exist (unique) coordinates $\lambda_1^{(v)}, \ldots, \lambda_m^{(v)}$ with $v = \sum_{i=1}^m \lambda_i^{(v)} v_i$ (cp. Lemma 9.27). We define the map $f : \mathcal{V} \to \mathcal{W}$ by

$$f(v) := \sum_{i=1}^m \lambda_i^{(v)} w_i \quad \text{for all } v \in \mathcal{V}.$$

By definition, $f(v_i) = w_i$ for $i = 1, \ldots, m$.

We next show that f is linear. For every $\lambda \in K$ we have $\lambda v = \sum_{i=1}^m (\lambda \lambda_i^{(v)}) v_i$, and hence

$$f(\lambda v) = \sum_{i=1}^m (\lambda \lambda_i^{(v)}) w_i = \lambda \sum_{i=1}^m \lambda_i^{(v)} w_i = \lambda f(v).$$

If $u = \sum_{i=1}^{m} \lambda_i^{(u)} v_i \in \mathcal{V}$, then $v + u = \sum_{i=1}^{m} (\lambda_i^{(v)} + \lambda_i^{(u)}) v_i$, and hence

$$f(v+u) = \sum_{i=1}^{m} (\lambda_i^{(v)} + \lambda_i^{(u)}) w_i = \sum_{i=1}^{m} \lambda_i^{(v)} w_i + \sum_{i=1}^{m} \lambda_i^{(u)} w_i = f(v) + f(u).$$

Thus, $f \in \mathcal{L}(\mathcal{V}, \mathcal{W})$.

Suppose that $g \in \mathcal{L}(\mathcal{V}, \mathcal{W})$ also satisfies $g(v_i) = w_i$ for $i = 1, \ldots, m$. Then for every $v = \sum_{i=1}^{m} \lambda_i^{(v)} v_i$ we have

$$f(v) = f\left(\sum_{i=1}^{m} \lambda_i^{(v)} v_i\right) = \sum_{i=1}^{m} \lambda_i^{(v)} f(v_i) = \sum_{i=1}^{m} \lambda_i^{(v)} w_i$$

$$= \sum_{i=1}^{m} \lambda_i^{(v)} g(v_i) = g\left(\sum_{i=1}^{m} \lambda_i^{(v)} v_i\right) = g(v),$$

and hence $f = g$, so that f is indeed uniquely determined. $\square$

Theorem 10.4 shows that the map $f \in \mathcal{L}(\mathcal{V}, \mathcal{W})$ is uniquely determined by the images of f at the given basis vectors of $\mathcal{V}$. Note that the image vectors $w_1, \ldots, w_m \in \mathcal{W}$ may be linearly dependent, and that $\mathcal{W}$ may be infinite dimensional.

In Definition 2.13 we have introduced the image and pre-image of a map. We next recall these definitions for completeness and introduce the kernel of a linear map.

Definition 10.5 If $\mathcal{V}$ and $\mathcal{W}$ are K-vector spaces and $f \in \mathcal{L}(\mathcal{V}, \mathcal{W})$, then the *kernel* and the *image* of f are defined by

$$\ker(f) := \{v \in \mathcal{V} \mid f(v) = 0\} \quad \text{and} \quad \text{im}(f) := \{f(v) \in \mathcal{W} \mid v \in \mathcal{V}\}.$$

For $w \in \mathcal{W}$ the *pre-image* of w in the space $\mathcal{V}$ is defined by

$$f^{-1}(w) := f^{-1}(\{w\}) = \{v \in \mathcal{V} \mid f(v) = w\}.$$

The kernel of a linear map is sometimes called the *null space* (or *nullspace*) of the map, and some authors use the notation $\text{null}(f)$ instead of $\ker(f)$.

Note that the pre-image $f^{-1}(w)$ is a set, and that f^{-1} here does *not* mean the inverse map of f (cp. Definition 2.13). In particular, we have $f^{-1}(0) = \ker(f)$, and if $w \notin \text{im}(f)$, then $f^{-1}(w) = \emptyset$.

Example 10.6 For $A \in K^{n,m}$ and the corresponding map $A \in \mathcal{L}(K^{m,1}, K^{n,1})$ from (1) in Example 10.2 we have

$$\ker(A) = \{x \in K^{m,1} \mid Ax = 0\} \quad \text{and} \quad \text{im}(A) = \{Ax \in K^{n,1} \mid x \in K^{m,1}\}.$$

10.1 Basic Definitions and Properties

Note that $\ker(A) = \mathcal{L}(A, 0)$ (cp. Definition 6.1). Let $a_j \in K^{n,1}$ denote the jth column of A, $j = 1, \ldots, m$. For $x = [x_1, \ldots, x_m]^T \in K^{m,1}$ we then can write

$$Ax = \sum_{j=1}^{m} x_j a_j.$$

Clearly, $0 \in \ker(A)$. Moreover, we see from the representation of Ax that $\ker(A) = \{0\}$ if and only if the columns of A are linearly independent. The set $\operatorname{im}(A)$ is given by the linear combinations of the columns of A, i.e., $\operatorname{im}(A) = \operatorname{span}\{a_1, \ldots, a_m\}$.

Lemma 10.7 *If V and W are K-vector spaces, then for every $f \in \mathcal{L}(V, W)$ the following assertions hold:*

(1) $f(0) = 0$ and $f(-v) = -f(v)$ for all $v \in V$.
(2) If f is an isomorphism, then $f^{-1} \in \mathcal{L}(W, V)$.
(3) $\ker(f)$ is a subspace of V and $\operatorname{im}(f)$ is a subspace of W.
(4) f is surjective if and only if $\operatorname{im}(f) = W$.
(5) f is injective if and only if $\ker(f) = \{0\}$.
(6) If f is injective and if $v_1, \ldots, v_m \in V$ are linearly independent, then $f(v_1), \ldots, f(v_m) \in W$ are linearly independent.
(7) If $v_1, \ldots, v_m \in V$ are linearly dependent, then $f(v_1), \ldots, f(v_m) \in W$ are linearly dependent, or, equivalently, if $f(v_1), \ldots, f(v_m) \in W$ are linearly independent, then $v_1, \ldots, v_m \in V$ are linearly independent.
(8) If $w \in \operatorname{im}(f)$ and if $u \in f^{-1}(w)$ is arbitrary, then

$$f^{-1}(w) = u + \ker(f) := \{u + v \mid v \in \ker(f)\}.$$

Proof

(1) We have $f(0_V) = f(0_K \cdot 0_V) = 0_K \cdot f(0_V) = 0_V$ as well as $f(v) + f(-v) = f(v + (-v)) = f(0_V) = 0_W$ for all $v \in V$.
(2) The existence of the inverse map $f^{-1} : W \to V$ is guaranteed by Theorem 2.21, so we just have to show that f^{-1} is linear. If $w_1, w_2 \in W$, then there exist uniquely determined $v_1, v_2 \in V$ with $w_1 = f(v_1)$ and $w_2 = f(v_2)$. Hence,

$$f^{-1}(w_1 + w_2) = f^{-1}(f(v_1) + f(v_2)) = f^{-1}(f(v_1 + v_2)) = v_1 + v_2$$
$$= f^{-1}(w_1) + f^{-1}(w_2).$$

Moreover, for every $\lambda \in K$ we have

$$f^{-1}(\lambda w_1) = f^{-1}(\lambda f(v_1)) = f^{-1}(f(\lambda v_1)) = \lambda v_1 = \lambda f^{-1}(w_1).$$

(3) and (4) are obvious from the corresponding definitions.

(5) Let f be injective and $v \in \ker(f)$, i.e., $f(v) = 0$. From (1) we know that $f(0) = 0$. Since $f(v) = f(0)$, the injectivity of f yields $v = 0$. Suppose now that $\ker(f) = \{0\}$ and let $u, v \in \mathcal{V}$ with $f(u) = f(v)$. Then $f(u - v) = 0$, i.e., $u - v \in \ker(f)$, which implies $u - v = 0$, i.e., $u = v$.

(6) Let $\sum_{i=1}^{m} \lambda_i f(v_i) = 0$. The linearity of f yields

$$f\left(\sum_{i=1}^{m} \lambda_i v_i\right) = 0, \quad \text{i.e.,} \quad \sum_{i=1}^{m} \lambda_i v_i \in \ker(f).$$

Since f is injective, we have $\sum_{i=1}^{m} \lambda_i v_i = 0$ by (5), and hence $\lambda_1 = \cdots = \lambda_m = 0$ due to the linear independence of $v_1, \ldots, v_m$. Thus, $f(v_1), \ldots, f(v_m)$ are linearly independent.

(7) If $v_1, \ldots, v_m$ are linearly dependent, then $\sum_{i=1}^{m} \lambda_i v_i = 0$ for some $\lambda_1, \ldots, \lambda_m \in K$ that are not all equal to zero. Applying f on both sides and using the linearity yields $\sum_{i=1}^{m} \lambda_i f(v_i) = 0$, hence $f(v_1), \ldots, f(v_m)$ are linearly dependent.

(8) Let $w \in \text{im}(f)$ and $u \in f^{-1}(w)$.

If $v \in f^{-1}(w)$, then $f(v) = f(u)$, and thus $f(v-u) = 0$, i.e., $v-u \in \ker(f)$ or $v \in u + \ker(f)$. This shows that $f^{-1}(w) \subseteq u + \ker(f)$.

If, on the other hand, $v \in u + \ker(f)$, then $f(v) = f(u) = w$, i.e., $v \in f^{-1}(w)$. This shows that $u + \ker(f) \subseteq f^{-1}(w)$. □

Example 10.8 Consider a matrix $A \in K^{n,m}$ and the corresponding map $A \in \mathcal{L}(K^{m,1}, K^{n,1})$ from (1) in Example 10.2. For a given $b \in K^{n,1}$ we have

$$A^{-1}(b) = \{\widehat{x} \in K^{m,1} \mid A\widehat{x} = b\} = \mathscr{L}(A, b).$$

If $b \notin \text{im}(A)$, then $\mathscr{L}(A, b) = \emptyset$ (case (1) in Theorem 6.5). Now suppose that $b \in \text{im}(A)$ and let $\widehat{x} \in \mathscr{L}(A, b)$ be arbitrary. Then (8) in Lemma 10.7 yields

$$\mathscr{L}(A, b) = \widehat{x} + \ker(A),$$

which is the assertion of Lemma 6.2. If $\ker(A) = \{0\}$, i.e., the columns of A are linearly independent, then $|\mathscr{L}(A, b)| = 1$ (case (2) in Theorem 6.5). If $\ker(A) \neq \{0\}$, i.e., the columns of A are linearly dependent, then $|\mathscr{L}(A, b)| > 1$ (case (3) in Theorem 6.5). If in this case $\{w_1, \ldots, w_\ell\}$ is a basis of $\ker(A)$, then

$$\mathscr{L}(A, b) = \left\{\widehat{x} + \sum_{i=1}^{\ell} \lambda_i w_i \,\Big|\, \lambda_1, \ldots, \lambda_\ell \in K\right\}.$$

Thus, the solutions of $Ax = b$ depend of $\ell \leq m$ parameters.

10.1 Basic Definitions and Properties

The following result, which gives an important dimension formula for linear maps, is also known as the *rank-nullity theorem*: The dimension of the image of f is equal to the rank of a matrix associated with f (cp. Theorem 10.23 below), and the dimension of the kernel (or null space) of f is sometimes called the *nullity*[1] of f.

Theorem 10.9 *Let $\mathcal{V}$ and $\mathcal{W}$ be K-vector spaces and let $\mathcal{V}$ be finite dimensional. Then for every $f \in \mathcal{L}(\mathcal{V}, \mathcal{W})$ we have the dimension formula*

$$\dim(\mathcal{V}) = \dim(\operatorname{im}(f)) + \dim(\ker(f)).$$

Proof Let $v_1, \ldots, v_n \in \mathcal{V}$. If $f(v_1), \ldots, f(v_n) \in \mathcal{W}$ are linearly independent, then by (7) in Lemma 10.7 also $v_1, \ldots, v_n$ are linearly independent, and thus $\dim(\operatorname{im}(f)) \leq \dim(\mathcal{V})$. Since $\ker(f) \subseteq \mathcal{V}$, we have $\dim(\ker(f)) \leq \dim(\mathcal{V})$, so that $\operatorname{im}(f)$ and $\ker(f)$ are both finite dimensional.

Let $\{w_1, \ldots, w_r\}$ and $\{v_1, \ldots, v_k\}$ be bases of $\operatorname{im}(f)$ and $\ker(f)$, respectively, and let $u_1 \in f^{-1}(w_1), \ldots, u_r \in f^{-1}(w_r)$. We will show that $\{u_1, \ldots, u_r, v_1, \ldots, v_k\}$ is a basis of $\mathcal{V}$, which then implies the assertion.

If $v \in \mathcal{V}$, then by Lemma 9.27 there exist (unique) coordinates $\mu_1, \ldots, \mu_r \in K$ with $f(v) = \sum_{i=1}^{r} \mu_i w_i$. Let $\widetilde{v} := \sum_{i=1}^{r} \mu_i u_i$, then $f(\widetilde{v}) = f(v)$, and hence $v - \widetilde{v} \in \ker(f)$, which gives $v - \widetilde{v} = \sum_{i=1}^{k} \lambda_i v_i$ for some (unique) coordinates $\lambda_1, \ldots, \lambda_k \in K$. Therefore,

$$v = \widetilde{v} + \sum_{i=1}^{k} \lambda_i v_i = \sum_{i=1}^{r} \mu_i u_i + \sum_{i=1}^{k} \lambda_i v_i,$$

and thus $v \in \operatorname{span}\{u_1, \ldots, u_r, v_1, \ldots, v_k\}$. Since $\{u_1, \ldots, u_r, v_1, \ldots, v_k\} \subset \mathcal{V}$, we have

$$\mathcal{V} = \operatorname{span}\{u_1, \ldots, u_r, v_1, \ldots, v_k\},$$

and it remains to show that $u_1, \ldots, u_r, v_1, \ldots, v_k$ are linearly independent. If

$$\sum_{i=1}^{r} \alpha_i u_i + \sum_{i=1}^{k} \beta_i v_i = 0,$$

then

$$0 = f(0) = f\left(\sum_{i=1}^{r} \alpha_i u_i + \sum_{i=1}^{k} \beta_i v_i\right) = \sum_{i=1}^{r} \alpha_i f(u_i) = \sum_{i=1}^{r} \alpha_i w_i$$

---

[1] This term was introduced in 1884 by James Joseph Sylvester (1814–1897).

and thus $\alpha_1 = \cdots = \alpha_r = 0$, because $w_1, \ldots, w_r$ are linearly independent. Finally, the linear independence of $v_1, \ldots, v_k$ implies that $\beta_1 = \cdots = \beta_k = 0$. □

Example 10.10

(1) For the linear map

$$f : \mathbb{Q}^{3,1} \to \mathbb{Q}^{2,1}, \quad \begin{bmatrix} \alpha_1 \\ \alpha_2 \\ \alpha_3 \end{bmatrix} \mapsto \begin{bmatrix} 1 & 0 & 1 \\ 1 & 0 & 1 \end{bmatrix} \begin{bmatrix} \alpha_1 \\ \alpha_2 \\ \alpha_3 \end{bmatrix} = \begin{bmatrix} \alpha_1 + \alpha_3 \\ \alpha_1 + \alpha_3 \end{bmatrix},$$

we have

$$\operatorname{im}(f) = \left\{ \begin{bmatrix} \alpha \\ \alpha \end{bmatrix} \,\middle|\, \alpha \in \mathbb{Q} \right\}, \quad \ker(f) = \left\{ \begin{bmatrix} \alpha_1 \\ \alpha_2 \\ -\alpha_1 \end{bmatrix} \,\middle|\, \alpha_1, \alpha_2 \in \mathbb{Q} \right\}.$$

Hence $\dim(\operatorname{im}(f)) = 1$ and $\dim(\ker(f)) = 2$, so that indeed $\dim(\operatorname{im}(f)) + \dim(\ker(f)) = \dim(\mathbb{Q}^{3,1})$.

(2) If $A \in K^{n,m}$ and $A \in \mathcal{L}(K^{m,1}, K^{n,1})$ are as in (1) in Example 10.2, then

$$m = \dim(K^{m,1}) = \dim(\ker(A)) + \dim(\operatorname{im}(A)).$$

Thus, $\dim(\operatorname{im}(A)) = m$ if and only if $\dim(\ker(A)) = 0$. This holds if and only if $\ker(A) = \{0\}$, i.e., if and only if the columns of A are linearly independent (cp. Example 10.6). If, on the other hand, $\dim(\operatorname{im}(A)) < m$, then $\dim(\ker(A)) = m - \dim(\operatorname{im}(A)) > 0$, and thus $\ker(A) \neq \{0\}$. In this case the columns of A are linearly dependent, since there exists an $x \in K^{m,1} \setminus \{0\}$ with $Ax = 0$.

Corollary 10.11 *If V and W are K-vector spaces with $\dim(V) = \dim(W) \in \mathbb{N}$ and if $f \in \mathcal{L}(V, W)$, then the following are equivalent:*

(1) *f is injective.*
(2) *f is surjective.*
(3) *f is bijective.*

Proof If (3) holds, then (1) and (2) hold by definition. We now show that (3) is implied by (1) as well as by (2).

If f is injective, then $\ker(f) = \{0\}$ (cp. (5) in Lemma 10.7) and the dimension formula of Theorem 10.9 yields $\dim(W) = \dim(V) = \dim(\operatorname{im}(f))$. Thus, $\operatorname{im}(f) = W$ (cp. Lemma 9.32), so that f is also surjective.

If f is surjective, i.e., $\operatorname{im}(f) = \mathcal{W}$, then the dimension formula and $\dim(\mathcal{W}) = \dim(\mathcal{V})$ yield

$$\dim(\ker(f)) = \dim(\mathcal{V}) - \dim(\operatorname{im}(f)) = \dim(\mathcal{W}) - \dim(\operatorname{im}(f)) = 0.$$

Thus, $\ker(f) = \{0\}$, so that f is also injective. $\square$

Using Theorem 10.9 we can also characterize when two finite dimensional vector spaces are isomorphic.

Corollary 10.12 *Two finite dimensional K-vector spaces $\mathcal{V}$ and $\mathcal{W}$ are isomorphic if and only if $\dim(\mathcal{V}) = \dim(\mathcal{W})$.*

Proof If $\mathcal{V} \cong \mathcal{W}$, then there exists a bijective map $f \in \mathcal{L}(\mathcal{V}, \mathcal{W})$. By (4) and (5) in Lemma 10.7 we have $\operatorname{im}(f) = \mathcal{W}$ and $\ker(f) = \{0\}$, and the dimension formula of Theorem 10.9 yields

$$\dim(\mathcal{V}) = \dim(\operatorname{im}(f)) + \dim(\ker(f)) = \dim(\mathcal{W}) + \dim(\{0\}) = \dim(\mathcal{W}).$$

Let now $\dim(\mathcal{V}) = \dim(\mathcal{W})$. We need to show that there exists a bijective $f \in \mathcal{L}(\mathcal{V}, \mathcal{W})$. Let $\{v_1, \ldots, v_n\}$ and $\{w_1, \ldots, w_n\}$ be bases of $\mathcal{V}$ and $\mathcal{W}$. By Theorem 10.4 there exists a unique $f \in \mathcal{L}(\mathcal{V}, \mathcal{W})$ with $f(v_i) = w_i$, $i = 1, \ldots, n$. If $v = \lambda_1 v_1 + \ldots + \lambda_n v_n \in \ker(f)$, then

$$0 = f(v) = f(\lambda_1 v_1 + \ldots + \lambda_n v_n) = \lambda_1 f(v_1) + \ldots + \lambda_n f(v_n)$$
$$= \lambda_1 w_1 + \ldots + \lambda_n w_n.$$

Since $w_1, \ldots, w_n$ are linearly independent, we have $\lambda_1 = \cdots = \lambda_n = 0$, hence $v = 0$ and $\ker(f) = \{0\}$. Thus, f is injective. Moreover, the dimension formula yields $\dim(\mathcal{V}) = \dim(\operatorname{im}(f)) = \dim(\mathcal{W})$ and, therefore, $\operatorname{im}(f) = \mathcal{W}$ (cp. Lemma 9.32), so that f is also surjective. $\square$

Example 10.13

(1) The vector spaces $K^{n,m}$ and $K^{m,n}$ both have the dimension $n \cdot m$ and are therefore isomorphic. An isomorphism is given by the linear map $A \mapsto A^T$.
(2) The $\mathbb{R}$-vector spaces $\mathbb{R}^{1,2}$ and $\mathbb{C} = \{x + \mathbf{i}y \mid x, y \in \mathbb{R}\}$ both have the dimension 2 and are therefore isomorphic. An isomorphism is given by the linear map $[x, y] \mapsto x + \mathbf{i}y$.
(3) The vector spaces $\mathbb{Q}[t]_{\leq 2}$ and $\mathbb{Q}^{1,3}$ both have dimension 3 and are therefore isomorphic. An isomorphism is given by the linear map $\alpha_2 t^2 + \alpha_1 t + \alpha_0 \mapsto [\alpha_2, \alpha_1, \alpha_0]$.

Although Mathematics is a formal and exact science, where smallest details matter, one sometimes uses an "abuse of notation" in order to simplify the presentation. We have used this for example in the inductive existence proof of the echelon form in Theorem 5.2. There we kept, for simplicity, the indices of the larger matrix $A^{(1)}$ in the smaller matrix $A^{(2)} = [a_{ij}^{(2)}]$. The matrix $A^{(2)}$ had, of course, an entry in position $(1, 1)$, but this entry was denoted by $a_{22}^{(2)}$ rather than $a_{11}^{(2)}$. Keeping the indices in the induction made the argument much less technical, while the proof itself remained formally correct. Another example is using the same notation for a given matrix $A \in K^{n,m}$ and the corresponding linear map $A \in \mathcal{L}(K^{n,1}, K^{m,1})$. We could have denoted the linear map differently, e.g. by f_A, but using the same notation is simpler and not misleading.

An abuse of notation should always be justified and should not be confused with a "misuse" of notation. In the field of Linear Algebra a justification is often given by an isomorphism that identifies vector spaces with each other. For example, the constant polynomials over a field K, i.e., polynomials of the form αt^0 with $\alpha \in K$, are often written simply as α, i.e., as elements of the field itself. This is justified since $K[t]_{\leq 0}$ and K are isomorphic K-vector spaces (of dimension 1). We already used this identification above. Similarly, we have identified the vector space $\mathcal{V}$ with $\mathcal{V}^1$ and written just v instead of (v) in Sect. 9.3. Another common example in the literature is the notation K^n that in our text denotes the set of n-tuples with elements from K, but which is often used for the (matrix) sets of the "column vectors" $K^{n,1}$ or the "row vectors" $K^{1,n}$. The actual meaning then should be clear from the context. An attentive reader can significantly benefit from the simplifications due to such abuses of notation.

The dimension formula for linear maps in Theorem 10.9 can be generalized to the composition of maps as follows.

Theorem 10.14 *If $\mathcal{V}, \mathcal{W}, \mathcal{X}$ are K-vector spaces, and $f \in \mathcal{L}(\mathcal{V}, \mathcal{W})$ and $g \in \mathcal{L}(\mathcal{W}, \mathcal{X})$, then $g \circ f \in \mathcal{L}(\mathcal{V}, \mathcal{X})$. Moreover, if $\mathrm{im}(f) \subseteq \mathcal{W}$ is finite dimensional, then*

$$\dim(\mathrm{im}(g \circ f)) = \dim(\mathrm{im}(f)) - \dim(\mathrm{im}(f) \cap \ker(g)).$$

Proof We first show that $g \circ f \in \mathcal{L}(\mathcal{V}, \mathcal{X})$. For any $u, v \in \mathcal{V}$ and $\lambda, \mu \in K$ we have

$$(g \circ f)(\lambda u + \mu v) = g(f(\lambda u + \mu v)) = g(\lambda f(u) + \mu f(v))$$
$$= \lambda g(f(u)) + \mu g(f(v)) = \lambda (g \circ f)(u) + \mu (g \circ f)(v).$$

Now let $\mathrm{im}(f) \subseteq \mathcal{W}$ be finite dimensional, and let $\widetilde{g} := g|_{\mathrm{im}(f)}$ be the restriction of g to the image of f, i.e., the map

$$\widetilde{g} \in \mathcal{L}(\mathrm{im}(f), \mathcal{X}), \quad v \mapsto g(v).$$

Applying Theorem 10.9 to $\widetilde{g}$ yields

$$\dim(\operatorname{im}(f)) = \dim(\operatorname{im}(\widetilde{g})) + \dim(\ker(\widetilde{g})).$$

Now

$$\operatorname{im}(\widetilde{g}) = \{g(v) \in \mathcal{X} \mid v \in \operatorname{im}(f)\} = \operatorname{im}(g \circ f)$$

and

$$\ker(\widetilde{g}) = \{v \in \operatorname{im}(f) \mid \widetilde{g}(v) = 0\} = \operatorname{im}(f) \cap \ker(g),$$

imply the assertion. □

Note that Theorem 10.23 with $\mathcal{V} = \mathcal{W}$, $f = \operatorname{Id}_\mathcal{V}$, and $g \in \mathcal{L}(\mathcal{V}, \mathcal{X})$ gives $\dim(\operatorname{im}(g)) = \dim(\mathcal{V}) - \dim(\ker(g))$, which is equivalent to Theorem 10.9.

10.2 Linear Maps and Matrices

Let $\mathcal{V}$ and $\mathcal{W}$ be finite dimensional K-vector spaces with bases $\{v_1, \ldots, v_m\}$ and $\{w_1, \ldots, w_n\}$, respectively, and let $f \in \mathcal{L}(\mathcal{V}, \mathcal{W})$. By Lemma 9.27, for every $f(v_j) \in \mathcal{W}$, $j = 1, \ldots, m$, there exist (unique) coordinates $a_{ij} \in K$, $i = 1, \ldots, n$, with

$$f(v_j) = a_{1j} w_1 + \ldots + a_{nj} w_n.$$

We define $A := [a_{ij}] \in K^{n,m}$ and write, similarly to (9.3), the m equations for the vectors $f(v_j)$ as

$$(f(v_1), \ldots, f(v_m)) = (w_1, \ldots, w_n) A. \tag{10.1}$$

The matrix A is determined uniquely by f and the given bases of $\mathcal{V}$ and $\mathcal{W}$.

If $v = \lambda_1 v_1 + \ldots + \lambda_m v_m \in \mathcal{V}$, then

$$f(v) = f(\lambda_1 v_1 + \ldots + \lambda_m v_m) = \lambda_1 f(v_1) + \ldots + \lambda_m f(v_m)$$

$$= (f(v_1), \ldots, f(v_m)) \begin{bmatrix} \lambda_1 \\ \vdots \\ \lambda_m \end{bmatrix}$$

$$= ((w_1, \ldots, w_n) A) \begin{bmatrix} \lambda_1 \\ \vdots \\ \lambda_m \end{bmatrix}$$

$$= (w_1, \ldots, w_n) \left(A \begin{bmatrix} \lambda_1 \\ \vdots \\ \lambda_m \end{bmatrix} \right).$$

The coordinates of $f(v)$ with respect to the given basis of $\mathcal{W}$ are therefore given by

$$A \begin{bmatrix} \lambda_1 \\ \vdots \\ \lambda_m \end{bmatrix}.$$

Thus, we can compute the coordinates of $f(v)$ simply by multiplying the coordinates of v with A. This motivates the following definition.

Definition 10.15 The uniquely determined matrix in (10.1) is called the *matrix representation* of $f \in \mathcal{L}(\mathcal{V}, \mathcal{W})$ with respect to the bases $B_1 = \{v_1, \ldots, v_m\}$ of $\mathcal{V}$ and $B_2 = \{w_1, \ldots, w_n\}$ of $\mathcal{W}$. We denote this matrix by $[f]_{B_1, B_2}$.

The construction of the matrix representation and Definition 10.15 can be consistently extended to the case that (at least) one of the K-vector spaces has dimension zero. If, for instance, $m = \dim(\mathcal{V}) \in \mathbb{N}$ and $\mathcal{W} = \{0\}$, then $f(v_j) = 0$ for every basis vector v_j of $\mathcal{V}$. Thus, every vector $f(v_j)$ is an empty linear combination of vector of the basis $\emptyset$ of $\mathcal{W}$. The matrix representation of f then is an empty matrix of size $0 \times m$. If also $\mathcal{V} = \{0\}$, then the matrix representation of f is an empty matrix of size 0×0.

There are many different notations for the matrix representation of linear maps in the literature. The notation should reflect that the matrix depends on the linear map f and the given bases B_1 and B_2. Examples of alternative notations are $[f]_{B_2}^{B_1}$ and $M(f)_{B_1, B_2}$ (where "M" means "matrix").

Example 10.16

(1) For a given matrix $A = [a_{ij}] \in K^{n,m}$, consider $f \in \mathcal{L}(K^{m,1}, K^{n,1})$ with $x \mapsto Ax$. If $B_1 = \{e_1^{(m)}, \ldots, e_m^{(m)}\}$ and $B_2 = \{e_1^{(n)}, \ldots, e_n^{(n)}\}$ are the standard bases of $K^{m,1}$, respectively $K^{n,1}$ (cp. (1) in Example 9.14), then for $j = 1, \ldots, m$ we have

$$f(e_j^{(m)}) = A e_j^{(m)} = \begin{bmatrix} a_{1j} \\ \vdots \\ a_{nj} \end{bmatrix} = \sum_{i=1}^{n} a_{ij} e_i^{(n)},$$

i.e., $[f]_{B_1, B_2} = A$. The matrix representation of the map $x \mapsto Ax$ with respect to the standard bases is therefore the matrix A itself.

(2) Consider the vector space $\mathbb{Q}[t]_{\leq 1}$ with the bases $B_1 = \{1, t\}$ and $B_2 = \{t + 1, t - 1\}$. Then the linear map

$$f : \mathbb{Q}[t]_{\leq 1} \to \mathbb{Q}[t]_{\leq 1}, \quad \alpha_1 t + \alpha_0 \mapsto 2\alpha_1 t + \alpha_0,$$

has the matrix representations

$$[f]_{B_1,B_1} = \begin{bmatrix} 1 & 0 \\ 0 & 2 \end{bmatrix}, \quad [f]_{B_1,B_2} = \begin{bmatrix} \frac{1}{2} & 1 \\ -\frac{1}{2} & 1 \end{bmatrix},$$

$$[f]_{B_2,B_2} = \begin{bmatrix} \frac{3}{2} & \frac{1}{2} \\ \frac{1}{2} & \frac{3}{2} \end{bmatrix}, \quad [f]_{B_2,B_1} = \begin{bmatrix} 1 & -1 \\ 2 & 2 \end{bmatrix}.$$

Theorem 10.17 *Let $\mathcal{V}$ and $\mathcal{W}$ be finite dimensional K-vector spaces with bases B_1 and B_2, respectively. Then the map*

$$\mathcal{L}(\mathcal{V}, \mathcal{W}) \to K^{n,m}, \quad f \mapsto [f]_{B_1,B_2},$$

is an isomorphism. Hence $\mathcal{L}(\mathcal{V}, \mathcal{W}) \cong K^{n,m}$ and $\dim(\mathcal{L}(\mathcal{V}, \mathcal{W})) = \dim(K^{n,m}) = n \cdot m$.

Proof We denote the map $f \mapsto [f]_{B_1,B_2}$ by φ, i.e., $\varphi(f) = [f]_{B_1,B_2}$. We first show that this map is linear. Let Let $B_1 = \{v_1, \ldots, v_m\}$, $B_2 = \{w_1, \ldots, w_n\}$, $f, g \in \mathcal{L}(\mathcal{V}, \mathcal{W})$, $\varphi(f) = [f_{ij}]$, $\varphi(g) = [g_{ij}]$, and $\lambda, \mu \in K$. For $j = 1, \ldots, m$ we have

$$(\lambda f + \mu g)(v_j) = \lambda f(v_j) + \mu g(v_j) = \lambda \sum_{i=1}^{n} f_{ij} w_i + \mu \sum_{i=1}^{n} g_{ij} w_i$$

$$= \sum_{i=1}^{n} (\lambda f_{ij} + \mu g_{ij}) w_i$$

and hence $\varphi(\lambda f + \mu g) = [\lambda f_{ij} + \mu g_{ij}] = \lambda [f_{ij}] + \mu [g_{ij}] = \lambda \varphi(f) + \mu \varphi(g)$.

It remains to show that φ is bijective. If $f \in \ker(\varphi)$, i.e., $\varphi(f) = 0 \in K^{n,m}$, then $f(v_j) = 0$ for $j = 1, \ldots, m$. Thus, $f(v) = 0$ for all $v \in \mathcal{V}$, so that $f = 0$ (the zero map) and φ is injective (cp. (5) in Lemma 10.7). If, on the other hand, $A = [a_{ij}] \in K^{n,m}$ is arbitrary, then according to Theorem 10.4 there exists exactly one linear map $f \in \mathcal{L}(\mathcal{V}, \mathcal{W})$ with $f(v_j) = \sum_{i=1}^{n} a_{ij} w_i$, $j = 1, \ldots, m$. For this map we have $\varphi(f) = A$, and hence φ is also surjective (cp. (4) in Lemma 10.7).

Corollary 10.12 now shows that $\dim(\mathcal{L}(\mathcal{V}, \mathcal{W})) = \dim(K^{n,m}) = n \cdot m$ (cp. also (1) in Example 9.23). □

Theorem 10.17 shows, in particular, that $f, g \in \mathcal{L}(\mathcal{V}, \mathcal{W})$ satisfy $f = g$ if and only if $[f]_{B_1, B_2} = [g]_{B_1, B_2}$ holds for given bases B_1 of $\mathcal{V}$ and B_2 of $\mathcal{W}$. Thus, we can prove the equality of linear maps via the equality of their matrix representations.

We now consider the map from the elements of a finite dimensional vector space to their coordinates with respect to a given basis.

Lemma 10.18 *If $B = \{v_1, \ldots, v_n\}$ is a basis of a K-vector space $\mathcal{V}$, then the map*

$$\Phi_B : \mathcal{V} \to K^{n,1}, \quad v = \lambda_1 v_1 + \ldots + \lambda_n v_n \mapsto \Phi_B(v) := \begin{bmatrix} \lambda_1 \\ \vdots \\ \lambda_n \end{bmatrix},$$

is an isomorphism, called the coordinate map of $\mathcal{V}$ with respect to the basis B.

Proof The linearity of Φ_B is clear. Moreover, we obviously have $\Phi_B(\mathcal{V}) = K^{n,1}$, i.e., Φ_B is surjective. If $v \in \ker(\Phi_B)$, i.e., $\lambda_1 = \cdots = \lambda_n = 0$, then $v = 0$, so that $\ker(\Phi_B) = \{0\}$ and Φ_B is also injective (cp. (5) in Lemma 10.7). $\square$

Example 10.19 In the vector space $K[t]_{\leq n}$ with the basis $B = \{t^0, t^1, \ldots, t^n\}$ we have

$$\Phi_B(\alpha_n t^n + \alpha_{n-1} t^{n-1} + \ldots + \alpha_1 t + \alpha_0) = \begin{bmatrix} \alpha_0 \\ \alpha_1 \\ \vdots \\ \alpha_n \end{bmatrix} \in K^{n+1,1}.$$

For the linear map

$$f : K[t]_{\leq n} \to K[t]_{\leq n},$$
$$\alpha_n t^n + \alpha_{n-1} t^{n-1} + \ldots + \alpha_1 t + \alpha_0 \mapsto \alpha_0 t^n + \alpha_1 t^{n-1} + \ldots + \alpha_{n-1} t + \alpha_n,$$

we obtain

$$[f]_{B,B} = \begin{bmatrix} & & 1 \\ & \cdot^{\cdot^{\cdot}} & \\ 1 & & \end{bmatrix} \in K^{n+1, n+1}.$$

Thus, the matrix $[f]_{B,B}$ is a permuation matrix.

10.2 Linear Maps and Matrices

If B_1 and B_2 are bases of the finite dimensional vector spaces V and W, respectively, then we can illustrate the meaning and the construction of the matrix representation $[f]_{B_1,B_2}$ of $f \in \mathcal{L}(V,W)$ in the following *commutative diagram*:

$$\begin{array}{ccc} V & \xrightarrow{f} & W \\ \Phi_{B_1} \downarrow & & \downarrow \Phi_{B_2} \\ K^{m,1} & \xrightarrow{[f]_{B_1,B_2}} & K^{n,1} \end{array} \qquad (10.2)$$

We see that different compositions of maps yield the same result. In particular, we have

$$f = \Phi_{B_2}^{-1} \circ [f]_{B_1,B_2} \circ \Phi_{B_1}, \qquad (10.3)$$

where the matrix $[f]_{B_1,B_2} \in K^{n,m}$ is interpreted as a linear map from $K^{m,1}$ to $K^{n,1}$, and we use that the coordinate map Φ_{B_2} is bijective and hence invertible. In the same way we obtain

$$\Phi_{B_2} \circ f = [f]_{B_1,B_2} \circ \Phi_{B_1},$$

i.e.,

$$\Phi_{B_2}(f(v)) = [f]_{B_1,B_2} \Phi_{B_1}(v) \quad \text{for all } v \in V. \qquad (10.4)$$

In words, the coordinates of $f(v)$ with respect to the basis B_2 of W are given by the product of $[f]_{B_1,B_2}$ and the coordinates of v with respect to the basis B_1 of V.

An important special case is obtained for $V = W$, hence in particular $m = n$, and $f = \mathrm{Id}_V$, the identity on V. We then obtain

$$(v_1, \ldots, v_n) = (w_1, \ldots, w_n)[\mathrm{Id}_V]_{B_1,B_2}, \qquad (10.5)$$

so that $[\mathrm{Id}_V]_{B_1,B_2}$ is exactly the matrix P in (9.4), i.e., the coordinate transformation matrix in Theorem 9.30. On the other hand,

$$(w_1, \ldots, w_n) = (v_1, \ldots, v_n)\,[\mathrm{Id}_V]_{B_2,B_1},$$

and thus

$$\left([\mathrm{Id}_V]_{B_1,B_2}\right)^{-1} = [\mathrm{Id}_V]_{B_2,B_1}.$$

We next show that the consecutive application of linear maps corresponds to the multiplication of their matrix representations.

Theorem 10.20 Let V, W, X be finite dimensional K-vector spaces with respective bases B_1, B_2, B_3. If $f \in \mathcal{L}(V, W)$ and $g \in \mathcal{L}(W, X)$, then

$$[g \circ f]_{B_1, B_3} = [g]_{B_2, B_3} [f]_{B_1, B_2}.$$

Proof We know from Theorem 10.14 that $h := g \circ f \in \mathcal{L}(V, X)$. Now let $B_1 = \{v_1, \ldots, v_m\}$, $B_2 = \{w_1, \ldots, w_n\}$ and $B_3 = \{x_1, \ldots, x_s\}$. If $[f]_{B_1, B_2} = [f_{ij}]$ and $[g]_{B_2, B_3} = [g_{ij}]$, then for $j = 1, \ldots, m$ we have

$$h(v_j) = g(f(v_j)) = g\left(\sum_{k=1}^{n} f_{kj} w_k\right) = \sum_{k=1}^{n} f_{kj} g(w_k) = \sum_{k=1}^{n} f_{kj} \sum_{i=1}^{s} g_{ik} x_i$$

$$= \sum_{i=1}^{s} \left(\sum_{k=1}^{n} f_{kj} g_{ik}\right) x_i = \sum_{i=1}^{s} \underbrace{\left(\sum_{k=1}^{n} g_{ik} f_{kj}\right)}_{=: h_{ij}} x_i.$$

By definition of the matrix representation we have $[h]_{B_1, B_3} = [h_{ij}]$, and hence $[h]_{B_1, B_3} = [g_{ij}][f_{ij}] = [g]_{B_2, B_3} [f]_{B_1, B_2}$. □

Using this theorem we can study how a change of the bases affects the matrix representation of a linear map.

Corollary 10.21 Let V and W be finite dimensional K-vector spaces with bases $B_1, \widetilde{B}_1$ of V and $B_2, \widetilde{B}_2$ of W. If $f \in \mathcal{L}(V, W)$, then

$$[f]_{B_1, B_2} = [\mathrm{Id}_W]_{\widetilde{B}_2, B_2} [f]_{\widetilde{B}_1, \widetilde{B}_2} [\mathrm{Id}_V]_{B_1, \widetilde{B}_1}. \tag{10.6}$$

In particular, the matrices $[f]_{B_1, B_2}$ and $[f]_{\widetilde{B}_1, \widetilde{B}_2}$ are equivalent.

Proof Applying Theorem 10.20 twice to the identity $f = \mathrm{Id}_W \circ f \circ \mathrm{Id}_V$ yields

$$[f]_{B_1, B_2} = [(\mathrm{Id}_W \circ f) \circ \mathrm{Id}_V]_{B_1, B_2}$$
$$= [\mathrm{Id}_W \circ f]_{\widetilde{B}_1, B_2} [\mathrm{Id}_V]_{B_1, \widetilde{B}_1}$$
$$= [\mathrm{Id}_W]_{\widetilde{B}_2, B_2} [f]_{\widetilde{B}_1, \widetilde{B}_2} [\mathrm{Id}_V]_{B_1, \widetilde{B}_1}.$$

The matrices $[f]_{B_1, B_2}$ and $[f]_{\widetilde{B}_1, \widetilde{B}_2}$ are equivalent, since both $[\mathrm{Id}_W]_{\widetilde{B}_2, B_2}$ and $[\mathrm{Id}_V]_{B_1, \widetilde{B}_1}$ are invertible. □

If $V = W$, $B_1 = B_2$, and $\widetilde{B}_1 = \widetilde{B}_2$, then (10.6) becomes

$$[f]_{B_1, B_1} = [\mathrm{Id}_V]_{\widetilde{B}_1, B_1} [f]_{\widetilde{B}_1, \widetilde{B}_1} [\mathrm{Id}_V]_{B_1, \widetilde{B}_1} = ([\mathrm{Id}_V]_{B_1, \widetilde{B}_1})^{-1} [f]_{\widetilde{B}_1, \widetilde{B}_1} [\mathrm{Id}_V]_{B_1, \widetilde{B}_1}.$$

10.2 Linear Maps and Matrices

Thus, the matrix representations $[f]_{B_1,B_1}$ and $[f]_{\tilde{B}_1,\tilde{B}_1}$ of the endomorphism $f \in \mathcal{L}(\mathcal{V}, \mathcal{V})$ are *similar* (cp. Definition 8.10).

The following commutative diagram illustrates Corollary 10.21:

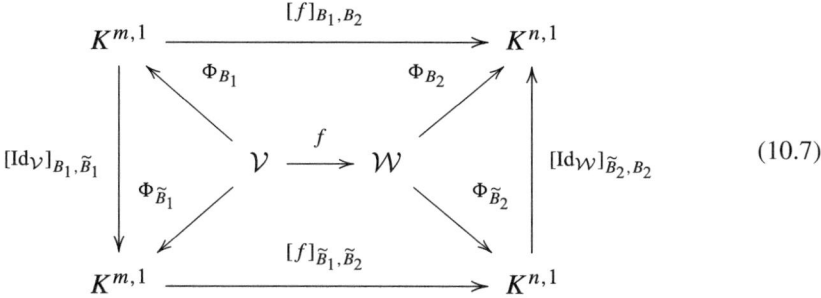
(10.7)

Analogously to (10.3) we have

$$f = \Phi_{B_2}^{-1} \circ [f]_{B_1,B_2} \circ \Phi_{B_1} = \Phi_{\tilde{B}_2}^{-1} \circ [f]_{\tilde{B}_1,\tilde{B}_2} \circ \Phi_{\tilde{B}_1}.$$

Example 10.22 For the following bases of the vector space $\mathcal{V} = \mathbb{Q}^{2,2}$,

$$B_1 = \left\{ \begin{bmatrix} 1 & 0 \\ 0 & 0 \end{bmatrix}, \begin{bmatrix} 0 & 1 \\ 0 & 0 \end{bmatrix}, \begin{bmatrix} 0 & 0 \\ 1 & 0 \end{bmatrix}, \begin{bmatrix} 0 & 0 \\ 0 & 1 \end{bmatrix} \right\},$$

$$B_2 = \left\{ \begin{bmatrix} 1 & 0 \\ 0 & 1 \end{bmatrix}, \begin{bmatrix} 1 & 0 \\ 0 & 0 \end{bmatrix}, \begin{bmatrix} 1 & 1 \\ 0 & 0 \end{bmatrix}, \begin{bmatrix} 0 & 0 \\ 1 & 0 \end{bmatrix} \right\},$$

we have the coordinate transformation matrices

$$[\mathrm{Id}_{\mathcal{V}}]_{B_1,B_2} = \begin{bmatrix} 0 & 0 & 0 & 1 \\ 1 & -1 & 0 & -1 \\ 0 & 1 & 0 & 0 \\ 0 & 0 & 1 & 0 \end{bmatrix}$$

and

$$[\mathrm{Id}_{\mathcal{V}}]_{B_2,B_1} = ([\mathrm{Id}_{\mathcal{V}}]_{B_1,B_2})^{-1} = \begin{bmatrix} 1 & 1 & 1 & 0 \\ 0 & 0 & 1 & 0 \\ 0 & 0 & 0 & 1 \\ 1 & 0 & 0 & 0 \end{bmatrix}.$$

The coordinate maps are

$$\Phi_{B_1}\left(\begin{bmatrix} a_{11} & a_{12} \\ a_{21} & a_{22} \end{bmatrix}\right) = \begin{bmatrix} a_{11} \\ a_{12} \\ a_{21} \\ a_{22} \end{bmatrix}, \quad \Phi_{B_2}\left(\begin{bmatrix} a_{11} & a_{12} \\ a_{21} & a_{22} \end{bmatrix}\right) = \begin{bmatrix} a_{22} \\ a_{11} - a_{12} - a_{22} \\ a_{12} \\ a_{21} \end{bmatrix},$$

and we can easily verify that

$$\Phi_{B_2}\left(\begin{bmatrix} a_{11} & a_{12} \\ a_{21} & a_{22} \end{bmatrix}\right) = ([\mathrm{Id}_\mathcal{V}]_{B_1, B_2} \circ \Phi_{B_1})\left(\begin{bmatrix} a_{11} & a_{12} \\ a_{21} & a_{22} \end{bmatrix}\right).$$

Theorem 10.23 *Let $\mathcal{V}$ and $\mathcal{W}$ be K-vector spaces with $\dim(\mathcal{V}) = m$ and $\dim(\mathcal{W}) = n$, respectively, and let $f \in \mathcal{L}(\mathcal{V}, \mathcal{W})$. Then there exist bases B_1 of $\mathcal{V}$ and B_2 of $\mathcal{W}$ such that*

$$[f]_{B_1, B_2} = \begin{bmatrix} I_r & 0 \\ 0 & 0 \end{bmatrix} \in K^{n,m},$$

where $0 \leq r = \dim(\mathrm{im}(f)) \leq \min\{n, m\}$. Furthermore, $r = \mathrm{rank}(F)$, where F is the matrix representation of f with respect to arbitrary bases of $\mathcal{V}$ and $\mathcal{W}$, and we define $\mathrm{rank}(f) := \mathrm{rank}(F) = \dim(\mathrm{im}(f))$.

Proof Let $\widetilde{B}_1 = \{\widetilde{v}_1, \ldots, \widetilde{v}_m\}$ and $\widetilde{B}_2 = \{\widetilde{w}_1, \ldots, \widetilde{w}_n\}$ be two arbitrary bases of $\mathcal{V}$ and $\mathcal{W}$, respectively. Let $r := \mathrm{rank}([f]_{\widetilde{B}_1, \widetilde{B}_2})$. Then by Theorem 5.12 there exist invertible matrices $Q \in K^{n,n}$ and $Z \in K^{m,m}$ with

$$Q[f]_{\widetilde{B}_1, \widetilde{B}_2} Z = \begin{bmatrix} I_r & 0 \\ 0 & 0 \end{bmatrix}, \tag{10.8}$$

where $r = \mathrm{rank}([f]_{\widetilde{B}_1, \widetilde{B}_2}) \leq \min\{n, m\}$. Since Q and Z are invertible, we can define two new bases $B_1 = \{v_1, \ldots, v_m\}$ and $B_2 = \{w_1, \ldots, w_n\}$ of $\mathcal{V}$ and $\mathcal{W}$ via

$$(v_1, \ldots, v_m) := (\widetilde{v}_1, \ldots, \widetilde{v}_m) Z,$$

$$(w_1, \ldots, w_n) := (\widetilde{w}_1, \ldots, \widetilde{w}_n) Q^{-1}, \quad \text{hence} \quad (\widetilde{w}_1, \ldots, \widetilde{w}_n) = (w_1, \ldots, w_n) Q.$$

Then, by construction,

$$Z = [\mathrm{Id}_\mathcal{V}]_{B_1, \widetilde{B}_1} \quad \text{and} \quad Q = [\mathrm{Id}_\mathcal{W}]_{\widetilde{B}_2, B_2}.$$

From (10.8) and Corollary 10.21 we obtain

$$\begin{bmatrix} I_r & 0 \\ 0 & 0 \end{bmatrix} = [\mathrm{Id}_{\mathcal{W}}]_{\widetilde{B}_2, B_2} \, [f]_{\widetilde{B}_1, \widetilde{B}_2} \, [\mathrm{Id}_{\mathcal{V}}]_{B_1, \widetilde{B}_1} = [f]_{B_1, B_2} .$$

We thus have found bases B_1 and B_2 that yield the desired matrix representation of f. Every other choice of bases leads, by Corollary 10.21, to an equivalent matrix which therefore also has rank r. It remains to show that $r = \dim(\mathrm{im}(f))$.

The structure of the matrix $[f]_{B_1, B_2}$ shows that

$$f(v_j) = \begin{cases} w_j, & 1 \leq j \leq r, \\ 0, & r+1 \leq j \leq m. \end{cases}$$

Therefore, $v_{r+1}, \ldots, v_m \in \ker(f)$, which implies that $\dim(\ker(f)) \geq m-r$. On the other hand, $w_1, \ldots, w_r \in \mathrm{im}(f)$ and thus $\dim(\mathrm{im}(f)) \geq r$. Theorem 10.9 yields

$$\dim(\mathcal{V}) = m = \dim(\mathrm{im}(f)) + \dim(\ker(f)),$$

and hence $\dim(\ker(f)) = m - r$ and $\dim(\mathrm{im}(f)) = r$. □

Example 10.24 For $A \in K^{n,m}$ and the corresponding map $A \in \mathcal{L}(K^{m,1}, K^{n,1})$ we have $\mathrm{im}(A) = \mathrm{span}\{a_1, \ldots, a_m\}$ (cp. Example 10.6). Thus, $\mathrm{rank}(A) = \dim(\mathrm{im}(A))$ is equal to the maximal number of linearly independent columns of A. Since $\mathrm{rank}(A) = \mathrm{rank}(A^T)$ (cp. (5) in Theorem 5.12), this number is equal to the maximal number of linearly independent rows of A.

If we interpret matrices $A \in K^{n,m}$ and $B \in K^{s,n}$ as linear maps, then Theorem 10.14 implies the equation

$$\mathrm{rank}(BA) = \mathrm{rank}(A) - \dim(\mathrm{im}(A) \cap \ker(B)).$$

For the special case $K = \mathbb{Q}$ and $B = A^T$ we have the following result.

Corollary 10.25 *If $A \in \mathbb{Q}^{n,m}$, then $\mathrm{rank}(A^T A) = \mathrm{rank}(A)$.*

Proof Let $w = [\omega_1, \ldots, \omega_n]^T \in \mathrm{im}(A) \cap \ker(A^T)$. Then $w = Ay$ for a vector $y \in \mathbb{Q}^{m,1}$. Multiplying this equation from the left by A^T, and using that $w \in \ker(A^T)$, we obtain $0 = A^T w = A^T A y$, which implies

$$0 = y^T A^T A y = w^T w = \sum_{j=1}^n \omega_j^2.$$

Since this holds only for $w = 0$, we have $\mathrm{im}(A) \cap \ker(A^T) = \{0\}$. □

It is obvious from the proof of Corollary 10.25, that $\text{rank}(A^T A) = \text{rank}(A)$ also holds for each $A \in \mathbb{R}^{n,m}$.

Theorem 10.23 is a first example of a general strategy that we will use several times in the following chapters:

By choosing appropriate bases, the matrix representation should reveal a desired information about a linear map in an efficient way.

In Theorem 10.23 this information is the rank of the linear map f, i.e., the dimension of its image.

Exercises

(In the following exercises K is an arbitrary field.)

10.1 Consider the linear map on $\mathbb{R}^{3,1}$ given by the matrix $A = \begin{bmatrix} 2 & 0 & 1 \\ 2 & 1 & 0 \\ 4 & 1 & 1 \end{bmatrix} \in \mathbb{R}^{3,3}$.

Determine $\ker(A)$, $\dim(\ker(A))$ and $\dim(\text{im}(A))$.

10.2 Construct a map $f \in \mathcal{L}(\mathcal{V}, \mathcal{W})$ such that for linearly independent vectors $v_1, \ldots, v_r \in \mathcal{V}$ the images $f(v_1), \ldots, f(v_r) \in \mathcal{W}$ are linearly dependent.

10.3 The map

$$f : \mathbb{R}[t]_{\leq n} \to \mathbb{R}[t]_{\leq n-1},$$
$$\alpha_n t^n + \alpha_{n-1} t^{n-1} + \ldots + \alpha_1 t + \alpha_0 \mapsto n\alpha_n t^{n-1}$$
$$+ (n-1)\alpha_{n-1} t^{n-2} + \ldots + 2\alpha_2 t + \alpha_1,$$

is called the *derivative* of the polynomial $p \in \mathbb{R}[t]_{\leq n}$ with respect to the variable t. Show that f is linear and determine $\ker(f)$ and $\text{im}(f)$.

10.4 Construct a map $f \in \mathcal{L}(K[t], K[t])$ with the following properties:
(1) $f(pq) = (f(p))q + p(f(q))$ for all $p, q \in K[t]$.
(2) $f(t) = 1$.
Is this map uniquely determined by these properties or are there further maps with the same properties?

10.5 Let $\alpha \in K$ and $A \in K^{n,n}$. Show that the maps

$$K[t] \to K, \quad p \mapsto p(\alpha), \quad \text{and} \quad K[t] \to K^{m,m}, \quad p \mapsto p(A),$$

are linear and justify the name *evaluation homomorphism* for this map.

10.6 Let $A \in K^{n,n}$. Show that $f : K^{n,n} \to K^{n,n}$, $B \mapsto AB - BA$ is linear. Determine $\dim(\text{im}(f))$ and $\dim(\ker(f))$ for $A = \begin{bmatrix} 0 & 1 \\ 0 & 0 \end{bmatrix}$.

10.7 Let $A \in K^{n,n}$ with $n \geq 2$. Show the following assertions:
   (a) If $\text{rank}(A) = n - 1$, then there exists a minor $A(j, i) \in K^{n-1,n-1}$ with $\text{rank}(A(j, i)) = n - 1$.
   (b) If $\text{rank}(A) = n$, then $\text{rank}(\text{adj}(A)) = n$.
   (c) If $\text{rank}(A) = n - 1$, then $\text{rank}(\text{adj}(A)) = 1$.
   (d) If $\text{rank}(A) \leq n - 2$, then $\text{rank}(\text{adj}(A)) = 0$.
   (*Hints:* You can use that the rank of a matrix is equal to the maximal number of linear independent columns, respectively rows (cp. Example 10.24). Furthermore, Theorem 10.9 can be helpful.)

10.8 Let $S \in GL_n(K)$. Show that the map $f : K^{n,n} \to K^{n,n}$, $A \mapsto S^{-1}AS$ is an isomorphism.

10.9 Let $f \in \mathcal{L}(K^{n,n}, K^{n,n})$ with $\det(f(A)) = \det(A)$ for all $A \in K^{n,n}$. Show that f is bijective. Show further that the set of all these maps f is a subgroup of $(\{g \in \mathcal{L}(K^{n,n}, K^{n,n}) \mid g \text{ bijective}\}, \circ)$.

10.10 Let K be a field with $1 + 1 \neq 0$ and let $A \in K^{n,n}$. Consider the map
$$f : K^{n,1} \to K, \quad x \mapsto x^T A x.$$
Is f a linear map? Show that $f = 0$ if and only if $A + A^T = 0$.

10.11 For the bases $B_1 = \left\{ \begin{bmatrix} 1 \\ 0 \\ 0 \end{bmatrix}, \begin{bmatrix} 0 \\ 1 \\ 0 \end{bmatrix}, \begin{bmatrix} 0 \\ 0 \\ 1 \end{bmatrix} \right\}$ of $\mathbb{R}^{3,1}$ and $B_2 = \left\{ \begin{bmatrix} 1 \\ 0 \end{bmatrix}, \begin{bmatrix} 0 \\ 1 \end{bmatrix} \right\}$ of $\mathbb{R}^{2,1}$, let $f \in \mathcal{L}(\mathbb{R}^{3,1}, \mathbb{R}^{2,1})$ have the matrix representation $[f]_{B_1, B_2} = \begin{bmatrix} 0 & 2 & 3 \\ 1 & -2 & 0 \end{bmatrix}$.

   (a) Determine $[f]_{\widetilde{B}_1, \widetilde{B}_2}$ for the bases $\widetilde{B}_1 = \left\{ \begin{bmatrix} 2 \\ 1 \\ -1 \end{bmatrix}, \begin{bmatrix} 1 \\ 0 \\ 3 \end{bmatrix}, \begin{bmatrix} -1 \\ 2 \\ 1 \end{bmatrix} \right\}$ of $\mathbb{R}^{3,1}$ and $\widetilde{B}_2 = \left\{ \begin{bmatrix} 1 \\ 1 \end{bmatrix}, \begin{bmatrix} 1 \\ -1 \end{bmatrix} \right\}$ of $\mathbb{R}^{2,1}$.
   (b) Determine the coordinates of $f([4, 1, 3]^T)$ with respect to the basis $\widetilde{B}_2$.

10.12 Let $\mathcal{V}$ be a $\mathbb{Q}$-vector space with the basis $B_1 = \{v_1, \ldots, v_n\}$ and let $f \in \mathcal{L}(\mathcal{V}, \mathcal{V})$ be defined by
$$f(v_j) = \begin{cases} v_j + v_{j+1}, & j = 1, \ldots, n-1, \\ v_1 + v_n, & j = n. \end{cases}$$

   (a) Determine $[f]_{B_1, B_1}$.
   (b) Let $B_2 = \{w_1, \ldots, w_n\}$ with $w_j = jv_{n+1-j}$, $j = 1, \ldots, n$. Show that B_2 is a basis of $\mathcal{V}$. Determine the coordinate transformation matrices $[\text{Id}_\mathcal{V}]_{B_1, B_2}$ and $[\text{Id}_\mathcal{V}]_{B_2, B_1}$, as well as the matrix representations $[f]_{B_1, B_2}$ and $[f]_{B_2, B_2}$.

10.13 Let $\mathcal{V}$ be a finite dimensional K-vector space, $f \in \mathcal{L}(\mathcal{V}, \mathcal{V})$, and P_f the characteristic polynomial of f. Show that $P_f(0) \neq 0$ holds if and only if f is an automorphism.

10.14 Can you extend Theorem 10.20 consistently to the case $\mathcal{W} = \{0\}$? What are the properties of the matrices $[g \circ f]_{B_1, B_3}$, $[g]_{B_2, B_3}$ and $[f]_{B_1, B_2}$?

10.15 Let $\mathcal{V}, \mathcal{W}$ be finite dimensional K-vector spaces with bases B_1, B_2, and let $f \in \mathcal{L}(\mathcal{V}, \mathcal{W})$ be bijective. Show that then $[f^{-1}]_{B_2, B_1} = ([f]_{B_1, B_2})^{-1}$.

10.16 Let $\mathcal{V}, \mathcal{W}$ be finite dimensional K-vector spaces, $\mathcal{U} \subseteq \mathcal{W}$ a subspace, and $f \in \mathcal{L}(\mathcal{V}, \mathcal{W})$. Show the following assertions:
  (a) $f^{-1}(\mathcal{U})$ is a subspace of $\mathcal{V}$.
  (b) $\dim(f^{-1}(\mathcal{U})) = \dim(\mathcal{U} \cap \mathrm{im}(f)) + \dim(\ker(f))$.

10.17 Consider the map

$$f : \mathbb{R}[t]_{\leq n} \to \mathbb{R}[t]_{\leq n+1},$$

$$\alpha_n t^n + \alpha_{n-1} t^{n-1} + \ldots + \alpha_1 t + \alpha_0 \mapsto \frac{1}{n+1} \alpha_n t^{n+1}$$

$$+ \frac{1}{n} \alpha_{n-1} t^n + \ldots + \frac{1}{2} \alpha_1 t^2 + \alpha_0 t.$$

  (a) Show that f is linear. Determine $\ker(f)$ and $\mathrm{im}(f)$.
  (b) Choose bases B_1, B_2 in the two vector spaces and verify that for your choice $\mathrm{rank}([f]_{B_1, B_2}) = \dim(\mathrm{im}(f))$ holds.

10.18 Let $\alpha_1, \ldots, \alpha_n \in \mathbb{R}$, $n \geq 2$, be pairwise distinct numbers and let n polynomials in $\mathbb{R}[t]$ be defined by

$$p_j = \prod_{\substack{k=1 \\ k \neq j}}^{n} \left(\frac{1}{\alpha_j - \alpha_k} (t - \alpha_k) \right), \quad j = 1, \ldots, n.$$

  (a) Show that the set $B = \{p_1, \ldots, p_n\}$ is a basis of $\mathbb{R}[t]_{\leq n-1}$. (This basis is called the *Lagrange basis*[2] of $\mathbb{R}[t]_{\leq n-1}$.)
  (b) Show that the corresponding coordinate map is given by

$$\Phi_B : \mathbb{R}[t]_{\leq n-1} \to \mathbb{R}^{n,1}, \quad p \mapsto \begin{bmatrix} p(\alpha_1) \\ \vdots \\ p(\alpha_n) \end{bmatrix}.$$

(*Hint:* You can use Exercise 7.8 (b).)

---

[2] Joseph-Louis de Lagrange (1736–1813).

(c) Let $\beta_1, \ldots, \beta_n \in \mathbb{R}$. Show that the *Lagrange interpolation problem*

$$p(\alpha_j) = \beta_j, \quad j = 1, \ldots, n,$$

has a unique solution $p \in \mathbb{R}[t]_{\leq n-1}$, and write this solution as a linear combination of polynomials of the Lagrange basis.

10.19 Verify different paths in the commutative diagram (10.7) for the vector spaces and bases of Example 10.22 and linear map $f : \mathbb{Q}^{2,2} \to \mathbb{Q}^{2,2}$, $A \mapsto FA$ with

$$F = \begin{bmatrix} 1 & 1 \\ -1 & 1 \end{bmatrix}.$$

Linear Forms and Bilinear Forms 11

In this chapter we study different classes of maps between one or two K-vector spaces and the one dimensional K-vector space defined by the field K itself. These maps play an important role in many areas of Mathematics, including Analysis, Functional Analysis and the solution of differential equations. They will also be essential for the further developments in this book: Using bilinear and sesquilinear forms, which are introduced in this chapter, we will define and study Euclidean and unitary vector spaces in Chap. 12. Linear forms and dual spaces will be used in the existence proof of the Jordan canonical form in Chap. 16.

11.1 Linear Forms and Dual Spaces

We start with the set of linear maps from a K-vector space to the vector space K.

Definition 11.1 If $\mathcal{V}$ is a K-vector space, then $f \in \mathcal{L}(\mathcal{V}, K)$ is called a *linear form* on $\mathcal{V}$. The K-vector space $\mathcal{V}^* := \mathcal{L}(\mathcal{V}, K)$ is called the *dual space* of $\mathcal{V}$.

A linear form is sometimes called a *linear functional* or a *one-form*, which stresses that it (linearly) maps into a one dimensional vector space.

Example 11.2 If $\mathcal{V}$ is the $\mathbb{R}$-vector space of the continuous and real valued functions on the real interval $[0, 1]$ and if $\gamma \in [0, 1]$, then the two maps

$$f_1 : \mathcal{V} \to \mathbb{R}, \quad g \mapsto g(\gamma),$$

$$f_2 : \mathcal{V} \to \mathbb{R}, \quad g \mapsto \int_\alpha^\beta g(x)dx,$$

are linear forms on $\mathcal{V}$.

If $\dim(\mathcal{V}) = n$, then $\dim(\mathcal{V}^*) = n$ by Theorem 10.17. Let $B_1 = \{v_1, \ldots, v_n\}$ be a basis of $\mathcal{V}$ and let $B_2 = \{1\}$ be a basis of the K-vector space K. If $f \in \mathcal{V}^*$, then $f(v_i) = \alpha_i$ for some $\alpha_i \in K$, $i = 1, \ldots, n$, and

$$[f]_{B_1, B_2} = [\alpha_1, \ldots, \alpha_n] \in K^{1,n}.$$

For an element $v = \sum_{i=1}^{n} \lambda_i v_i \in \mathcal{V}$ we have

$$f(v) = f\left(\sum_{i=1}^{n} \lambda_i v_i\right) = \sum_{i=1}^{n} \lambda_i f(v_i) = \sum_{i=1}^{n} \lambda_i \alpha_i = \underbrace{[\alpha_1, \ldots, \alpha_n]}_{\in K^{1,n}} \underbrace{\begin{bmatrix} \lambda_1 \\ \vdots \\ \lambda_n \end{bmatrix}}_{\in K^{n,1}}$$

$$= [f]_{B_1, B_2} \, \Phi_{B_1}(v),$$

where we have identified the isomorphic vector spaces K and $K^{1,1}$ with each other.

For a given basis of a finite dimensional vector space $\mathcal{V}$ we will now construct a special, uniquely determined basis of the dual space $\mathcal{V}^*$.

Theorem 11.3 *If $\mathcal{V}$ is K-vector space with the basis $B = \{v_1, \ldots, v_n\}$, then there exists a unique basis $B^* = \{v_1^*, \ldots, v_n^*\}$ of $\mathcal{V}^*$ such that*

$$v_i^*(v_j) = \delta_{ij}, \quad i, j = 1, \ldots, n,$$

which is called the dual basis *of B.*

Proof By Theorem 10.4, a unique linear map from $\mathcal{V}$ to K can be constructed by prescribing its images at the given basis B. Thus, for each $i = 1, \ldots, n$, there exists a unique map $v_i^* \in \mathcal{L}(\mathcal{V}, K)$ with $v_i^*(v_j) = \delta_{ij}$, $j = 1, \ldots, n$.

It remains to show that $B^* := \{v_1^*, \ldots, v_n^*\}$ is a basis of $\mathcal{V}^*$. If $\lambda_1, \ldots, \lambda_n \in K$ are such that

$$\sum_{i=1}^{n} \lambda_i v_i^* = 0_{\mathcal{V}^*} \in \mathcal{V}^*,$$

then

$$0 = 0_{\mathcal{V}^*}(v_j) = \sum_{i=1}^{n} \lambda_i v_i^*(v_j) = \lambda_j, \quad j = 1, \ldots, n.$$

Thus, $v_1^*, \ldots, v_n^*$ are linearly independent, and $\dim(\mathcal{V}^*) = n$ implies that B^* is a basis of $\mathcal{V}^*$ (cp. Lemma 9.22). $\square$

11.1 Linear Forms and Dual Spaces

Let $B = \{v_1, \ldots, v_n\}$ and $B^* = \{v_1^*, \ldots, v_n^*\}$ be as in Theorem 11.3, and let

$$v = \sum_{j=1}^{n} \lambda_j v_j \in \mathcal{V}$$

be arbitrary. Then

$$v_i^*(v) = \sum_{j=1}^{n} \lambda_j v_i^*(v_j) = \lambda_i, \quad i = 1, \ldots, n,$$

i.e., the coordinates of v with respect to the basis B are obtained from the application of the elements of the dual basis B^* to v. If, on the other hand,

$$f = \sum_{i=1}^{n} \mu_i v_i^* \in \mathcal{V}^*$$

is arbitrary, then

$$f(v_j) = \sum_{i=1}^{n} \mu_i v_i^*(v_j) = \mu_j, \quad j = 1, \ldots, n,$$

i.e., the coordinates of f with respect to the dual basis B^* are obtained from the application of f to the elements of the basis B. Therefore, the corresponding coordinate maps are given by

$$\Phi_B : \mathcal{V} \to K^{n,1}, \quad v \mapsto \begin{bmatrix} v_1^*(v) \\ \vdots \\ v_n^*(v) \end{bmatrix}, \quad \text{and}$$

$$\Phi_{B^*} : \mathcal{V}^* \to K^{n,1}, \quad f \mapsto \begin{bmatrix} f(v_1) \\ \vdots \\ f(v_n) \end{bmatrix}.$$

Example 11.4

(1) Consider $\mathcal{V} = K^{n,1}$ with the canonical basis $B = \{e_1, \ldots, e_n\}$. If $\{e_1^*, \ldots, e_n^*\}$ is the dual basis of B, then $e_i^*(e_j) = \delta_{ij}$, which shows that $[e_i^*]_{B,\{1\}} = e_i^T \in K^{1,n}$, $i = 1, \ldots, n$.

(2) For $n \in \mathbb{N}$ we consider the $(n+1)$-dimensional real vector space $\mathcal{V} = \mathbb{R}[t]_{\leq n}$. Let $\alpha_1, \ldots, \alpha_{n+1} \in \mathbb{R}$ be pairwise distinct, and let $v_i^* \in \mathcal{V}^*$ be defined through the evaluation of $p \in \mathcal{V}$ at the point α_i, i.e.,

$$v_i^*(p) = p(\alpha_i), \quad i = 1, \ldots, n+1$$

(cp. Example 11.2). We now show that $v_1^*, \ldots v_{n+1}^* \in \mathcal{V}^*$ are linearly independent. Let $\lambda_1, \ldots, \lambda_{n+1} \in \mathbb{R}$ with

$$\sum_{i=1}^{n+1} \lambda_i v_i^* = 0 \in \mathcal{V}^*.$$

Inserting the polynomial t^j for $j = 0, 1, \ldots, n$ on both sides of this equation yields

$$\sum_{i=1}^{n+1} \lambda_i \alpha_i^j = 0, \quad j = 0, 1, \ldots, n.$$

We can write these $n+1$ equations as linear system

$$\begin{bmatrix} 1 & 1 & \cdots & 1 \\ \alpha_1 & \alpha_2 & \cdots & \alpha_{n+1} \\ \vdots & \vdots & & \vdots \\ \alpha_1^n & \alpha_2^n & \cdots & \alpha_{n+1}^n \end{bmatrix} \begin{bmatrix} \lambda_1 \\ \lambda_2 \\ \vdots \\ \lambda_{n+1} \end{bmatrix} = \begin{bmatrix} 0 \\ 0 \\ \vdots \\ 0 \end{bmatrix}.$$

Since $\alpha_1, \ldots, \alpha_{n+1}$ are pairwise distinct, the Vandermonde matrix on the left side is invertible (cp. Exercise 7.8), and hence $\lambda_1 = \lambda_2 = \cdots = \lambda_{n+1} = 0$. Since the $n+1$ linear forms $v_1^*, \ldots v_{n+1}^* \in \mathcal{V}^*$ are linearly independent and $\dim(\mathcal{V}^*) = n+1$, we see that $B^* := \{v_1^*, \ldots, v_{n+1}^*\}$ is a basis of $\mathcal{V}^*$.

The Lagrange polynomials

$$p_j = \prod_{\substack{k=1 \\ k \neq j}}^{n+1} \left(\frac{1}{\alpha_j - \alpha_k}(t - \alpha_k) \right) \in \mathcal{V}, \quad j = 1, \ldots, n+1,$$

form a basis $B := \{p_1, \ldots, p_{n+1}\}$ of $\mathcal{V}$ (cp. Exercise 10.18). We have $v_i^*(p_j) = p_j(\alpha_i) = \delta_{ij}$ for $i, j = 1, \ldots, n+1$. Hence, the basis B^* of $\mathcal{V}^*$ is the dual basis of B. If $p \in \mathcal{V}$ is arbitrary, then

$$p = \sum_{j=1}^{n} v_j^*(p) p_j = \sum_{j=1}^{n} p(\alpha_j) p_j.$$

11.1 Linear Forms and Dual Spaces

Let V be an n-dimensional K-vector space with the basis $B_1 = \{v_1, \ldots, v_n\}$, and let $B_1^* = \{v_1^*, \ldots, v_n^*\}$ be the dual basis of B_1. If $B_2 = \{w_1, \ldots, w_n\}$ is another basis of V, then

$$(w_1, \ldots, w_n) = (v_1, \ldots, v_n)A, \quad \text{where} \quad A = [a_{ij}] = [\mathrm{Id}_V]_{B_2, B_1} \in GL_n(K).$$

Using the matrix $A^{-T} \in GL_n(K)$ we define the vectors $w_1^*, \ldots, w_n^* \in V^*$ by

$$(w_1^*, \ldots, w_n^*) := (v_1^*, \ldots, v_n^*) A^{-T}.$$

Since A^{-T} is invertible, these vectors are linearly independent. Hence they form another basis $B_2^* := \{w_1^*, \ldots, w_n^*\}$ of V^*. Let $A^{-T} = [b_{ij}]$, then $v_k^*(v_\ell) = \delta_{k\ell}$ yields

$$w_i^*(w_j) = \sum_{k=1}^n b_{ki} v_k^*\left(\sum_{\ell=1}^n a_{\ell j} v_\ell\right) = \sum_{k=1}^n a_{kj} b_{ki} \quad \text{(entry } (j,i) \text{ of } A^T A^{-T} = I_n)$$
$$= \delta_{ij}.$$

Thus, B_2^* is the dual basis of B_2, and we have $[\mathrm{Id}_{V^*}]_{B_2^*, B_1^*} = A^{-T} = ([\mathrm{Id}_V]_{B_1, B_2})^T$.

Definition 11.5 Let V and W be K-vector spaces with their respective dual spaces V^* and W^*, and let $f \in \mathcal{L}(V, W)$. Then

$$f^* : W^* \to V^*, \quad h \mapsto f^*(h) := h \circ f,$$

is called the *dual map* of f.

We next derive some properties of the dual map.

Lemma 11.6 *If V, W and X are K-vector spaces, then the following assertions hold:*

(1) *If $f \in \mathcal{L}(V, W)$, then the dual map f^* is linear, hence $f^* \in \mathcal{L}(W^*, V^*)$.*
(2) *If $f \in \mathcal{L}(V, W)$ and $g \in \mathcal{L}(W, X)$, then $(g \circ f)^* \in \mathcal{L}(X^*, V^*)$ and $(g \circ f)^* = f^* \circ g^*$.*
(3) *If $f \in \mathcal{L}(V, W)$ is bijective, then $f^* \in \mathcal{L}(W^*, V^*)$ is bijective and $(f^*)^{-1} = (f^{-1})^*$.*

Proof

(1) If $h_1, h_2 \in \mathcal{W}^*$, $\lambda_1, \lambda_2 \in K$, then

$$f^*(\lambda_1 h_1 + \lambda_2 h_2) = (\lambda_1 h_1 + \lambda_2 h_2) \circ f = (\lambda_1 h_1) \circ f + (\lambda_2 h_2) \circ f$$
$$= \lambda_1(h_1 \circ f) + \lambda_2(h_2 \circ f) = \lambda_1 f^*(h_1) + \lambda_2 f^*(h_2).$$

(2) and (3) are exercises. □

As the following theorem shows, the concepts of the dual map and the transposed matrix are closely related.

Theorem 11.7 *Let $\mathcal{V}$ and $\mathcal{W}$ be finite dimensional K-vector spaces with bases B_1 and B_2, respectively, and let B_1^* and B_2^* be the corresponding dual bases. If $f \in \mathcal{L}(\mathcal{V}, \mathcal{W})$, then*

$$[f^*]_{B_2^*, B_1^*} = ([f]_{B_1, B_2})^T.$$

Proof Let $B_1 = \{v_1, \ldots, v_m\}$ and $B_2 = \{w_1, \ldots, w_n\}$, as well as $B_1^* = \{v_1^*, \ldots, v_m^*\}$ and $B_2^* = \{w_1^*, \ldots, w_n^*\}$. Let $[f]_{B_1, B_2} = [a_{ij}] \in K^{n,m}$, i.e.,

$$f(v_j) = \sum_{i=1}^{n} a_{ij} w_i, \quad j = 1, \ldots, m,$$

and $[f^*]_{B_2^*, B_1^*} = [b_{ij}] \in K^{m,n}$, i.e.,

$$f^*\left(w_j^*\right) = \sum_{i=1}^{m} b_{ij} v_i^*, \quad j = 1, \ldots, n.$$

For every pair (k, ℓ) with $1 \leq k \leq n$ and $1 \leq \ell \leq m$ we then have

$$a_{k\ell} = \sum_{i=1}^{n} a_{i\ell} w_k^*(w_i) = w_k^*\left(\sum_{i=1}^{n} a_{i\ell} w_i\right) = w_k^*(f(v_\ell)) = f^*\left(w_k^*\right)(v_\ell)$$
$$= \left(\sum_{i=1}^{m} b_{ik} v_i^*\right)(v_\ell) = \sum_{i=1}^{m} b_{ik} v_i^*(v_\ell)$$
$$= b_{\ell k},$$

where we have used the definition of the dual map as well as $w_k^*(w_i) = \delta_{ki}$ and $v_i^*(v_\ell) = \delta_{i\ell}$. □

11.1 Linear Forms and Dual Spaces

For $\mathcal{V} = \mathcal{W}$, $f = \text{Id}_{\mathcal{V}}$, and $f^* = \text{Id}_{\mathcal{V}^*}$ this theorem yields the equation

$$[\text{Id}_{\mathcal{V}^*}]_{B_2^*, B_1^*} = ([\text{Id}_{\mathcal{V}}]_{B_1, B_2})^T,$$

which we have already shown above. Applied to matrices, Theorem 11.7 and Lemma 11.6 yield the following rules known from Chap. 4:

$$(AB)^T = B^T A^T \quad \text{for } A \in K^{n,m} \text{ and } B \in K^{m,\ell}, \text{ and}$$

$$(A^{-1})^T = (A^T)^{-1} \quad \text{for } A \in GL_n(K).$$

Because of the close relationship between the transposed matrix and the dual map, some authors call the dual map f^* the *transpose* of the linear map f.

Example 11.8 For the two bases of $\mathbb{R}^{2,1}$,

$$B_1 = \left\{ v_1 = \begin{bmatrix} 1 \\ 0 \end{bmatrix}, v_2 = \begin{bmatrix} 0 \\ 2 \end{bmatrix} \right\}, \quad B_2 = \left\{ w_1 = \begin{bmatrix} 1 \\ 0 \end{bmatrix}, w_2 = \begin{bmatrix} 1 \\ 1 \end{bmatrix} \right\},$$

the elements of the corresponding dual bases are given by

$$v_1^* : \mathbb{R}^{2,1} \to \mathbb{R}, \quad \begin{bmatrix} \alpha_1 \\ \alpha_2 \end{bmatrix} \mapsto \alpha_1 + 0, \quad v_2^* : \mathbb{R}^{2,1} \to \mathbb{R}, \quad \begin{bmatrix} \alpha_1 \\ \alpha_2 \end{bmatrix} \mapsto 0 + \frac{1}{2}\alpha_2,$$

$$w_1^* : \mathbb{R}^{2,1} \to \mathbb{R}, \quad \begin{bmatrix} \alpha_1 \\ \alpha_2 \end{bmatrix} \mapsto \alpha_1 - \alpha_2, \quad w_2^* : \mathbb{R}^{2,1} \to \mathbb{R}, \quad \begin{bmatrix} \alpha_1 \\ \alpha_2 \end{bmatrix} \mapsto 0 + \alpha_2.$$

The matrix representations of these maps are

$$[v_1^*]_{B_1, \{1\}} = [1, 0], \quad [v_2^*]_{B_1, \{1\}} = [0, 1], \quad [w_1^*]_{B_2, \{1\}} = [1, 0],$$

$$[w_2^*]_{B_2, \{1\}} = [0, 1].$$

For the linear map

$$f : \mathbb{R}^{2,1} \to \mathbb{R}^{2,1}, \quad \begin{bmatrix} \alpha_1 \\ \alpha_2 \end{bmatrix} \mapsto \begin{bmatrix} \alpha_1 + \alpha_2 \\ 3\alpha_2 \end{bmatrix},$$

we have

$$[f]_{B_1, B_2} = \begin{bmatrix} 1 & -4 \\ 0 & 6 \end{bmatrix}, \quad [f^*]_{B_2^*, B_1^*} = \begin{bmatrix} 1 & 0 \\ -4 & 6 \end{bmatrix}.$$

11.2 Bilinear Forms

Bilinear forms map pairs of K-vector spaces to the vector space K. They play an important role in the classification of solution sets of quadratic equations (the so-called quadrics), and in the solution of partial differential equations.

Definition 11.9 Let $\mathcal{V}$ and $\mathcal{W}$ be K-vector spaces. A map $\beta : \mathcal{V} \times \mathcal{W} \to K$ is called a *bilinear form on* $\mathcal{V} \times \mathcal{W}$, when

(1) $\beta(v_1 + v_2, w) = \beta(v_1, w) + \beta(v_2, w)$,
(2) $\beta(v, w_1 + w_2) = \beta(v, w_1) + \beta(v, w_2)$,
(3) $\beta(\lambda v, w) = \beta(v, \lambda w) = \lambda \beta(v, w)$,

hold for all $v, v_1, v_2 \in \mathcal{V}$, $w, w_1, w_2 \in \mathcal{W}$, and $\lambda \in K$.

A bilinear form β is called *non-degenerate in the first variable*, if $\beta(v, w) = 0$ for all $w \in \mathcal{W}$ implies that $v = 0$. Analogously, it is called *non-degenerate in the second variable*, if $\beta(v, w) = 0$ for all $v \in \mathcal{V}$ implies that $w = 0$. If β is non-degenerate in both variables, then β is called *non-degenerate* and the spaces $\mathcal{V}, \mathcal{W}$ are called a *dual pair* with respect to β.

If $\mathcal{V} = \mathcal{W}$, then β is called a *bilinear form on* $\mathcal{V}$. If additionally $\beta(v, w) = \beta(w, v)$ holds for all $v, w \in \mathcal{V}$, then β is called *symmetric*. Otherwise, β is called *nonsymmetric*.

Example 11.10

(1) If $A \in K^{n,m}$, then

$$\beta : K^{m,1} \times K^{n,1} \to K, \quad (v, w) \mapsto w^T A v,$$

is a bilinear form on $K^{m,1} \times K^{n,1}$ that is non-degenerate if and only if $n = m$ and $A \in GL_n(K)$ (cp. Exercise 11.13).
(2) The bilinear form

$$\beta : \mathbb{R}^{2,1} \times \mathbb{R}^{2,1} \to \mathbb{R}, \quad (x, y) \mapsto y^T \begin{bmatrix} 1 & 1 \\ 1 & 1 \end{bmatrix} x,$$

is degenerate in both variables: For $\widehat{x} = [1, -1]^T$, we have $\beta(\widehat{x}, y) = 0$ for all $y \in \mathbb{R}^{2,1}$; for $\widehat{y} = [1, -1]^T$ we have $\beta(x, \widehat{y}) = 0$ for all $x \in \mathbb{R}^{2,1}$. The set of all $x = [x_1, x_2]^T \in \mathbb{R}^{2,1}$ with $\beta(x, x) = 1$ is equal to the solution set of the quadratic equation in two variables $x_1^2 + 2x_1 x_2 + x_2^2 = 1$, or $(x_1 + x_2)^2 = 1$, for $x_1, x_2 \in \mathbb{R}$. Geometrically, this set is given by the two straight lines $x_1 + x_2 = 1$ and $x_1 + x_2 = -1$ in the Cartesian coordinate system of $\mathbb{R}^2$.

11.2 Bilinear Forms

(3) If $\mathcal{V}$ is the $\mathbb{R}$-vector space of the continuous and real valued functions on the real interval $[0, 1]$, then the map

$$\beta : \mathcal{V} \times \mathcal{V} \to \mathbb{R}, \quad (f, g) \mapsto \int_0^1 f(x) g(x) dx,$$

is a symmetric bilinear form on $\mathcal{V}$ (cp. Exercise 11.14)

(4) If $\mathcal{V}$ is a K-vector space, then

$$\beta : \mathcal{V} \times \mathcal{V}^* \to K, \quad (v, f) \mapsto f(v),$$

is a bilinear form on $\mathcal{V} \times \mathcal{V}^*$, since

$$\beta(v_1 + v_2, f) = f(v_1 + v_2) = f(v_1) + f(v_2) = \beta(v_1, f) + \beta(v_2, f),$$
$$\beta(v, f_1 + f_2) = (f_1 + f_2)(v) = f_1(v) + f_2(v) = \beta(v, f_1) + \beta(v, f_2),$$
$$\beta(\lambda v, f) = f(\lambda v) = \lambda f(v) = \lambda \beta(v, f) = (\lambda f)(v) = \beta(v, \lambda f),$$

hold for all $v, v_1, v_2 \in \mathcal{V}$, $f, f_1, f_2 \in \mathcal{V}^*$ and $\lambda \in K$. This bilinear form is non-degenerate and thus $\mathcal{V}, \mathcal{V}^*$ are a dual pair with respect to β (cp. Exercise 11.15 for the case $\dim(\mathcal{V}) \in \mathbb{N}$).

Let $\mathcal{V}$ and $\mathcal{W}$ be K-vector spaces with bases $B_1 = \{v_1, \ldots, v_m\}$ and $B_2 = \{w_1, \ldots, w_n\}$, respectively. Let β be a bilinear form on $\mathcal{V} \times \mathcal{W}$, and let $v = \sum_{j=1}^m \lambda_j v_j \in \mathcal{V}$ and $w = \sum_{i=1}^n \mu_i w_i \in \mathcal{W}$. Then we have

$$\beta(v, w) = \sum_{j=1}^m \sum_{i=1}^n \lambda_j \mu_i \beta(v_j, w_i) = \sum_{i=1}^n \mu_i \sum_{j=1}^m \beta(v_j, w_i) \lambda_j$$
$$= \left(\Phi_{B_2}(w)\right)^T [b_{ij}] \Phi_{B_1}(v), \tag{11.1}$$

where $b_{ij} := \beta(v_j, w_i)$, and where we have used the coordinate map from Lemma 10.18. This motivates the following definition.

Definition 11.11 The matrix constructed in (11.1), i.e.,

$$[\beta]_{B_1 \times B_2} := [b_{ij}] \in K^{n,m}, \quad b_{ij} := \beta(v_j, w_i),$$

is called the *matrix representation* of β with respect to the bases B_1 and B_2.

Example 11.12 If $B_1 = \{e_1^{(m)}, \ldots, e_m^{(m)}\}$ and $B_2 = \{e_1^{(n)}, \ldots, e_n^{(n)}\}$ are the canonical bases of $K^{m,1}$ and $K^{n,1}$, respectively, and if β is the bilinear form from (1) in Example 11.10 with $A = [a_{ij}] \in K^{n,m}$, then $[\beta]_{B_1 \times B_2} = [b_{ij}]$, where

$$b_{ij} = \beta(e_j^{(m)}, e_i^{(n)}) = (e_i^{(n)})^T A e_j^{(m)} = a_{ij},$$

and hence $[\beta]_{B_1 \times B_2} = A$.

Using the notation of Definition 11.11 we define the two maps

$$\Phi_{B_1,B_2} : \mathcal{V} \times \mathcal{W} \to K^{m,1} \times K^{n,1}, \quad (v, w) \mapsto (\Phi_{B_1}(v), \Phi_{B_2}(w)),$$

$$\varphi_{\beta,B_1,B_2} : K^{m,1} \times K^{n,1} \to K, \quad (x, y) \mapsto y^T [\beta]_{B_1 \times B_2} x.$$

For all $v \in \mathcal{V}$ and $w \in \mathcal{W}$, we then have

$$\beta(v, w) = (\varphi_{\beta,B_1,B_2} \circ \Phi_{B_1,B_2})(v, w).$$

Analogously to the commutative diagram (10.2) for the matrix representation of a linear map, we obtain the following diagram:

$$\begin{array}{ccc} \mathcal{V} \times \mathcal{W} & \xrightarrow{\beta} & K \\ {\scriptstyle \Phi_{B_1,B_2}} \downarrow & \nearrow {\scriptstyle \varphi_{\beta,B_1,B_2}} & \\ K^{m,1} \times K^{n,1} & & \end{array}$$

Remark 11.13 The matrix representation of a bilinear form could also be constructed and defined "the other way around", i.e., instead of (11.1) one computes

$$\beta(v, w) = \sum_{i=1}^{m} \sum_{j=1}^{n} \lambda_i \mu_j \beta(v_i, w_j) = \sum_{i=1}^{m} \lambda_i \sum_{j=1}^{n} \beta(v_i, w_j) \mu_j$$

$$= (\Phi_{B_1}(v))^T [\widetilde{b}_{ij}] \Phi_{B_2}(w)$$

with $\widetilde{b}_{ij} := \beta(v_i, w_j)$ and, therefore, $[\widetilde{b}_{ij}] \in K^{m,n}$. To be consistent, in this case one would define the bilinear form in Example 11.10 (1) via $A \in K^{m,n}$ and

$$\beta : K^{m,1} \times K^{n,1} \to K, \quad (v, w) \mapsto v^T A w.$$

The difference between both definitions is purely formal. The reason for our construction in (11.1) and choice in Definition 11.11 is a desired property of the sesquilinear forms $s : \mathcal{V} \times \mathcal{W} \to \mathbb{C}$, that we will consider in the next section.

11.2 Bilinear Forms

These maps are supposed to be linear in the first and semi-linear in the second component (cp. Definition 11.19). This property is satisfied by $s(v, w) = w^H A v$ (cp. Example 11.24), but not by $s(v, w) = v^H A w$. To keep the development consistent, this requires choosing $\beta(v, w) = w^T A v$ in Example 11.10 (1), and setting $b_{ij} := \beta(v_j, w_i)$ in Definition 11.11. In the end it is a matter of taste, because we could also define sesquilinear forms as linear in the second and semi-linear in the first component. This would lead to $s(v, w) = v^H A w$, and would then require $\beta(v, w) = v^T A w$ to be consistent. In order to have linearity in the first and semi-linearity in the second component of a sesquilinear form, but nevertheless have v "on the left" and w "on the right", some authors use $s(v, w) = v^T A \overline{w}$ as their standard example, where $\overline{w} := [\overline{w}_1, \ldots, \overline{w}_n]^T$.

The bilinear forms on $\mathcal{V} \times \mathcal{W}$ with pointwise addition and scalar multiplication form the K-vector space $\text{Bil}(\mathcal{V}, \mathcal{W})$ (cp. Exercise 11.11), and we obtain the following result analogously to Theorem 10.17.

Theorem 11.14 *Let $\mathcal{V}$ and $\mathcal{W}$ be two finite dimensional K-vector spaces with bases $B_1 = \{v_1, \ldots, v_m\}$, respectively $B_2 = \{w_1, \ldots, w_n\}$. Then the map*

$$\text{Bil}(\mathcal{V}, \mathcal{W}) \to K^{n,m}, \quad \beta \mapsto [\beta]_{B_1 \times B_2},$$

is an isomorphism. Hence, $\text{Bil}(\mathcal{V}, \mathcal{W}) \cong K^{n,m}$ and $\dim(\text{Bil}(\mathcal{V}, \mathcal{W})) = \dim(K^{n,m}) = n \cdot m$.

Proof Analogously to the proof of Theorem 10.17, we denote the map $\beta \mapsto [\beta]_{B_1 \times B_2}$ by φ, and thus $\varphi(\beta) = [\beta]_{B_1 \times B_2}$. The linearity of this map follows immediately from the pointwise definition of the operations in the vector space $\text{Bil}(\mathcal{V}, \mathcal{W})$, since for all $\beta_1, \beta_2 \in \text{Bil}(\mathcal{V}, \mathcal{W})$ and $\lambda_1, \lambda_2 \in K$ we have

$$(\lambda_1 \cdot \beta_1 + \lambda_2 \cdot \beta_2)(v_j, w_i) = \lambda_1 \beta_1(v_j, w_i) + \lambda_2 \beta_2(v_j, w_i),$$

and thus

$$\varphi(\lambda_1 \cdot \beta_1 + \lambda_2 \cdot \beta_2) = [\lambda_1 \beta_1 + \lambda_2 \beta_2]_{B_1 \times B_2} = \lambda_1 [\beta_1]_{B_1 \times B_2} + \lambda_2 [\beta_2]_{B_1 \times B_2}$$
$$= \lambda_1 \varphi(\beta_1) + \lambda_2 \varphi(\beta_2).$$

Suppose that $\beta \in \ker(\varphi)$, i.e., $\varphi(\beta) = 0 \in K^{n,m}$ and, therefore, $\beta(v_j, w_i) = 0$ for all $j = 1, \ldots, m$ and $i = 1, \ldots, n$. Then $\beta = 0 \in \text{Bil}(\mathcal{V}, \mathcal{W})$, and hence φ is injective (cp. (5) in Lemma 10.7). If, on the other hand, $B = [b_{ij}] \in K^{n,m}$ is arbitrary, then for $v = \sum_{j=1}^{m} \lambda_j v_j \in \mathcal{V}$ and $w = \sum_{i=1}^{n} \mu_i w_i \in \mathcal{W}$ we define the map $\beta : \mathcal{V} \times \mathcal{W} \to K$ via

$$\beta(v, w) := \left(\Phi_{B_2}(w)\right)^T [b_{ij}] \Phi_{B_1}(v).$$

It is easy to see that $\beta \in \mathrm{Bil}(\mathcal{V}, \mathcal{W})$, and that $\beta(v_j, w_i) = \left(e_i^{(n)}\right)^T B e_j^{(m)} = b_{ij}$, and thus also $B = [\beta]_{B_1 \times B_2} = \varphi(\beta)$. Hence φ is surjective, and thus bijective.

Corollary 10.12 implies that $\dim(\mathrm{Bil}(\mathcal{V}, \mathcal{W})) = \dim(K^{n,m}) = n \cdot m$. □

By Theorem 11.14 we also have $\mathrm{Bil}(\mathcal{V}, \mathcal{W}) \cong \mathcal{L}(\mathcal{V}, \mathcal{W})$ for finite dimensional K-vector spaces.

The following result shows that symmetric bilinear forms have symmetric matrix representations.

Lemma 11.15 *For a bilinear form β on a finite dimensional vector space $\mathcal{V}$ the following statements are equivalent:*

(1) β is symmetric.
(2) For every basis B of $\mathcal{V}$ the matrix $[\beta]_{B \times B}$ is symmetric.
(3) There exists a basis B of $\mathcal{V}$ such that $[\beta]_{B \times B}$ is symmetric.

Proof Exercise. □

We will now analyze the effect of a basis change on the matrix representation of a bilinear form.

Theorem 11.16 *Let $\mathcal{V}$ and $\mathcal{W}$ be finite dimensional K-vector spaces with bases $B_1, \widetilde{B}_1$ of $\mathcal{V}$ and $B_2, \widetilde{B}_2$ of $\mathcal{W}$. If β is a bilinear form on $\mathcal{V} \times \mathcal{W}$, then*

$$[\beta]_{B_1 \times B_2} = \left([\mathrm{Id}_\mathcal{W}]_{B_2, \widetilde{B}_2}\right)^T [\beta]_{\widetilde{B}_1 \times \widetilde{B}_2} [\mathrm{Id}_\mathcal{V}]_{B_1, \widetilde{B}_1}.$$

Proof Let $B_1 = \{v_1, \ldots, v_m\}$, $\widetilde{B}_1 = \{\widetilde{v}_1, \ldots, \widetilde{v}_m\}$, $B_2 = \{w_1, \ldots, w_n\}$, $\widetilde{B}_2 = \{\widetilde{w}_1, \ldots, \widetilde{w}_n\}$, and

$$(v_1, \ldots, v_m) = (\widetilde{v}_1, \ldots, \widetilde{v}_m) P, \quad \text{where} \quad P = [p_{ij}] = [\mathrm{Id}_\mathcal{V}]_{B_1, \widetilde{B}_1},$$

$$(w_1, \ldots, w_n) = (\widetilde{w}_1, \ldots, \widetilde{w}_n) Q, \quad \text{where} \quad Q = [q_{ij}] = [\mathrm{Id}_\mathcal{W}]_{B_2, \widetilde{B}_2}.$$

With $[\beta]_{\widetilde{B}_1 \times \widetilde{B}_2} = [\widetilde{b}_{ij}]$, where $\widetilde{b}_{ij} = \beta(\widetilde{v}_j, \widetilde{w}_i)$, we then have

$$\beta(v_j, w_i) = \beta\left(\sum_{k=1}^m p_{kj} \widetilde{v}_k, \sum_{\ell=1}^n q_{\ell i} \widetilde{w}_\ell\right) = \sum_{\ell=1}^n q_{\ell i} \sum_{k=1}^m \beta(\widetilde{v}_k, \widetilde{w}_\ell) p_{kj}$$

$$= \sum_{\ell=1}^n q_{\ell i} \sum_{k=1}^m \widetilde{b}_{\ell k} p_{kj}$$

$$= \begin{bmatrix} q_{1i} \\ \vdots \\ q_{ni} \end{bmatrix}^T [\beta]_{\widetilde{B}_1 \times \widetilde{B}_2} \begin{bmatrix} p_{1j} \\ \vdots \\ p_{mj} \end{bmatrix},$$

which implies that $[\beta]_{B_1 \times B_2} = Q^T [\beta]_{\widetilde{B}_1 \times \widetilde{B}_2} P$, and hence the assertion follows. □

If $\mathcal{V} = \mathcal{W}$ and B_1, B_2 are two bases of $\mathcal{V}$, then we obtain the following special case of Theorem 11.16:

$$[\beta]_{B_1 \times B_1} = \left([\mathrm{Id}_\mathcal{V}]_{B_1, B_2}\right)^T [\beta]_{B_2 \times B_2} [\mathrm{Id}_\mathcal{V}]_{B_1, B_2}.$$

The two matrix representations $[\beta]_{B_1 \times B_1}$ and $[\beta]_{B_2 \times B_2}$ of β in this case are *congruent*, which we formally define as follows.

Definition 11.17 If for two matrices $A, B \in K^{n,n}$ there exists a matrix $Z \in GL_n(K)$ with $B = Z^T A Z$, then A and B are called *congruent*.

Lemma 11.18 *Congruence is an equivalence relation on the set $K^{n,n}$.*

Proof Exercise. □

11.3 Sesquilinear Forms

In this section we introduce another special class of forms on *complex* vector spaces.

Definition 11.19 Let $\mathcal{V}$ and $\mathcal{W}$ be $\mathbb{C}$-vector spaces. A map $s : \mathcal{V} \times \mathcal{W} \to \mathbb{C}$ is called a *sesquilinear form* on $\mathcal{V} \times \mathcal{W}$, when

(1) $s(v_1 + v_2, w) = s(v_1, w) + s(v_2, w)$,
(2) $s(\lambda v, w) = \lambda s(v, w)$,
(3) $s(v, w_1 + w_2) = s(v, w_1) + s(v, w_2)$,
(4) $s(v, \lambda w) = \bar{\lambda} s(v, w)$,

hold for all $v, v_1, v_2 \in \mathcal{V}$, $w, w_1, w_2 \in \mathcal{W}$ and $\lambda \in \mathbb{C}$.

A sesquilinear form s is called *non-degenerate in the first variable*, if $s(v, w) = 0$ for all $w \in \mathcal{W}$ implies that $v = 0$. Analogously, it is called *non-degenerate in the second variable*, if $s(v, w) = 0$ for all $v \in \mathcal{V}$ implies that $w = 0$. If s is non-degenerate in both variables, then s is called *non-degenerate*.

If $\mathcal{V} = \mathcal{W}$, then s is called a *sesquilinear form on $\mathcal{V}$*. If additionally $s(v, w) = \overline{s(w, v)}$ holds for all $v, w \in \mathcal{V}$, then s is called *Hermitian*.[1]

---

[1] Charles Hermite (1822–1901).

The prefix *sesqui* is Latin and means "one and a half". Note that a sesquilinear form is linear in the first variable and *semilinear* ("half linear") in the second variable.

The following result characterizes Hermitian sesquilinear forms.

Lemma 11.20 *A sesquilinear form on the $\mathbb{C}$-vector space $\mathcal{V}$ is Hermitian if and only if $s(v, v) \in \mathbb{R}$ for all $v \in \mathcal{V}$.*

Proof If s is Hermitian then, in particular, $s(v, v) = \overline{s(v, v)}$ for all $v \in \mathcal{V}$, and thus $s(v, v) \in \mathbb{R}$.

If, on the other hand, $v, w \in \mathcal{V}$, then by definition

$$s(v + w, v + w) = s(v, v) + s(v, w) + s(w, v) + s(w, w), \tag{11.2}$$

$$s(v + \mathbf{i}w, v + \mathbf{i}w) = s(v, v) + \mathbf{i}s(w, v) - \mathbf{i}s(v, w) + s(w, w). \tag{11.3}$$

The first equation implies that $s(v, w) + s(w, v) \in \mathbb{R}$, since $s(v + w, v + w), s(v, v), s(w, w) \in \mathbb{R}$ by assumption. The second equation implies analogously that $\mathbf{i}s(w, v) - \mathbf{i}s(v, w) \in \mathbb{R}$. Therefore,

$$s(v, w) + s(w, v) = \overline{s(v, w)} + \overline{s(w, v)},$$

$$-\mathbf{i}s(v, w) + \mathbf{i}s(w, v) = \overline{\mathbf{i}s(v, w)} - \overline{\mathbf{i}s(w, v)}.$$

Multiplying the second equation with $\mathbf{i}$ and adding the resulting equation to the first we obtain $s(v, w) = \overline{s(w, v)}$. $\square$

Corollary 11.21 *For a sesquilinear form s on the $\mathbb{C}$-vector space $\mathcal{V}$ we have*

$$2s(v, w) = s(v + w, v + w) + \mathbf{i}s(v + \mathbf{i}w, v + \mathbf{i}w) - (\mathbf{i} + 1)(s(v, v) + s(w, w)).$$

for all $v, w \in \mathcal{V}$.

Proof The result follows from multiplication of (11.3) with $\mathbf{i}$ and adding the result to (11.2). $\square$

Corollary 11.21 shows that a sesquilinear form on a $\mathbb{C}$-vector space $\mathcal{V}$ is uniquely determined by the values of $s(v, v)$ for all $v \in \mathcal{V}$.

Definition 11.22 The *Hermitian transpose* of $A = [a_{ij}] \in \mathbb{C}^{n,m}$ is the matrix

$$A^H := [\overline{a_{ij}}]^T \in \mathbb{C}^{m,n}.$$

If $A = A^H$, then A is called *Hermitian*.

11.3 Sesquilinear Forms

If a matrix A has real entries, then obviously $A^H = A^T$. Thus, a real symmetric matrix is also Hermitian. If $A = [a_{ij}] \in \mathbb{C}^{n,n}$ is Hermitian, then in particular $a_{ii} = \overline{a}_{ii}$ for $i = 1, \ldots, n$, i.e., Hermitian matrices have real diagonal entries.

The Hermitian transposition satisfies similar rules as the (usual) transposition (cp. Lemma 4.6).

Lemma 11.23 *For* $A, \widehat{A} \in \mathbb{C}^{n,m}$, $B \in \mathbb{C}^{m,\ell}$ *and* $\lambda \in \mathbb{C}$ *the following assertions hold:*

(1) $(A^H)^H = A$.
(2) $(A + \widehat{A})^H = A^H + \widehat{A}^H$.
(3) $(\lambda A)^H = \overline{\lambda} A^H$.
(4) $(AB)^H = B^H A^H$.

Proof Exercise. □

If $A \in \mathbb{C}^{n,n}$ is invertible, then analogously to (1) in Lemma 4.10 we have that A^H is invertible with $(A^H)^{-1} = (A^{-1})^H$. We denote the inverse of A^H with A^{-H}.

Example 11.24 For $A \in \mathbb{C}^{n,m}$ the map

$$s : \mathbb{C}^{m,1} \times \mathbb{C}^{n,1} \to \mathbb{C}, \quad (v, w) \mapsto w^H A v,$$

is a sesquilinear form.

The matrix representation of a sesquilinear form is defined analogously to the matrix representation of bilinear forms (cp. Definition 11.11).

Definition 11.25 Let $\mathcal{V}$ and $\mathcal{W}$ be $\mathbb{C}$-vector spaces with bases $B_1 = \{v_1, \ldots, v_m\}$ and $B_2 = \{w_1, \ldots, w_n\}$, respectively. If s is a sesquilinear form on $\mathcal{V} \times \mathcal{W}$, then

$$[s]_{B_1 \times B_2} = [b_{ij}] \in \mathbb{C}^{n,m}, \quad b_{ij} := s(v_j, w_i),$$

is called the *matrix representation* of s with respect to the bases B_1 and B_2.

Analogously to (11.1) we have

$$s(v, w) = \left(\Phi_{B_2}(w)\right)^H [s]_{B_1 \times B_2} \Phi_{B_1}(v),$$

for all $v \in \mathcal{V}$ and $w \in \mathcal{W}$.

Example 11.26 If $B_1 = \{e_1^{(m)}, \ldots, e_m^{(m)}\}$ and $B_2 = \{e_1^{(n)}, \ldots, e_n^{(n)}\}$ are the canonical bases of $\mathbb{C}^{m,1}$ and $\mathbb{C}^{n,1}$, respectively, and s is the sesquilinear form of Example 11.24 with $A = [a_{ij}] \in \mathbb{C}^{n,m}$, then $[s]_{B_1 \times B_2} = [b_{ij}]$ with

$$b_{ij} = s(e_j^{(m)}, e_i^{(n)}) = (e_i^{(n)})^H A e_j^{(m)} = \overline{(e_i^{(n)})^T A e_j^{(m)}} = \overline{a_{ij}}$$

and, hence, $[s]_{B_1 \times B_2} = \overline{A}$.

Analogously to Theorem 11.16, we have the following result for the change of bases in sesquilinear forms.

Theorem 11.27 *Let $\mathcal{V}$ and $\mathcal{W}$ be two finite dimensional $\mathbb{C}$-vector spaces with bases $B_1, \widetilde{B}_1$ of $\mathcal{V}$ and $B_2, \widetilde{B}_2$ of $\mathcal{W}$. If s is a sesquilinear form on $\mathcal{V} \times \mathcal{W}$, then*

$$[s]_{B_1 \times B_2} = \overline{\left([\mathrm{Id}_{\mathcal{W}}]_{B_2, \widetilde{B}_2}\right)^H} [s]_{\widetilde{B}_1 \times \widetilde{B}_2} [\mathrm{Id}_{\mathcal{V}}]_{B_1, \widetilde{B}_1}.$$

Proof Exercise. □

If $\mathcal{V} = \mathcal{W}$ and B_1, B_2 are two bases of $\mathcal{V}$, then

$$[s]_{B_1 \times B_1} = \overline{\left([\mathrm{Id}_{\mathcal{V}}]_{B_1, B_2}\right)^H} [s]_{B_2 \times B_2} [\mathrm{Id}_{\mathcal{V}}]_{B_1, B_2}.$$

This motivates the following definition (cp. Definition 11.17).

Definition 11.28 If, for two matrices $A, B \in \mathbb{C}^{n,n}$, there exists a matrix $Z \in GL_n(\mathbb{C})$ with $B = Z^H A Z$, then A and B are called *complex congruent*.

Lemma 11.29 *The relation of complex congruence is an equivalence relation on the set $\mathbb{C}^{n,n}$.*

Proof Exercise. □

Exercises

(In the following exercises K is an arbitrary field.)

11.1 Let $\mathcal{V}$ be a finite dimensional K-vector space and $v \in \mathcal{V}$. Show that $f(v) = 0$ for all $f \in \mathcal{V}^*$ if and only if $v = 0$.

11.2 Let $\mathcal{V}$ be a finite dimensional K-vector space and $f \in \mathcal{V}^* \setminus \{0\}$. Show that then $\dim(\ker(f)) = \dim(\mathcal{V}) - 1$.

11.3 Consider the basis $B = \{10, t - 1, t^2 - t\}$ of the 3-dimensional vector space $\mathbb{R}[t]_{\leq 2}$. Compute the dual basis B^* to B.

11.4 Let V be an n-dimensional K-vector space, and let $\{v_1^*, \ldots, v_n^*\}$ be a basis of V^*. Prove or disprove: There exists a unique basis $\{v_1, \ldots, v_n\}$ of V with $v_i^*(v_j) = \delta_{ij}$.

11.5 Let V and W be K-vector spaces, and let

$$\varphi : V^* \times W^* \to (V \times W)^*, \quad (f, g) \mapsto h_{f,g},$$
$$\text{where} \quad h_{f,g} : (v, w) \mapsto f(v) + g(w).$$

Show that φ is an isomorphism.

11.6 Let V be a finite dimensional K-vector space, let V^* be the dual space of V, and let $V^{**} := (V^*)^*$ be the dual space of V^*. (The vector space V^{**} is called the *bidual space* of V.) Consider the map

$$\Phi : V \to V^{**}, \quad v \mapsto \varphi_v,$$

where for all $v \in V$ the map φ_v is defined by $\varphi_v : V^* \to K,\; f \mapsto f(v)$. Show that Φ is an isomorphism.

11.7 Let V be a finite dimensional K-vector space and let $f, g \in V^*$ with $f \neq 0$. Show that $g = \lambda f$ for a $\lambda \in K \setminus \{0\}$ holds if and only if $\ker(f) = \ker(g)$. Is it possible to omit the assumption $f \neq 0$?

11.8 Let V be a K-vector space and let $\mathcal{U}$ be a subspace of V. The set

$$\mathcal{U}^0 := \{ f \in V^* \mid f(u) = 0 \text{ for all } u \in \mathcal{U} \}$$

is called the *annihilator* of $\mathcal{U}$. Show the following assertions:
(a) $\mathcal{U}^0$ is a subspace of V^*.
(b) If V is finite dimensional and $\mathcal{U}_1, \mathcal{U}_2$ are subspaces of V, then we have

$$(\mathcal{U}_1 + \mathcal{U}_2)^0 = \mathcal{U}_1^0 \cap \mathcal{U}_2^0 \quad \text{and} \quad (\mathcal{U}_1 \cap \mathcal{U}_2)^0 = \mathcal{U}_1^0 + \mathcal{U}_2^0,$$

and if $\mathcal{U}_1 \subseteq \mathcal{U}_2$, then $\mathcal{U}_2^0 \subseteq \mathcal{U}_1^0$. (Can you show this also for infinite dimensional vector spaces?)
(c) If W is a K-vector space and $f \in \mathcal{L}(V, W)$, then $\ker(f^*) = (\text{im}(f))^0$.

11.9 Prove Lemma 11.6 (2) and (3).

11.10 Let V be a finite dimensional K-vector space and $f, g \in \mathcal{L}(V, V)$. Show that $f \circ g = g \circ f$ holds if and only if $f^* \circ g^* = g^* \circ f^*$.

11.11 Let V and W be K-vector spaces. Show that the set $\text{Bil}(V, W)$ of all bilinear forms on $V \times W$ with the operations

$$+ : (\beta_1 + \beta_2)(v, w) := \beta_1(v, w) + \beta_2(v, w),$$
$$\cdot : (\lambda \cdot \beta)(v, w) := \lambda \beta(v, w),$$

is a K-vector space.

11.12 Let $\mathcal{V}$ and $\mathcal{W}$ be K-vector spaces with bases $\{v_1, \ldots, v_m\}$ and $\{w_1, \ldots, w_n\}$ and corresponding dual bases $\{v_1^*, \ldots, v_m^*\}$ and $\{w_1^*, \ldots, w_n^*\}$, respectively. For $i = 1, \ldots, m$ and $j = 1, \ldots, n$ let

$$\beta_{ij} : \mathcal{V} \times \mathcal{W} \to K, \quad (v, w) \mapsto v_i^*(v) w_j^*(w).$$

(a) Show that β_{ij} is a bilinear form on $\mathcal{V} \times \mathcal{W}$.
(b) Show that the set $\{\beta_{ij} \mid i = 1, \ldots, m, \ j = 1, \ldots, n\}$ is a basis of $\text{Bil}(\mathcal{V}, \mathcal{W})$ (cp. Exercise 11.11) and determine the dimension of this space.

11.13 Show that the map β in Example 11.10 (1) is a bilinear form, and that β is non-degenerate if and only if $n = m$ and $A \in GL_n(K)$.

11.14 Show that the map β in Example 11.10 (3) is a symmetric bilinear form. Is β non-degenerate?

11.15 Let $\mathcal{V}$ be a finite dimensional K-vector space. Show that $\mathcal{V}, \mathcal{V}^*$ are a dual pair with respect to the bilinear form β in Example 11.10 (4), i.e., that β is non-degenerate.

11.16 Let $\mathcal{V}$ be a finite dimensional K-vector space and let $\mathcal{U} \subseteq \mathcal{V}$ and $\mathcal{W} \subseteq \mathcal{V}^*$ be subspaces with $\dim(\mathcal{U}) = \dim(\mathcal{W}) \geq 1$. Prove or disprove: The spaces $\mathcal{U}, \mathcal{W}$ form a dual pair with respect to the bilinear form $\beta : \mathcal{U} \times \mathcal{W} \to K$, $(v, h) \mapsto h(v)$.

11.17 Let $\mathcal{V}$ and $\mathcal{W}$ be finite dimensional K-vector spaces with the bases B_1 and B_2, respectively, and let $\beta \in \text{Bil}(\mathcal{V}, \mathcal{W})$.
(a) Show that the following statements are equivalent:
   (1) $[\beta]_{B_1 \times B_2}$ is not invertible.
   (2) β is degenerate in the second variable.
   (3) β is degenerate in the first variable.
(b) Conclude from (a): β is non-degenerate if and only if $[\beta]_{B_1 \times B_2}$ is invertible.

11.18 Let $\beta : \mathbb{R}^{2,1} \times \mathbb{R}^{2,1} \to \mathbb{R}, (x, y) \mapsto \det([x, y])$.
(a) Show that β is a bilinear form on $\mathbb{R}^{2,1}$. Is β degenerate?
(b) Determine $[\beta]_{B \times B}$ for the standard basis $B = \{e_1, e_2\}$ of $\mathbb{R}^{2,1}$.
(c) Determine $[\beta]_{\widetilde{B} \times \widetilde{B}}$ for the basis $\widetilde{B} = \{[1, 1]^T, [-1, 1]^T\}$ of $\mathbb{R}^{2,1}$ using $[\beta]_{B \times B}$ from (b) and Theorem 11.16.

11.19 Prove Lemma 11.15.

11.20 Prove Lemma 11.18.

11.21 For a bilinear form β on a K-vector space $\mathcal{V}$, the map $q_\beta : \mathcal{V} \to K$, $v \mapsto \beta(v, v)$, is called the *quadratic form* induced by β. Show the following assertion:

If $1 + 1 \neq 0$ in K and β is symmetric, then $\beta(v, w) = \frac{1}{2}(q_\beta(v + w) - q_\beta(v) - q_\beta(w))$ holds for all $v, w \in \mathcal{V}$.

Exercises

11.22 Show that a sesquilinear form s on a $\mathbb{C}$-vector space $\mathcal{V}$ satisfies the *polarization identity*

$$s(v, w) = \frac{1}{4}\big(s(v + w, v + w) - s(v - w, v - w) + \mathbf{i}s(v + \mathbf{i}w, v + \mathbf{i}w) \\ - \mathbf{i}s(v - \mathbf{i}w, v - \mathbf{i}w)\big)$$

for all $v, w \in \mathcal{V}$.

11.23 Consider the following maps from $\mathbb{C}^{3,1} \times \mathbb{C}^{3,1}$ to $\mathbb{C}$:
  (a) $\beta_1(x, y) = 3x_1\overline{x}_1 + 3y_1\overline{y}_1 + x_2\overline{y}_3 - x_3\overline{y}_2$,
  (b) $\beta_2(x, y) = x_1\overline{y}_2 + x_2\overline{y}_3 + x_3\overline{y}_1$,
  (c) $\beta_3(x, y) = x_1y_2 + x_2y_3 + x_3y_1$,
  (d) $\beta_4(x, y) = 3x_1\overline{y}_1 + x_1\overline{y}_2 + x_2\overline{y}_1 + 2\mathbf{i}x_2\overline{y}_3 - 2\mathbf{i}x_3\overline{y}_2 + x_3\overline{y}_3$.

  Which of these are bilinear forms or sesquilinear forms on $\mathbb{C}^{3,1}$? Test whether the bilinear form is symmetric or the sesquilinear form is Hermitian, and derive the corresponding matrix representations with respect to the canonical basis $B_1 = \{e_1, e_2, e_3\}$ and the basis $B_2 = \{e_1, e_1 + \mathbf{i}e_2, e_2 + \mathbf{i}e_3\}$.

11.24 Prove Lemma 11.23.

11.25 Let $A \in \mathbb{C}^{n,n}$ be Hermitian. Show that

$$s : \mathbb{C}^{n,1} \times \mathbb{C}^{n,1} \to \mathbb{C}, \quad (v, w) \mapsto w^H A v,$$

is a Hermitian sesquilinear form on $\mathbb{C}^{n,1}$.

11.26 Let $\mathcal{V}$ be a finite dimensional $\mathbb{C}$-vector space with the basis B, and let s be a sesquilinear form on $\mathcal{V}$. Show that s is Hermitian if and only if $[s]_{B \times B}$ is Hermitian.

11.27 Let $\mathcal{V}$ be a finite dimensional $\mathbb{C}$-vector space and let s be a sesquilinear form on $\mathcal{V}$. For all $v, w \in \mathcal{V}$ we decompose $s(v, w) \in \mathbb{C}$ in its real and imaginary parts, and we write

$$s(v, w) = f(v, w) + \mathbf{i}g(v, w)$$

with $f(v, w) \in \mathbb{R}$ and $g(v, w) \in \mathbb{R}$.
  (a) Show that the maps $f : \mathcal{V} \times \mathcal{V} \to \mathbb{R}$ and $g : \mathcal{V} \times \mathcal{V} \to \mathbb{R}$ defined in this way are bilinear forms on the $\mathbb{R}$-vector space $\mathcal{V}$.
  (b) Show that the following statements are equivalent: (1) s is Hermitian. (2) f is symmetric. (3) g is *anti-symmetric*, i.e., $g(v, w) = -g(w, v)$ holds for all $v, w \in \mathcal{V}$. (An anti-symmetric bilinear form is also called *skew-symmetric*.)

11.28 Show the following assertions for $A, B \in \mathbb{C}^{n,n}$:
  (a) If $A^H = -A$, then the eigenvalues of A are purely imaginary.
  (b) If $A^H = -A$, then $\text{trace}(A^2) \leq 0$ and $(\text{trace}(A))^2 \leq 0$.
  (c) If $A^H = A$ and $B^H = B$, then $\text{trace}((AB)^2) \leq \text{trace}(A^2B^2)$.

11.29 Prove Theorem 11.27.

11.30 Prove Lemma 11.29.

Euclidean and Unitary Vector Spaces 12

In this chapter we study vector spaces over the fields $\mathbb{R}$ and $\mathbb{C}$. Using the definition of bilinear and sesquilinear forms, we introduce scalar products on such vector spaces. Scalar products allow the extension of well-known concepts from elementary geometry, such as length and angles, to abstract real and complex vector spaces. This, in particular, leads to the idea of orthogonality and to orthonormal bases of vector spaces. As an example for the importance of these concepts in many applications we study least-squares approximations.

12.1 Scalar Products and Norms

We start with the definition of a scalar product and the Euclidean or unitary vector spaces.

Definition 12.1 Let $\mathcal{V}$ be a K-vector space, where either $K = \mathbb{R}$ or $K = \mathbb{C}$. A map

$$\langle \cdot, \cdot \rangle : \mathcal{V} \times \mathcal{V} \to K, \quad (v, w) \mapsto \langle v, w \rangle,$$

is called a *scalar product* on $\mathcal{V}$, if the following properties hold:

(1) If $K = \mathbb{R}$, then $\langle \cdot, \cdot \rangle$ is a symmetric bilinear form.
    If $K = \mathbb{C}$, then $\langle \cdot, \cdot \rangle$ is an Hermitian sesquilinear form.
(2) $\langle \cdot, \cdot \rangle$ is *positive definite*, i.e., $\langle v, v \rangle \geq 0$ holds for all $v \in \mathcal{V}$, with equality if and only if $v = 0$.

An $\mathbb{R}$-vector space with a scalar product is called a *Euclidean vector space*,[1] and a $\mathbb{C}$-vector space with a scalar product is called a *unitary vector space*.

---
[1] Euclid of Alexandria (approx. 300 BC).

Scalar products are sometimes called *inner products*. Note that $\langle v, v \rangle$ is nonnegative and *real* also when $\mathcal{V}$ is a $\mathbb{C}$-vector space.

In both the Euclidean and unitary case we have $\langle 0, w \rangle = \langle 0 + 0, w \rangle = 2\langle 0, w \rangle$ and, therefore, $\langle 0, w \rangle = 0$ for all $w \in \mathcal{V}$. In the same way it follows that $\langle v, 0 \rangle = 0$ for all $v \in \mathcal{V}$.

A scalar product is always non-degenerate. If for a given $v \in \mathcal{V}$ the equation $\langle v, w \rangle = 0$ holds for all $w \in \mathcal{V}$, then, in particular, $\langle v, v \rangle = 0$, and the positive definiteness yields $v = 0$. If $\langle v, w \rangle = 0$ for a given $w \in \mathcal{V}$ and all $v \in \mathcal{V}$, then $\langle w, w \rangle = 0$, and hence $w = 0$.

It is easy to see that a subspace $\mathcal{U}$ of a Euclidean or unitary vector space $\mathcal{V}$ is again a Euclidean or unitary vector space, respectively, when the scalar product on the space $\mathcal{V}$ is restricted to the subspace $\mathcal{U}$.

Example 12.2

(1) A scalar product on $\mathbb{R}^{n,1}$ is given by

$$\langle v, w \rangle := w^T v.$$

It is called the *standard scalar product of* $\mathbb{R}^{n,1}$.

(2) A scalar product on $\mathbb{C}^{n,1}$ is given by

$$\langle v, w \rangle := w^H v.$$

It is called the *standard scalar product of* $\mathbb{C}^{n,1}$.

(3) For both $K = \mathbb{R}$ and $K = \mathbb{C}$,

$$\langle A, B \rangle := \text{trace}(B^H A)$$

is a scalar product on $K^{n,m}$.

(4) A scalar product on the vector space of the continuous and real valued functions on the real interval $[0, 1]$ is given by

$$\langle f, g \rangle := \int_0^1 f(x)g(x)dx.$$

We will now show how to use the Euclidean or unitary structure of a vector space in order to introduce geometric concepts such as the length of a vector or the angle between vectors.

As a motivation of a general concept of length we have the *absolute value* of real numbers, i.e., the map $|\cdot| : \mathbb{R} \to \mathbb{R}, x \mapsto |x|$. This map has the following properties:

(1) $|\lambda x| = |\lambda| \cdot |x|$ for all $\lambda, x \in \mathbb{R}$.

12.1 Scalar Products and Norms

(2) $|x| \geq 0$ for all $x \in \mathbb{R}$, with equality if and only if $x = 0$.
(3) $|x + y| \leq |x| + |y|$ for all $x, y \in \mathbb{R}$.

These properties are generalized to real or complex vector spaces as follows.

Definition 12.3 Let $\mathcal{V}$ be a K-vector space, where either $K = \mathbb{R}$ or $K = \mathbb{C}$. A map

$$\| \cdot \| : \mathcal{V} \to \mathbb{R}, \quad v \mapsto \|v\|,$$

is called a *norm* on $\mathcal{V}$, when for all $v, w \in \mathcal{V}$ and $\lambda \in K$ the following properties hold:

(1) $\|\lambda v\| = |\lambda| \cdot \|v\|$.
(2) $\|v\| \geq 0$, with equality if and only if $v = 0$.
(3) $\|v + w\| \leq \|v\| + \|w\|$ (triangle inequality).

A K-vector space on which a norm is defined is called a *normed space*.

Example 12.4

(1) If $\langle \cdot, \cdot \rangle$ is the standard scalar product on $\mathbb{R}^{n,1}$, then

$$\|v\| := \langle v, v \rangle^{1/2} = (v^T v)^{1/2}$$

defines a norm that is called the *Euclidean norm of* $\mathbb{R}^{n,1}$.

(2) If $\langle \cdot, \cdot \rangle$ is the standard scalar product on $\mathbb{C}^{n,1}$, then

$$\|v\| := \langle v, v \rangle^{1/2} = (v^H v)^{1/2}$$

defines a norm that is called the *Euclidean norm of* $\mathbb{C}^{n,1}$. (This is common terminology, although the space itself is unitary and not Euclidean.)

(3) For both $K = \mathbb{R}$ and $K = \mathbb{C}$,

$$\|A\|_F := (\text{trace}(A^H A))^{1/2} = \Big(\sum_{i=1}^{n} \sum_{j=1}^{m} |a_{ij}|^2 \Big)^{1/2}$$

is a norm on $K^{n,m}$ that is called the *Frobenius norm*[2] *of* $K^{n,m}$. For $m = 1$ the Frobenius norm is equal to the Euclidean norm of $K^{n,1}$. Moreover, the Frobenius norm of $K^{n,m}$ is equal to the Euclidean norm of $K^{nm,1}$ (or K^{nm}), if we identify these vector spaces via an isomorphism.
Obviously, we have $\|A\|_F = \|A^T\|_F = \|A^H\|_F$ for all $A \in K^{n,m}$.

---

[2] Ferdinand Georg Frobenius (1849–1917).

(4) If $\mathcal{V}$ is the vector space of the continuous and real valued functions on the real interval [0, 1], then

$$\|f\| := \langle f, f \rangle^{1/2} = \left(\int_0^1 (f(x))^2 dx \right)^{1/2}$$

is a norm on $\mathcal{V}$ that is called the L^2-norm.

(5) Let $K = \mathbb{R}$ or $K = \mathbb{C}$, and let $p \in \mathbb{R}$, $p \geq 1$ be given. Then for $v = [v_1, \ldots, v_n]^T \in K^{n,1}$ the *p-norm of $K^{n,1}$* is defined by

$$\|v\|_p := \left(\sum_{i=1}^n |v_i|^p \right)^{1/p}. \tag{12.1}$$

For $p = 2$ this is the Euclidean norm on $K^{n,1}$. For this norm we typically omit the index 2 and write $\|\cdot\|$ instead of $\|\cdot\|_2$ (as in (1) and (2) above). Taking the limit $p \to \infty$ in (12.1), we obtain the ∞-*norm of $K^{n,1}$*, given by

$$\|v\|_\infty = \max_{1 \leq i \leq n} |v_i|.$$

The following figures illustrate the unit circle in $\mathbb{R}^{2,1}$ with respect to the *p*-norm, i.e., the set of all $v \in \mathbb{R}^{2,1}$ with $\|v\|_p = 1$, for $p = 1$, $p = 2$ and $p = \infty$:

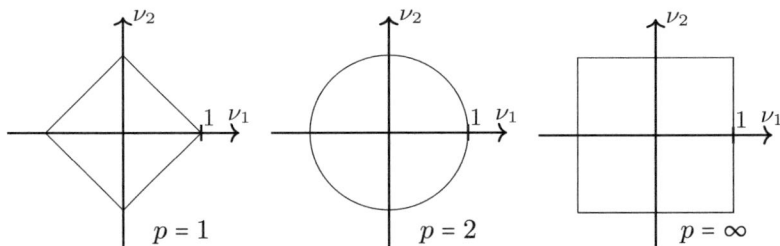

(6) For $K = \mathbb{R}$ or $K = \mathbb{C}$ the associated *p-norm of $K^{n,m}$* is defined by

$$\|A\|_p := \sup_{v \in K^{m,1} \setminus \{0\}} \frac{\|Av\|_p}{\|v\|_p}.$$

Here we use the *p*-norm of $K^{m,1}$ in the denominator and the *p*-norm of $K^{n,1}$ in the numerator. The notation sup means *supremum*, i.e., the least upper bound that is known from Analysis. One can show that the supremum is attained by a vector v, and thus we may write max instead of sup in the definition above.

12.1 Scalar Products and Norms

In particular, for $A = [a_{ij}] \in K^{n,m}$ we have

$$\|A\|_1 = \max_{1 \leq j \leq m} \sum_{i=1}^{n} |a_{ij}|,$$

$$\|A\|_\infty = \max_{1 \leq i \leq n} \sum_{j=1}^{m} |a_{ij}|.$$

These norms are called *maximum column sum* and *maximum row sum norm* of $K^{n,m}$, respectively. We easily see that $\|A\|_1 = \|A^T\|_\infty = \|A^H\|_\infty$ and $\|A\|_\infty = \|A^T\|_1 = \|A^H\|_1$. However, for the matrix

$$A = \begin{bmatrix} 1/2 & -1/4 \\ -1/2 & 2/3 \end{bmatrix} \in \mathbb{R}^{2,2}$$

we have $\|A\|_1 = 1$ and $\|A\|_\infty = 7/6$. Thus, this matrix A satisfies $\|A\|_1 < \|A\|_\infty$ and $\|A^T\|_\infty < \|A^T\|_1$. The 2-norm of matrices will be considered further in Chap. 19.

The norms in the above examples (1)–(4) have the form $\|v\| = \langle v, v \rangle^{1/2}$, where $\langle \cdot, \cdot \rangle$ is a given scalar product. We will show now that the map $v \mapsto \langle v, v \rangle^{1/2}$ always defines a norm. Our proof is based on the following theorem.

Theorem 12.5 *If $\mathcal{V}$ is a Euclidean or unitary vector space with the scalar product $\langle \cdot, \cdot \rangle$, then*

$$|\langle v, w \rangle|^2 \leq \langle v, v \rangle \cdot \langle w, w \rangle \quad \text{for all } v, w \in \mathcal{V}, \tag{12.2}$$

with equality if and only if v, w are linearly dependent.

Proof The inequality is trivial for $w = 0$. Thus, let $w \neq 0$ and let

$$\lambda := \frac{\langle v, w \rangle}{\langle w, w \rangle}.$$

Then

$$0 \leq \langle v - \lambda w, v - \lambda w \rangle = \langle v, v \rangle - \overline{\lambda} \langle v, w \rangle - \lambda \langle w, v \rangle - \lambda(-\overline{\lambda})\langle w, w \rangle$$

$$= \langle v, v \rangle - \overline{\frac{\langle v, w \rangle}{\langle w, w \rangle}} \langle v, w \rangle - \frac{\langle v, w \rangle}{\langle w, w \rangle} \overline{\langle v, w \rangle} + \frac{|\langle v, w \rangle|^2}{\langle w, w \rangle^2} \langle w, w \rangle$$

$$= \langle v, v \rangle - \frac{|\langle v, w \rangle|^2}{\langle w, w \rangle},$$

which implies (12.2).

If v, w are linearly dependent, then $v = \lambda w$ for a scalar λ, and hence

$$|\langle v, w\rangle|^2 = |\langle \lambda w, w\rangle|^2 = |\lambda \langle w, w\rangle|^2 = |\lambda|^2 |\langle w, w\rangle|^2 = \lambda \overline{\lambda} \langle w, w\rangle \langle w, w\rangle$$
$$= \langle \lambda w, \lambda w\rangle \langle w, w\rangle = \langle v, v\rangle \langle w, w\rangle.$$

On the other hand, let $|\langle v, w\rangle|^2 = \langle v, v\rangle \langle w, w\rangle$. If $w = 0$, then v, w are linearly dependent. If $w \neq 0$, then we define λ as above and get

$$\langle v - \lambda w, v - \lambda w\rangle = \langle v, v\rangle - \frac{|\langle v, w\rangle|^2}{\langle w, w\rangle} = 0.$$

Since the scalar product is positive definite, we have $v - \lambda w = 0$, and thus v, w are linearly dependent. □

The inequality (12.2) is called *Cauchy-Schwarz inequality*.[3] It is an important tool in Analysis, in particular in the estimation of approximation and interpolation errors.

Corollary 12.6 *If $\mathcal{V}$ is a Euclidean or unitary vector space with the scalar product $\langle \cdot, \cdot \rangle$, then the map*

$$\| \cdot \| : \mathcal{V} \to \mathbb{R}, \quad v \mapsto \|v\| := \langle v, v\rangle^{1/2},$$

is a norm on $\mathcal{V}$ that is called the norm induced by the scalar product.

Proof We have to prove the three defining properties of the norm. Since $\langle \cdot, \cdot \rangle$ is positive definite, we have $\|v\| \geq 0$, with equality if and only if $v = 0$. If $v \in \mathcal{V}$ and $\lambda \in K$ (where in the Euclidean case $K = \mathbb{R}$ and in the unitary case $K = \mathbb{C}$), then

$$\|\lambda v\|^2 = \langle \lambda v, \lambda v\rangle = \lambda \overline{\lambda} \langle v, v\rangle = |\lambda|^2 \|v\|^2,$$

and hence $\|\lambda v\| = |\lambda| \|v\|$. In order to show the triangle inequality, we use the Cauchy-Schwarz inequality and the fact that $\text{Re}(z) \leq |z|$ for every complex number z. For all $v, w \in \mathcal{V}$ we have

$$\|v + w\|^2 = \langle v + w, v + w\rangle = \langle v, v\rangle + \langle v, w\rangle + \langle w, v\rangle + \langle w, w\rangle$$
$$= \langle v, v\rangle + \langle v, w\rangle + \overline{\langle v, w\rangle} + \langle w, w\rangle$$
$$= \|v\|^2 + 2\,\text{Re}(\langle v, w\rangle) + \|w\|^2$$

---
[3] Augustin Louis Cauchy (1789–1857) and Hermann Amandus Schwarz (1843–1921).

$$\leq \|v\|^2 + 2|\langle v, w\rangle| + \|w\|^2$$
$$\leq \|v\|^2 + 2\|v\|\,\|w\| + \|w\|^2$$
$$= (\|v\| + \|w\|)^2,$$

and thus $\|v + w\| \leq \|v\| + \|w\|$. □

12.2 Orthogonality

We will now use the scalar product to introduce angles between vectors. As motivation we consider the Euclidean vector space $\mathbb{R}^{2,1}$ with the standard scalar product and the induced Euclidean norm $\|v\| = \langle v, v\rangle^{1/2}$. The Cauchy-Schwarz inequality shows that

$$-1 \leq \frac{\langle v, w\rangle}{\|v\|\,\|w\|} \leq 1 \quad \text{for all } v, w \in \mathbb{R}^{2,1} \setminus \{0\}.$$

If $v, w \in \mathbb{R}^{2,1} \setminus \{0\}$, then the angle between v and w is the uniquely determined real number $\varphi \in [0, \pi]$ with

$$\cos(\varphi) = \frac{\langle v, w\rangle}{\|v\|\,\|w\|}.$$

The vectors v, w are orthogonal if $\varphi = \pi/2$, so that $\cos(\varphi) = 0$. Thus, v, w are orthogonal if and only if $\langle v, w\rangle = 0$.

An elementary calculation now leads to the *cosine theorem for triangles*:

$$\|v - w\|^2 = \langle v - w, v - w\rangle = \langle v, v\rangle - 2\langle v, w\rangle + \langle w, w\rangle$$
$$= \|v\|^2 + \|w\|^2 - 2\|v\|\,\|w\|\cos(\varphi).$$

If v, w are orthogonal, i.e., $\langle v, w\rangle = 0$, then the cosine theorem implies the *Pythagorean theorem*:[4]

$$\|v - w\|^2 = \|v\|^2 + \|w\|^2.$$

---

[4] Pythagoras of Samos (approx. 570–500 BC).

The following figures illustrate the cosine theorem and the Pythagorean theorem for vectors in $\mathbb{R}^{2,1}$:

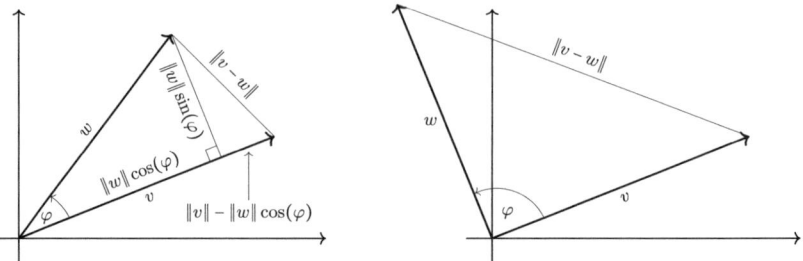

In the following definition we generalize the ideas of angles and orthogonality.

Definition 12.7 Let $\mathcal{V}$ be a Euclidean or unitary vector space with the scalar product $\langle \cdot, \cdot \rangle$.

(1) In the Euclidean case, the *angle* between two vectors $v, w \in \mathcal{V} \setminus \{0\}$ is the uniquely determined real number $\varphi \in [0, \pi]$ with

$$\cos(\varphi) = \frac{\langle v, w \rangle}{\|v\| \|w\|}.$$

(2) Two vectors $v, w \in \mathcal{V}$ are called *orthogonal* (with respect to the given scalar product $\langle \cdot, \cdot \rangle$), if $\langle v, w \rangle = 0$. We then also write $v \perp w$.
(3) A basis $\{v_1, \ldots, v_n\}$ of $\mathcal{V}$ is called an *orthogonal basis*, if

$$\langle v_i, v_j \rangle = 0, \quad i, j = 1, \ldots, n \text{ and } i \neq j.$$

If, furthermore,

$$\|v_i\| = 1, \quad i = 1, \ldots, n,$$

where $\|v\| = \langle v, v \rangle^{1/2}$ is the norm induced by the scalar product, then $\{v_1, \ldots, v_n\}$ is called an *orthonormal basis* of $\mathcal{V}$. (For an orthonormal basis we therefore have $\langle v_i, v_j \rangle = \delta_{ij}$.)

We now prove some immediate consequences of the orthogonality of vectors.

Lemma 12.8 *Let $\mathcal{V}$ be a Euclidean or unitary vector space with the scalar product $\langle \cdot, \cdot \rangle$ and the induced norm $\| \cdot \| = \langle \cdot, \cdot \rangle^{1/2}$.*

(1) If $v_1, \ldots, v_k \in \mathcal{V} \setminus \{0\}$ are pairwise orthogonal, i.e., $v_i \perp v_j = 0$ holds for all $i \neq j$, then these vectors are linearly independent.

12.2 Orthogonality

(2) *If $v_1, \ldots, v_k \in \mathcal{V}$ are pairwise orthogonal, then $\|\sum_{i=1}^{k} v_i\|^2 = \sum_{i=1}^{k} \|v_i\|^2$. (Generalized Pythagorean theorem.)*

Proof

(1) If $v_1, \ldots, v_k \in \mathcal{V} \setminus \{0\}$ are pairwise orthogonal and $0 = \sum_{i=1}^{k} \lambda_i v_i$, then

$$0 = \left\langle \sum_{i=1}^{k} \lambda_i v_i, \, v_j \right\rangle = \sum_{i=1}^{k} \lambda_i \langle v_i, v_j \rangle = \lambda_j \langle v_j, v_j \rangle, \quad j = 1, \ldots, k.$$

Now $\langle v_j, v_j \rangle \neq 0$ yields $\lambda_j = 0$ for all $j = 1, \ldots, k$.

(2) If $v_1, \ldots, v_k \in \mathcal{V}$ are pairwise orthogonal, then

$$\left\| \sum_{i=1}^{k} v_i \right\|^2 = \left\langle \sum_{i=1}^{k} v_i, \sum_{j=1}^{k} v_j \right\rangle = \sum_{i=1}^{k} \sum_{j=1}^{k} \langle v_i, v_j \rangle = \sum_{i=1}^{k} \langle v_i, v_i \rangle = \sum_{i=1}^{k} \|v_i\|^2,$$

where we have used that $\langle v_i, v_j \rangle = 0$ for $i \neq j$. □

Note that the terms in Definition 12.7 are defined with respect to the given scalar product. Different scalar products yield different angles between vectors. In particular, the orthogonality of two given vectors may be lost when we consider a different scalar product.

We now show that every finite dimensional Euclidean or unitary vector space has an orthonormal basis.

Theorem 12.9 *Let $\mathcal{V}$ be a Euclidean or unitary vector space with the basis $\{v_1, \ldots, v_n\}$. Then there exists an orthonormal basis $\{u_1, \ldots, u_n\}$ of $\mathcal{V}$ with*

$$\text{span}\{u_1, \ldots, u_k\} = \text{span}\{v_1, \ldots, v_k\}, \quad k = 1, \ldots, n.$$

Proof We give the proof by induction on $\dim(\mathcal{V}) = n$. If $n = 1$, then we set $u_1 := \|v_1\|^{-1} v_1$. Then $\|u_1\| = 1$, and $\{u_1\}$ is an orthonormal basis of $\mathcal{V}$ with $\text{span}\{u_1\} = \text{span}\{v_1\}$.

Let the assertion hold for an $n \geq 1$. Let $\dim(\mathcal{V}) = n+1$ and let $\{v_1, \ldots, v_{n+1}\}$ be a basis of $\mathcal{V}$. Then $\mathcal{V}_n := \text{span}\{v_1, \ldots, v_n\}$ is an n-dimensional subspace of $\mathcal{V}$. By the induction hypothesis there exists an orthonormal basis $\{u_1, \ldots, u_n\}$ of $\mathcal{V}_n$ with $\text{span}\{u_1, \ldots, u_k\} = \text{span}\{v_1, \ldots, v_k\}$ for $k = 1, \ldots, n$. We define

$$\widehat{u}_{n+1} := v_{n+1} - \sum_{k=1}^{n} \langle v_{n+1}, u_k \rangle u_k, \quad u_{n+1} := \|\widehat{u}_{n+1}\|^{-1} \widehat{u}_{n+1}.$$

Since $v_{n+1} \notin \mathcal{V}_n = \operatorname{span}\{u_1, \ldots, u_n\}$, we must have $\widehat{u}_{n+1} \neq 0$, and Lemma 9.18 yields $\operatorname{span}\{u_1, \ldots, u_{n+1}\} = \operatorname{span}\{v_1, \ldots, v_{n+1}\}$.

For $j = 1, \ldots, n$ we have

$$\begin{aligned}
\langle u_{n+1}, u_j \rangle &= \langle \|\widehat{u}_{n+1}\|^{-1} \widehat{u}_{n+1}, u_j \rangle \\
&= \|\widehat{u}_{n+1}\|^{-1} \left(\langle v_{n+1}, u_j \rangle - \sum_{k=1}^{n} \langle v_{n+1}, u_k \rangle \langle u_k, u_j \rangle \right) \\
&= \|\widehat{u}_{n+1}\|^{-1} \left(\langle v_{n+1}, u_j \rangle - \langle v_{n+1}, u_j \rangle \right) \\
&= 0.
\end{aligned}$$

Finally, $\langle u_{n+1}, u_{n+1} \rangle = \|\widehat{u}_{n+1}\|^{-2} \langle \widehat{u}_{n+1}, \widehat{u}_{n+1} \rangle = 1$ which completes the proof. □

The proof of Theorem 12.9 shows how a given basis $\{v_1, \ldots, v_n\}$ can be *orthonormalized*, i.e., transformed into an orthonormal basis $\{u_1, \ldots, u_n\}$ with

$$\operatorname{span}\{u_1, \ldots, u_k\} = \operatorname{span}\{v_1, \ldots, v_k\}, \quad k = 1, \ldots, n.$$

The resulting algorithm is called the *Gram-Schmidt method*:[5]

Algorithm 12.10 *Given a basis $\{v_1, \ldots, v_n\}$ of $\mathcal{V}$.*

(1) Set $u_1 := \|v_1\|^{-1} v_1$.
(2) For $j = 1, \ldots, n-1$ set

$$\widehat{u}_{j+1} := v_{j+1} - \sum_{k=1}^{j} \langle v_{j+1}, u_k \rangle u_k,$$

$$u_{j+1} := \|\widehat{u}_{j+1}\|^{-1} \widehat{u}_{j+1}.$$

A slight reordering and combination of steps in the Gram-Schmidt method yields

$$\underbrace{(v_1, v_2, \ldots, v_n)}_{\in \mathcal{V}^n} = \underbrace{(u_1, u_2, \ldots, u_n)}_{\in \mathcal{V}^n} \begin{bmatrix} \|v_1\| & \langle v_2, u_1 \rangle & \cdots & \langle v_n, u_1 \rangle \\ & \|\widehat{u}_2\| & \ddots & \vdots \\ & & \ddots & \langle v_n, u_{n-1} \rangle \\ & & & \|\widehat{u}_n\| \end{bmatrix}.$$

---

[5] Jørgen Pedersen Gram (1850–1916) and Erhard Schmidt (1876–1959).

12.2 Orthogonality

The upper triangular matrix on the right hand side is the coordinate transformation matrix from the basis $\{v_1, \ldots, v_n\}$ to the basis $\{u_1, \ldots, u_n\}$ of $\mathcal{V}$ (cp. Theorem 9.30 or (10.5)). Thus, we have shown the following result.

Theorem 12.11 *If $\mathcal{V}$ is a finite dimensional Euclidean or unitary vector space with a given basis B_1, then the Gram-Schmidt method applied to B_1 yields an orthonormal basis B_2 of $\mathcal{V}$, such that $[\text{Id}_\mathcal{V}]_{B_1, B_2}$ is an invertible upper triangular matrix.*

This theorem has the following simple but useful consequence.

Corollary 12.12 *Let $\mathcal{V}$ be an n-dimensional Euclidean or unitary vector space, and let $\mathcal{U} \subset \mathcal{V}$ be an m-dimensional subspace of $\mathcal{V}$, where $1 \leq m < n$. If $\mathcal{U}$ has the orthonormal basis $\{u_1, \ldots, u_m\}$, then $\mathcal{V}$ has an orthonormal basis of the form $\{u_1, \ldots, u_m, u_{m+1}, \ldots, u_n\}$.*

Proof Using Theorem 9.15 we can extend the orthonormal basis $\{u_1, \ldots, u_m\}$ of $\mathcal{U}$ to a basis $\{u_1, \ldots, u_m, v_{m+1}, \ldots, v_n\}$ of $\mathcal{V}$. If we apply the Gram-Schmidt method to this basis, we obtain the orthonormal basis $\{u_1, \ldots, u_m, u_{m+1}, \ldots, u_n\}$ of $\mathcal{V}$. □

Consider an m-dimensional subspace of $\mathbb{R}^{n,1}$ or $\mathbb{C}^{n,1}$ with the standard scalar product $\langle \cdot, \cdot \rangle$, and write the m vectors of an orthonormal basis $\{q_1, \ldots, q_m\}$ as columns of a matrix, $Q := [q_1, \ldots, q_m]$. Then we obtain in the real case

$$Q^T Q = [q_i^T q_j] = [\langle q_j, q_i \rangle] = [\delta_{ji}] = I_m,$$

and analogously in the complex case

$$Q^H Q = [q_i^H q_j] = [\langle q_j, q_i \rangle] = [\delta_{ji}] = I_m.$$

If, on the other hand, $Q^T Q = I_m$ or $Q^H Q = I_m$ for a matrix $Q \in \mathbb{R}^{n,m}$ or $Q \in \mathbb{C}^{n,m}$, respectively, then the m columns of Q form an orthonormal basis (with respect to the standard scalar product) of the m-dimensional subspace $\text{span}\{q_1, \ldots, q_m\}$ of $\mathbb{R}^{n,1}$ or $\mathbb{C}^{n,1}$, respectively. A "matrix version" of Theorem 12.11 can therefore be formulated as follows.

Corollary 12.13 *Let $K = \mathbb{R}$ or $K = \mathbb{C}$ and let $A \in K^{n,m}$ have $\text{rank}(A) = m$. Then there exists a matrix $Q \in K^{n,m}$ with its m columns being orthonormal with respect to the standard scalar product of $K^{n,1}$, i.e., $Q^T Q = I_m$ for $K = \mathbb{R}$ or $Q^H Q = I_m$ for $K = \mathbb{C}$, and an upper triangular matrix $R \in GL_m(K)$, such that*

$$A = QR. \tag{12.3}$$

The factorization (12.3) is called a *QR-decomposition* of the matrix A. The QR-decomposition has many applications in Numerical Mathematics (cp. Example 12.17 below). It is of particular importance in this context that the multiplication with a matrix that has orthonormal columns with respect to the standard scalar products, does not change the Euclidean norm of a vector. More precisely, we have the following result.

Lemma 12.14 *Let $K = \mathbb{R}$ or $K = \mathbb{C}$ and let $Q \in K^{n,m}$ be a matrix with orthonormal columns with respect to the standard scalar product of $K^{n,1}$. Then $\|Qv\| = \|v\|$ holds for all $v \in K^{m,1}$. (Here $\|\cdot\|$ is the Euclidean norm of $K^{n,1}$ and of $K^{m,1}$.)*

Proof We have

$$\|Qv\|^2 = \langle Qv, Qv \rangle = v^H(Q^H Q)v = v^H v = \langle v, v \rangle = \|v\|^2.$$

(For $K = \mathbb{R}$ we can replace "H" by "T".) □

We now introduce two important classes of matrices.

Definition 12.15

(1) A matrix $Q \in \mathbb{R}^{n,n}$ whose columns form an orthonormal basis with respect to the standard scalar product of $\mathbb{R}^{n,1}$ is called *orthogonal*.
(2) A matrix $Q \in \mathbb{C}^{n,n}$ whose columns form an orthonormal basis with respect to the standard scalar product of $\mathbb{C}^{n,1}$ is called *unitary*.

A matrix $Q = [q_1, \ldots, q_n] \in \mathbb{R}^{n,n}$ is therefore orthogonal if and only if

$$Q^T Q = [q_i^T q_j] = [\langle q_j, q_i \rangle] = [\delta_{ji}] = I_n.$$

In particular, an orthogonal matrix Q is invertible with $Q^{-1} = Q^T$ (cp. Corollary 7.19). The equation $QQ^T = I_n$ means that the n rows of Q form an orthonormal basis of $\mathbb{R}^{1,n}$ (with respect to the scalar product $\langle v, w \rangle := wv^T$).

Analogously, a unitary matrix $Q \in \mathbb{C}^{n,n}$ is invertible with $Q^{-1} = Q^H$ and $Q^H Q = I_n = QQ^H$. The n columns of Q form an orthonormal basis of $\mathbb{C}^{1,n}$.

Lemma 12.16 *The sets $\mathcal{O}(n)$ of orthogonal and $\mathcal{U}(n)$ of unitary $n \times n$ matrices form subgroups of $GL_n(\mathbb{R})$ and $GL_n(\mathbb{C})$, respectively.*

Proof We consider only $\mathcal{O}(n)$; the proof for $\mathcal{U}(n)$ is analogous.

Since every orthogonal matrix is invertible, we have that $\mathcal{O}(n) \subset GL_n(\mathbb{R})$. The identity matrix I_n is orthogonal, and hence $I_n \in \mathcal{O}(n) \neq \emptyset$. If $Q \in \mathcal{O}(n)$, then also $Q^T = Q^{-1} \in \mathcal{O}(n)$, since $(Q^T)^T Q^T = QQ^T = I_n$. Finally, if $Q_1, Q_2 \in \mathcal{O}(n)$, then

$$(Q_1 Q_2)^T (Q_1 Q_2) = Q_2^T (Q_1^T Q_1) Q_2 = Q_2^T Q_2 = I_n,$$

and thus $Q_1 Q_2 \in \mathcal{O}(n)$. □

12.2 Orthogonality

We have already seen that the set of the $n \times n$ permutation matrices forms a subgroup of $GL_n(\mathbb{R})$ respectively $GL_n(\mathbb{C})$ (cp. Theorem 4.16). Since the permutation matrices are real orthogonal as well as complex unitary, the group of the $n \times n$ permutation matrices is also a subgroup of $\mathcal{O}(n)$ respectively $\mathcal{U}(n)$.

Example 12.17 In many applications measurements or samples lead to a data set that is represented by tuples $(\tau_i, \mu_i) \in \mathbb{R}^2$, $i = 1, \ldots, m$. Here $\tau_1 < \cdots < \tau_m$, are the pairwise distinct measurement points and $\mu_1, \ldots, \mu_m$ are the corresponding measurements. In order to approximate the given data set by a simple model, one can try to construct a polynomial p of small degree so that the values $p(\tau_1), \ldots, p(\tau_m)$ are as close as possible to the measurements $\mu_1, \ldots, \mu_m$.

The simplest case is a real polynomial of degree (at most) 1. Geometrically, this corresponds to the construction of a straight line in $\mathbb{R}^2$ that has a minimal distance to the given points, as shown in the figure below (cp. Sect. 1.4). There are many possibilities to measure the distance. In the following we will describe one of them in more detail and use the Gram-Schmidt method, or the QR-decomposition, for the construction of the straight line. In Statistics this method is called *linear regression*.

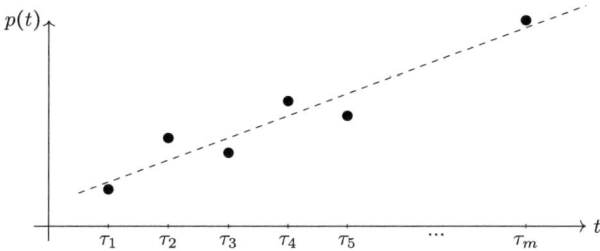

A real polynomial of degree (at most) 1 has the form $p = \alpha t + \beta$ and we are looking for coefficients $\alpha, \beta \in \mathbb{R}$ with

$$p(\tau_i) = \alpha \tau_i + \beta \approx \mu_i, \quad i = 1, \ldots, m.$$

Using matrices we can write this problem as

$$\begin{bmatrix} \tau_1 & 1 \\ \vdots & \vdots \\ \tau_m & 1 \end{bmatrix} \begin{bmatrix} \alpha \\ \beta \end{bmatrix} \approx \begin{bmatrix} \mu_1 \\ \vdots \\ \mu_m \end{bmatrix} \quad \text{or} \quad A \begin{bmatrix} \alpha \\ \beta \end{bmatrix} \approx y.$$

As mentioned above, there are different possibilities for interpreting the symbol "$\approx$". In particular, there are different norms in which we can measure the distance between the given values $\mu_1, \ldots, \mu_m$ and the polynomial values $p(\tau_1), \ldots, p(\tau_m)$.

Here we will use the Euclidean norm $\|\cdot\|$ and consider the minimization problem

$$\min_{\alpha,\beta\in\mathbb{R}} \left\| A\begin{bmatrix}\alpha\\\beta\end{bmatrix} - y \right\|.$$

The matrix A has rank 2, since the entries in its first column are pairwise distinct, while all entries in its second column are equal. Let

$$A = [q_1, q_2]R$$

be a QR-decomposition. We extend the vectors $q_1, q_2 \in \mathbb{R}^{m,1}$ to an orthonormal basis $\{q_1, q_2, q_3, \ldots, q_m\}$ of $\mathbb{R}^{m,1}$. Then $Q = [q_1, \ldots, q_m] \in \mathbb{R}^{m,m}$ is an orthogonal matrix and

$$\min_{\alpha,\beta\in\mathbb{R}} \left\| A\begin{bmatrix}\alpha\\\beta\end{bmatrix} - y \right\| = \min_{\alpha,\beta\in\mathbb{R}} \left\| [q_1,q_2]R\begin{bmatrix}\alpha\\\beta\end{bmatrix} - y \right\|$$

$$= \min_{\alpha,\beta\in\mathbb{R}} \left\| Q\begin{bmatrix}R\\0_{m-2,2}\end{bmatrix}\begin{bmatrix}\alpha\\\beta\end{bmatrix} - y \right\|$$

$$= \min_{\alpha,\beta\in\mathbb{R}} \left\| Q\left(\begin{bmatrix}R\\0_{m-2,2}\end{bmatrix}\begin{bmatrix}\alpha\\\beta\end{bmatrix} - Q^T y\right) \right\|$$

$$= \min_{\alpha,\beta\in\mathbb{R}} \left\| \begin{bmatrix}R\begin{bmatrix}\alpha\\\beta\end{bmatrix}\\0\\\vdots\\0\end{bmatrix} - \begin{bmatrix}q_1^T y\\q_2^T y\\q_3^T y\\\vdots\\q_m^T y\end{bmatrix} \right\|.$$

Here we have used that $QQ^T = I_m$ and $\|Qv\| = \|v\|$ for all $v \in \mathbb{R}^{m,1}$. The upper triangular matrix R is invertible and thus the minimization problem is solved by

$$\begin{bmatrix}\widehat{\alpha}\\\widehat{\beta}\end{bmatrix} = R^{-1}\begin{bmatrix}q_1^T y\\q_2^T y\end{bmatrix}.$$

Using the definition of the Euclidean norm, we can write the minimizing property of the polynomial $\widehat{p} := \widehat{\alpha} t + \widehat{\beta}$ as

$$\left\| A\begin{bmatrix}\widehat{\alpha}\\\widehat{\beta}\end{bmatrix} - y \right\|^2 = \sum_{i=1}^{m} (\widehat{p}(\tau_i) - \mu_i)^2$$

$$= \min_{\alpha,\beta\in\mathbb{R}} \left(\sum_{i=1}^{m} ((\alpha\tau_i + \beta) - \mu_i)^2 \right).$$

12.2 Orthogonality

Since the polynomial $\widehat{p}$ minimizes the sum of squares of the distances between the measurements μ_i and the polynomial values $\widehat{p}(\tau_i)$, this polynomial yields a *least squares approximation* of the measurement values.

Consider the example from Sect. 1.4. In the four quarters of a year, a company has profits of 10, 8, 9, 11 million Euros. Under the assumption that the profit grows linearly, i.e., like a straight line, the goal is to estimate the profit in the last quarter of the following year. The given data leads to the approximation problem

$$\begin{bmatrix} 1 & 1 \\ 2 & 1 \\ 3 & 1 \\ 4 & 1 \end{bmatrix} \begin{bmatrix} \alpha \\ \beta \end{bmatrix} \approx \begin{bmatrix} 10 \\ 8 \\ 9 \\ 11 \end{bmatrix} \quad \text{or} \quad A \begin{bmatrix} \alpha \\ \beta \end{bmatrix} \approx y.$$

The numerical computation of a QR-decomposition of A yields

$$\begin{bmatrix} \widehat{\alpha} \\ \widehat{\beta} \end{bmatrix} = \underbrace{\begin{bmatrix} \sqrt{30} & \frac{1}{3}\sqrt{30} \\ 0 & \frac{1}{3}\sqrt{6} \end{bmatrix}^{-1}}_{= R^{-1}} \underbrace{\begin{bmatrix} \frac{1}{\sqrt{30}} & \frac{2}{\sqrt{30}} & \frac{3}{\sqrt{30}} & \frac{4}{\sqrt{30}} \\ \frac{2}{\sqrt{6}} & \frac{1}{\sqrt{6}} & 0 & -\frac{1}{\sqrt{6}} \end{bmatrix}}_{= Q^T} \begin{bmatrix} 10 \\ 8 \\ 9 \\ 11 \end{bmatrix} = \begin{bmatrix} 0.4 \\ 8.5 \end{bmatrix},$$

and the resulting profit estimate for the last quarter of the following year is $\widehat{p}(8) = 11.7$, i.e., 11.7 million Euros.

> **MATLAB-Minute 7**
>
> In Example 12.17 one could imagine that the profit grows quadratically instead of linearly. Determine, analogously to the procedure in Example 12.17, a polynomial $\widehat{p} = \widehat{\alpha}t^2 + \widehat{\beta}t + \widehat{\gamma}$ that solves the least squares problem
>
> $$\sum_{i=1}^{4}(\widehat{p}(\tau_i) - \mu_i)^2 = \min_{\alpha,\beta,\gamma \in \mathbb{R}} \left(\sum_{i=1}^{4} \left((\alpha\tau_i^2 + \beta\tau_i + \gamma) - \mu_i \right)^2 \right).$$
>
> Use the MATLAB command qr for computing a QR-decomposition (why is A invertible?), and determine the estimated profit in the last quarter of the following year.

We will now analyze the properties of orthonormal bases in more detail.

Lemma 12.18 *If $\mathcal{V}$ is a Euclidean or unitary vector space with the scalar product $\langle \cdot, \cdot \rangle$ and the orthonormal basis $\{u_1, \ldots, u_n\}$, then*

$$v = \sum_{i=1}^{n} \langle v, u_i \rangle u_i$$

for all $v \in \mathcal{V}$.

Proof For every $v \in \mathcal{V}$ there exist uniquely determined coordinates $\lambda_1, \ldots, \lambda_n$ with $v = \sum_{i=1}^{n} \lambda_i u_i$. For every $j = 1, \ldots, n$ we then have $\langle v, u_j \rangle = \sum_{i=1}^{n} \lambda_i \langle u_i, u_j \rangle = \lambda_j$. □

The coordinates $\langle v, u_i \rangle$, $i = 1, \ldots, n$, of v with respect to an orthonormal basis $\{u_1, \ldots, u_n\}$ are often called the *Fourier coefficients*[6] of v with respect to this basis. The representation $v = \sum_{i=1}^{n} \langle v, u_i \rangle u_i$ is called the (abstract) *Fourier expansion* of v in the given orthonormal basis.

Corollary 12.19 *If $\mathcal{V}$ is a Euclidean or unitary vector space with the scalar product $\langle \cdot, \cdot \rangle$ and the orthonormal basis $\{u_1, \ldots, u_n\}$, then the following assertions hold:*

(1) $\langle v, w \rangle = \sum_{i=1}^{n} \langle v, u_i \rangle \langle u_i, w \rangle = \sum_{i=1}^{n} \langle v, u_i \rangle \overline{\langle w, u_i \rangle}$ *for all $v, w \in \mathcal{V}$ (Parseval's identity*[7]*).*
(2) $\langle v, v \rangle = \sum_{i=1}^{n} |\langle v, u_i \rangle|^2$ *for all $v \in \mathcal{V}$ (Bessel's identity*[8]*).*

Proof

(1) We have $v = \sum_{i=1}^{n} \langle v, u_i \rangle u_i$, and thus

$$\langle v, w \rangle = \left\langle \sum_{i=1}^{n} \langle v, u_i \rangle u_i, w \right\rangle = \sum_{i=1}^{n} \langle v, u_i \rangle \langle u_i, w \rangle = \sum_{i=1}^{n} \langle v, u_i \rangle \overline{\langle w, u_i \rangle}.$$

(2) is a special case of (1) for $v = w$. □

By Bessel's identity, every vector $v \in \mathcal{V}$ satisfies

$$\|v\|^2 = \langle v, v \rangle = \sum_{i=1}^{n} |\langle v, u_i \rangle|^2 \geq \max_{1 \leq i \leq n} |\langle v, u_i \rangle|^2,$$

---

[6] Jean Baptiste Joseph Fourier (1768–1830).
[7] Marc-Antoine Parseval (1755–1836).
[8] Friedrich Wilhelm Bessel (1784–1846).

12.2 Orthogonality

where $\|\cdot\|$ is the norm induced by the scalar product. The absolute value of each coordinate of v with respect to an orthonormal basis of $\mathcal{V}$ is therefore bounded by the norm of v. This property does not hold for a general basis of $\mathcal{V}$.

Example 12.20 Consider $\mathcal{V} = \mathbb{R}^{2,1}$ with the standard scalar product and the Euclidean norm, then for every real $\varepsilon \neq 0$ the set

$$\left\{ \begin{bmatrix} 1 \\ 0 \end{bmatrix}, \begin{bmatrix} 1 \\ \varepsilon \end{bmatrix} \right\}$$

is a basis of $\mathcal{V}$. For every vector $v = [v_1, v_2]^T$ we then have

$$v = \left(v_1 - \frac{v_2}{\varepsilon} \right) \begin{bmatrix} 1 \\ 0 \end{bmatrix} + \frac{v_2}{\varepsilon} \begin{bmatrix} 1 \\ \varepsilon \end{bmatrix}.$$

If $|v_1|, |v_2|$ are moderate numbers and if $|\varepsilon|$ is (very) small, then $|v_1 - v_2/\varepsilon|$ and $|v_2/\varepsilon|$ are (very) large. In numerical algorithms such a situation can lead to significant problems (e.g. due to roundoff errors) that are avoided when orthonormal bases are used.

Definition 12.21 Let $\mathcal{V}$ be a Euclidean or unitary vector space with the scalar product $\langle \cdot, \cdot \rangle$, and let $\mathcal{U} \subseteq \mathcal{V}$ be a subspace. Then

$$\mathcal{U}^\perp := \{ v \in \mathcal{V} \mid \langle v, u \rangle = 0 \text{ for all } u \in \mathcal{U} \}$$

is called the *orthogonal complement of* $\mathcal{U}$ (in $\mathcal{V}$).

Lemma 12.22 *The orthogonal complement* $\mathcal{U}^\perp$ *is a subspace of* $\mathcal{V}$.

Proof Exercise. □

Lemma 12.23 *If* $\mathcal{V}$ *is an n-dimensional Euclidean or unitary vector space, and if* $\mathcal{U} \subseteq \mathcal{V}$ *is an m-dimensional subspace, then* $\dim(\mathcal{U}^\perp) = n - m$ *and* $\mathcal{V} = \mathcal{U} \oplus \mathcal{U}^\perp$.

Proof We know that $0 \leq m \leq n$ (cp. Lemma 9.32). If $m = 0$, then $\mathcal{U} = \{0\}$ and $\mathcal{U}^\perp = \mathcal{V}$, so that the assertion is trivial. If $m = n$, then $\mathcal{U} = \mathcal{V}$, and thus

$$\mathcal{U}^\perp = \mathcal{V}^\perp = \{ v \in \mathcal{V} \mid \langle v, u \rangle = 0 \text{ for all } u \in \mathcal{V} \} = \{0\},$$

so that the assertion is again trivial.

Thus, let $1 \leq m < n$. According to Theorem 12.9 there exists an orthonormal basis $\{u_1, \ldots, u_m\}$ of $\mathcal{U}$, and Corollary 12.12 yields an orthonormal basis $\{u_1, \ldots, u_m, u_{m+1}, \ldots, u_n\}$ of $\mathcal{V}$. Then $\text{span}\{u_{m+1}, \ldots, u_n\} \subseteq \mathcal{U}^\perp$ and therefore $\mathcal{V} = \mathcal{U} + \mathcal{U}^\perp$. If $w \in \mathcal{U} \cap \mathcal{U}^\perp$, then $\langle w, w \rangle = 0$, and hence $w = 0$, since the scalar product is positive definite. Thus, $\mathcal{U} \cap \mathcal{U}^\perp = \{0\}$, which implies that $\mathcal{V} = \mathcal{U} \oplus \mathcal{U}^\perp$ and $\dim(\mathcal{U}^\perp) = n - m$ (cp. Theorem 9.34). In particular, we have $\mathcal{U}^\perp = \text{span}\{u_{m+1}, \ldots, u_n\}$. □

In the following we consider another important class of endomorphisms.

Definition 12.24 Let $\mathcal{V}$ be a Euclidean or unitary vector space. A map $f \in \mathcal{L}(\mathcal{V}, \mathcal{V})$ is called a *projection* if $f^2 = f \circ f = f$.

Simple examples of projections are $f = 0 \in \mathcal{L}(\mathcal{V}, \mathcal{V})$ and $f = \text{Id}_\mathcal{V}$. If $f \in \mathcal{L}(\mathcal{V}, \mathcal{V})$ is a projection and $v \in \text{im}(f)$, then $v = f(w)$ for some $w \in \mathcal{V}$. We obtain

$$f(v) = f^2(w) = f(w) = v,$$

i.e., $f = \text{Id}_{\text{im}(f)}$. We therefore call f a *projection onto* $\text{im}(f)$.

Then it holds that

$$(\text{Id}_\mathcal{V} - f)^2 = (\text{Id}_\mathcal{V} - f) \circ (\text{Id}_\mathcal{V} - f) = \text{Id}_\mathcal{V} \circ (\text{Id}_\mathcal{V} - f) - f \circ (\text{Id}_\mathcal{V} - f)$$
$$= \text{Id}_\mathcal{V} - f - f + f^2 = \text{Id}_\mathcal{V} - f,$$

and thus, for every projection f, also $\text{Id}_\mathcal{V} - f$ is a projection. If $v \in \mathcal{V}$ and $w = (\text{Id}_\mathcal{V} - f)(v) \in \text{im}(f)$, then

$$f(w) = f((\text{Id}_\mathcal{V} - f)(v)) = f(v) - f^2(v) = 0,$$

i.e., $w \in \ker(f)$ and, therefore, $\text{im}(\text{Id}_\mathcal{V} - f) \subseteq \ker(f)$. If on the other hand $w \in \ker(f)$, then

$$w = w - f(w) = (\text{Id}_\mathcal{V} - f)(w),$$

i.e., $w \in \text{im}(\text{Id}_\mathcal{V} - f)$ and, hence, $\ker(f) \subseteq \text{im}(\text{Id}_\mathcal{V} - f)$. Since $\text{im}(\text{Id}_\mathcal{V} - f) = \ker(f)$, it follows that $\text{Id}_\mathcal{V} - f$ is a projection onto $\ker(f)$.

Theorem 12.25 *Let $\mathcal{V}$ be a Euclidean or unitary vector space and let $f \in \mathcal{L}(\mathcal{V}, \mathcal{V})$ be a projection, then*

$$\mathcal{V} = \text{im}(f) \oplus \ker(f).$$

Proof If $v \in \mathcal{V}$ is arbitrary, then $v = f(v) + (\text{Id}_\mathcal{V} - f)(v)$, and thus $\mathcal{V} = \text{im}(f) + \ker(f)$. If $v \in \text{im}(f) \cap \ker(f)$, then $f(v) = v$ and $f(v) = 0$, which implies that $v = 0$ and, hence, $\text{im}(f) \cap \ker(f) = \{0\}$. □

12.2 Orthogonality

If for a projection f we have $\text{im}(f) \perp \ker(f)$, i.e., $v \perp w$ for all $v \in \text{im}(f)$ and $w \in \ker(f)$, then this theorem gives an orthogonal decomposition,

$$\mathcal{V} = \mathcal{U} \oplus \mathcal{U}^\perp \quad \text{with} \quad \mathcal{U} = \text{im}(f).$$

This motivates the following definition.

Definition 12.26 Let $\mathcal{V}$ be a Euclidean or unitary vector space. A projection $f \in \mathcal{L}(\mathcal{V}, \mathcal{V})$ is called *orthogonal* if $\text{im}(f) \perp \ker(f)$.

We next prove the existence of an orthogonal projection onto an arbitrary subspace of a Euclidean or unitary vector space.

Theorem 12.27 *Let $\mathcal{V}$ be a finite dimensional Euclidean or unitary vector space, and let $\mathcal{U} \subseteq \mathcal{V}$ be a subspace. Then there exists a uniquely determined orthogonal projection $f_\mathcal{U} \in \mathcal{L}(\mathcal{V}, \mathcal{V})$ with $\mathcal{U} = \text{im}(f_\mathcal{U})$ and $\mathcal{U}^\perp = \ker(f_\mathcal{U})$. We call $f_\mathcal{U}$ the orthogonal projection onto $\mathcal{U}$.*

Proof If $\mathcal{U} = \{0\}$, then we set $f_\mathcal{U} = 0 \in \mathcal{L}(\mathcal{V}, \mathcal{V})$. Now let $\{0\} \neq \mathcal{U} \subseteq \mathcal{V}$. We set $m := \dim(\mathcal{U})$ and let $\{u_1, \ldots, u_m\}$ be an orthonormal basis of $\mathcal{U}$. If $m < n := \dim(\mathcal{V})$, then we extend this basis to an orthonormal basis $\{u_1, \ldots, u_m, u_{m+1}, \ldots, u_n\}$ of $\mathcal{V}$. We then define a map $f_\mathcal{U} \in \mathcal{L}(\mathcal{V}, \mathcal{V})$ via the images of the basis vectors (cp. Theorem 10.4),

$$f_\mathcal{U}(u_i) := \begin{cases} u_i, & i = 1, \ldots, m, \\ 0, & i = m+1, \ldots, n. \end{cases}$$

Then $f_\mathcal{U}^2(u_i) = f_\mathcal{U}(f_\mathcal{U}(u_i)) = f_\mathcal{U}(u_i)$ for $i = 1, \ldots, n$. Therefore, also $f_\mathcal{U}^2(v) = f_\mathcal{U}(v)$ for all $v \in \mathcal{V}$, and hence $f_\mathcal{U}^2 = f_\mathcal{U}$. Furthermore, it is immediate that $\text{im}(f_\mathcal{U}) = \mathcal{U}$ and

$$\ker(f_\mathcal{U}) = \text{span}\{u_{m+1}, \ldots, u_n\} = \mathcal{U}^\perp$$

(cp. the proof of Lemma 12.23). Thus, $f_\mathcal{U}$ is an orthogonal projection.

If $g \in \mathcal{L}(\mathcal{V}, \mathcal{V})$ is another orthogonal projection with $\text{im}(g) = \mathcal{U}$ and $\ker(g) = \mathcal{U}^\perp$, then, in particular, $g(u_i) = f_\mathcal{U}(u_i)$ for $i = 1, \ldots, n$, which implies that $g = f_\mathcal{U}$. □

The following figure illustrates the orthogonal projection of a vector $v \in \mathcal{V}$ onto a subspace $\mathcal{U} \subseteq \mathcal{V}$. We have $v = f_\mathcal{U}(v) + (\text{Id}_\mathcal{V} - f_\mathcal{U})(v)$ and $f_\mathcal{U}(v) \perp (\text{Id}_\mathcal{V} - f_\mathcal{U})(v)$:

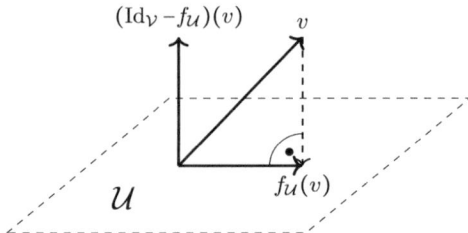

Let $\{u_1, \ldots, u_m, u_{m+1}, \ldots, u_n\}$, as in the proof of Theorem 12.27, be an orthonormal basis of $\mathcal{V} = \mathcal{U} \oplus \mathcal{U}^\perp$. By Lemma 12.18 we know that every $v \in \mathcal{V}$ can be written as $v = \sum_{i=1}^n \langle v, u_i \rangle u_i$. If $f_\mathcal{U}$ is the orthogonal projection onto $\mathcal{U}$, then

$$f_\mathcal{U}(v) = \sum_{i=1}^n \langle v, u_i \rangle f_\mathcal{U}(u_i) = \sum_{i=1}^m \langle v, u_i \rangle u_i. \tag{12.4}$$

The following result shows that the orthogonal projection onto $\mathcal{U}$ yields the *best approximation* of a given vector in $\mathcal{V}$ by vectors in $\mathcal{U}$.

Theorem 12.28 *Let $\mathcal{V}$ be a finite dimensional Euclidean or unitary vector space with the scalar product $\langle \cdot, \cdot \rangle$ and induced norm $\|\cdot\| = \langle \cdot, \cdot \rangle^{1/2}$. Let $\mathcal{U} \subseteq \mathcal{V}$ be a subspace and $f_\mathcal{U} \in \mathcal{L}(\mathcal{V}, \mathcal{V})$ the orthogonal projection onto $\mathcal{U}$. Then*

$$\|v - f_\mathcal{U}(v)\| = \min_{u \in \mathcal{U}} \|v - u\|$$

for every $v \in \mathcal{V}$.

Proof Let $v \in \mathcal{V}$ be given. Then $v = f_\mathcal{U}(v) + (\text{Id}_\mathcal{V} - f_\mathcal{U})(v)$ with $f_\mathcal{U}(v) \in \mathcal{U}$ and $(\text{Id}_\mathcal{V} - f_\mathcal{U})(v) \in \mathcal{U}^\perp$. Using (2) in Lemma 12.8, it follows that

$$\|v - u\|^2 = \| \underbrace{(v - f_\mathcal{U}(v))}_{\in \mathcal{U}^\perp} + \underbrace{(f_\mathcal{U}(v) - u)}_{\in \mathcal{U}} \|^2 = \|v - f_\mathcal{U}(v)\|^2 + \|f_\mathcal{U}(v) - u\|^2$$
$$\geq \|v - f_\mathcal{U}(v)\|^2$$

for every $u \in \mathcal{U}$. Hence, $\|v - f_\mathcal{U}(v)\| \leq \|v - u\|$ for all $u \in \mathcal{U}$, and equality holds if and only if $u = f_\mathcal{U}(v)$. □

12.3 The Vector Product in $\mathbb{R}^{3,1}$

Example 12.29 In Example 12.17 we have considered the minimization problem, respectively least-squares problem, to determine for given $A = [a_1, a_2] \in \mathbb{R}^{m,2}$ of rank 2 and $y \in \mathbb{R}^{m,1}$,

$$\min_{\alpha, \beta \in \mathbb{R}} \left\| A \begin{bmatrix} \alpha \\ \beta \end{bmatrix} - y \right\|$$

in the Euclidean norm. Consider $\mathcal{V} = \mathbb{R}^{m,1}$ with the standard scalar product and the induced Euclidean norm, and let $\mathcal{U} = \text{span}\{a_1, a_2\} \subseteq \mathcal{V}$. Using the orthogonal projection $f_\mathcal{U}$ onto $\mathcal{U}$ and Theorem 12.28, we obtain

$$\min_{\alpha, \beta \in \mathbb{R}} \left\| A \begin{bmatrix} \alpha \\ \beta \end{bmatrix} - y \right\| = \min_{u \in \mathcal{U}} \|y - u\| = \|y - f_\mathcal{U}(y)\|.$$

The solution of the minimization problem in Example 12.17 using the QR-decomposition $A = [q_1, q_2]R$ shows that

$$f_\mathcal{U}(y) = AR^{-1} \begin{bmatrix} q_1^T y \\ q_2^T y \end{bmatrix} = \langle y, q_1 \rangle q_1 + \langle y, q_2 \rangle q_2.$$

This is a special case of (12.4).

12.3 The Vector Product in $\mathbb{R}^{3,1}$

In this section we consider a further product on the vector space $\mathbb{R}^{3,1}$ that is frequently used in Physics and Electrical Engineering.

Definition 12.30 The *vector product* or *cross product* in $\mathbb{R}^{3,1}$ is the map $\mathbb{R}^{3,1} \times \mathbb{R}^{3,1} \to \mathbb{R}^{3,1}$ with

$$(v, w) \mapsto v \times w := [v_2 w_3 - v_3 w_2, \; v_3 w_1 - v_1 w_3, \; v_1 w_2 - v_2 w_1]^T,$$

where $v = [v_1, v_2, v_3]^T$ and $w = [w_1, w_2, w_3]^T$.

In contrast to the scalar product, the vector product of two elements of the vector space $\mathbb{R}^{3,1}$ is not a scalar but again a vector in $\mathbb{R}^{3,1}$. Using the canonical basis vectors of $\mathbb{R}^{3,1}$,

$$e_1 = [1, 0, 0]^T, \quad e_2 = [0, 1, 0]^T, \quad e_3 = [0, 0, 1]^T,$$

we can write the vector product as

$$v \times w = \det\left(\begin{bmatrix} v_2 & \omega_2 \\ v_3 & \omega_3 \end{bmatrix}\right) e_1 - \det\left(\begin{bmatrix} v_1 & \omega_1 \\ v_3 & \omega_3 \end{bmatrix}\right) e_2 + \det\left(\begin{bmatrix} v_1 & \omega_1 \\ v_2 & \omega_2 \end{bmatrix}\right) e_3.$$

Lemma 12.31 *The vector product is linear in both components and for all $v, w \in \mathbb{R}^{3,1}$ the following properties hold:*

(1) $v \times w = -w \times v$, i.e., the vector product is anti-commutative *or* alternating.
(2) $\|v \times w\|^2 = \|v\|^2 \|w\|^2 - \langle v, w \rangle^2$, where $\langle \cdot, \cdot \rangle$ is the standard scalar product and $\|\cdot\|$ the Euclidean norm of $\mathbb{R}^{3,1}$.
(3) $\langle v, v \times w \rangle = \langle w, v \times w \rangle = 0$, where $\langle \cdot, \cdot \rangle$ is the standard scalar product of $\mathbb{R}^{3,1}$.

Proof Exercise. □

By (2) and the Cauchy-Schwarz inequality (12.2), it follows that $v \times w = 0$ holds if and only if v, w are linearly dependent. From (3) we obtain

$$\langle \lambda v + \mu w, v \times w \rangle = \lambda \langle v, v \times w \rangle + \mu \langle w, v \times w \rangle = 0,$$

for arbitrary $\lambda, \mu \in \mathbb{R}$. If v, w are linearly independent, then the product $v \times w$ is orthogonal to the plane through the origin spanned by v and w in $\mathbb{R}^{3,1}$, i.e.,

$$v \times w \in \{\lambda v + \mu w \mid \lambda, \mu \in \mathbb{R}\}^{\perp}.$$

Geometrically, there are two possibilities:

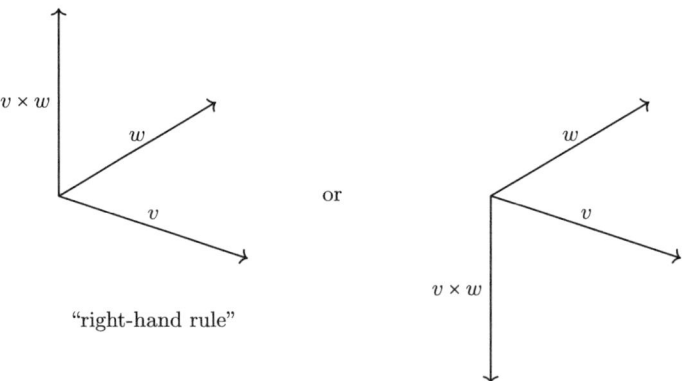

"right-hand rule"

The positions of the three vectors $v, w, v \times w$ on the left side of this figure correspond to the "right-handed orientation" of the usual coordinate system of $\mathbb{R}^{3,1}$, where the canonical basis vectors e_1, e_2, e_3 are associated with thumb, index finger and middle finger of the right hand. This motivates the name *right-hand rule*. In order to explain this in detail, one needs to introduce the concept of *orientation*, which we omit here.

If $\varphi \in [0, \pi]$ is the angle between the vectors v, w, then

$$\langle v, w \rangle = \|v\| \|w\| \cos(\varphi)$$

(cp. Definition 12.7) and we can write (2) in Lemma 12.31 as

$$\|v \times w\|^2 = \|v\|^2 \|w\|^2 - \|v\|^2 \|w\|^2 \cos^2(\varphi) = \|v\|^2 \|w\|^2 \sin^2(\varphi),$$

so that

$$\|v \times w\| = \|v\| \|w\| \sin(\varphi).$$

A geometric interpretation of this equation is the following: *The norm of the vector product of v and w is equal to the area of the parallelogram spanned by v and w.* This interpretation is illustrated in the following figure:

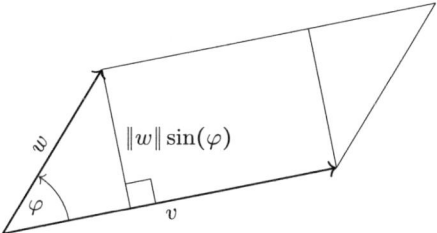

Exercises

12.1 Let $\mathcal{V}$ be a finite dimensional real or complex vector space. Show that there exists a scalar product on $\mathcal{V}$.

12.2 Show that the maps defined in Example 12.2 are scalar products on the corresponding vector spaces.

12.3 Show that the map

$$\langle \cdot, \cdot \rangle : \mathbb{C}[t]_{\leq n} \times \mathbb{C}[t]_{\leq n} \to \mathbb{C}, \quad (p, q) \mapsto \langle p, q \rangle := \sum_{j=0}^{n} \alpha_j \overline{\beta}_j,$$

for $p = \alpha_n t^n + \ldots + \alpha_1 t + \alpha_0$ and $q = \beta_n t^n + \ldots + \beta_1 t + \beta_0$ is a scalar product on $\mathbb{C}[t]_{\leq n}$.

12.4 Let $\langle \cdot, \cdot \rangle$ be an arbitrary scalar product on $\mathbb{R}^{n,1}$. Show that there exists a matrix $A \in \mathbb{R}^{n,n}$ with $\langle v, w \rangle = w^T A v$ for all $v, w \in \mathbb{R}^{n,1}$.

12.5 Let $\mathcal{V}$ be a finite dimensional $\mathbb{R}$- or $\mathbb{C}$-vector space. Let s_1 and s_2 be scalar products on $\mathcal{V}$ with the following property: If $v, w \in \mathcal{V}$ satisfy $s_1(v, w) = 0$, then also $s_2(v, w) = 0$. Prove or disprove: There exists a real scalar $\lambda > 0$ with $s_1(v, w) = \lambda s_2(v, w)$ for all $v, w \in \mathcal{V}$.

12.6 Show that the maps defined in Example 12.4 are norms on the corresponding vector spaces.

12.7 Let

$$A_1 = \begin{bmatrix} 1 & -1 \\ a & b \end{bmatrix} \in \mathbb{R}^{2,2} \quad \text{and} \quad A_2 = \begin{bmatrix} 1 & -\mathbf{i} \\ c & d \end{bmatrix} \in \mathbb{C}^{2,2}.$$

Determine for which choices of $a, b \in \mathbb{R}$ and $c, d \in \mathbb{C}$ the matrices A_1 resp. A_2 are matrix representations of scalar products on $\mathbb{R}^{2,1}$ resp. $\mathbb{C}^{2,1}$.

12.8 Determine a scalar product $\langle \cdot, \cdot \rangle$ on $\mathbb{R}[t]_{\leq 2}$, so that the three polynomials

$$p_1 = t^2 + t + 1, \quad p_2 = t^2 + 2, \quad p_3 = 1,$$

form an orthonormal basis of $\mathbb{R}[t]_{\leq 2}$. Next determine $\langle p, q \rangle$ for arbitrary polynomials $p = \alpha_2 t^2 + \alpha_1 t + \alpha_0$ and $q = \beta_2 t^2 + \beta_1 t + \beta_0$ in $\mathbb{R}[t]_{\leq 2}$.

12.9 Show that

$$\|A\|_1 = \max_{1 \leq j \leq m} \sum_{i=1}^{n} |a_{ij}| \quad \text{and} \quad \|A\|_\infty = \max_{1 \leq i \leq n} \sum_{j=1}^{m} |a_{ij}|$$

for all $A = [a_{ij}] \in K^{n,m}$, where $K = \mathbb{R}$ or $K = \mathbb{C}$ (cp. (6) in Example 12.4).

12.10 Sketch for the matrix A from (6) in Example 12.4 and $p \in \{1, 2, \infty\}$, the sets $\{Av \mid v \in \mathbb{R}^{2,1}, \|v\|_p = 1\} \subset \mathbb{R}^{2,1}$.

12.11 Let $K = \mathbb{R}$ or $K = \mathbb{C}$, and let $\|\cdot\|_n$ and $\|\cdot\|_m$ be norms on $K^{n,1}$ respectively $K^{m,1}$. Show that then

$$\|A\|_{n,m} := \max_{x \neq 0} \frac{\|Ax\|_n}{\|x\|_m}$$

defines a norm on $K^{n,m}$.

12.12 Let $\mathcal{V}$ be a Euclidean or unitary vector space and let $\|\cdot\|$ be the norm induced by a scalar product on $\mathcal{V}$. Show that $\|\cdot\|$ satisfies the *parallelogram identity*

$$\|v + w\|^2 + \|v - w\|^2 = 2(\|v\|^2 + \|w\|^2)$$

for all $v, w \in \mathcal{V}$.

12.13 Let $\mathcal{V}$ be a K-vector space ($K = \mathbb{R}$ or $K = \mathbb{C}$) with the scalar product $\langle \cdot, \cdot \rangle$ and the induced norm $\|\cdot\|$. Show that $v, w \in \mathcal{V}$ are orthogonal with respect to $\langle \cdot, \cdot \rangle$ if and only if $\|v + \lambda w\| = \|v - \lambda w\|$ for all $\lambda \in K$.

12.14 Let $\mathcal{V}$ be a K-vector space ($K = \mathbb{R}$ or $K = \mathbb{C}$) with the scalar product $\langle \cdot, \cdot \rangle$ and the induced norm $\|\cdot\|$. Show that $v, w \in \mathcal{V}$ are orthogonal with respect to $\langle \cdot, \cdot \rangle$ if and only if $\|v\| \leq \|v + \lambda w\|$ for all $\lambda \in K$.

Exercises

12.15 Does there exist a scalar product $\langle \cdot, \cdot \rangle$ on $\mathbb{C}^{n,1}$, such that the 1-norm of $\mathbb{C}^{n,1}$ (cp. (5) in Example 12.4) is the induced norm by this scalar product?

12.16 Let X be a nonempty set. A map $d : X \times X \to \mathbb{R}$ is called a *metric* on X, if for all $x, y, z \in X$ the following conditions hold:
   (1) $d(x, y) \geq 0$ with equality if and only if $x = y$.
   (2) $d(x, y) = d(y, x)$.
   (3) $d(x, y) \leq d(x, z) + d(z, y)$.
   (a) Let $\mathcal{V}$ be a Euclidean or unitary vector space with a norm $\|\cdot\|$. Show that
   $$d : \mathcal{V} \times \mathcal{V} \to \mathbb{R}, \quad (x, y) \mapsto \|x - y\|,$$
   is a metric on $\mathcal{V}$.
   (b) Show that each of the following maps is a metric:
   $$d_1 : \mathbb{R}^2 \times \mathbb{R}^2 \to \mathbb{R}, \quad ((x_1, x_2), (y_1, y_2)) \mapsto \max\{|x_1 - y_1|, |x_2 - y_2|\},$$
   $$d_2 : \mathbb{R} \times \mathbb{R} \to \mathbb{R}, \quad (x, y) \mapsto \frac{|x - y|}{|x - y| + 1},$$
   $$d_3 : \mathbb{R} \times \mathbb{R} \to \mathbb{R}, \quad (x, y) \mapsto \min\{|x - y|, 1\}.$$

12.17 Show that the inequality
$$\left(\sum_{i=1}^n \alpha_i \beta_i \right)^2 \leq \left(\sum_{i=1}^n (\gamma_i \alpha_i)^2 \right) \cdot \sum_{i=1}^n \left(\frac{\beta_i}{\gamma_i} \right)^2$$
holds for arbitrary $\alpha_1, \ldots, \alpha_n, \beta_1, \ldots, \beta_n \in \mathbb{R}$ and $\gamma_1, \ldots, \gamma_n \in \mathbb{R} \setminus \{0\}$.

12.18 Let $\mathcal{V}$ be a finite dimensional Euclidean or unitary vector space with the scalar product $\langle \cdot, \cdot \rangle$. Let $f : \mathcal{V} \to \mathcal{V}$ be a map with $\langle f(v), f(w) \rangle = \langle v, w \rangle$ for all $v, w \in \mathcal{V}$. Show that f is an isomorphism.

12.19 Let $\mathcal{V}$ be a unitary vector space and suppose that $f \in \mathcal{L}(\mathcal{V}, \mathcal{V})$ satisfies $\langle f(v), v \rangle = 0$ for all $v \in \mathcal{V}$. Prove or disprove that $f = 0$.
Does the same statement also hold for Euclidean vector spaces?

12.20 Let $D = \text{diag}(d_1, \ldots, d_n) \in \mathbb{R}^{n,n}$ with $d_1, \ldots, d_n > 0$. Show that $\langle v, w \rangle = w^T D v$ is a scalar product on $\mathbb{R}^{n,1}$. Analyze which properties of a scalar product are violated if at least one of the d_i is zero, or when all d_i are nonzero but have different signs.

12.21 Orthonormalize the following basis of the vector space $\mathbb{C}^{2,2}$ with respect to the scalar product $\langle A, B \rangle = \text{trace}(B^H A)$:
$$\left\{ \begin{bmatrix} 1 & 0 \\ 0 & 0 \end{bmatrix}, \begin{bmatrix} 1 & 0 \\ 0 & 1 \end{bmatrix}, \begin{bmatrix} 1 & 1 \\ 0 & 1 \end{bmatrix}, \begin{bmatrix} 1 & 1 \\ 1 & 1 \end{bmatrix} \right\}.$$

12.22 Let $Q \in \mathbb{R}^{n,n}$ be an orthogonal or let $Q \in \mathbb{C}^{n,n}$ be a unitary matrix. What are the possible values of $\det(Q)$?

12.23 Let $u \in \mathbb{R}^{n,1} \setminus \{0\}$ and let

$$H(u) = I_n - 2\frac{1}{u^T u} u u^T \in \mathbb{R}^{n,n}.$$

Show that the n columns of $H(u)$ form an orthonormal basis of $\mathbb{R}^{n,1}$ with respect to the standard scalar product. (Matrices of this form are called *Householder matrices*.[9] We will study them in more detail in Example 18.15.)

12.24 Prove Lemma 12.22.

12.25 Let $\mathcal{V}$ be a Euclidean or unitary vector space and let $\mathcal{U} \subseteq \mathcal{V}$ be a subspace. Show that $\mathcal{U} \subseteq (\mathcal{U}^\perp)^\perp$. Show, furthermore, that equality holds if $\mathcal{V}$ is finite dimensional.

12.26 Let

$$[v_1, v_2, v_3] = \begin{bmatrix} \frac{1}{\sqrt{2}} & 0 & \frac{1}{\sqrt{2}} \\ -\frac{1}{\sqrt{2}} & 0 & \frac{1}{\sqrt{2}} \\ 0 & 0 & 0 \end{bmatrix} \in \mathbb{R}^{3,3}.$$

Analyze whether the vectors v_1, v_2, v_3 are orthonormal with respect to the standard scalar product and compute the orthogonal complement of $\operatorname{span}\{v_1, v_2, v_3\}$.

12.27 Let $\mathcal{V}$ be a Euclidean or unitary vector space with the scalar product $\langle \cdot, \cdot \rangle$, let $u_1, \ldots, u_k \in \mathcal{V}$ and let $\mathcal{U} = \operatorname{span}\{u_1, \ldots, u_k\}$. Show that for $v \in \mathcal{V}$ we have $v \in \mathcal{U}^\perp$ if and only if $\langle v, u_j \rangle = 0$ for $j = 1, \ldots, k$.

12.28 In the unitary vector space $\mathbb{C}^{4,1}$ with the standard scalar product let $v_1 = [-1, \mathbf{i}, 0, 1]^T$ and $v_2 = [\mathbf{i}, 0, 2, 0]^T$ be given. Determine an orthonormal basis of $\operatorname{span}\{v_1, v_2\}^\perp$.

12.29 Let $\mathcal{V}$ be a finite dimensional Euclidean or unitary vector space and let $\mathcal{U}, \mathcal{W} \subseteq \mathcal{V}$ be two subspaces with $\mathcal{V} = \mathcal{U} \oplus \mathcal{W}$. Show that there exists a uniquely determined projection $f \in \mathcal{L}(\mathcal{V}, \mathcal{V})$ with $\operatorname{im}(f) = \mathcal{U}$ and $\ker(f) = \mathcal{W}$.

12.30 Consider $\mathcal{V} = \mathbb{R}^{3,1}$ with the standard scalar product and $\mathcal{U} = \operatorname{span}\{[2, 0, 0]^T, [0, -1, 2]^T\} \subseteq \mathcal{V}$. Determine the orthogonal projection $f_\mathcal{U}(v)$ for $v = [1, 2, 0]^T \in \mathcal{V}$ as well as $\|v - f_\mathcal{U}(v)\|_2$.

12.31 Let $\mathcal{V}$ and $\mathcal{W}$ be Euclidean vector spaces with scalar products $\langle \cdot, \cdot \rangle_\mathcal{V}$ resp. $\langle \cdot, \cdot \rangle_\mathcal{W}$, and the induced norms $\|\cdot\|_\mathcal{V}$ resp. $\|\cdot\|_\mathcal{W}$. A map $f \in \mathcal{L}(\mathcal{V}, \mathcal{W})$ is called an *isometry* if $\|f(v)\|_\mathcal{W} = \|v\|_\mathcal{V}$ for all $v \in \mathcal{V}$. Show the following assertions:

(a) An isometry $f \in \mathcal{L}(\mathcal{V}, \mathcal{W})$ is injective.

---

[9] Alston Scott Householder (1904–1993), pioneer of Numerical Linear Algebra.

(b) A map $f \in \mathcal{L}(\mathcal{V}, \mathcal{W})$ is an isometry if and only if $\langle f(v_1), f(v_2) \rangle_\mathcal{W} = \langle v_1, v_2 \rangle_\mathcal{V}$ for all $v_1, v_2 \in \mathcal{V}$.

(c) Let $f \in \mathcal{L}(\mathcal{V}, \mathcal{W})$ be an isometry, where $\dim(\mathcal{V}) = \dim(\mathcal{W}) = n$, and let $v_1, \ldots, v_n \in \mathcal{V}$. The set $\{v_1, \ldots, v_n\}$ is an orthonormal basis of $\mathcal{V}$ if and only if $\{f(v_1), \ldots, f(v_n)\}$ is an orthonormal basis of $\mathcal{W}$.

12.32 Let $\mathcal{V}$ be a finite dimensional Euclidean vector space with the scalar product $\langle \cdot, \cdot \rangle$ and induced norm $\|\cdot\|$, and let $u \in \mathcal{V}$ with $\|u\| = 1$. Show that the linear map $f : \mathcal{V} \to \mathcal{V}$, $v \mapsto v - 2\langle v, u \rangle u$, is a bijective isometry.

12.33 Prove Lemma 12.31.

Adjoints of Linear Maps 13

In this chapter we introduce adjoints of linear maps. In some sense these represent generalizations of the (Hermitian) transposes of a matrices. A matrix is symmetric (or Hermitian) if it is equal to its (Hermitian) transpose. In an analogous way, an endomorphism is selfadjoint if it is equal to its adjoint endomorphism. The sets of symmetric (or Hermitian) matrices and of selfadjoint endomorphisms form certain vector spaces which will play a key role in our proof of the Fundamental Theorem of Algebra in Chap. 15. Special properties of selfadjoint endomorphisms will be studied in Chap. 18.

13.1 Adjoints in Finite Dimensional K-vector Spaces

In Chap. 12 we have considered Euclidean and unitary vector spaces, and hence vector spaces over the fields $\mathbb{R}$ and $\mathbb{C}$. Now let $\mathcal{V}$ and $\mathcal{W}$ be vector spaces over a general field K, and let β be a bilinear form on $\mathcal{V} \times \mathcal{W}$.

For every *fixed* vector $v \in \mathcal{V}$, the map

$$\beta_v : \mathcal{W} \to K, \quad w \mapsto \beta(v, w),$$

is a linear form on $\mathcal{W}$. Thus, we can assign to every $v \in \mathcal{V}$ a vector $\beta_v \in \mathcal{W}^*$, which defines the map

$$\beta^{(1)} : \mathcal{V} \to \mathcal{W}^*, \quad v \mapsto \beta_v. \tag{13.1}$$

Analogously, we define the map

$$\beta^{(2)} : \mathcal{W} \to \mathcal{V}^*, \quad w \mapsto \beta_w, \tag{13.2}$$

where $\beta_w : \mathcal{V} \to K$ is defined by $v \mapsto \beta(v, w)$ for every $w \in \mathcal{W}$.

Lemma 13.1 *The maps $\beta^{(1)}$ and $\beta^{(2)}$ defined in (13.1) and (13.2), respectively, are linear, i.e., $\beta^{(1)} \in \mathcal{L}(\mathcal{V}, \mathcal{W}^*)$ and $\beta^{(2)} \in \mathcal{L}(\mathcal{W}, \mathcal{V}^*)$. If $\dim(\mathcal{V}) = \dim(\mathcal{W}) \in \mathbb{N}$ and β is non-degenerate (cp. Definition 11.9), then $\beta^{(1)}$ and $\beta^{(2)}$ are bijective and thus isomorphisms.*

Proof We prove the assertion only for the map $\beta^{(1)}$; the proof for $\beta^{(2)}$ is analogous.

We first show the linearity. Let $v_1, v_2 \in \mathcal{V}$ and $\lambda_1, \lambda_2 \in K$. For every $w \in \mathcal{W}$ we then have

$$\beta^{(1)}(\lambda_1 v_1 + \lambda_2 v_2)(w) = \beta(\lambda_1 v_1 + \lambda_2 v_2, w)$$
$$= \lambda_1 \beta(v_1, w) + \lambda_2 \beta(v_2, w)$$
$$= \lambda_1 \beta^{(1)}(v_1)(w) + \lambda_2 \beta^{(1)}(v_2)(w)$$
$$= \bigl(\lambda_1 \beta^{(1)}(v_1) + \lambda_2 \beta^{(1)}(v_2)\bigr)(w),$$

and hence $\beta^{(1)}(\lambda_1 v_1 + \lambda_2 v_2) = \lambda_1 \beta^{(1)}(v_1) + \lambda_2 \beta^{(1)}(v_2)$. Therefore, $\beta^{(1)} \in \mathcal{L}(\mathcal{V}, \mathcal{W}^*)$.

Let now $\dim(\mathcal{V}) = \dim(\mathcal{W}) \in \mathbb{N}$ and let β be non-degenerate. We show that $\beta^{(1)} \in \mathcal{L}(\mathcal{V}, \mathcal{W}^*)$ is injective. By (5) in Lemma 10.7, this holds if and only if $\ker(\beta^{(1)}) = \{0\}$. If $v \in \ker(\beta^{(1)})$, then $\beta^{(1)}(v) = \beta_v = 0 \in \mathcal{W}^*$, and thus

$$\beta_v(w) = \beta(v, w) = 0 \quad \text{for all } w \in \mathcal{W}.$$

Since β is non-degenerate, we have $v = 0$. Finally, $\dim(\mathcal{V}) = \dim(\mathcal{W})$ and $\dim(\mathcal{W}) = \dim(\mathcal{W}^*)$ imply that $\dim(\mathcal{V}) = \dim(\mathcal{W}^*)$ so that $\beta^{(1)}$ is bijective (cp. Corollary 10.11). □

We next discuss the existence of the adjoint map.

Theorem 13.2 *If $\mathcal{V}$ and $\mathcal{W}$ are K-vector spaces with $\dim(\mathcal{V}) = \dim(\mathcal{W}) \in \mathbb{N}$ and β is a non-degenerate bilinear form on $\mathcal{V} \times \mathcal{W}$, then the following assertions hold:*

(1) For every $f \in \mathcal{L}(\mathcal{V}, \mathcal{V})$ there exists a uniquely determined $g \in \mathcal{L}(\mathcal{W}, \mathcal{W})$ with

$$\beta(f(v), w) = \beta(v, g(w)) \quad \text{for all } v \in \mathcal{V} \text{ and } w \in \mathcal{W}.$$

The map g is called the right adjoint of f with respect to β.

(2) For every $h \in \mathcal{L}(\mathcal{W}, \mathcal{W})$ there exists a uniquely determined $k \in \mathcal{L}(\mathcal{V}, \mathcal{V})$ with

$$\beta(v, h(w)) = \beta(k(v), w) \quad \text{for all } v \in \mathcal{V} \text{ and } w \in \mathcal{W}.$$

The map k is called the left adjoint of h with respect to β.

13.1 Adjoints in Finite Dimensional K-vector Spaces

Proof We only show (1); the proof of (2) is analogous.

Let $\mathcal{V}^*$ be the dual space of $\mathcal{V}$, let $f^* \in \mathcal{L}(\mathcal{V}^*, \mathcal{V}^*)$ be the dual map of f, and let $\beta^{(2)} \in \mathcal{L}(\mathcal{W}, \mathcal{V}^*)$ be as in (13.2). Since β is non-degenerate, $\beta^{(2)}$ is bijective by Lemma 13.1. Define

$$g := (\beta^{(2)})^{-1} \circ f^* \circ \beta^{(2)} \in \mathcal{L}(\mathcal{W}, \mathcal{W}).$$

Then, for all $v \in \mathcal{V}$ and $w \in \mathcal{W}$,

$$\begin{aligned}
\beta(v, g(w)) &= \beta(v, ((\beta^{(2)})^{-1} \circ f^* \circ \beta^{(2)})(w)) \\
&= \beta^{(2)}\big(((\beta^{(2)})^{-1} \circ f^* \circ \beta^{(2)})(w)\big)(v) \\
&= \beta^{(2)}\big((\beta^{(2)})^{-1}(f^*(\beta^{(2)}(w)))\big)(v) \\
&= \big(\beta^{(2)} \circ (\beta^{(2)})^{-1} \circ \beta^{(2)}(w) \circ f\big)(v) \\
&= \beta^{(2)}(w)(f(v)) \\
&= \beta(f(v), w).
\end{aligned}$$

(Recall that the dual map satisfies $f^*(\beta^{(2)}(w)) = \beta^{(2)}(w) \circ f$.)

It remains to show the uniqueness of g. Let $\widetilde{g} \in \mathcal{L}(\mathcal{W}, \mathcal{W})$ with $\beta(v, \widetilde{g}(w)) = \beta(f(v), w)$ for all $v \in \mathcal{V}$ and $w \in \mathcal{W}$. Then $\beta(v, \widetilde{g}(w)) = \beta(v, g(w))$, and hence

$$\beta(v, (\widetilde{g} - g)(w)) = 0 \quad \text{for all } v \in \mathcal{V} \text{ and } w \in \mathcal{W}.$$

Since β is non-degenerate in the second variable, we have $(\widetilde{g} - g)(w) = 0$ for all $w \in \mathcal{W}$, so that $g = \widetilde{g}$. $\square$

Example 13.3 Let $\mathcal{V} = \mathcal{W} = K^{n,1}$ and $\beta(v, w) = w^T B v$ with a matrix $B \in GL_n(K)$, so that β is non-degenerate (cp. (1) in Example 11.10). We consider the linear map $f : \mathcal{V} \to \mathcal{V}$, $v \mapsto Fv$, with a matrix $F \in K^{n,n}$, and the linear map $h : \mathcal{W} \to \mathcal{W}$, $w \mapsto Hw$, with a matrix $H \in K^{n,n}$. Then

$$\begin{aligned}
\beta_v &: \mathcal{W} \to K, & w &\mapsto w^T(Bv), \\
\beta^{(1)} &: \mathcal{V} \to \mathcal{W}^*, & v &\mapsto (Bv)^T, \\
\beta^{(2)} &: \mathcal{W} \to \mathcal{V}^*, & w &\mapsto w^T B,
\end{aligned}$$

where we have identified the isomorphic vector spaces $\mathcal{W}^*$ and $K^{1,n}$, respectively $\mathcal{V}^*$ and $K^{1,n}$, with each other. If $g \in \mathcal{L}(\mathcal{W}, \mathcal{W})$ is the right adjoint of f with respect to β, then

$$\beta(f(v), w) = w^T B f(v) = w^T B F v = \beta(v, g(w)) = g(w)^T B v$$

for all $v \in \mathcal{V}$ and $w \in \mathcal{W}$. If we represent the linear map g via the multiplication with a matrix $G \in K^{n,n}$, i.e., $g(w) = Gw$, then $w^T BFv = w^T G^T Bv$ for all $v, w \in K^{n,1}$. Hence $BF = G^T B$. Since B is invertible, the unique right adjoint is given by $G = (BFB^{-1})^T = B^{-T} F^T B^T$.

Analogously, for the left adjoint $k \in \mathcal{L}(\mathcal{V}, \mathcal{V})$ of h with respect to β we obtain the equation

$$\beta(v, h(w)) = (h(w))^T Bv = w^T H^T Bv = \beta(k(v), w) = w^T Bk(v)$$

for all $v \in \mathcal{V}$ and $w \in \mathcal{W}$. With $k(v) = Lv$ for a matrix $L \in K^{n,n}$, we obtain $H^T B = BL$ and hence $L = B^{-1} H^T B$.

If in this example $f = h$ and $B = B^T$, i.e., β is symmetric, then $G = B^{-T} F^T B^T = B^{-1} H^T B = L$, and thus also $g = k$.

The last observation in this example can be generalized. If $\mathcal{V}$ is finite dimensional and β is a non-degenerate bilinear form on $\mathcal{V}$, then by Theorem 13.2 every $f \in \mathcal{L}(\mathcal{V}, \mathcal{V})$ has a unique right adjoint g and a unique left adjoint k, such that

$$\beta(f(v), w) = \beta(v, g(w)) \quad \text{and} \quad \beta(v, f(w)) = \beta(k(v), w) \tag{13.3}$$

for all $v, w \in \mathcal{V}$. If β is symmetric, i.e., if $\beta(v, w) = \beta(w, v)$ holds for all $v, w \in \mathcal{V}$, then (13.3) yields

$$\beta(v, g(w)) = \beta(f(v), w) = \beta(w, f(v)) = \beta(k(w), v) = \beta(v, k(w)).$$

Therefore, $\beta(v, (g-k)(w)) = 0$ for all $v, w \in \mathcal{V}$, and hence $g = k$, since β is non-degenerate. Thus, we have proved the following result.

Corollary 13.4 *If β is a symmetric and non-degenerate bilinear form on a finite dimensional K-vector space $\mathcal{V}$, then for every $f \in \mathcal{L}(\mathcal{V}, \mathcal{V})$ there exists a unique $g \in \mathcal{L}(\mathcal{V}, \mathcal{V})$ with*

$$\beta(f(v), w) = \beta(v, g(w)) \quad \text{and} \quad \beta(v, f(w)) = \beta(g(v), w)$$

for all $v, w \in \mathcal{V}$.

13.2 Adjoints in Finite Dimensional Euclidean and Unitary Vector Spaces

We know from Chap. 12 that a scalar product on a Euclidean vector space is a symmetric and non-degenerate bilinear form. Therefore, Corollary 13.4 yields the following result.

13.2 Adjoints in Finite Dimensional Euclidean and Unitary Vector Spaces

Corollary 13.5 *If $\mathcal{V}$ is a finite dimensional Euclidean vector space with the scalar product $\langle \cdot, \cdot \rangle$, then for every $f \in \mathcal{L}(\mathcal{V}, \mathcal{V})$ there exists a unique $f^{ad} \in \mathcal{L}(\mathcal{V}, \mathcal{V})$ with*

$$\langle f(v), w \rangle = \langle v, f^{ad}(w) \rangle \quad \text{and} \quad \langle v, f(w) \rangle = \langle f^{ad}(v), w \rangle \tag{13.4}$$

for all $v, w \in \mathcal{V}$. The map f^{ad} is called the adjoint of f (with respect to $\langle \cdot, \cdot \rangle$).

In order to determine whether a given map $g \in \mathcal{L}(\mathcal{V}, \mathcal{V})$ is the unique adjoint of $f \in \mathcal{L}(\mathcal{V}, \mathcal{V})$, only one of the two conditions in (13.4) have to be verified: If for $f, g \in \mathcal{L}(\mathcal{V}, \mathcal{V})$ the equation

$$\langle f(v), w \rangle = \langle v, g(w) \rangle$$

holds for all $v, w \in \mathcal{V}$, then also

$$\langle v, f(w) \rangle = \langle f(w), v \rangle = \langle w, g(v) \rangle = \langle g(v), w \rangle$$

for all $v, w \in \mathcal{V}$, where we have used the symmetry of the scalar product. Similarly, if $\langle v, f(w) \rangle = \langle g(v), w \rangle$ holds for all $v, w \in \mathcal{V}$, then also $\langle f(v), w \rangle = \langle v, g(w) \rangle$ for all $v, w \in \mathcal{V}$.

Example 13.6 Consider the Euclidean vector space $\mathbb{R}^{3,1}$ with the scalar product

$$\langle v, w \rangle = w^T B v, \quad \text{where} \quad B = \begin{bmatrix} 1 & 0 & 0 \\ 0 & 2 & 0 \\ 0 & 0 & 1 \end{bmatrix},$$

and the linear map

$$f : \mathbb{R}^{3,1} \to \mathbb{R}^{3,1}, \quad v \mapsto Fv, \quad \text{where} \quad F = \begin{bmatrix} 1 & 2 & 2 \\ 1 & 0 & 1 \\ 2 & 0 & 0 \end{bmatrix}.$$

For all $v, w \in \mathbb{R}^{3,1}$ we then have

$$\langle f(v), w \rangle = w^T B F v = w^T B F B^{-1} B v = (B^{-T} F^T B^T w)^T B v = \langle v, f^{ad}(w) \rangle$$

(cp. Example 13.3), and thus

$$f^{ad} : \mathbb{R}^{3,1} \to \mathbb{R}^{3,1}, \quad v \mapsto B^{-1} F^T B v = \begin{bmatrix} 1 & 2 & 2 \\ 1 & 0 & 0 \\ 2 & 2 & 0 \end{bmatrix} v,$$

where we have used that B is symmetric.

We now show that uniquely determined adjoint maps also exist in the unitary case. However, we cannot conclude this directly from Corollary 13.4, since a scalar product on a $\mathbb{C}$-vector space is not a symmetric bilinear form, but a Hermitian sesquilinear form. In order to show the existence of the adjoint map in the unitary case we construct it explicitly. This construction works also in the Euclidean case.

Let $\mathcal{V}$ be a unitary vector space with the scalar product $\langle \cdot, \cdot \rangle$ and let $\{u_1, \ldots, u_n\}$ be an orthonormal basis of $\mathcal{V}$. For a given $f \in \mathcal{L}(\mathcal{V}, \mathcal{V})$ we define the map

$$g : \mathcal{V} \to \mathcal{V}, \quad v \mapsto \sum_{i=1}^n \langle v, f(u_i) \rangle u_i.$$

If $v, w \in \mathcal{V}$ and $\lambda, \mu \in \mathbb{C}$, then

$$g(\lambda v + \mu w) = \sum_{i=1}^n \langle \lambda v + \mu w, f(u_i) \rangle u_i = \sum_{i=1}^n \left(\lambda \langle v, f(u_i) \rangle u_i + \mu \langle v, f(u_i) \rangle u_i \right)$$
$$= \lambda g(v) + \mu g(w),$$

and hence $g \in \mathcal{L}(\mathcal{V}, \mathcal{V})$. Let now $v = \sum_{i=1}^n \lambda_i u_i \in \mathcal{V}$ and $w \in \mathcal{V}$, then

$$\langle v, g(w) \rangle = \left\langle \sum_{i=1}^n \lambda_i u_i, \sum_{j=1}^n \langle w, f(u_j) \rangle u_j \right\rangle = \sum_{i=1}^n \sum_{j=1}^n \lambda_i \overline{\langle w, f(u_j) \rangle} \langle u_i, u_j \rangle$$
$$= \sum_{i=1}^n \lambda_i \overline{\langle w, f(u_i) \rangle} = \sum_{i=1}^n \lambda_i \langle f(u_i), w \rangle$$
$$= \langle f(v), w \rangle.$$

Furthermore,

$$\langle v, f(w) \rangle = \overline{\langle f(w), v \rangle} = \overline{\langle w, g(v) \rangle} = \langle g(v), w \rangle$$

for all $v, w \in \mathcal{V}$. If $\widetilde{g} \in \mathcal{L}(\mathcal{V}, \mathcal{V})$ satisfies $\langle f(v), w \rangle = \langle v, \widetilde{g}(w) \rangle$ for all $v, w \in \mathcal{V}$, then $\langle v, (g - \widetilde{g})(w) \rangle = 0$ for all $v, w \in \mathcal{V}$, and in particular

$$\langle (g - \widetilde{g})(w), (g - \widetilde{g})(w) \rangle = 0 \quad \text{for all } w \in \mathcal{V}.$$

Since the scalar product is positive definite, we obtain $(g - \widetilde{g})(w) = 0$ for all $w \in \mathcal{V}$, and hence $g = \widetilde{g}$. We can therefore formulate the following result analogously to Corollary 13.5.

13.2 Adjoints in Finite Dimensional Euclidean and Unitary Vector Spaces

Corollary 13.7 *If $\mathcal{V}$ is a finite dimensional unitary vector space with the scalar product $\langle \cdot, \cdot \rangle$, then for every $f \in \mathcal{L}(\mathcal{V}, \mathcal{V})$ there exists a unique $f^{ad} \in \mathcal{L}(\mathcal{V}, \mathcal{V})$ with*

$$\langle f(v), w \rangle = \langle v, f^{ad}(w) \rangle \quad \text{and} \quad \langle v, f(w) \rangle = \langle f^{ad}(v), w \rangle \tag{13.5}$$

for all $v, w \in \mathcal{V}$. The map f^{ad} is called the adjoint of f (with respect to $\langle \cdot, \cdot \rangle$).

As in the Euclidean case, again the validity of one of the two equations in (13.5) for all $v, w \in \mathcal{V}$ implies the validity of the other for all $v, w \in \mathcal{V}$.

Example 13.8 Consider the unitary vector space $\mathbb{C}^{3,1}$ with the scalar product

$$\langle v, w \rangle = w^H B v, \quad \text{where} \quad B = \begin{bmatrix} 1 & 0 & 0 \\ 0 & 2 & 0 \\ 0 & 0 & 1 \end{bmatrix},$$

and the linear map

$$f : \mathbb{C}^{3,1} \to \mathbb{C}^{3,1}, \quad v \mapsto Fv, \quad \text{where} \quad F = \begin{bmatrix} 1 & 2\mathbf{i} & 2 \\ \mathbf{i} & 0 & -\mathbf{i} \\ 2 & 0 & 3\mathbf{i} \end{bmatrix}.$$

For all $v, w \in \mathbb{C}^{3,1}$ we then have

$$\langle f(v), w \rangle = w^H B F v = w^H B F B^{-1} B v = (B^{-H} F^H B^H w)^H B v$$
$$= \langle v, f^{ad}(w) \rangle,$$

and thus

$$f^{ad} : \mathbb{C}^{3,1} \to \mathbb{C}^{3,1}, \quad v \mapsto B^{-1} F^H B v = \begin{bmatrix} 1 & -2\mathbf{i} & 2 \\ -\mathbf{i} & 0 & 0 \\ 2 & 2\mathbf{i} & -3\mathbf{i} \end{bmatrix} v,$$

where we have used that B is real and symmetric.

Example 13.9 In this example we will use some results from Analysis to show that endomorphisms on infinite dimensional vector spaces do not necessarily have adjoints. For this we consider the Euclidean vector space $\mathcal{V} = \mathbb{R}[t]$ with the scalar product

$$\langle p, q \rangle = \int_0^1 p(t) q(t) dt$$

(cp. (4) in Example 12.2). Let $f \in \mathcal{L}(\mathcal{V}, \mathcal{V})$ be the usual derivative of polynomials, i.e., $f(p) = p'$ for all $p \in \mathcal{V}$ (cp. also Exercise 10.3 for the derivative on $\mathbb{R}[t]_{\leq n}$). Suppose that for f there would exist a map $f^{ad} \in \mathcal{L}(\mathcal{V}, \mathcal{V})$ that satisfies the equations in (13.4). Using partial integration, for all polynomials $p, q \in \mathcal{V}$ we obtain the equation

$$\langle p, f^{ad}(q)\rangle = \langle f(p), q\rangle = \int_0^1 p'(t)q(t)dt$$

$$= p(1)q(1) - p(0)q(0) - \int_0^1 p(t)q'(t)dt$$

$$= p(1)q(1) - p(0)q(0) - \langle p, f(q)\rangle,$$

which yields

$$\langle p, (f + f^{ad})(q)\rangle = p(1)q(1) - p(0)q(0).$$

For an arbitrary $q \in \mathcal{V}$ we can choose $p = (t-1)^2 t^2 (f + f^{ad})(q) \in \mathcal{V}$. Then $p(1) = p(0) = 0$, and hence

$$0 = \langle (t-1)^2 t^2 (f + f^{ad})(q), (f + f^{ad})(q)\rangle = \int_0^1 (t-1)^2 t^2 ((f + f^{ad})(q(t)))^2 dt$$

for all $q \in \mathcal{V}$. The integral on the right hand side of a continuous and nonnegative real function can be zero only if $(f + f^{ad})(q) = 0 \in \mathcal{V}$. Since this has to hold for all $q \in \mathcal{V}$ we obtain $f + f^{ad} = 0 \in \mathcal{L}(\mathcal{V}, \mathcal{V})$. Thus,

$$0 = \langle p, (f + f^{ad})(q)\rangle = p(1)q(1) - p(0)q(0)$$

for all polynomials $p, q \in \mathcal{V}$, which is obviously wrong. A counterexample are the polynomials $p = 1$ and $q = t$, for which $p(1)q(1) - p(0)q(0) = 1$. This shows that the map f^{ad} cannot exist.

This construction cannot be made for the finite dimensional vector space $\mathcal{V} = \mathbb{R}[t]_{\leq n}$ with the same scalar product, because in this case every $f \in \mathcal{L}(\mathcal{V}, \mathcal{V})$ has a unique adjoint map $f^{ad} \in \mathcal{L}(\mathcal{V}, \mathcal{V})$. As an exercise, discuss which step in the construction above does not work in the finite dimensional case.

We next prove some properties of the adjoint map.

Lemma 13.10 *Let $\mathcal{V}$ be a finite dimensional Euclidean or unitary vector space.*

(1) *If $f_1, f_2 \in \mathcal{L}(\mathcal{V}, \mathcal{V})$ and $\lambda_1, \lambda_2 \in K$ (where $K = \mathbb{R}$ in the Euclidean and $K = \mathbb{C}$ in the unitary case), then*

$$(\lambda_1 f_1 + \lambda_2 f_2)^{ad} = \overline{\lambda}_1 f_1^{ad} + \overline{\lambda}_2 f_2^{ad}.$$

13.2 Adjoints in Finite Dimensional Euclidean and Unitary Vector Spaces

In the Euclidean case the map $f \mapsto f^{ad}$ is therefore linear, and in the unitary case semilinear.

(2) We have $(\mathrm{Id}_\mathcal{V})^{ad} = \mathrm{Id}_\mathcal{V}$.
(3) For every $f \in \mathcal{L}(\mathcal{V}, \mathcal{V})$ we have $(f^{ad})^{ad} = f$.
(4) If $f_1, f_2 \in \mathcal{L}(\mathcal{V}, \mathcal{V})$, then $(f_2 \circ f_1)^{ad} = f_1^{ad} \circ f_2^{ad}$.
(5) For every $f \in \mathcal{L}(\mathcal{V}, \mathcal{V})$ we have $\ker(f^{ad}) = \mathrm{im}(f)^\perp$ and $\ker(f) = \mathrm{im}(f^{ad})^\perp$.

Proof

(1) If $v, w \in \mathcal{V}$ and $\lambda_1, \lambda_2 \in K$, then

$$\langle (\lambda_1 f_1 + \lambda_2 f_2)(v), w \rangle = \lambda_1 \langle f_1(v), w \rangle + \lambda_2 \langle f_2(v), w \rangle$$
$$= \lambda_1 \left\langle v, f_1^{ad}(w) \right\rangle + \lambda_2 \left\langle v, f_2^{ad}(w) \right\rangle$$
$$= \left\langle v, \overline{\lambda_1} f_1^{ad}(w) + \overline{\lambda_2} f_2^{ad}(w) \right\rangle$$
$$= \left\langle v, \left(\overline{\lambda_1} f_1^{ad} + \overline{\lambda_2} f_2^{ad} \right)(w) \right\rangle,$$

and thus $(\lambda_1 f_1 + \lambda_2 f_2)^{ad} = \overline{\lambda_1} f_1^{ad} + \overline{\lambda_2} f_2^{ad}$.

(2) For all $v, w \in \mathcal{V}$ we have $\langle \mathrm{Id}_\mathcal{V}(v), w \rangle = \langle v, w \rangle = \langle v, \mathrm{Id}_\mathcal{V}(w) \rangle$, and thus $(\mathrm{Id}_\mathcal{V})^{ad} = \mathrm{Id}_\mathcal{V}$.

(3) For all $v, w \in \mathcal{V}$ we have $\langle f^{ad}(v), w \rangle = \langle v, f(w) \rangle$, and thus $(f^{ad})^{ad} = f$.

(4) For all $v, w \in \mathcal{V}$ we have

$$\langle (f_2 \circ f_1)(v), w \rangle = \langle f_2(f_1(v)), w \rangle = \left\langle f_1(v), f_2^{ad}(w) \right\rangle = \left\langle v, f_1^{ad}\left(f_2^{ad}(w)\right) \right\rangle$$
$$= \left\langle v, \left(f_1^{ad} \circ f_2^{ad}\right)(w) \right\rangle,$$

and thus $(f_2 \circ f_1)^{ad} = f_1^{ad} \circ f_2^{ad}$.

(5) Let $w \in \ker(f^{ad})$, then $f^{ad}(w) = 0$ and hence

$$0 = \langle v, f^{ad}(w) \rangle = \langle f(v), w \rangle$$

for all $v \in \mathcal{V}$, and therefore $w \in \mathrm{im}(f)^\perp$. If, on the other hand, $w \in \mathrm{im}(f)^\perp$, then

$$0 = \langle f(v), w \rangle = \langle v, f^{ad}(w) \rangle$$

for all $v \in \mathcal{V}$. Thus, in particular, $\langle f^{ad}(w), f^{ad}(w) \rangle = 0$, and therefore $f^{ad}(w) = 0$, i.e., $w \in \ker(f^{ad})$. This shows the equation $\ker(f^{ad}) = \mathrm{im}(f)^\perp$. From this and $(f^{ad})^{ad} = f$ we get $\ker(f) = \ker((f^{ad})^{ad}) = \mathrm{im}(f^{ad})^\perp$. □

Example 13.11 Consider the unitary vector space $\mathbb{C}^{3,1}$ with the standard scalar product and the linear map

$$f : \mathbb{C}^{3,1} \to \mathbb{C}^{3,1}, \quad v \mapsto Fv, \quad \text{with} \quad F = \begin{bmatrix} 1 & i & i \\ i & 0 & 0 \\ 1 & 0 & 0 \end{bmatrix}.$$

Then

$$f^{ad} : \mathbb{C}^{3,1} \to \mathbb{C}^{3,1}, \quad v \mapsto F^H v, \quad \text{with} \quad F^H = \begin{bmatrix} 1 & -i & 1 \\ -i & 0 & 0 \\ -i & 0 & 0 \end{bmatrix}$$

(cp. Example 13.8 with $B = I_3$). The matrices F and F^H have rank 2. Therefore, $\dim(\ker(f)) = \dim(\ker(f^{ad})) = 1$. A simple calculation shows that

$$\ker(f) = \text{span}\left\{ \begin{bmatrix} 0 \\ 1 \\ -1 \end{bmatrix} \right\} \quad \text{and} \quad \ker(f^{ad}) = \text{span}\left\{ \begin{bmatrix} 0 \\ 1 \\ i \end{bmatrix} \right\}.$$

The dimension formula for linear maps implies that $\dim(\text{im}(f)) = \dim(\text{im}(f^{ad})) = 2$. From the matrices F and F^H we can see that

$$\text{im}(f) = \text{span}\left\{ \begin{bmatrix} 1 \\ i \\ 1 \end{bmatrix}, \begin{bmatrix} 1 \\ 0 \\ 0 \end{bmatrix} \right\} \quad \text{and} \quad \text{im}(f^{ad}) = \text{span}\left\{ \begin{bmatrix} 1 \\ -i \\ -i \end{bmatrix}, \begin{bmatrix} 1 \\ 0 \\ 0 \end{bmatrix} \right\}.$$

The equations $\ker(f^{ad}) = \text{im}(f)^\perp$ and $\ker(f) = \text{im}(f^{ad})^\perp$ can be verified by direct computation.

We now study the relation between the matrix representations of an endomorphism and its adjoint. Let $\mathcal{V}$ be a finite dimensional unitary vector space with the scalar product $\langle \cdot, \cdot \rangle$ and let $f \in \mathcal{L}(\mathcal{V}, \mathcal{V})$. For an orthonormal basis $B = \{u_1, \ldots, u_n\}$ of $\mathcal{V}$ let $[f]_{B,B} = [a_{ij}] \in \mathbb{C}^{n,n}$, i.e.,

$$f(u_j) = \sum_{k=1}^{n} a_{kj} u_k, \quad j = 1, \ldots, n,$$

and hence

$$\langle f(u_j), u_i \rangle = \left\langle \sum_{k=1}^{n} a_{kj} u_k, u_i \right\rangle = a_{ij}, \quad i, j = 1, \ldots, n.$$

13.2 Adjoints in Finite Dimensional Euclidean and Unitary Vector Spaces

If $[f^{ad}]_{B,B} = [b_{ij}] \in \mathbb{C}^{n,n}$, i.e.,

$$f^{ad}(u_j) = \sum_{k=1}^{n} b_{kj} u_k, \quad j = 1, \ldots, n,$$

then

$$b_{ij} = \langle f^{ad}(u_j), u_i \rangle = \langle u_j, f(u_i) \rangle = \overline{\langle f(u_i), u_j \rangle} = \overline{a}_{ji}.$$

Thus, $[f^{ad}]_{B,B} = ([f]_{B,B})^H$. The same holds for a finite dimensional Euclidean vector space, but then we can omit the complex conjugation. Therefore, we have shown the following result.

Theorem 13.12 *If $\mathcal{V}$ is a finite dimensional Euclidean or unitary vector space with the orthonormal basis B and $f \in \mathcal{L}(\mathcal{V}, \mathcal{V})$, then*

$$[f^{ad}]_{B,B} = ([f]_{B,B})^H.$$

(In the Euclidean case $([f]_{B,B})^H = ([f]_{B,B})^T$.)

An important special class are the selfadjoint endomorphisms.

Definition 13.13 Let $\mathcal{V}$ be a finite dimensional Euclidean or unitary vector space. An endomorphism $f \in \mathcal{L}(\mathcal{V}, \mathcal{V})$ is called *selfadjoint* when $f = f^{ad}$.

Trivial examples of selfadjoint endomorphism in $\mathcal{L}(\mathcal{V}, \mathcal{V})$ are $f = 0$ and $\text{Id}_\mathcal{V}$. Using Theorem 13.12 we obtain the following result about the matrix representation of selfadjoint endomorphisms.

Corollary 13.14

(1) If $\mathcal{V}$ is a finite dimensional Euclidean vector space, $f \in \mathcal{L}(\mathcal{V}, \mathcal{V})$ is selfadjoint and B is an orthonormal basis of $\mathcal{V}$, then $[f]_{B,B}$ is a symmetric matrix.
(2) If $\mathcal{V}$ is a finite dimensional unitary vector space, $f \in \mathcal{L}(\mathcal{V}, \mathcal{V})$ is selfadjoint and B is an orthonormal basis of $\mathcal{V}$, then $[f]_{B,B}$ is an Hermitian matrix.

The selfadjoint endomorphisms again form a vector space. However, we have to be careful to use the appropriate field over which this vector space is defined. In particular, the set of selfadjoint endomorphisms on a unitary vector space $\mathcal{V}$ does *not* form a $\mathbb{C}$-vector space. If $f = f^{ad} \in \mathcal{L}(\mathcal{V}, \mathcal{V}) \setminus \{0\}$, then $(\mathbf{i}f)^{ad} = -\mathbf{i}f^{ad} = -\mathbf{i}f \neq \mathbf{i}f$ (cp. (1) in Lemma 13.10). Similarly, the Hermitian matrices in $\mathbb{C}^{n,n}$ do *not* form a $\mathbb{C}$-vector space. If $A = A^H \in \mathbb{C}^{n,n} \setminus \{0\}$ is Hermitian, then $(\mathbf{i}A)^H = -\mathbf{i}A^H = -\mathbf{i}A \neq \mathbf{i}A$.

Lemma 13.15

(1) If $\mathcal{V}$ is an n-dimensional Euclidean vector space, then the set of selfadjoint endomorphisms $\{f \in \mathcal{L}(\mathcal{V},\mathcal{V}) \mid f = f^{ad}\}$ forms an $\mathbb{R}$-vector space of dimension $n(n+1)/2$.

(2) If $\mathcal{V}$ is an n-dimensional unitary vector space, then the set of selfadjoint endomorphisms $\{f \in \mathcal{L}(\mathcal{V},\mathcal{V}) \mid f = f^{ad}\}$ forms an $\mathbb{R}$-vector space of dimension n^2.

Proof Exercise. □

A matrix $A \in \mathbb{C}^{n,n}$ with $A = A^T$ is called *complex symmetric*. Unlike the Hermitian matrices, the complex symmetric matrices form a $\mathbb{C}$-vector space.

Lemma 13.16 *The set of complex symmetric matrices in $\mathbb{C}^{n,n}$ forms a $\mathbb{C}$-vector space of dimension $n(n+1)/2$.*

Proof Exercise. □

Lemmas 13.15 and 13.16 will be used in Chap. 15 in our proof of the Fundamental Theorem of Algebra.

Exercises

13.1 Let $\beta(v,w) = w^T Bv$ with $B = \text{diag}(1,-1)$ be defined for $v, w \in \mathbb{R}^{2,1}$. Consider the linear maps $f : \mathbb{R}^{2,1} \to \mathbb{R}^{2,1}$, $v \mapsto Fv$, and $h : \mathbb{R}^{2,1} \to \mathbb{R}^{2,1}$, $w \mapsto Hw$, where

$$F = \begin{bmatrix} 1 & 2 \\ 0 & 1 \end{bmatrix} \in \mathbb{R}^{2,2}, \quad H = \begin{bmatrix} 1 & 0 \\ 1 & 1 \end{bmatrix} \in \mathbb{R}^{2,2}.$$

Determine β_v, $\beta^{(1)}$ and $\beta^{(2)}$ as in (13.1)–(13.2) as well as the right adjoint of f and the left adjoint of h with respect to β.

13.2 Let $(\mathcal{V}, \langle \cdot, \cdot \rangle_\mathcal{V})$ and $(\mathcal{W}, \langle \cdot, \cdot \rangle_\mathcal{W})$ be two finite dimensional Euclidean vector spaces and let $f \in \mathcal{L}(\mathcal{V}, \mathcal{W})$. Show that there exists a unique $g \in \mathcal{L}(\mathcal{W}, \mathcal{V})$ with $\langle f(v), w \rangle_\mathcal{W} = \langle v, g(w) \rangle_\mathcal{V}$ for all $v \in \mathcal{V}$ and $w \in \mathcal{W}$.

13.3 Let $\langle v, w \rangle = w^T Bv$ for all $v, w \in \mathbb{R}^{2,1}$ with

$$B = \begin{bmatrix} 2 & 1 \\ 1 & 1 \end{bmatrix} \in \mathbb{R}^{2,2}.$$

(a) Show that $\langle v, w \rangle = w^T Bv$ is a scalar product on $\mathbb{R}^{2,1}$.

(b) Using this scalar product, determine the adjoint map f^{ad} of $f : \mathbb{R}^{2,1} \to \mathbb{R}^{2,1}$, $v \mapsto Fv$, with $F \in \mathbb{R}^{2,2}$.

(c) Investigate which properties F needs to satisfy so that f is selfadjoint.

13.4 Let $n \geq 2$ and
$$f : \mathbb{R}^{n,1} \to \mathbb{R}^{n,1}, \quad [x_1, \ldots, x_n]^T \mapsto [0, x_1, \ldots, x_{n-1}]^T.$$
Determine the adjoint f^{ad} of f with respect to the standard scalar product of $\mathbb{R}^{n,1}$.

13.5 Show that the two maps
$$f_1 : \mathbb{C}[t]_{\leq n} \to \mathbb{C}[t]_{\leq n}, \quad p(t) \mapsto p(-t),$$
$$f_2 : \mathbb{C}[t]_{\leq n} \to \mathbb{C}[t]_{\leq n}, \quad p(t) \mapsto t^n p(t^{-1}),$$
are linear and determine f_1^{ad} and f_2^{ad} with respect to the scalar product in Exercise 12.3.

13.6 Let V be a finite dimensional Euclidean or unitary vector space and let $f \in \mathcal{L}(V, V)$. Show that $\ker(f^{ad} \circ f) = \ker(f)$ and $\text{im}(f^{ad} \circ f) = \text{im}(f^{ad})$.

13.7 Let V be a finite dimensional Euclidean or unitary vector space, let $\mathcal{U} \subseteq V$ be a subspace and let $f \in \mathcal{L}(V, V)$ with $f(\mathcal{U}) \subseteq \mathcal{U}$. Show that then $f^{ad}(\mathcal{U}^\perp) \subseteq \mathcal{U}^\perp$.

13.8 Let $A \in \mathbb{C}^{n,n}$ and $b \in \mathbb{C}^{n,1}$. Show that the linear system of equations $Ax = b$ has a solution, i.e., $\mathscr{L}(A, b) \neq \emptyset$, if and only if $b \in \mathscr{L}(A^H, 0)^\perp$.

13.9 Let V be a finite dimensional Euclidean or unitary vector space and let $f, g \in \mathcal{L}(V, V)$ be selfadjoint. Show that $f \circ g$ is selfadjoint if and only if f and g commute, i.e., $f \circ g = g \circ f$.

13.10 Let V be a finite dimensional unitary vector space and let $f \in \mathcal{L}(V, V)$. Show that f is selfadjoint if and only if $\langle f(v), v \rangle \in \mathbb{R}$ holds for all $v \in V$.

13.11 Let V be a finite dimensional Euclidean or unitary vector space, and let $f \in \mathcal{L}(V, V)$ be a projection. Show the following assertions:

(a) f is an orthogonal projection onto $\text{im}(f)$ if and only if f is selfadjoint.

(b) Let $u \in V$ with $\|u\| = 1$, where $\|\cdot\|$ is the norm induced by the scalar product. Then $f \in \mathcal{L}(V, V)$, $v \mapsto \langle v, u \rangle u$, is an orthogonal projection. Determine $\ker(f)$ and $\text{im}(f)$.

13.12 Let V be a finite dimensional Euclidean or unitary vector space and let $f, g \in \mathcal{L}(V, V)$. Show that $g^{ad} \circ f = 0 \in \mathcal{L}(V, V)$ holds if and only if $\langle v, w \rangle = 0$ for all $v \in \text{im}(f)$ and $w \in \text{im}(g)$.

13.13 Prove Lemma 13.15.

13.14 Prove Lemma 13.16.

13.15 Let V be a finite dimensional Euclidean vector space. Show the following assertions:

(a) For $v \in V$ the map $f_v : V \to \mathbb{R}$, $w \mapsto \langle w, v \rangle$, is linear, i.e., $f_v \in V^*$.

(b) The map $\Phi : \mathcal{V} \to \mathcal{V}^*$, $v \mapsto f_v$, is an isomorphism. (This map is called *Fréchet-Riesz isomorphism*.[1])
(c) If $\mathcal{U} \subseteq \mathcal{V}$ is a subspace, then $\Phi(\mathcal{U}^\perp) = \mathcal{U}^0$ (cp. Exercise 11.8).
(d) For $f \in \mathcal{L}(\mathcal{V}, \mathcal{V})$ we have $f^{ad} = \Phi^{-1} \circ f^* \circ \Phi$.

13.16 For two polynomials $p, q \in \mathbb{R}[t]_{\leq n}$ let

$$\langle p, q \rangle := \int_{-1}^{1} p(t) q(t) \, dt.$$

(a) Show that this defines a scalar product on $\mathbb{R}[t]_{\leq n}$.
(b) Consider the map

$$f : \mathbb{R}[t]_{\leq n} \to \mathbb{R}[t]_{\leq n}, \quad p = \sum_{i=0}^{n} \alpha_i t^i \mapsto \sum_{i=1}^{n} i \alpha_i t^{i-1},$$

and determine f^{ad}, $\ker(f^{ad})$, $\mathrm{im}(f)$, $\ker(f^{ad})^\perp$ and $\mathrm{im}(f)^\perp$.

---
[1] Maurice René Fréchet (1878–1973) and Frigyes Riesz (1880–1956).

Eigenvalues of Endomorphisms 14

In previous chapters we have already studied eigenvalues and eigenvectors of matrices. In this chapter we generalize these concepts to endomorphisms, and we investigate when endomorphisms on finite dimensional vector spaces can be represented by diagonal matrices or (upper) triangular matrices. From such representations we easily can read off important information about the endomorphism, in particular its eigenvalues.

14.1 Basic Definitions and Properties

We first consider an arbitrary vector space and then concentrate on the finite dimensional case.

Definition 14.1 Let $\mathcal{V}$ be a K-vector space and $f \in \mathcal{L}(\mathcal{V}, \mathcal{V})$. If $\lambda \in K$ and $v \in \mathcal{V} \setminus \{0\}$ satisfy

$$f(v) = \lambda v,$$

then λ is called an *eigenvalue* of f, and v is called an *eigenvector* of f corresponding to λ.

By definition, $v = 0$ cannot be an eigenvector, but an eigenvalue $\lambda = 0$ may occur (cp. the example following Definition 8.7).

The equation $f(v) = \lambda v$ can be written as

$$0 = \lambda v - f(v) = (\lambda \operatorname{Id}_\mathcal{V} - f)(v).$$

Hence, $\lambda \in K$ is an eigenvalue of f if and only if

$$\ker(\lambda \, \mathrm{Id}_\mathcal{V} - f) \neq \{0\}.$$

We already know that the kernel of an endomorphism on $\mathcal{V}$ forms a subspace of $\mathcal{V}$ (cp. Lemma 10.7). This holds, in particular, for $\ker(\lambda \, \mathrm{Id}_\mathcal{V} - f)$.

Definition 14.2 If $\mathcal{V}$ is a K-vector space and $\lambda \in K$ is an eigenvalue of $f \in \mathcal{L}(\mathcal{V}, \mathcal{V})$, then the subspace

$$\mathcal{V}_f(\lambda) := \ker(\lambda \, \mathrm{Id}_\mathcal{V} - f)$$

is called the *eigenspace* of f corresponding to λ and

$$g(\lambda, f) := \dim(\mathcal{V}_f(\lambda))$$

is called the *geometric multiplicity* of the eigenvalue λ.

By definition, the eigenspace $\mathcal{V}_f(\lambda)$ is spanned by all eigenvectors of f corresponding to the eigenvalue λ. If $\mathcal{V}_f(\lambda)$ is finite dimensional, then $g(\lambda, f) = \dim(\mathcal{V}_f(\lambda))$ is equal to the maximal number of linearly independent eigenvectors of f corresponding to λ.

Definition 14.3 Let $\mathcal{V}$ be a K-vector space, let $\mathcal{U} \subseteq \mathcal{V}$ be a subspace, and let $f \in \mathcal{L}(\mathcal{V}, \mathcal{V})$. If $f(\mathcal{U}) \subseteq \mathcal{U}$, i.e., if $f(u) \in \mathcal{U}$ holds for all $u \in \mathcal{U}$, then $\mathcal{U}$ is called an *f-invariant subspace* of $\mathcal{V}$.

An important example of f-invariant subspaces are the eigenspaces of f.

Lemma 14.4 *If $\mathcal{V}$ is a K-vector space and $\lambda \in K$ is an eigenvalue of $f \in \mathcal{L}(\mathcal{V}, \mathcal{V})$, then $\mathcal{V}_f(\lambda)$ is an f-invariant subspace of $\mathcal{V}$.*

Proof For every $v \in \mathcal{V}_f(\lambda)$ we have $f(v) = \lambda v \in \mathcal{V}_f(\lambda)$. □

We now consider finite dimensional vector spaces and discuss the relationship between the eigenvalues of f and the eigenvalues of a matrix representation of f with respect to a given basis.

Lemma 14.5 *If $\mathcal{V}$ is a finite dimensional K-vector space and $f \in \mathcal{L}(\mathcal{V}, \mathcal{V})$, then the following statements are equivalent:*

(1) $\lambda \in K$ is an eigenvalue of f.
(2) $\lambda \in K$ is an eigenvalue of the matrix $[f]_{B,B}$ for every basis B of $\mathcal{V}$.

14.1 Basic Definitions and Properties

Proof Let $\lambda \in K$ be an eigenvalue of f and let $B = \{v_1, \ldots, v_n\}$ be an arbitrary basis of $\mathcal{V}$. If $v \in \mathcal{V}$ is an eigenvector of f corresponding to the eigenvalue λ, then $f(v) = \lambda v$ and there exist (unique) coordinates $\mu_1, \ldots, \mu_n \in K$, not all equal to zero, with $v = \sum_{j=1}^n \mu_j v_j$. Using (10.4) we obtain

$$[f]_{B,B} \begin{bmatrix} \mu_1 \\ \vdots \\ \mu_n \end{bmatrix} = \Phi_B(f(v)) = \Phi_B(\lambda v) = \lambda \Phi_B(v) = \lambda \begin{bmatrix} \mu_1 \\ \vdots \\ \mu_n \end{bmatrix},$$

and thus λ is an eigenvalue of $[f]_{B,B}$.

If, on the other hand, $[f]_{B,B}[\mu_1, \ldots, \mu_n]^T = \lambda[\mu_1, \ldots, \mu_n]^T$ with $[\mu_1, \ldots, \mu_n]^T \neq 0$ for a given (arbitrary) basis $B = \{v_1, \ldots, v_n\}$ of $\mathcal{V}$, then we set $v := \sum_{j=1}^n \mu_j v_j$. Then $v \neq 0$ and

$$f(v) = f\left(\sum_{j=1}^n \mu_j v_j\right) = \sum_{j=1}^n \mu_j f(v_j) = (v_1, \ldots, v_n) \left([f]_{B,B} \begin{bmatrix} \mu_1 \\ \vdots \\ \mu_n \end{bmatrix}\right)$$

$$= (v_1, \ldots, v_n) \left(\lambda \begin{bmatrix} \mu_1 \\ \vdots \\ \mu_n \end{bmatrix}\right) = \lambda v,$$

i.e., λ is an eigenvalue of f. □

If $v \in \mathcal{V} \setminus \{0\}$ is an eigenvector corresponding to the eigenvalue λ of f, then the proof of Lemma 14.5 shows that $\Phi_B(v)$ (for every basis B of $\mathcal{V}$) is an eigenvector corresponding to the eigenvalue λ of $[f]_{B,B}$.

Lemma 14.5 implies that the eigenvalues of f are the roots of the characteristic polynomial of the matrix $[f]_{B,B}$ (cp. Theorem 8.8). This, however, does *not* hold in general for a matrix representation of the form $[f]_{B,\widetilde{B}}$, where B and $\widetilde{B}$ are two *different* bases of $\mathcal{V}$. In general, the two matrices

$$[f]_{B,\widetilde{B}} = [\mathrm{Id}_\mathcal{V}]_{B,\widetilde{B}} [f]_{B,B} \quad \text{and} \quad [f]_{B,B}$$

do not have the same eigenvalues.

Example 14.6 Consider the vector space $\mathbb{R}^{2,1}$ with the bases

$$B = \left\{ \begin{bmatrix} 1 \\ 0 \end{bmatrix}, \begin{bmatrix} 0 \\ 1 \end{bmatrix} \right\}, \quad \widetilde{B} = \left\{ \begin{bmatrix} 1 \\ -1 \end{bmatrix}, \begin{bmatrix} 1 \\ 1 \end{bmatrix} \right\}.$$

Then the endomorphism

$$f : \mathbb{R}^{2,1} \to \mathbb{R}^{2,1}, \quad v \mapsto Fv, \quad \text{where} \quad F = \begin{bmatrix} 0 & 1 \\ 1 & 0 \end{bmatrix},$$

has the matrix representations

$$[f]_{B,B} = \begin{bmatrix} 0 & 1 \\ 1 & 0 \end{bmatrix} \quad \text{and} \quad [f]_{B,\widetilde{B}} = \frac{1}{2}\begin{bmatrix} -1 & 1 \\ 1 & 1 \end{bmatrix}.$$

We have $\det(tI_2 - [f]_{B,B}) = t^2 - 1$, and thus f has the eigenvalues -1 and 1. On the other hand, the characteristic polynomial of $[f]_{B,\widetilde{B}}$ is $t^2 - \frac{1}{2}$, so that this matrix has the eigenvalues $-1/\sqrt{2}$ and $1/\sqrt{2}$.

For two different bases B and $\widetilde{B}$ of $\mathcal{V}$ the matrices $[f]_{B,B}$ and $[f]_{\widetilde{B},\widetilde{B}}$ are similar (cp. the discussion following Corollary 10.21). In Theorem 8.11 we have shown that similar matrices have the same characteristic polynomial. This justifies the following definition.

Definition 14.7 If $n \in \mathbb{N}$, $\mathcal{V}$ is an n-dimensional K-vector space with the basis B and $f \in \mathcal{L}(\mathcal{V}, \mathcal{V})$, then

$$P_f := \det(tI_n - [f]_{B,B}) \in K[t]$$

is called the *characteristic polynomial* of f.

The characteristic polynomial P_f is always a monic polynomial with

$$\deg(P_f) = n = \dim(\mathcal{V}).$$

As we have discussed before, P_f is independent of the choice of the basis of $\mathcal{V}$. A scalar $\lambda \in K$ is an eigenvalue of f if and only if λ is a root of P_f, i.e., $P_f(\lambda) = 0$. As shown in Example 8.9, in real vector spaces with dimensions at least two, there exist endomorphisms that do not have eigenvalues.

If λ is a root of P_f, then $P_f = (t - \lambda) \cdot q$ for a monic polynomial $q \in K[t]$, i.e., the *linear factor* $t - \lambda$ divides the polynomial P_f; we will show this formally in Corollary 15.5 below. If also $q(\lambda) = 0$, then $q = (t - \lambda) \cdot \widetilde{q}$ for a monic polynomial $\widetilde{q} \in K[t]$, and thus $P_f = (t - \lambda)^2 \cdot \widetilde{q}$. We can continue until $P_f = (t - \lambda)^d \cdot g$ for a monic polynomial $g \in K[t]$ with $g(\lambda) \neq 0$. This leads to the following definition.

Definition 14.8 Let $\mathcal{V}$ be a finite dimensional K-vector space, and let $f \in \mathcal{L}(\mathcal{V}, \mathcal{V})$ have the eigenvalue $\lambda \in K$. If the characteristic polynomial of f has the form

$$P_f = (t - \lambda)^d \cdot g$$

for some monic polynomial $g \in K[t]$ with $g(\lambda) \neq 0$, then d is called the *algebraic multiplicity* of the eigenvalue λ of f. It is denoted by $a(\lambda, f)$.

If $\lambda_1, \ldots, \lambda_k$ are the *pairwise distinct* eigenvalues of f with corresponding algebraic multiplicities $a(\lambda_1, f), \ldots, a(\lambda_k, f)$, and if $\dim(\mathcal{V}) = n$, then

$$a(\lambda_1, f) + \ldots + a(\lambda_k, f) \leq \deg(P_f) \leq n$$

(cp. Corollary 15.7).

Example 14.9 The endomorphism $f : \mathbb{R}^{4,1} \to \mathbb{R}^{4,1}$, $v \mapsto Fv$ with

$$F = \begin{bmatrix} 1 & 2 & 3 & 4 \\ 0 & 1 & 2 & 3 \\ 0 & 0 & 0 & 1 \\ 0 & 0 & -1 & 0 \end{bmatrix} \in \mathbb{R}^{4,4},$$

has the characteristic polynomial $P_f = (t-1)^2(t^2+1)$. The only real root of P_f is $\lambda_1 = 1$, and $a(\lambda_1, f) = 2 < 4 = \dim(\mathbb{R}^{4,1})$.

To determine the geometric multiplicity $g(\lambda_1, f)$ we compute

$$\dim(\text{im}(\lambda_1 \, \text{Id}_{\mathbb{R}^{4,1}} - f)) = \text{rank}(1 \cdot I_4 - F) = \text{rank}\left(\begin{bmatrix} 0 & -2 & -3 & -4 \\ 0 & 0 & -2 & -3 \\ 0 & 0 & 1 & -1 \\ 0 & 0 & 1 & -1 \end{bmatrix}\right) = 3.$$

With the dimension formula for linear maps (Theorem 10.9) we then obtain

$$4 = \dim(\text{im}(\lambda_1 \, \text{Id}_{\mathbb{R}^{4,1}} - f)) + \dim(\ker(\lambda_1 \, \text{Id}_{\mathbb{R}^{4,1}} - f)) = 3 + g(\lambda_1, f),$$

which implies that $g(\lambda_1, f) = 1$.

Lemma 14.10 *If $\mathcal{V}$ is a finite dimensional K-vector space and $f \in \mathcal{L}(\mathcal{V}, \mathcal{V})$, then*

$$g(\lambda, f) \leq a(\lambda, f)$$

for every eigenvalue λ of f.

Proof Let $\lambda \in K$ be an eigenvalue of f with geometric multiplicity $m = g(\lambda, f)$. Then there exist m linear independent eigenvectors $v_1, \ldots, v_m \in \mathcal{V}$ of f corresponding to the eigenvalue λ. If $m = \dim(\mathcal{V})$, then these m eigenvectors form a basis B of $\mathcal{V}$. If $m < \dim(\mathcal{V}) = n$, then we can extend the m eigenvectors to a basis $B = \{v_1, \ldots, v_m, v_{m+1}, \ldots, v_n\}$ of $\mathcal{V}$.

We have $f(v_j) = \lambda v_j$ for $j = 1, \ldots, m$ and, therefore,

$$[f]_{B,B} = \begin{bmatrix} \lambda I_m & Z_1 \\ 0 & Z_2 \end{bmatrix}$$

for two matrices $Z_1 \in K^{m,n-m}$ and $Z_2 \in K^{n-m,n-m}$. Using (1) in Lemma 7.10 we obtain

$$P_f = \det(t I_n - [f]_{B,B}) = (t - \lambda)^m \cdot \det(t I_{n-m} - Z_2),$$

which implies $a(\lambda, f) \geq m = g(\lambda, f)$. □

In the following we will try to find a basis of $\mathcal{V}$, so that the eigenvalues of a given endomorphism f can be read off easily from its matrix representation. The easiest forms of matrices in this sense are diagonal and triangular matrices, since their eigenvalues are just their diagonal entries.

14.2 Diagonalization

In this section we will analyze when a given endomorphism has a diagonal matrix representation. We formally define this property as follows.

Definition 14.11 Let $\mathcal{V}$ be a finite dimensional K-vector space. An endomorphism $f \in \mathcal{L}(\mathcal{V}, \mathcal{V})$ is called *diagonalizable*, if there exists a basis B of $\mathcal{V}$, such that $[f]_{B,B}$ is a diagonal matrix.

Accordingly, a matrix $A \in K^{n,n}$ is diagonalizable when there exists a matrix $S \in GL_n(K)$ with $A = SDS^{-1}$ for a diagonal matrix $D \in K^{n,n}$.

In order to analyze the diagonalizablility, we begin with a sufficient condition for the linear independence of eigenvectors. This condition also holds when $\mathcal{V}$ is infinite dimensional.

Lemma 14.12 Let $\mathcal{V}$ be a K-vector space and $f \in \mathcal{L}(\mathcal{V}, \mathcal{V})$. If $\lambda_1, \ldots, \lambda_k \in K$, $k \geq 2$, are pairwise distinct eigenvalues of f with corresponding eigenvectors $v_1, \ldots, v_k \in \mathcal{V}$, then $v_1, \ldots, v_k$ are linearly independent.

Proof We prove the assertion by induction on k. Let $k = 2$ and let v_1, v_2 be eigenvectors of f corresponding to the eigenvalues $\lambda_1 \neq \lambda_2$. Let $\mu_1, \mu_2 \in K$ with $\mu_1 v_1 + \mu_2 v_2 = 0$. Applying f on both sides of this equation as well as multiplying the equation with λ_2 yields the two equations

$$\mu_1 \lambda_1 v_1 + \mu_2 \lambda_2 v_2 = 0,$$
$$\mu_1 \lambda_2 v_1 + \mu_2 \lambda_2 v_2 = 0.$$

14.2 Diagonalization

Subtracting the second equation from the first, we get $\mu_1(\lambda_1 - \lambda_2)v_1 = 0$. Since $\lambda_1 \neq \lambda_2$ and $v_1 \neq 0$, we have $\mu_1 = 0$. Then from $\mu_1 v_1 + \mu_2 v_2 = 0$ we also obtain $\mu_2 = 0$, since $v_2 \neq 0$. Thus, v_1 and v_2 are linearly independent.

The proof of the inductive step is analogous. We assume that the assertion holds for some $k \geq 2$. Let $\lambda_1, \ldots, \lambda_{k+1}$ be pairwise distinct eigenvalues of f with corresponding eigenvectors $v_1, \ldots, v_{k+1}$, and let $\mu_1, \ldots, \mu_{k+1} \in K$ satisfy

$$\mu_1 v_1 + \ldots + \mu_k v_k + \mu_{k+1} v_{k+1} = 0.$$

Applying f to this equation yields

$$\mu_1 \lambda_1 v_1 + \ldots + \mu_k \lambda_k v_k + \mu_{k+1} \lambda_{k+1} v_{k+1} = 0,$$

while a multiplication with λ_{k+1} gives

$$\mu_1 \lambda_{k+1} v_1 + \ldots + \mu_k \lambda_{k+1} v_k + \mu_{k+1} \lambda_{k+1} v_{k+1} = 0.$$

Subtracting this equation from the previous one we get

$$\mu_1(\lambda_1 - \lambda_{k+1})v_1 + \ldots + \mu_k(\lambda_k - \lambda_{k+1})v_k = 0.$$

Since $\lambda_1, \ldots, \lambda_{k+1}$ are pairwise distinct and $v_1, \ldots, v_k$ are linearly independent, by the induction hypothesis we obtain $\mu_1 = \cdots = \mu_k = 0$. But then $\mu_{k+1} v_{k+1} = 0$ implies that also $\mu_{k+1} = 0$, so that $v_1, \ldots, v_{k+1}$ are linearly independent. □

Using this result we next show that the sum of eigenspaces corresponding to pairwise distinct eigenvalues is direct (cp. Theorem 9.36).

Lemma 14.13 *Let $\mathcal{V}$ be a K-vector space and $f \in \mathcal{L}(\mathcal{V}, \mathcal{V})$. If $\lambda_1, \ldots, \lambda_k \in K$, $k \geq 2$, are pairwise distinct eigenvalues of f, then the corresponding eigenspaces satisfy*

$$\mathcal{V}_f(\lambda_i) \cap \sum_{\substack{j=1 \\ j \neq i}}^{k} \mathcal{V}_f(\lambda_j) = \{0\}$$

for all $i = 1, \ldots, k$.

Proof Let i be fixed and let

$$v \in \mathcal{V}_f(\lambda_i) \cap \sum_{\substack{j=1 \\ j \neq i}}^{k} \mathcal{V}_f(\lambda_j).$$

In particular, we have $v = \sum_{j \neq i} v_j$ for some $v_j \in \mathcal{V}_f(\lambda_j)$, $j \neq i$. Then $-v + \sum_{j \neq i} v_j = 0$, and the linear independence of eigenvectors corresponding to pairwise distinct eigenvalues (cp. Lemma 14.12) implies $v = 0$. □

The following theorem gives necessary and sufficient conditions for the diagonalizability of an endomorphism on a finite dimensional vector space.

Theorem 14.14 *If $\mathcal{V}$ is a finite dimensional K-vector space and $f \in \mathcal{L}(\mathcal{V}, \mathcal{V})$, then the following statements are equivalent:*

(1) f is diagonalizable.
(2) There exists a basis of $\mathcal{V}$ consisting of eigenvectors of f.
(3) The characteristic polynomial P_f decomposes into $n = \dim(\mathcal{V})$ linear factors over K, i.e.,

$$P_f = (t - \lambda_1) \cdot \ldots \cdot (t - \lambda_n)$$

with the eigenvalues $\lambda_1, \ldots, \lambda_n \in K$ of f, and for every eigenvalue λ_j we have $g(\lambda_j, f) = a(\lambda_j, f)$.

Proof

(1) ⇔ (2): If $f \in \mathcal{L}(\mathcal{V}, \mathcal{V})$ is diagonalizable, then there exists a basis $B = \{v_1, \ldots, v_n\}$ of $\mathcal{V}$ and scalars $\lambda_1, \ldots, \lambda_n \in K$ with

$$[f]_{B,B} = \begin{bmatrix} \lambda_1 & & \\ & \ddots & \\ & & \lambda_n \end{bmatrix}, \tag{14.1}$$

and hence $f(v_j) = \lambda_j v_j$, $j = 1, \ldots, n$. The scalars $\lambda_1, \ldots, \lambda_n$ are thus eigenvalues of f, and the corresponding eigenvectors are $v_1, \ldots, v_n$.
If, on the other hand, there exists a basis $B = \{v_1, \ldots, v_n\}$ of $\mathcal{V}$ consisting of eigenvectors of f, then $f(v_j) = \lambda_j v_j$, $j = 1, \ldots, n$, for scalars $\lambda_1, \ldots, \lambda_n \in K$ (the corresponding eigenvalues), and hence $[f]_{B,B}$ has the form (14.1).

(2) ⇒ (3): Let $B = \{v_1, \ldots, v_n\}$ be a basis of $\mathcal{V}$ consisting of eigenvectors of f, and let $\lambda_1, \ldots, \lambda_n \in K$ be the corresponding eigenvalues. Then $[f]_{B,B}$ has the form (14.1) and hence

$$P_f = (t - \lambda_1) \cdot \ldots \cdot (t - \lambda_n),$$

so that P_f decomposes into linear factors over K.
We still have to show that $g(\lambda_j, f) = a(\lambda_j, f)$ for every eigenvalue λ_j. The eigenvalue λ_j has the algebraic multiplicity $m_j := a(\lambda_j, f)$ if and only if λ_j occurs m_j times on the diagonal of the (diagonal) matrix $[f]_{B,B}$. This holds if and only if exactly m_j vectors of the basis B are eigenvectors of f corresponding

14.2 Diagonalization

to the eigenvalue λ_j. Each of these m_j linearly independent vectors is an element of the eigenspace $\mathcal{V}_f(\lambda_j)$ and, hence,

$$\dim(\mathcal{V}_f(\lambda_j)) = g(\lambda_j, f) \geq m_j = a(\lambda_j, f).$$

From Lemma 14.10 we know that $g(\lambda_j, f) \leq a(\lambda_j, f)$, and thus $g(\lambda_j, f) = a(\lambda_j, f)$.

(3) $\Rightarrow$ (2): Let $\widetilde{\lambda}_1, \ldots, \widetilde{\lambda}_k$ be the pairwise distinct eigenvalues of f with corresponding geometric and algebraic multiplicities $g(\widetilde{\lambda}_j, f)$ and $a(\widetilde{\lambda}_j, f)$, $j = 1, \ldots, k$, respectively. Since P_f decomposes into linear factors, we have

$$\sum_{j=1}^{k} a(\widetilde{\lambda}_j, f) = n = \dim(\mathcal{V}).$$

Now $g(\widetilde{\lambda}_j, f) = a(\widetilde{\lambda}_j, f)$, $j = 1, \ldots, k$, implies that

$$\sum_{j=1}^{k} g(\widetilde{\lambda}_j, f) = n = \dim(\mathcal{V}).$$

By Lemma 14.13 we obtain (cp. also Theorem 9.36)

$$\mathcal{V}_f(\widetilde{\lambda}_1) \oplus \ldots \oplus \mathcal{V}_f(\widetilde{\lambda}_k) = \mathcal{V}.$$

If we select bases of the respective eigenspaces $\mathcal{V}_f(\widetilde{\lambda}_j)$, $j = 1, \ldots, k$, then we get a basis of $\mathcal{V}$ that consists of eigenvectors of f. $\square$

Theorem 14.14 and Lemma 14.12 imply an important sufficient condition for diagonalizability.

Corollary 14.15 *If $\mathcal{V}$ is an n-dimensional K-vector space and $f \in \mathcal{L}(\mathcal{V}, \mathcal{V})$ has n pairwise distinct eigenvalues, then f is diagonalizable.*

This corollary also shows that every matrix $A \in K^{n,n}$ with n pairwise distinct eigenvalues is diagonalizable.

The condition of having $n = \dim(\mathcal{V})$ pairwise distinct eigenvalues is, however, not necessary for the diagonalizability of an endomorphism. A simple counterexample is the identity $\mathrm{Id}_\mathcal{V}$, which has the n-fold eigenvalue 1, while $[\mathrm{Id}_\mathcal{V}]_{B,B} = I_n$ holds for every basis B of $\mathcal{V}$. On the other hand, there exist endomorphisms with multiple eigenvalues that are not diagonalizable.

Example 14.16 The endomorphism

$$f : \mathbb{R}^{2,1} \to \mathbb{R}^{2,1}, \quad v \mapsto Fv \quad \text{with} \quad F = \begin{bmatrix} 1 & 1 \\ 0 & 1 \end{bmatrix},$$

has the characteristic polynomial $(t-1)^2$ and thus only has the eigenvalue 1. We have $\ker(\mathcal{V}_f(1)) = \operatorname{span}\{[1, 0]^T\}$ and thus $g(1, f) = 1 < a(1, f) = 2$. By Theorem 14.14, f is not diagonalizable.

Example 14.17 In this example we consider an interesting application of the diagonalization of matrices. The sequence of *Fibonacci numbers*[1] is defined by

$$f_0 = 0, \quad f_1 = 1 \quad \text{and} \quad f_{k+1} = f_k + f_{k-1}, \quad k = 1, 2, \ldots.$$

If we add the trivial equation $f_k = f_k$, $k = 1, 2, \ldots$, and formulate the two equations in matrix form, we obtain the iteration

$$\begin{bmatrix} f_{k+1} \\ f_k \end{bmatrix} = \begin{bmatrix} 1 & 1 \\ 1 & 0 \end{bmatrix} \begin{bmatrix} f_k \\ f_{k-1} \end{bmatrix} = \begin{bmatrix} 1 & 1 \\ 1 & 0 \end{bmatrix}^2 \begin{bmatrix} f_{k-1} \\ f_{k-2} \end{bmatrix} = \cdots = \begin{bmatrix} 1 & 1 \\ 1 & 0 \end{bmatrix}^k \begin{bmatrix} f_1 \\ f_0 \end{bmatrix}, \quad k = 1, 2, \ldots.$$

The (symmetric) matrix $A = \begin{bmatrix} 1 & 1 \\ 1 & 0 \end{bmatrix} \in \mathbb{R}^{2,2}$ has the characteristic polynomial $P_A = t^2 - t - 1$, so that the eigenvalues of A are given by

$$\lambda_1 := \frac{1 + \sqrt{5}}{2} \quad \text{and} \quad \lambda_2 := \frac{1 - \sqrt{5}}{2}.$$

The value $(1 + \sqrt{5})/2 \approx 1.618$ of λ_1 called the *golden ratio*.[2]

The matrix A has two different eigenvalues and therefore is diagonalizable. To diagonalize A we need eigenvectors corresponding to λ_1 and λ_2. These can be obtained by solving the homogeneous linear systems $(\lambda_j I_2 - A) s_j = 0$, $j = 1, 2$. A detailed observation shows that

$$(\lambda_j I_2 - A) \begin{bmatrix} \lambda_j \\ 1 \end{bmatrix} = \begin{bmatrix} \lambda_j - 1 & -1 \\ -1 & \lambda_j \end{bmatrix} \begin{bmatrix} \lambda_j \\ 1 \end{bmatrix} = \begin{bmatrix} \lambda_j^2 - \lambda_j - 1 \\ -\lambda_j + \lambda_j \end{bmatrix} = \begin{bmatrix} 0 \\ 0 \end{bmatrix},$$

---

[1] Leonardo di Pisa, called Fibonacci (ca. 1170–1240).

[2] Two (positive real) numbers $a > b > 0$ are in the *golden ratio*, if their ratio a/b is equal to $(a+b)/a$, i.e., equal to the ratio of their sum to the larger of the two numbers. Multiplying the equation $a/b = (a+b)/a$ by ab yields $a^2 - ab - b^2 = 0$, and solving this quadratic equation for a gives $a = b(1 \pm \sqrt{5})/2$. Only the positive solution $a = b(1 + \sqrt{5})/2$ is relevant, and hence the golden ratio is $a/b = (1 + \sqrt{5})/2$.

which gives the diagonalization $A = SDS^{-1}$ with $S = [s_1, s_2] \in GL_2(\mathbb{R})$, $s_j = [\lambda_j, 1]^T$ for $j = 1, 2$, and $D = \text{diag}(\lambda_1, \lambda_2)$. The inverse of S is given by

$$S^{-1} = \frac{1}{\lambda_1 - \lambda_2} \begin{bmatrix} 1 & -\lambda_2 \\ -1 & \lambda_1 \end{bmatrix}.$$

The diagonalization of A yields $A^2 = (SDS^{-1})(SDS^{-1}) = SD^2S^{-1}$, and inductively $A^k = SD^kS^{-1}$ for all $k \geq 1$. (This also holds for $k = 0$, since $I_2 = A^0 = SD^0S^{-1}$.) For the Fibonacci numbers we obtain

$$\begin{bmatrix} f_{k+1} \\ f_k \end{bmatrix} = A^k \begin{bmatrix} f_1 \\ f_0 \end{bmatrix} = SD^kS^{-1} \begin{bmatrix} 1 \\ 0 \end{bmatrix} = \frac{1}{\lambda_1 - \lambda_2} \begin{bmatrix} \lambda_1 & \lambda_2 \\ 1 & 1 \end{bmatrix} \begin{bmatrix} \lambda_1^k & 0 \\ 0 & \lambda_2^k \end{bmatrix} \begin{bmatrix} 1 \\ -1 \end{bmatrix}$$
$$= \frac{1}{\lambda_1 - \lambda_2} \begin{bmatrix} \lambda_1^{k+1} - \lambda_2^{k+1} \\ \lambda_1^k - \lambda_2^k \end{bmatrix}, \quad k = 1, 2, \ldots,$$

and hence, in particular,

$$f_k = \frac{\lambda_1^k - \lambda_2^k}{\lambda_1 - \lambda_2} = \frac{1}{\sqrt{5}} \left(\left(\frac{1 + \sqrt{5}}{2} \right)^k - \left(\frac{1 - \sqrt{5}}{2} \right)^k \right), \quad k = 1, 2, \ldots.$$

This explicit representation of the Fibonacci numbers is also called the *Moivre-Binet formula*.[3] A similar formula can be constructed not only for the initial vector $[1, 0]^T$, but for every initial vector $[f_1, f_0]^T$ of the iteration. This procedure can also be applied to every recursive sequence of the form $f_n = c_1 f_{n-1} + \ldots + c_k f_{n-k}$, as long as the corresponding matrix is diagonalizable.

14.3 Triangulation and Schur's Theorem

If the property $g(\lambda_j, f) = a(\lambda_j, f)$ does not hold for every eigenvalue λ_j of f, then f is not diagonalizable. However, as long as the characteristic polynomial P_f decomposes into linear factors, we can find a special basis B such that $[f]_{B,B}$ is a triangular matrix.

Theorem 14.18 *If $\mathcal{V}$ is a finite dimensional K-vector space and $f \in \mathcal{L}(\mathcal{V}, \mathcal{V})$, then the following statements are equivalent:*

(1) The characteristic polynomial P_f decomposes into linear factors over K.
(2) There exists a basis B of $\mathcal{V}$ such that $[f]_{B,B}$ is upper triangular, i.e., f can be triangulated.

---
[3] Abraham de Moivre (1667–1754) and Jacques Philippe Marie Binet (1786–1856).

Proof

(2) $\Rightarrow$ (1): If $n = \dim(\mathcal{V})$ and $[f]_{B,B} = [r_{ij}] \in K^{n,n}$ is upper triangular, then $P_f = (t - r_{11}) \cdot \ldots \cdot (t - r_{nn})$.

(1) $\Rightarrow$ (2): We show the assertion by induction on $n = \dim(\mathcal{V})$. The case $n = 1$ is trivial, since then $[f]_{B,B} \in K^{1,1}$.

Suppose that the assertion holds for an $n \geq 1$, and let $\dim(\mathcal{V}) = n + 1$. By assumption,

$$P_f = (t - \lambda_1) \cdot \ldots \cdot (t - \lambda_{n+1}),$$

where $\lambda_1, \ldots, \lambda_{n+1} \in K$ are the eigenvalues of f. Let $v_1 \in \mathcal{V}$ be an eigenvector corresponding to the eigenvalue λ_1. We extend this vector to a basis $B = \{v_1, w_2, \ldots, w_{n+1}\}$ of $\mathcal{V}$. With $B_{\mathcal{W}} := \{w_2, \ldots, w_{n+1}\}$ and $\mathcal{W} := \text{span } B_{\mathcal{W}}$ we have $\mathcal{V} = \text{span}\{v_1\} \oplus \mathcal{W}$ and

$$[f]_{B,B} = \begin{bmatrix} \lambda_1 & a_{12} & \cdots & a_{1,n+1} \\ 0 & a_{22} & \cdots & a_{2,n+1} \\ \vdots & \vdots & \ddots & \vdots \\ 0 & a_{n+1,2} & \cdots & a_{n+1,n+1} \end{bmatrix}.$$

We define $h \in \mathcal{L}(\mathcal{W}, \text{span}\{v_1\})$ and $g \in \mathcal{L}(\mathcal{W}, \mathcal{W})$ by

$$h(w_j) := a_{1j} v_1 \quad \text{and} \quad g(w_j) := \sum_{k=2}^{n+1} a_{kj} w_k, \quad j = 2, \ldots, n+1.$$

Then $f(w) = h(w) + g(w)$ for all $w \in \mathcal{W}$, and

$$[f]_{B,B} = \begin{bmatrix} \lambda_1 & [h]_{B_{\mathcal{W}}, \{v_1\}} \\ 0 & [g]_{B_{\mathcal{W}}, B_{\mathcal{W}}} \end{bmatrix}.$$

Consequently,

$$(t - \lambda_1) P_g = P_f = (t - \lambda_1) \cdot \ldots \cdot (t - \lambda_{n+1}),$$

and hence $P_g = (t - \lambda_2) \cdot \ldots \cdot (t - \lambda_{n+1})$. Now $\dim(\mathcal{W}) = n$ and the characteristic polynomial of $g \in \mathcal{L}(\mathcal{W}, \mathcal{W})$ decomposes into linear factors. By the induction hypothesis there exists a basis $\widehat{B}_{\mathcal{W}} = \{\widehat{w}_2, \ldots, \widehat{w}_{n+1}\}$ of $\mathcal{W}$ such that $[g]_{\widehat{B}_{\mathcal{W}}, \widehat{B}_{\mathcal{W}}}$ upper triangular. Thus, for the basis $B_1 := \{v_1, \widehat{w}_2, \ldots, \widehat{w}_{n+1}\}$ the matrix $[f]_{B_1, B_1}$ is upper triangular. $\square$

A "matrix version" of this theorem reads as follows: The characteristic polynomial P_A of $A \in K^{n,n}$ decomposes into linear factors over K if and only if A can be

14.3 Triangulation and Schur's Theorem

triangulated, i.e., there exists a matrix $S \in GL_n(K)$ with $A = SRS^{-1}$ for an upper triangular matrix $R \in K^{n,n}$.

Corollary 14.19 *Let $\mathcal{V}$ be a finite dimensional Euclidean or unitary vector space and $f \in \mathcal{L}(\mathcal{V}, \mathcal{V})$. If P_f decomposes over $\mathbb{R}$ (in the Euclidean case case) or $\mathbb{C}$ (in the unitary case) into linear factors, then there exists an orthonormal basis B of $\mathcal{V}$, such that $[f]_{B,B}$ is upper triangular.*

Proof If P_f decomposes into linear factors, then by Theorem 14.18 there exists a basis B_1 of $\mathcal{V}$, such that $[f]_{B_1,B_1}$ is upper triangular. Applying the Gram-Schmidt method to the basis B_1, we obtain an orthonormal basis B_2 of $\mathcal{V}$, such that $[\text{Id}_\mathcal{V}]_{B_1,B_2}$ is upper triangular (cp. Theorem 12.11). Then

$$[f]_{B_2,B_2} = [\text{Id}_\mathcal{V}]_{B_1,B_2}[f]_{B_1,B_1}[\text{Id}_\mathcal{V}]_{B_2,B_1} = ([\text{Id}_\mathcal{V}]_{B_2,B_1})^{-1}[f]_{B_1,B_1}[\text{Id}_\mathcal{V}]_{B_2,B_1}.$$

The invertible upper triangular matrices form a group with respect to the matrix multiplication (cp. Theorem 4.13). Thus, all matrices in the product on the right hand side are upper triangular, and hence $[f]_{B_2,B_2}$ is upper triangular. □

Example 14.20 Consider the Euclidean vector space $\mathbb{R}[t]_{\leq 1}$ with the scalar product $\langle p, q \rangle = \int_0^1 p(t)q(t)\, dt$ (cp. (4) in Example 12.2), and the endomorphism

$$f : \mathbb{R}[t]_{\leq 1} \to \mathbb{R}[t]_{\leq 1}, \quad \alpha_1 t + \alpha_0 \mapsto 2\alpha_1 t + \alpha_0.$$

We have $f(1) = 1$ and $f(t) = 2t$, i.e., the polynomials 1 and t are eigenvectors of f corresponding to the (distinct) eigenvalues 1 and 2. Thus, $\widehat{B} = \{1, t\}$ is a basis of $\mathbb{R}[t]_{\leq 1}$, and $[f]_{\widehat{B},\widehat{B}}$ is a diagonal matrix. Note that $\widehat{B}$ is not an orthonormal basis, since in particular $\langle 1, t \rangle \neq 0$.

Since P_f decomposes into linear factors, Corollary 14.19 guarantees the existence of an orthonormal basis B for which $[f]_{B,B}$ is upper triangular. In the proof of the implication (1) $\Rightarrow$ (2) of Theorem 14.18 we choose any eigenvector of f, and then proceed inductively in order to obtain the triangulation of f. In this example, let us use $q_1 = 1$ as the first vector. This vector is an eigenvector of f with norm 1 corresponding to the eigenvalue 1. If $q_2 \in \mathbb{R}[t]_{\leq 1}$ is a vector with norm 1 and $\langle q_1, q_2 \rangle = 0$, then $B = \{q_1, q_2\}$ is an orthonormal basis for which $[f]_{B,B}$ is an upper triangular matrix. We construct the vector q_2 by orthogonalizing t against q_1 using the Gram-Schmidt method:

$$\widehat{q_2} = t - \langle t, q_1 \rangle q_1 = t - \frac{1}{2},$$

$$\|\widehat{q_2}\| = \left\langle t - \frac{1}{2}, t - \frac{1}{2} \right\rangle^{1/2} = \frac{1}{\sqrt{12}},$$

$$q_2 = \|\widehat{q_2}\|^{-1} \widehat{q_2} = \sqrt{12}\, t - \sqrt{3}.$$

This leads to the triangulation

$$[f]_{B,B} = \begin{bmatrix} 1 & \sqrt{3} \\ 0 & 2 \end{bmatrix} \in \mathbb{R}^{2,2}.$$

We could also choose $q_1 = \sqrt{3}t$, which is an eigenvector of f with norm 1 corresponding to the eigenvalue 2. Orthogonalizing the vector 1 against q_1 leads to the second basis vector $q_2 = -3t + 2$. With the corresponding basis B_1 we obtain the triangulation

$$[f]_{B_1,B_1} = \begin{bmatrix} 2 & -\sqrt{3} \\ 0 & 1 \end{bmatrix} \in \mathbb{R}^{2,2}.$$

This example shows that in the triangulation of f the elements above the diagonal can be different for different orthonormal bases. Only the diagonal elements are (except for their order) uniquely determined, since they are the eigenvalues of f. A more detailed statement about the uniqueness is given in Lemma 14.23.

In the next chapter we will prove the Fundamental Theorem of Algebra, which states that every non-constant polynomial over $\mathbb{C}$ decomposes into linear factors. This result has the following corollary, which is known as *Schur's theorem*.[4]

Corollary 14.21 *If $\mathcal{V}$ is a finite dimensional unitary vector space, then every endomorphism on $\mathcal{V}$ can be unitarily triangulated, i.e., for each $f \in \mathcal{L}(\mathcal{V}, \mathcal{V})$ there exists an orthonormal basis B of $\mathcal{V}$, such that $[f]_{B,B}$ is upper triangular. The matrix $[f]_{B,B}$ is called a Schur form of f.*

If $\mathcal{V}$ is the unitary vector space $\mathbb{C}^{n,1}$ with the standard scalar product, then we obtain the following "matrix version" of Corollary 14.21.

Corollary 14.22 *If $A \in \mathbb{C}^{n,n}$, then there exists a unitary matrix $Q \in \mathbb{C}^{n,n}$ with $A = QRQ^H$ for an upper triangular matrix $R \in \mathbb{C}^{n,n}$. The matrix R is called a Schur form of A.*

The following result shows that a Schur form of a matrix $A \in \mathbb{C}^{n,n}$ with n pairwise distinct eigenvalues is "almost unique".

Lemma 14.23 *Let $A \in \mathbb{C}^{n,n}$ have n pairwise distinct eigenvalues, and let $R_1, R_2 \in \mathbb{C}^{n,n}$ be two Schur forms of A. If the diagonals of R_1 and R_2 are equal then $R_1 = QR_2Q^H$ for a unitary diagonal matrix $Q \in \mathbb{C}^{n,n}$.*

Proof Exercise. □

---

[4] Issai Schur (1875–1941).

A survey of the results on unitary similarity of matrices can be found in the article [14].

MATLAB-Minute 8

Consider for $n \geq 2$ the matrix

$$A = \begin{bmatrix} 1 & 2 & 3 & \cdots & n \\ 1 & 3 & 4 & \cdots & n+1 \\ 1 & 4 & 5 & \cdots & n+2 \\ \vdots & \vdots & \vdots & & \vdots \\ 1 & n+1 & n+2 & \cdots & 2n-1 \end{bmatrix} \in \mathbb{C}^{n,n}.$$

Compute a Schur form of A using the command [Q,R]=schur(A) for $n = 2, 3, 4, \ldots 10$. What are the eigenvalues of A? Formulate a conjecture about the rank of A for general n. Can you prove your conjecture?

Exercises

(In the following exercises K is an arbitrary field.)

14.1 Let $\mathcal{V}$ be a vector space and let $f \in \mathcal{L}(\mathcal{V}, \mathcal{V})$ have the eigenvalue $\lambda \in K$. Show that $\text{im}(\lambda \, \text{Id}_\mathcal{V} - f)$ is an f-invariant subspace.

14.2 Let $\mathcal{V}$ be a K-vector space, $f \in \mathcal{L}(\mathcal{V}, \mathcal{V})$ and $\lambda \in K$. Prove or disprove: A subspace $\mathcal{U} \subseteq \mathcal{V}$ is f-invariant, if it is $(f - \lambda \, \text{Id}_\mathcal{V})$-invariant.

14.3 Let $\mathcal{V}$ be a finite dimensional K-vector space and let $f \in \mathcal{L}(\mathcal{V}, \mathcal{V})$ be bijective. Show that f and f^{-1} have the same invariant subspaces.

14.4 Let $\mathcal{V}$ be a K-Vector space, let $f \in \mathcal{L}(\mathcal{V}, \mathcal{V})$ be bijective and let $\lambda \in K$ be an eigenvalue of f. Show the following assertions:
 (a) $\lambda \neq 0$.
 (b) λ^{-1} is an eigenvalue of f^{-1}.
 (c) $\mathcal{V}_f(\lambda) = \mathcal{V}_{f^{-1}}(\lambda^{-1})$.

14.5 Let $\mathcal{V}$ be an n-dimensional K-vector space, let $f \in \mathcal{L}(\mathcal{V}, \mathcal{V})$, and let $\mathcal{U}$ be an m-dimensional f-invariant subspace of $\mathcal{V}$. Show that a basis B of $\mathcal{V}$ exists such that

$$[f]_{B,B} = \begin{bmatrix} A_1 & A_2 \\ 0 & A_3 \end{bmatrix}$$

for some matrices $A_1 \in K^{m,m}$, $A_2 \in K^{m,n-m}$ and $A_3 \in K^{n-m,n-m}$.

14.6 Let $K \in \{\mathbb{R}, \mathbb{C}\}$ and $f : K^{4,1} \to K^{4,1}$, $v \mapsto Fv$, with

$$F = \begin{bmatrix} 2 & 1 & 3 & 4 \\ -1 & 0 & 2 & 3 \\ 0 & 0 & 1 & 1 \\ 0 & 0 & -1 & 0 \end{bmatrix}.$$

Compute P_f and determine for $K = \mathbb{R}$ and $K = \mathbb{C}$ the eigenvalues of f with their algebraic and geometric multiplicities, as well as the associated eigenspaces.

14.7 Let $\mathcal{V}$ be a finite dimensional K-vector space, $f \in \mathcal{L}(\mathcal{V}, \mathcal{V})$ and $\mathcal{V} = \mathcal{U}_1 \oplus \mathcal{U}_2$, where $\mathcal{U}_1, \mathcal{U}_2$ are f-invariant subspaces of $\mathcal{V}$. Let, furthermore, $f_j := f|_{\mathcal{U}_j} \in \mathcal{L}(\mathcal{U}_j, \mathcal{U}_j)$, $j = 1, 2$.
   (a) For every $v \in \mathcal{V}$ there exist unique $u_1 \in \mathcal{U}_1$ and $u_2 \in \mathcal{U}_2$ with $v = u_1 + u_2$. Show that then also $f(v) = f(u_1) + f(u_2) = f_1(u_1) + f_2(u_2)$.
   (We write this as $f = f_1 \oplus f_2$ and call f the *direct sum* of f_1 and f_2 with respect to the decomposition $\mathcal{V} = \mathcal{U}_1 \oplus \mathcal{U}_2$.)
   (b) Show that $\text{rank}(f) = \text{rank}(f_1) + \text{rank}(f_2)$ and $P_f = P_{f_1} \cdot P_{f_2}$.
   (c) Show that $a(\lambda, f) = a(\lambda, f_1) + a(\lambda, f_2)$ for all $\lambda \in K$.
   (Here we set $a(\lambda, h) = 0$, if λ is not an eigenvalue of $h \in \mathcal{L}(\mathcal{V}, \mathcal{V})$.)
   (d) Show that $g(\lambda, f) = g(\lambda, f_1) + g(\lambda, f_2)$ for all $\lambda \in K$.
   (Here we set $g(\lambda, h) = \dim(\ker(\lambda \, \text{Id}_\mathcal{V} - h))$ even if λ is not an eigenvalue of $h \in \mathcal{L}(\mathcal{V}, \mathcal{V})$.)
   (e) Show that $p(f) = p(f_1) \oplus p(f_2)$ for all $p \in K[t]$.

14.8 Consider the vector space $\mathbb{R}[t]_{\leq n}$ with the standard basis $\{1, t, \ldots, t^n\}$ and the endomorphism

$$f : \mathbb{R}[t]_{\leq n} \to \mathbb{R}[t]_{\leq n}, \quad \sum_{i=0}^{n} \alpha_i t^i \mapsto \sum_{i=2}^{n} i(i-1) \alpha_i t^{i-2} = \frac{d^2}{dt^2} p.$$

Compute P_f, the eigenvalues of f with their algebraic and geometric multiplicities, and examine whether f is diagonalizable or not. What changes if one considers as map the kth derivative (for $k = 3, 4, \ldots, n$)?

14.9 Examine whether the following matrices

$$A = \begin{bmatrix} 0 & 1 \\ -1 & 0 \end{bmatrix} \in \mathbb{Q}^{2,2}, \quad B = \begin{bmatrix} 1 & 0 & 0 \\ -1 & 2 & 0 \\ -1 & 1 & 1 \end{bmatrix} \in \mathbb{Q}^{3,3}, \quad C = \begin{bmatrix} 3 & 1 & 0 & -2 \\ 0 & 2 & 0 & 0 \\ 2 & 2 & 2 & -4 \\ 0 & 0 & 0 & 2 \end{bmatrix} \in \mathbb{Q}^{4,4}$$

are diagonalizable.

14.10 Show that the matrix

$$A = \begin{bmatrix} \alpha & -1 \\ 0 & \beta \end{bmatrix} \in K^{2,2}$$

is diagonalizable if and only if $\alpha \neq \beta$.

14.11 Is the set of all diagonalizable and invertible matrices a subgroup of $GL_n(K)$?

14.12 Show that the endomorphism $f \in \mathcal{L}(\mathbb{R}^{n,n}, \mathbb{R}^{n,n})$, $A \mapsto A^T$, is diagonalizable.

14.13 Let $n \in \mathbb{N}_0$. Consider the $\mathbb{R}$-vector space $\mathbb{R}[t]_{\leq n}$ and the map

$$f : \mathbb{R}[t]_{\leq n} \to \mathbb{R}[t]_{\leq n}, \quad p(t) \mapsto p(t+1) - p(t).$$

Show that f is linear. For which n is f diagonalizable?

14.14 Let $f \in \mathcal{L}(\mathbb{C}[t]_{\leq 2}, \mathbb{C}[t]_{\leq 2})$ with $f(p) := p(-1)t^2 + p(1)t + p(0)$. Determine P_f, the eigenvalues and eigenspaces of f, and the corresponding algebraic and geometric multiplicities. Is f diagonalizable?

14.15 Let $\mathcal{V}$ be an $\mathbb{R}$-vector space with the basis $\{v_1, \ldots, v_n\}$. Examine whether the following endomorphisms are diagonalizable or not:
(a) $f(v_j) = v_j + v_{j+1}$, $j = 1, \ldots, n-1$, and $f(v_n) = v_n$,
(b) $f(v_j) = jv_j + v_{j+1}$, $j = 1, \ldots, n-1$, and $f(v_n) = nv_n$.

14.16 Let $\mathcal{V}$ be a finite dimensional Euclidean vector space and let $f \in \mathcal{L}(\mathcal{V}, \mathcal{V})$ with $f + f^{ad} = 0 \in \mathcal{L}(\mathcal{V}, \mathcal{V})$. Show that $f \neq 0$ if and only if f is not diagonalizable.

14.17 Let $\mathcal{V}$ be a finite dimensional Euclidean or unitary vector space and let $f \in \mathcal{L}(\mathcal{V}, \mathcal{V})$ be a projection. Show that there exists a basis B of $\mathcal{V}$ such that

$$[f]_{B,B} = \begin{bmatrix} I_k & \\ & 0_{n-k} \end{bmatrix},$$

where $k = \dim(\text{im}(f))$ and $n = \dim(\mathcal{V})$. In particular, we obtain $P_f = (t-1)^k t^{n-k}$ und $\lambda \in \{0, 1\}$ for each eigenvalue λ of f.

14.18 Let $\mathcal{V}$ be a $\mathbb{C}$-vector space and let $f \in \mathcal{L}(\mathcal{V}, \mathcal{V})$ with $f^2 = -\text{Id}_\mathcal{V}$. Determine all possible eigenvalues of f.

14.19 Let $\mathcal{V}$ be a finite dimensional vector space and $f \in \mathcal{L}(\mathcal{V}, \mathcal{V})$. Show that $P_f(f) = 0 \in \mathcal{L}(\mathcal{V}, \mathcal{V})$.

14.20 Let $\mathcal{V}$ be a finite dimensional K-vector space, let $f \in \mathcal{L}(\mathcal{V}, \mathcal{V})$ and

$$p = (t - \mu_1) \cdot \ldots \cdot (t - \mu_m) \in K[t]_{\leq m}.$$

Show that $p(f)$ is bijective if and only if $\mu_1, \ldots, \mu_m$ are not eigenvalues of f.

14.21 Let $\mathcal{V}$ be a finite dimensional unitary vector space. Let $f \in \mathcal{L}(\mathcal{V}, \mathcal{V})$ be normal, i.e., f satisfies $f \circ f^{ad} = f^{ad} \circ f$.
  (a) Show that if $\lambda \in \mathbb{C}$ is an eigenvalue of f, then $\mathcal{V}_f(\lambda)^\perp$ is an f-invariant subspace.
  (b) Show (using (a)) that f is diagonalizable. (*Hint:* Show by induction on $\dim(\mathcal{V})$, that $\mathcal{V}$ is the direct sum of the eigenspaces of f.)
  (c) Show (using (a) or (b)), that f is even *unitarily diagonalizable*, i.e., there exists an orthonormal basis B of $\mathcal{V}$ such that $[f]_{B,B}$ is a diagonal matrix.
  (d) Let $g \in \mathcal{L}(\mathcal{V}, \mathcal{V})$ be unitarily diagonalizable. Show that g is normal.
  (This shows that an endomorphism on a finite dimensional unitary vector space is normal if and only if it is unitarily diagonalizable. We will give a different proof of this result in Theorem 18.2; cp. also Exercise 15.22.)

14.22 Determine conditions for the entries of the matrices

$$A = \begin{bmatrix} \alpha & \beta \\ \gamma & \delta \end{bmatrix} \in \mathbb{R}^{2,2},$$

such that A is diagonalizable or can be triangulated.

14.23 Determine an endomorphism on $\mathbb{R}[t]_{\leq 3}$ that is not diagonalizable and that cannot be triangulated.

14.24 Let $\mathcal{V}$ be a vector space with $\dim(\mathcal{V}) = n$. Show that $f \in \mathcal{L}(\mathcal{V}, \mathcal{V})$ can be triangulated if and only if there exist subspaces $\mathcal{V}_0, \mathcal{V}_1, \ldots, \mathcal{V}_n$ of $\mathcal{V}$ with
  (a) $\mathcal{V}_j \subset \mathcal{V}_{j+1}$ for $j = 0, 1, \ldots, n-1$,
  (b) $\dim(\mathcal{V}_j) = j$ for $j = 0, 1, \ldots, n$, and
  (c) $\mathcal{V}_j$ is f-invariant for $j = 0, 1, \ldots, n$.

14.25 Prove Lemma 14.23.

Polynomials and the Fundamental Theorem of Algebra 15

In this chapter we discuss polynomials in more detail. We consider the division of polynomials and derive classical results from polynomial algebra, including the factorization into irreducible factors. We also prove the Fundamental Theorem of Algebra, which states that every non-constant polynomial over the complex numbers has a least one complex root. This implies that every complex matrix and every endomorphism on a (finite dimensional) complex vector space has at least one eigenvalue.

15.1 Polynomials

Let us recall some of the most important terms in the context of polynomials. If K is a field, then

$$p = \alpha_0 + \alpha_1 t + \ldots + \alpha_n t^n \text{ with } n \in \mathbb{N}_0 \text{ and } \alpha_0, \alpha_1, \ldots \alpha_n \in K$$

is a polynomial over K in the variable t. The set $K[t]$ of all these polynomials forms a commutative ring with unit (cp. Example 3.15). If $\alpha_n \neq 0$, then $\deg(p) = n$ is called the *degree* of p. If $\alpha_n = 1$, then p is called *monic*. If $p = 0$, then $\deg(p) := -\infty$, and if $\deg(p) < 1$, then p is called *constant*.

Lemma 15.1 *For two polynomials $p, q \in K[t]$ the following assertions hold:*

(1) $\deg(p+q) \leq \max\{\deg(p), \deg(q)\}$.
(2) $\deg(p \cdot q) = \deg(p) + \deg(q)$.

Proof Exercise. □

We now introduce some concepts associated with the division of polynomials.

Definition 15.2 Let K be a field.

(1) If for two polynomials $p, s \in K[t]$ there exists a polynomial $q \in K[t]$ with $p = s \cdot q$, then s is called a *divisor* of p and we write $s|p$ (read this as "s divides p").
(2) Two polynomials $p, s \in K[t]$ are called *coprime*, if $q|p$ and $q|s$ for some $q \in K[t]$ always imply that q is constant.
(3) A non-constant polynomial $p \in K[t]$ is called *irreducible* (over K), if $p = s \cdot q$ for two polynomials $s, q \in K[t]$ implies that s or q are constant. If there exist two non-constant polynomials $s, q \in K[t]$ with $p = s \cdot q$, then p is called *reducible* (over K).

Note that the property of irreducibility is only defined for polynomials of degree at least 1. A polynomial of degree 1 is always irreducible. Whether a polynomial of degree at least 2 is irreducible may depend on the underlying field.

Example 15.3 The polynomial $2 - t^2 \in \mathbb{Q}[t]$ is irreducible, but the factorization

$$2 - t^2 = (\sqrt{2} - t) \cdot (\sqrt{2} + t)$$

shows that $2 - t^2 \in \mathbb{R}[t]$ is reducible. The polynomial $1 + t^2 \in \mathbb{R}[t]$ is irreducible, but using the imaginary unit $\mathbf{i}$ we have

$$1 + t^2 = (-\mathbf{i} + t) \cdot (\mathbf{i} + t),$$

so that $1 + t^2 \in \mathbb{C}[t]$ is reducible.

The next result concerns the *division with remainder* of polynomials.

Theorem 15.4 *If $p \in K[t]$ and $s \in K[t] \setminus \{0\}$, then there exist uniquely defined polynomials $q, r \in K[t]$ with*

$$p = s \cdot q + r \quad \text{and} \quad \deg(r) < \deg(s). \tag{15.1}$$

Proof We first show the existence of polynomials $q, r \in K[t]$ such that (15.1) holds.

If $\deg(s) = 0$, then $s = s_0$ for an $s_0 \in K \setminus \{0\}$ and (15.1) follows with $q := s_0^{-1} \cdot p$ and $r := 0$, where $\deg(r) < \deg(s)$.

We now assume that $\deg(s) \geq 1$. If $\deg(p) < \deg(s)$, then we set $q := 0$ and $r := p$. Then $p = s \cdot q + r$ with $\deg(r) < \deg(s)$.

15.1 Polynomials

Let $n := \deg(p) \geq m := \deg(s) \geq 1$. We prove (15.1) by induction on n. If $n = 1$, then $m = 1$. Hence $p = p_1 \cdot t + p_0$ with $p_1 \neq 0$ and $s = s_1 \cdot t + s_0$ with $s_1 \neq 0$. Therefore,

$$p = s \cdot q + r \quad \text{for} \quad q := p_1 s_1^{-1}, \quad r := p_0 - p_1 s_1^{-1} s_0,$$

where $\deg(r) < \deg(s)$.

Suppose that the assertion holds for an $n \geq 1$. Let two polynomials p and s with $n + 1 = \deg(p) \geq \deg(s) = m$ be given, and let $p_{n+1} (\neq 0)$ and $s_m (\neq 0)$ be the highest coefficients of p and s. If

$$h := p - p_{n+1} s_m^{-1} s \cdot t^{n+1-m} \in K[t],$$

then $\deg(h) < \deg(p) = n+1$. By the induction hypothesis there exist polynomials $\widetilde{q}, r \in K[t]$ with

$$h = s \cdot \widetilde{q} + r \quad \text{and} \quad \deg(r) < \deg(s).$$

It then follows that

$$p = s \cdot q + r \quad \text{with} \quad q := \widetilde{q} + p_{n+1} s_m^{-1} t^{n+1-m},$$

where $\deg(r) < \deg(s)$.

It remains to show the uniqueness. Suppose that (15.1) holds and that there exist polynomials $\widehat{q}, \widehat{r} \in K[t]$ with $p = s \cdot \widehat{q} + \widehat{r}$ and $\deg(\widehat{r}) < \deg(s)$. Then

$$r - \widehat{r} = s \cdot (\widehat{q} - q).$$

If $\widehat{r} - r \neq 0$, then $\widehat{q} - q \neq 0$ and thus

$$\deg(r - \widehat{r}) = \deg(s \cdot (\widehat{q} - q)) = \deg(s) + \deg(\widehat{q} - q) \geq \deg(s).$$

On the other hand, we also have

$$\deg(r - \widehat{r}) \leq \max\{\deg(r), \deg(\widehat{r})\} < \deg(s).$$

This is a contradiction, which shows that indeed $r = \widehat{r}$ and $q = \widehat{q}$. □

This theorem has some important consequences for the roots of polynomials. The first of these is known as the *Theorem of Ruffini*.[1]

Corollary 15.5 *If $\lambda \in K$ is a root of $p \in K[t]$, i.e., $p(\lambda) = 0$, then there exists a uniquely determined polynomial $q \in K[t]$ with $p = (t - \lambda) \cdot q$.*

---
[1] Paolo Ruffini (1765–1822).

Proof When we apply Theorem 15.4 to the polynomials p and $s = t - \lambda \neq 0$, then we get uniquely determined polynomials q and r with $\deg(r) < \deg(s) = 1$ and

$$p = (t - \lambda) \cdot q + r.$$

The polynomial r is constant and evaluating it at λ gives

$$0 = p(\lambda) = (\lambda - \lambda) \cdot q(\lambda) + r(\lambda) = r(\lambda),$$

which yields $r = 0$ and $p = (t - \lambda) \cdot q$. $\square$

In the trivial case $p = 0$ in Corollary 15.5 we have $p(\lambda) = 0$ for each $\lambda \in K$, and hence $q = 0$ for each $\lambda \in K$.

If a polynomial $p \in K[t]$ has at least degree 2 and a root $\lambda \in K$, then the linear factor $t - \lambda$ is a divisor of p and, in particular, p is reducible. The converse of this statement *does not hold*. For instance the polynomial $4 - 4t^2 + t^4 = (2-t^2) \cdot (2-t^2) \in \mathbb{Q}[t]$ is reducible, but it does not have a root in $\mathbb{Q}$.

Corollary 15.5 motivates the following definition (cp. Definition 14.8).

Definition 15.6 If $\lambda \in K$ is a root of $p \in K[t] \setminus \{0\}$, then its *multiplicity* is the uniquely determined nonnegative integer m, such that $p = (t - \lambda)^m \cdot q$ for a polynomial $q \in K[t]$ with $q(\lambda) \neq 0$.

In this definition we exclude the polynomial $p = 0$, since for this polynomial we obtain $q = 0$ in Corollary 15.5, so that a factorization of the form $p = (t - \lambda)^m \cdot q$ with $q(\lambda) \neq 0$ does not exist for any $\lambda \in K$.

Repeated application of Corollary 15.5 to a given polynomial $p \in K[t] \setminus \{0\}$ leads to the following result.

Corollary 15.7 *If $\lambda_1 \ldots, \lambda_k \in K$ are pairwise distinct roots of $p \in K[t] \setminus \{0\}$ with the corresponding multiplicities $m_1, \ldots, m_k$, then there exists a unique polynomial $q \in K[t]$ with*

$$p = (t - \lambda_1)^{m_1} \cdot \ldots \cdot (t - \lambda_k)^{m_k} \cdot q$$

and $q(\lambda_j) \neq 0$ for $j = 1, \ldots, k$. In particular, the sum of the multiplicities of all pairwise distinct roots of p is at most $\deg(p)$.

Now let $\lambda \in K$ be a root of $p \in K[t]$ and of the derivative $p' \in K[t]$. Applying Corollary 15.5 to p yields $p = (t - \lambda) \cdot q_1$ with a uniquely determined polynomial $q_1 \in K[t]$. Using the product rule for the derivative, i.e., $(fg)' = f'g + gf'$, we obtain

$$p' = 1 \cdot q_1 + (t - \lambda) \cdot q_1'.$$

15.1 Polynomials

This implies $0 = p'(\lambda) = q_1(\lambda)$. Applying Corollary 15.5 to q_1 yields $q_1 = (t-\lambda) \cdot q_2$ with a uniquely determined polynomial $q_2 \in K[t]$, and hence $p = (t-\lambda)^2 \cdot q_2$. Inductively we obtain the following result, where $p^{(j)}$ is the jth derivative of the polynomial p (with $p^{(0)} := p$).

Corollary 15.8 *Let $d \in \mathbb{N}$, $\lambda \in K$ and $p \in K[t]$ with $p^{(j)}(\lambda) = 0$ for $j = 0, 1, \ldots, d-1$. Then there exists a uniquely determined polynomial $q \in K[t]$ with $p = (t-\lambda)^d \cdot q$.*

The next result is known as the *Lemma of Bézout*.[2]

Lemma 15.9 *If $p, s \in K[t] \setminus \{0\}$ are coprime, then there exist polynomials $q_1, q_2 \in K[t]$ with*

$$p \cdot q_1 + s \cdot q_2 = 1.$$

Proof We may assume without loss of generality that $\deg(p) \geq \deg(s) \, (\geq 0)$, and we proceed by induction on $\deg(s)$.

If $\deg(s) = 0$, then $s = s_0$ for an $s_0 \in K \setminus \{0\}$, and thus

$$p \cdot q_1 + s \cdot q_2 = 1 \quad \text{with} \quad q_1 := 0, \quad q_2 := s_0^{-1}.$$

Suppose that the assertion holds for all polynomials $p, s \in K[t] \setminus \{0\}$ with $\deg(s) = n$ for an $n \geq 0$. Let $p, s \in K[t] \setminus \{0\}$ with $\deg(p) \geq \deg(s) = n+1$ be given. By Theorem 15.4 there exist polynomials q and r with

$$p = s \cdot q + r \quad \text{and} \quad \deg(r) < \deg(s).$$

Here we have $r \neq 0$, since by assumption p and s are coprime.

Suppose that there exists a non-constant polynomial $h \in K[t]$ that divides both s and r. Then h also divides p, in contradiction to the assumption that p and s are coprime. Thus, the polynomials s and r are coprime. Since $\deg(r) < \deg(s)$, we can apply the induction hypothesis to the polynomials $s, r \in K[t] \setminus \{0\}$. Hence there exist polynomials $\widetilde{q}_1, \widetilde{q}_2 \in K[t]$ with

$$s \cdot \widetilde{q}_1 + r \cdot \widetilde{q}_2 = 1.$$

From $r = p - s \cdot q$ we then get

$$1 = s \cdot \widetilde{q}_1 + (p - s \cdot q) \cdot \widetilde{q}_2 = p \cdot \widetilde{q}_2 + s \cdot (\widetilde{q}_1 - q \cdot \widetilde{q}_2),$$

which completes the proof. □

---
[2] Étienne Bézout (1730–1783).

Using the Lemma of Bézout we can easily prove the following result.

Lemma 15.10 *If $p \in K[t]$ is irreducible and a divisor of the product $s \cdot h$ of two polynomials $s, h \in K[t]$, then p divides at least one of the factors, i.e., $p|s$ or $p|h$.*

Proof If $s = 0$, then $p|s$, because every polynomial is a divisor of the zero polynomial.

If $s \neq 0$ and p is not a divisor of s, then p and s are coprime, since p is irreducible. By Lemma 15.9 there exist polynomials $q_1, q_2 \in K[t]$ with $p \cdot q_1 + s \cdot q_2 = 1$, and hence

$$h = h \cdot 1 = (q_1 \cdot h) \cdot p + q_2 \cdot (s \cdot h).$$

The polynomial p divides both terms on the right hand side, and thus also $p|h$. □

By recursive application of Lemma 15.10 we obtain the *Euclidean theorem*, which describes a prime factor decomposition in the ring of polynomials.

Theorem 15.11 *Every polynomial $p = \alpha_0 + \alpha_1 t + \ldots + \alpha_n t^n \in K[t] \setminus \{0\}$ has a unique (up to the ordering of the factors) decomposition*

$$p = \mu \cdot p_1 \cdot \ldots \cdot p_k$$

with $\mu \in K$ and monic irreducible polynomials $p_1, \ldots, p_k \in K[t]$.

Proof If $\deg(p) = 0$, and thus $p = \alpha_0$, then the assertion holds with $k = 0$ and $\mu = \alpha_0$.

Let $\deg(p) \geq 1$. If p is irreducible, then the assertion holds with $p_1 = \mu^{-1} p$ and $\mu = \alpha_n$. If p is reducible, then $p = p_1 \cdot p_2$ for two non-constant polynomials p_1 and p_2. These are either irreducible, or we can decompose them further. Every multiplicative decomposition of p that is obtained in this way has at most $\deg(p) = n$ non-constant factors. Suppose that

$$p = \mu \cdot p_1 \cdot \ldots \cdot p_k = \beta \cdot q_1 \cdot \ldots \cdot q_\ell \tag{15.2}$$

for some k, ℓ, where $1 \leq \ell \leq k \leq n$, $\mu, \beta \in K$, as well as monic irreducible polynomials $p_1, \ldots, p_k, q_1, \ldots, q_\ell \in K[t]$. Then $p_1|p$ and hence $p_1|q_j$ for some j. Since the polynomials p_1 and q_j are irreducible, we must have $p_1 = q_j$.

We may assume without loss of generality that $j = 1$ and cancel the polynomial $p_1 = q_1$ in the identity (15.2), which gives

$$\mu \cdot p_2 \cdot \ldots \cdot p_k = \beta \cdot q_2 \cdot \ldots \cdot q_\ell.$$

15.1 Polynomials

Proceeding analogously for the polynomials $p_2, \ldots, p_k$, we finally obtain $k = \ell$, $\mu = \beta$ and $p_j = q_j$ for $j = 1, \ldots, k$. □

Finally, we consider common divisors and multiples of polynomials.

Definition 15.12 Let $k \geq 2$ and $p_1, \ldots, p_k \in K[t] \setminus \{0\}$.

(1) A polynomial $q \in K[t]$ is called *common divisor* of $p_1, \ldots, p_k$, if $q | p_i$ for $i = 1, \ldots, k$. A monic common divisor q of $p_1, \ldots, p_k$ is called *greatest common divisor*, denoted by $\gcd(p_1, \ldots, p_k)$, if $s|q$ for every common divisor $s \in K[t]$ of $p_1, \ldots, p_k$.
(2) A polynomial $h \in K[t]$ is called *common multiple* of $p_1, \ldots, p_k$, if $p_i | h$ for $i = 1, \ldots, k$. A monic common multiple h of $p_1, \ldots, p_k$ is called *least common multiple*, denoted by $\text{lcm}(p_1, \ldots, p_k)$, if $h|s$ for every common multiple $s \in K[t]$ of $p_1, \ldots, p_k$.

The greatest common divisor and least common multiple exist and are unique.

Lemma 15.13 *If $k \geq 2$ and $p_1, \ldots, p_k \in K[t] \setminus \{0\}$, then there exist unique monic polynomials $\gcd(p_1, \ldots, p_k) \in K[t]$ and $\text{lcm}(p_1, \ldots, p_k) \in K[t]$.*

Proof Exercise. □

For the case $k = 2$ we obtain the following consequence of the Lemma of Bézout.

Corollary 15.14 *If $p_1, p_2 \in K[t] \setminus \{0\}$, then there exist polynomials $q_1, q_2 \in K[t]$ with*

$$p_1 \cdot q_1 + p_2 \cdot q_2 = \gcd(p_1, p_2).$$

Proof If $q := \gcd(p_1, p_2)$, then there exist coprime polynomials $s_1, s_2 \in K[t] \setminus \{0\}$ with $p_1 = q \cdot s_1$ and $p_2 = q \cdot s_2$. (If s_1 and s_2 would have a common non-constant divisor, then q would not be the greatest common divisor of p_1 and p_2.) Applying Lemma 15.9 to s_1 and s_2 yields that

$$s_1 \cdot q_1 + s_2 \cdot q_2 = 1$$

for some polynomials $q_1, q_2 \in K[t]$. Multiplication of this equation with q yields

$$q = (q \cdot s_1) \cdot q_1 + (q \cdot s_2) \cdot q_2 =: p_1 \cdot q_1 + p_2 \cdot q_2,$$

which proves the assertion. □

For $k = 2$ and $p_1, p_2 \in K[t] \setminus \{0\}$ we can explicitly determine $\gcd(p_1, p_2)$ and $\text{lcm}(p_1, p_2)$ using the prime factor decompositions of p_1, p_2 as in Theorem 15.11.

If in these decompositions we combine equal monic and irreducible polynomials, then we obtain decompositions of the form

$$p_1 = \mu_1 \cdot q_1^{c_1} \cdot \ldots \cdot q_\ell^{c_\ell} \quad \text{und} \quad p_2 = \mu_2 \cdot q_1^{d_1} \cdot \ldots \cdot q_\ell^{d_\ell}, \tag{15.3}$$

with $\mu_1, \mu_2 \in K \setminus \{0\}$, pairwise distinct irreducible polynomials $q_1, \ldots, q_\ell \in K[t]$, and $c_i, d_i \in \mathbb{N}_0$ for $i = 1, \ldots, \ell$. We set $m_i := \min\{c_i, d_i\}$ and $M_i := \max\{c_i, d_i\}$ for $i = 1, \ldots, \ell$. Then

$$\gcd(p_1, p_2) = q_1^{m_1} \cdot \ldots \cdot q_\ell^{m_\ell},$$
$$\operatorname{lcm}(p_1, p_2) = q_1^{M_1} \cdot \ldots \cdot q_\ell^{M_\ell}, \tag{15.4}$$
$$p_1 \cdot p_2 = \mu_1 \cdot \mu_2 \cdot \gcd(p_1, p_2) \cdot \operatorname{lcm}(p_1, p_2).$$

For given $p_1, p_2 \in K[t] \setminus \{0\}$ we can determine $\gcd(p_1, p_2)$ by the *Euclidean algorithm*, in which one iteratively applies division with remainder:

Algorithm 15.15 *Given $p_1, p_2 \in K[t] \setminus \{0\}$. Carry out the following steps for $n = 3, 4, \ldots$:*

(1) Divide with remainder p_{n-2} by p_{n-1}. By Theorem 15.4 this yields unique polynomials $q_n, r_n \in K[t]$ with $p_{n-2} = p_{n-1} \cdot q_n + r_n$ and $\deg(r_n) < \deg(p_{n-1})$.
(2) Set $p_n := r_n = p_{n-2} - p_{n-1} \cdot q_n$.
(3) If $p_n = 0$ then stop, otherwise continue.

For every $n \geq 3$ it holds that $\deg(p_n) = \deg(r_n) < \deg(p_{n-1})$. Therefore, Algorithm 15.15 stops after finitely many steps with $p_N := r_N = 0$ for some $N \in \mathbb{N}$. We then have $p_{N-1} \neq 0$ and, hence, $p_{N-1} = \mu \cdot \widehat{p}_{N-1}$ for some $\mu \in K \setminus \{0\}$ and a monic polynomial $\widehat{p}_{N-1} \in K[t]$.

Lemma 15.16 *If $p_1, p_2 \in K[t] \setminus \{0\}$ and if $\widehat{p}_{N-1}$ is the monic polynomial constructed by Algorithm 15.15, then $\widehat{p}_{N-1} = \gcd(p_1, p_2)$.*

Proof Exercise. □

15.2 The Fundamental Theorem of Algebra

We have seen above that the existence of roots of a polynomial depends on the field over which it is considered. The field $\mathbb{C}$ is special in this sense, since here the

15.2 The Fundamental Theorem of Algebra

Fundamental Theorem of Algebra[3] guarantees that every non-constant polynomial has a root. In order to use this theorem in our context, we first present an equivalent formulation in the language of Linear Algebra.

Theorem 15.17 *The following statements are equivalent:*

(1) Every non-constant polynomial $p \in \mathbb{C}[t]$ has a root in $\mathbb{C}$.
(2) If $\mathcal{V} \neq \{0\}$ is a finite dimensional $\mathbb{C}$-vector space, then every endomorphism $f \in \mathcal{L}(\mathcal{V}, \mathcal{V})$ has an eigenvector.

Proof

(1) $\Rightarrow$ (2): If $\mathcal{V} \neq \{0\}$ and $f \in \mathcal{L}(\mathcal{V}, \mathcal{V})$, then the characteristic polynomial $P_f \in \mathbb{C}[t]$ is non-constant, since $\deg(P_f) = \dim(\mathcal{V}) > 0$. Thus, P_f has a root in $\mathbb{C}$, which is an eigenvalue of f, so that f indeed has an eigenvector.

(2) $\Rightarrow$ (1): Let $p = \alpha_0 + \alpha_1 t + \ldots + \alpha_n t^n \in \mathbb{C}[t]$ be a non-constant polynomial with $\alpha_n \neq 0$. The roots of p are equal to the roots of the monic polynomial $\widehat{p} := \alpha_n^{-1} p$. Let $A \in \mathbb{C}^{n,n}$ be the companion matrix of $\widehat{p}$, then $P_A = \widehat{p}$ (cp. Lemma 8.4).

If $\mathcal{V}$ is an n-dimensional $\mathbb{C}$-vector space and B is an arbitrary basis of $\mathcal{V}$, then there exists a uniquely determined $f \in \mathcal{L}(\mathcal{V}, \mathcal{V})$ with $[f]_{B,B} = A$ (cp. Theorem 10.17). By assumption, f has an eigenvector and hence also an eigenvalue, so that $\widehat{p} = P_A$ has a root. □

The Fundamental Theorem of Algebra cannot be proven without tools from Analysis. In particular, one needs that polynomials are *continuous*. We will use the following standard result, which is based on the continuity of polynomials.

Lemma 15.18 *Every polynomial $p \in \mathbb{R}[t]$ with odd degree has a (real) root.*

Proof Let the highest coefficient of p be positive. Then

$$\lim_{t \to \infty} p(t) = +\infty, \quad \lim_{t \to -\infty} p(t) = -\infty.$$

Since the real function $p(t)$ is continuous, the *Intermediate Value Theorem* from Analysis implies the existence of a root of p. The argument in the case of a negative leading coefficient is analogous. □

Our proof of the Fundamental Theorem of Algebra below follows the presentation in the article [2]. The proof is by induction on the dimension of $\mathcal{V}$. However,

---

[3] Numerous proofs of this important result exist. Carl Friedrich Gauß (1777–1855) alone gave four different proofs, starting with the one in his dissertation from 1799, which contained however a gap. The history of this result is described in detail in the book [3].

we do not use the usual consecutive order, i.e., $\dim(\mathcal{V}) = 1, 2, 3, \ldots$, but an order that is based on the sets

$$M_j := \{2^m \cdot \ell \mid 0 \leq m \leq j-1, \ \ell \text{ odd}\} \subset \mathbb{N}, \quad j \in \mathbb{N}.$$

For instance,

$$M_1 = \{\ell \mid \ell \text{ odd}\} = \{1, 3, 5, 7, \ldots\}, \quad M_2 = M_1 \cup \{2, 6, 10, 14, \ldots\}.$$

Lemma 15.19

(1) If $\mathcal{V}$ is an $\mathbb{R}$-vector space and if $\dim(\mathcal{V})$ is odd, i.e., $\dim(\mathcal{V}) \in M_1$, then every $f \in \mathcal{L}(\mathcal{V}, \mathcal{V})$ has an eigenvector.
(2) Let K be a field and $j \in \mathbb{N}$. If for every K-vector space $\mathcal{V}$ with $\dim(\mathcal{V}) \in M_j$ every $f \in \mathcal{L}(\mathcal{V}, \mathcal{V})$ has an eigenvector, then two commuting $f_1, f_2 \in \mathcal{L}(\mathcal{V}, \mathcal{V})$ have a common eigenvector. That is, if $f_1 \circ f_2 = f_2 \circ f_1$, then there exists a vector $v \in \mathcal{V} \setminus \{0\}$ and two scalars $\lambda_1, \lambda_2 \in K$ with $f_1(v) = \lambda_1 v$ and $f_2(v) = \lambda_2 v$.
(3) If $\mathcal{V}$ is an $\mathbb{R}$-vector space and if $\dim(\mathcal{V})$ is odd, then two commuting $f_1, f_2 \in \mathcal{L}(\mathcal{V}, \mathcal{V})$ have a common eigenvector.

Proof

(1) For every $f \in \mathcal{L}(\mathcal{V}, \mathcal{V})$ the degree of $P_f \in \mathbb{R}[t]$ is odd. Hence Lemma 15.18 implies that P_f has a root, and therefore f has an eigenvector.
(2) We proceed by induction on $\dim(\mathcal{V})$, where $\dim(\mathcal{V})$ runs through the elements of M_j in increasing order. The set M_j is a proper subset of $\mathbb{N}$ consisting of natural numbers that are not divisible by 2^j and, in particular, 1 is the smallest element of M_j.

If $\dim(\mathcal{V}) = 1 \in M_j$, then by assumption two arbitrary $f_1, f_2 \in \mathcal{L}(\mathcal{V}, \mathcal{V})$ each have an eigenvector, i.e.,

$$f_1(v_1) = \lambda_1 v_1, \quad f_2(v_2) = \lambda_2 v_2.$$

Since $\dim(\mathcal{V}) = 1$, we have $v_1 = \alpha v_2$ for an $\alpha \in K \setminus \{0\}$. Thus,

$$f_2(v_1) = f_2(\alpha v_2) = \alpha f_2(v_2) = \lambda_2(\alpha v_2) = \lambda_2 v_1,$$

i.e., v_1 is a common eigenvector of f_1 and f_2.

Let now $\dim(\mathcal{V}) \in M_j$, and let the assertion be proven for all K-vector spaces whose dimensions is an element of M_j that is smaller than $\dim(\mathcal{V})$. Let $f_1, f_2 \in \mathcal{L}(\mathcal{V}, \mathcal{V})$ with $f_1 \circ f_2 = f_2 \circ f_1$. By assumption, f_1 has an eigenvector v_1 with corresponding eigenvalue λ_1. Let

$$\mathcal{U} := \operatorname{im}(\lambda_1 \operatorname{Id}_\mathcal{V} - f_1), \quad \mathcal{W} := \mathcal{V}_{f_1}(\lambda_1) = \ker(\lambda_1 \operatorname{Id}_\mathcal{V} - f_1).$$

15.2 The Fundamental Theorem of Algebra

The subspaces $\mathcal{U}$ and $\mathcal{W}$ of $\mathcal{V}$ are f_1-invariant, i.e., $f_1(\mathcal{U}) \subseteq \mathcal{U}$ and $f_1(\mathcal{W}) \subseteq \mathcal{W}$. For the space $\mathcal{W}$ we have shown this in Lemma 14.4 and for the space $\mathcal{U}$ this can be easily shown as well (cp. Exercise 14.1). The subspaces $\mathcal{U}$ and $\mathcal{W}$ are also f_2-invariant:

If $u \in \mathcal{U}$, then $u = (\lambda_1 \operatorname{Id}_\mathcal{V} - f_1)(v)$ for a $v \in \mathcal{V}$. Since f_1 and f_2 commute, we have

$$f_2(u) = (f_2 \circ (\lambda_1 \operatorname{Id}_\mathcal{V} - f_1))(v) = ((\lambda_1 \operatorname{Id}_\mathcal{V} - f_1) \circ f_2)(v)$$
$$= (\lambda_1 \operatorname{Id}_\mathcal{V} - f_1)(f_2(v)) \in \mathcal{U}.$$

If $w \in \mathcal{W}$, then

$$(\lambda_1 \operatorname{Id}_\mathcal{V} - f_1)(f_2(w)) = ((\lambda_1 \operatorname{Id}_\mathcal{V} - f_1) \circ f_2)(w) = (f_2 \circ (\lambda_1 \operatorname{Id}_\mathcal{V} - f_1))(w)$$
$$= f_2((\lambda_1 \operatorname{Id}_\mathcal{V} - f_1)(w)) = f_2(0) = 0,$$

hence $f_2(w) \in \mathcal{W}$.

We have $\dim(\mathcal{V}) = \dim(\mathcal{U}) + \dim(\mathcal{W})$ and since $\dim(\mathcal{V})$ is not divisible by 2^j, we see that $\dim(\mathcal{U})$ and $\dim(\mathcal{W})$ can not both be divisible by 2^j. Hence either $\dim(\mathcal{U}) \in M_j$ or $\dim(\mathcal{W}) \in M_j$.

If the corresponding subspace is a proper subspace of $\mathcal{V}$, then its dimension is an element of M_j that is smaller than $\dim(\mathcal{V})$. By the induction hypothesis then f_1 and f_2 have a common eigenvector in this subspace. Thus, f_1 and f_2 have a common eigenvector in $\mathcal{V}$.

If the corresponding subspace is equal to $\mathcal{V}$, then this must be the subspace $\mathcal{W}$, since $\dim(\mathcal{W}) \geq 1$. But if $\mathcal{V} = \mathcal{W}$, then every vector in $\mathcal{V} \setminus \{0\}$ is an eigenvector of f_1. By assumption also f_2 has an eigenvector, so that there exists at least one common eigenvector of f_1 and f_2.

(3) By (1) it follows that the assumption of (2) holds for $K = \mathbb{R}$ and $j = 1$, which means that (3) holds as well. □

We will now prove the Fundamental Theorem of Algebra in the formulation (2) of Theorem 15.17.

Theorem 15.20 *If $\mathcal{V} \neq \{0\}$ is a finite dimensional $\mathbb{C}$-vector space, then every $f \in \mathcal{L}(\mathcal{V}, \mathcal{V})$ has an eigenvector.*

Proof We prove the assertion by induction on $j = 1, 2, 3, \ldots$ and $\dim(\mathcal{V}) \in M_j$.

We start with $j = 1$ and thus by showing the assertion for all $\mathbb{C}$-vector spaces of odd dimension. Let $\mathcal{V}$ be an arbitrary $\mathbb{C}$-vector space with $n := \dim(\mathcal{V}) \in M_1$. Let $f \in \mathcal{L}(\mathcal{V}, \mathcal{V})$ and consider an arbitrary scalar product on $\mathcal{V}$ (such a scalar product always exists; cp. Exercise 12.1), as well as the set of self-adjoint maps with respect to this scalar product,

$$\mathcal{H} := \{g \in \mathcal{L}(\mathcal{V}, \mathcal{V}) \mid g = g^{ad}\}.$$

By Lemma 13.15 the set $\mathcal{H}$ forms an $\mathbb{R}$-vector space of dimension n^2. If we define $h_1, h_2 \in \mathcal{L}(\mathcal{H}, \mathcal{H})$ by

$$h_1(g) := \frac{1}{2}(f \circ g + g \circ f^{ad}), \quad h_2(g) := \frac{1}{2\mathbf{i}}(f \circ g - g \circ f^{ad})$$

for all $g \in \mathcal{H}$, then $h_1 \circ h_2 = h_2 \circ h_1$ (cp. Exercise 15.13). Since n is odd, also n^2 is odd. By (3) in Lemma 15.19, h_1 and h_2 have a common eigenvector. Hence, there exists a $\widetilde{g} \in \mathcal{H} \setminus \{0\}$ with

$$h_1(\widetilde{g}) = \lambda_1 \widetilde{g}, \quad h_2(\widetilde{g}) = \lambda_2 \widetilde{g} \quad \text{for some} \quad \lambda_1, \lambda_2 \in \mathbb{R}.$$

We have $(h_1 + \mathbf{i} h_2)(g) = f \circ g$ for all $g \in \mathcal{H}$ and therefore, in particular,

$$f \circ \widetilde{g} = (h_1 + \mathbf{i} h_2)(\widetilde{g}) = (\lambda_1 + \mathbf{i}\lambda_2)\widetilde{g}.$$

Since $\widetilde{g} \neq 0$, there exists a $v \in \mathcal{V}$ with $\widetilde{g}(v) \neq 0$. Then

$$f(\widetilde{g}(v)) = (\lambda_1 + \mathbf{i}\lambda_2)\,\widetilde{g}(v),$$

which shows that $\widetilde{g}(v) \in \mathcal{V}$ is an eigenvector of f, so that the proof for $j = 1$ is complete.

Assume now that for some $j \geq 1$ and every $\mathbb{C}$-vector space $\mathcal{V}$ with $\dim(\mathcal{V}) \in M_j$, every $f \in \mathcal{L}(\mathcal{V}, \mathcal{V})$ has an eigenvector. Then (2) in Lemma 15.19 implies that every two commuting $f_1, f_2 \in \mathcal{L}(\mathcal{V}, \mathcal{V})$ have a common eigenvector.

We have to show that for every $\mathbb{C}$-vector space $\mathcal{V}$ with $\dim(\mathcal{V}) \in M_{j+1}$, every $f \in \mathcal{L}(\mathcal{V}, \mathcal{V})$ has an eigenvector. Since

$$M_{j+1} = M_j \cup \{2^j q \mid q \text{ odd}\},$$

we only have to prove this for $\mathbb{C}$-vector spaces $\mathcal{V}$ with $n := \dim(\mathcal{V}) = 2^j q$ for odd q. Let $\mathcal{V}$ be such a vector space and let $f \in \mathcal{L}(\mathcal{V}, \mathcal{V})$ be given. We choose an arbitrary basis of $\mathcal{V}$ and denote the matrix representation of f with respect to this basis by $A \in \mathbb{C}^{n,n}$.

Let

$$\mathcal{S} := \{B \in \mathbb{C}^{n,n} \mid B = B^T\}$$

be the set of complex symmetric $n \times n$ matrices. If we define $h_1, h_2 \in \mathcal{L}(\mathcal{S}, \mathcal{S})$ by

$$h_1(B) := AB + BA^T, \quad h_2(B) := ABA^T$$

15.2 The Fundamental Theorem of Algebra

for all $B \in \mathcal{S}$, then $h_1 \circ h_2 = h_2 \circ h_1$ (cp. Exercise 15.14). By Lemma 13.16 the set $\mathcal{S}$ forms a $\mathbb{C}$-vector space of dimension $n(n+1)/2$. We have $n = 2^j q$ for an odd natural number q. Thus,

$$\frac{n(n+1)}{2} = \frac{2^j q \, (2^j q + 1)}{2} = 2^{j-1} q \cdot (2^j q + 1) \in M_j.$$

By the induction hypothesis, the commuting endomorphisms h_1 and h_2 have a common eigenvector. Hence there exists a $\widetilde{B} \in \mathcal{S} \setminus \{0\}$ with

$$h_1(\widetilde{B}) = \lambda_1 \widetilde{B}, \quad h_2(\widetilde{B}) = \lambda_2 \widetilde{B} \quad \text{for some} \quad \lambda_1, \lambda_2 \in \mathbb{C}.$$

In particular, we have $\lambda_1 \widetilde{B} = A\widetilde{B} + \widetilde{B}A^T$. Multiplying this equation from the left with A yields

$$\lambda_1 A \widetilde{B} = A^2 \widetilde{B} + A\widetilde{B}A^T = A^2 \widetilde{B} + h_2(\widetilde{B}) = A^2 \widetilde{B} + \lambda_2 \widetilde{B},$$

so that

$$\left(A^2 - \lambda_1 A + \lambda_2 I_n \right) \widetilde{B} = 0.$$

We now factorize $t^2 - \lambda_1 t + \lambda_2 = (t - \alpha)(t - \beta)$ with

$$\alpha = \frac{\lambda_1 + \sqrt{\lambda_1^2 - 4\lambda_2}}{2}, \quad \beta = \frac{\lambda_1 - \sqrt{\lambda_1^2 - 4\lambda_2}}{2},$$

where we have used that every complex number has a square root. Then

$$(A - \alpha I_n)(A - \beta I_n) \widetilde{B} = 0.$$

Since $\widetilde{B} \neq 0$, there exists a $v \in \mathbb{C}^{n,1}$ with $\widetilde{B}v \neq 0$. If $(A - \beta I_n)\widetilde{B}v = 0$, then $\widetilde{B}v$ is an eigenvector of A corresponding to the eigenvalue β. If $(A - \beta I_n)\widetilde{B}v \neq 0$, then $(A - \beta I_n)\widetilde{B}v$ is an eigenvector of A corresponding to the eigenvalue α. Since A has an eigenvector, also f has an eigenvector. $\square$

MATLAB-Minute 9

Compute the eigenvalues of the matrix

$$A = \begin{bmatrix} 1 & 2 & 3 & 4 & 5 \\ 1 & 2 & 4 & 3 & 5 \\ 2 & 3 & 4 & 1 & 5 \\ 5 & 1 & 4 & 2 & 3 \\ 4 & 2 & 3 & 1 & 5 \end{bmatrix} \in \mathbb{R}^{5,5}$$

using the command eig(A).

By definition a real matrix A can only have real eigenvalues. The reason for the occurrence of complex eigenvalues is that MATLAB interprets *every* matrix as a complex matrix. This means that within MATLAB *every* matrix can be unitarily triangulated, since every complex polynomial (of degree at least 1) decomposes into linear factors.

As a direct corollary of the Fundamental Theorem of Algebra and (2) in Lemma 15.19 we have the following result.

Corollary 15.21 *If $\mathcal{V} \neq \{0\}$ is a finite dimensional $\mathbb{C}$-vector space, then two commuting $f_1, f_2 \in \mathcal{L}(\mathcal{V}, \mathcal{V})$ have a common eigenvector.*

Example 15.22 The two complex 2×2 matrices

$$A = \begin{bmatrix} \mathbf{i} & 1 \\ 1 & \mathbf{i} \end{bmatrix} \quad \text{and} \quad B = \begin{bmatrix} 2\mathbf{i} & 1 \\ 1 & 2\mathbf{i} \end{bmatrix}$$

commute. The eigenvalues of A are $\pm 1 + \mathbf{i}$ and those of B are $\pm 2 + \mathbf{i}$. Hence A and B do not have a common eigenvalue, while $[1, 1]^T$ and $[-1, 1]^T$ are common eigenvectors of A and B.

Using Corollary 15.21, Schur's theorem (Corollary 14.21) can be generalized as follows.

Theorem 15.23 *If $\mathcal{V} \neq \{0\}$ is a finite dimensional unitary vector space and $f_1, f_2 \in \mathcal{L}(\mathcal{V}, \mathcal{V})$ commute, then f_1 and f_2 can be simultaneously unitarily triangulated, i.e., there exists an orthonormal basis B of $\mathcal{V}$, such that $[f_1]_{B,B}$ and $[f_2]_{B,B}$ are both upper triangular.*

Proof Exercise. □

Since every endomorphism (trivially) commutes with itself, the choice $f_1 = f_2 = f$ in Theorem 15.23 implies the assertion of Schur's theorem.

The "matrix version" of Theorem 15.23 is the following: If $A_1, A_2 \in \mathbb{C}^{n,n}$ satisfy $A_1 A_2 = A_2 A_1$, then there exists a unitary matrix $Q \in \mathbb{C}^{n,n}$ such that $Q^H A_1 Q = R_1$ and $Q^H A_2 Q = R_2$ are upper triangular.

The eigenvalues of the two commuting matrices A_1 and A_2 are on the diagonals of the upper triangular matrices R_1 and R_2, respectively. One easily sees that $A_1 A_2 = Q R_1 R_2 Q^H$. The matrices $A_1 A_2$ and $R_1 R_2$ are (unitarily) similar and, therefore, have the same eigenvalues. The product $R_1 R_2$ is upper triangular, and its diagonal entries are the products of the diagonal entries of R_1 and R_2, i.e., the eigenvalues of A_1 and A_2. This shows that for commuting matrices A_1 and A_2 the eigenvalues of the product $A_1 A_2$ are given by products of eigenvalues of A_1 and A_2.

One can show inductively that the eigenvalues of $p(A_1, A_2) = \sum_{i,j=0}^{n} \alpha_{ij} A_1^i A_2^j$, where $p \in K[t_1, t_2]$ is a bivariate polynomial (cp. Exercise 9.16), are given by $p(\lambda_j^{(1)}, \lambda_j^{(2)})$, where the $\lambda_j^{(i)}$, $j = 1, \ldots, n$, are the eigenvalues of A_i, $i = 1, 2$, in a specific order. This statement does not hold in general for non-commuting matrices (or endomorphisms).

Example 15.24 The matrices

$$A_1 = \begin{bmatrix} 1 & 1 \\ 0 & 1 \end{bmatrix} \quad \text{and} \quad A_2 = \begin{bmatrix} 0 & 0 \\ 1 & 0 \end{bmatrix}$$

do not commute. The matrix A_1 has only the eigenvalue 1 with $a(1, A_1) = 2$ and $g(1, A_1) = 1$. The matrix A_2 has only the eigenvalue 0 with $a(0, A_2) = 2$ and $g(0, A_2) = 1$. Hence, every product of eigenvalues of A_1 and A_2 is 0, while the products

$$A_1 A_2 = \begin{bmatrix} 1 & 0 \\ 1 & 0 \end{bmatrix} \quad \text{and} \quad A_2 A_1 = \begin{bmatrix} 0 & 0 \\ 1 & 1 \end{bmatrix}$$

have the eigenvalues 1 and 0 (each with algebraic and geometric multiplicity 1).

Exercises

(In the following exercises K is an arbitrary field.)

15.1 Prove Lemma 15.1.
15.2 Show the following assertions for $p_1, p_2, p_3 \in K[t]$:
  (a) $p_1 | (p_1 p_2)$.
  (b) $p_1 | p_2$ and $p_2 | p_3$ imply that $p_1 | p_3$.
  (c) $p_1 | p_2$ and $p_1 | p_3$ imply that $p_1 | (p_2 + p_3)$.
  (d) If $p_1 | p_2$ and $p_2 | p_1$, then there exists a $c \in K \setminus \{0\}$ with $p_1 = c p_2$.

15.3 Examine whether the following polynomials are irreducible:

$$p_1 = t^3 - t^2 + t - 1 \in \mathbb{Q}[t], \qquad p_4 = 4t^3 - 4t^2 - t + 1 \in \mathbb{Q}[t],$$
$$p_2 = t^3 - t^2 + t - 1 \in \mathbb{R}[t], \qquad p_5 = 4t^3 - 4t^2 - t + 1 \in \mathbb{R}[t],$$
$$p_3 = t^3 - t^2 + t - 1 \in \mathbb{C}[t], \qquad p_6 = t^3 - 2t^2 + t - 2 \in \mathbb{C}[t].$$

Determine the decompositions into irreducible factors.

15.4 Factorize the monic polynomials

$$p_1 = t^2 - 2, \quad p_2 = t^2 + 2, \quad p_3 = t^4 - 1,$$
$$p_4 = t^2 + t + 1, \quad p_5 = t^4 - t^2 - 2$$

into monic irreducible polynomials over the fields $K = \mathbb{Q}$, $K = \mathbb{R}$ and $K = \mathbb{C}$.

15.5 Show the following assertions for $p \in K[t]$:
(a) If $\deg(p) = 1$, then p is irreducible.
(b) If $\deg(p) \geq 2$ and p has a root, then p is not irreducible.
(c) If $\deg(p) \in \{2, 3\}$, then p is irreducible if and only if p does not have a root.

15.6 Determine a field K and a polynomial $p \in K[t]$ with $\deg(p) = 4$, such that p has no root and is not irreducible.

15.7 Let $A \in GL_n(\mathbb{C})$, $n \geq 2$, and let $\text{adj}(A) \in \mathbb{C}^{n,n}$ be the adjunct of A. Show that there exist $n-1$ matrices $A_j \in \mathbb{C}^{n,n}$ with $\det(-A_j) = \det(A)$, $j = 1, \ldots, n-1$, and

$$\text{adj}(A) = \prod_{j=1}^{n-1} A_j.$$

(*Hint:* Use P_A to construct a polynomial $p \in \mathbb{C}[t]_{\leq n-1}$ with $\text{adj}(A) = p(A)$ and express p as product of linear factors.)

15.8 Show that two polynomials $p, q \in \mathbb{C}[t] \setminus \{0\}$ have a common root if and only if there exist polynomials $r_1, r_2 \in \mathbb{C}[t]$ with $0 \leq \deg(r_1) < \deg(p)$ such that $0 \leq \deg(r_2) < \deg(q)$ and $p \cdot r_2 + q \cdot r_1 = 0$.

15.9 Prove Lemma 15.13.

15.10 Show that the equations in (15.4) hold for two polynomials $p_1, p_2 \in K[t] \setminus \{0\}$ that are factorized as in (15.3).

15.11 Prove Lemma 15.16.

15.12 Using the Euclidean algorithm determine $\gcd(p_1, p_2) \in \mathbb{Q}[t]$ for the polynomials $p_1 = t^4 - 8t^3 + 23t^2 - 28t + 12$ and $p_2 = t^3 - 9t^2 + 26t - 24$. Factorize $\gcd(p_1, p_2)$ into monic irreducible polynomials from $\mathbb{Q}[t]$.

15.13 Let V be a finite dimensional unitary vector space, $f \in \mathcal{L}(V, V)$, $\mathcal{H} = \{g \in \mathcal{L}(V, V) \mid g = g^{ad}\}$ and let

$$h_1 : \mathcal{H} \to \mathcal{L}(V, V), \quad g \mapsto \frac{1}{2}(f \circ g + g \circ f^{ad}),$$

$$h_2 : \mathcal{H} \to \mathcal{L}(V, V), \quad g \mapsto \frac{1}{2\mathrm{i}}(f \circ g - g \circ f^{ad}).$$

Show that $h_1, h_2 \in \mathcal{L}(\mathcal{H}, \mathcal{H})$ and $h_1 \circ h_2 = h_2 \circ h_1$.

15.14 Let $A \in \mathbb{C}^{n,n}$, $\mathcal{S} = \{B \in \mathbb{C}^{n,n} \mid B = B^T\}$ and let

$$h_1 : \mathcal{S} \to \mathbb{C}^{n,n}, \quad B \mapsto AB + BA^T,$$

$$h_2 : \mathcal{S} \to \mathbb{C}^{n,n}, \quad B \mapsto ABA^T.$$

Show that $h_1, h_2 \in \mathcal{L}(\mathcal{S}, \mathcal{S})$ and $h_1 \circ h_2 = h_2 \circ h_1$.

15.15 Let V be a $\mathbb{C}$-vector space, $f \in \mathcal{L}(V, V)$ and let $\mathcal{U} \neq \{0\}$ be a finite dimensional f-invariant subspace of V. Show that $\mathcal{U}$ contains at least one eigenvector of f.

15.16 Let $V \neq \{0\}$ be a finite dimensional K-vector space and let $f \in \mathcal{L}(V, V)$. Show the following assertions:
(a) If $K = \mathbb{C}$, then there exists an f-invariant subspace $\mathcal{U}$ of V with $\dim(\mathcal{U}) = 1$.
(b) If $K = \mathbb{R}$, then there exists an f-invariant subspace $\mathcal{U}$ of V with $\dim(\mathcal{U}) \in \{1, 2\}$.

15.17 Prove Theorem 15.23.

15.18 Construct an example showing that the condition $f \circ g = g \circ f$ in Theorem 15.23 is sufficient but not necessary for the simultaneous unitary triangulation of f and g.

15.19 Let $A \in K^{n,n}$ be a diagonal matrix with pairwise distinct diagonal entries and $B \in K^{n,n}$ with $AB = BA$. Show that in this case B is a diagonal matrix. What can you say about B, when the diagonal entries of A are not all pairwise distinct?

15.20 Show that the matrices

$$A = \begin{bmatrix} -1 & 1 \\ 1 & -1 \end{bmatrix} \quad \text{and} \quad B = \begin{bmatrix} 0 & 1 \\ 1 & 0 \end{bmatrix}$$

commute and determine a unitary matrix Q such that $Q^H A Q$ and $Q^H B Q$ are upper triangular.

15.21 Show the following assertions for $p \in K[t]$:
   (a) For all $A \in K^{n,n}$ and $S \in GL_n(K)$ we have $p(SAS^{-1}) = Sp(A)S^{-1}$.
   (b) For all $A, B, C \in K^{n,n}$ with $AB = CA$ we have $Ap(B) = p(C)A$.
   (c) If $K = \mathbb{C}$ and $A \in \mathbb{C}^{n,n}$, then there exists a unitary matrix Q, such that $Q^H A Q$ and $Q^H p(A) Q$ are upper triangular.

15.22 Let $\mathcal{V}$ be a finite dimensional unitary vector space, and let $f \in \mathcal{L}(\mathcal{V}, \mathcal{V})$ be *normal*, i.e., f satisfies $f \circ f^{ad} = f^{ad} \circ f$. Show, using Theorems 15.23 and 13.12, that f is unitarily diagonalizable. (Normal endomorphisms will be studied in detail in Chap. 18.)

16 The Jordan and the Frobenius Canonical Form

In this chapter we use the duality theory to analyze the properties of an endomorphism f on a finite dimensional vector space $\mathcal{V}$ in detail. We are particularly interested in the algebraic and geometric multiplicities of the eigenvalues of f and the characterization of the corresponding eigenspaces. Our strategy in this analysis is to decompose the vector space $\mathcal{V}$ into a direct sum of f-invariant subspaces so that, with appropriately chosen bases, the essential properties of f will be obvious from its matrix representation. The matrix representation that we derive is called the Jordan canonical form[1] of endomorphisms and matrices. It exists whenever the characteristic polynomial of f resp. A decomposes into linear factors over the given field. Because of its great importance, there have been many different derivations of this form using different mathematical tools since its discovery in the nineteenth century. Our approach using duality theory is based on an article by Vlastimil Pták (1925–1999) from 1956 [13]. Using the same strategy, in this chapter we also derive the Frobenius canonical form[2] of endomorphisms and matrices. Unlike the Jordan canonical form, the Frobenius canonical form exists even when f resp. A does not have eigenvalues.

16.1 Cyclic f-invariant Subspaces and Duality

Let $\mathcal{V}$ be a finite dimensional K-vector space. If $f \in \mathcal{L}(\mathcal{V}, \mathcal{V})$ and $v_0 \in \mathcal{V} \setminus \{0\}$, then there exists a uniquely defined smallest number $m \in \mathbb{N}$, such that the vectors

$$v_0, f(v_0), \ldots, f^{m-1}(v_0)$$

---

[1] Marie Ennemond Camille Jordan (1838–1922).
[2] Ferdinand Georg Frobenius (1849–1917).

are linearly independent and the vectors

$$v_0, f(v_0), \ldots, f^{m-1}(v_0), f^m(v_0)$$

are linearly dependent. Obviously $m \leq \dim(\mathcal{V})$, since at most $\dim(\mathcal{V})$ vectors of $\mathcal{V}$ can be linearly independent. The number m is called the *grade of v_0 with respect to f*. We denote this grade by $m(f, v_0)$. The vector $v_0 = 0$ is linearly dependent, and thus its grade is 0 (with respect to any f).

For $v_0 \neq 0$ we have $m(f, v_0) = 1$ if and only if the vectors $v_0, f(v_0)$ are linearly dependent. This holds if and only if v_0 is an eigenvector of f. If $v_0 \neq 0$ is not an eigenvector of f, then $m(f, v_0) \geq 2$.

For every $j \in \mathbb{N}$ we define the subspace

$$\mathcal{K}_j(f, v_0) := \mathrm{span}\{v_0, f(v_0), \ldots, f^{j-1}(v_0)\} \subseteq \mathcal{V}.$$

The space $\mathcal{K}_j(f, v_0)$ is called the *jth Krylov subspace*[3] of f and v_0.

Lemma 16.1 *If $\mathcal{V}$ is a finite dimensional K-vector space, $f \in \mathcal{L}(\mathcal{V}, \mathcal{V})$, and $v_0 \in \mathcal{V}$, then the following assertions hold:*

(1) If $m = m(f, v_0)$, then $\mathcal{K}_m(f, v_0)$ is an f-invariant subspace of $\mathcal{V}$, and

$$\mathrm{span}\{v_0\} = \mathcal{K}_1(f, v_0) \subset \mathcal{K}_2(f, v_0) \subset \cdots \subset \mathcal{K}_m(f, v_0) = \mathcal{K}_{m+j}(f, v_0)$$

for all $j \in \mathbb{N}$.

(2) If $m = m(f, v_0)$ and $\mathcal{U} \subseteq \mathcal{V}$ is an f-invariant subspace that contains the vector v_0, then $\mathcal{K}_m(f, v_0) \subseteq \mathcal{U}$. Thus, among all f-invariant subspaces of $\mathcal{V}$ that contain the vector v_0, the Krylov subspace $\mathcal{K}_m(f, v_0)$ is the one of smallest dimension.

(3) If $f^{m-1}(v_0) \neq 0$ and $f^m(v_0) = 0$ for an $m \in \mathbb{N}$, then $\dim(\mathcal{K}_j(f, v_0)) = j$ for $j = 1, \ldots, m$.

Proof

(1) Exercise.
(2) The assertion is trivial if $v_0 = 0$. Thus, let $v_0 \neq 0$ with $m = d(f, v_0) \geq 1$ and let $\mathcal{U} \subseteq \mathcal{V}$ be an f-invariant subspace that contains v_0. Then $\mathcal{U}$ also contains the vectors $f(v_0), \ldots, f^{m-1}(v_0)$, so that $\mathcal{K}_m(f, v_0) \subseteq \mathcal{U}$ and, in particular, $\dim(\mathcal{U}) \geq m = \dim(\mathcal{K}_m(f, v_0))$.
(3) Let $\gamma_0, \ldots, \gamma_{m-1} \in K$ with

$$0 = \gamma_0 v_0 + \ldots + \gamma_{m-1} f^{m-1}(v_0).$$

---

[3] Aleksey Nikolaevich Krylov (1863–1945).

16.1 Cyclic f-invariant Subspaces and Duality

If we apply f^{m-1} to both sides, then $0 = \gamma_0 f^{m-1}(v_0)$ and thus $\gamma_0 = 0$, since $f^{m-1}(v_0) \neq 0$. If $m > 1$, then we apply inductively f^{m-k} for $k = 2, \ldots, m$ and obtain $\gamma_1 = \cdots = \gamma_{m-1} = 0$. Thus, the vectors $v_0, \ldots, f^{m-1}(v_0)$ are linearly independent, which implies that $\dim(\mathcal{K}_j(f, v_0)) = j$ for $j = 1, \ldots, m$. □

The vectors $v_0, f(v_0), \ldots, f^{m-1}(v_0)$ form, by construction, a basis of the Krylov subspace $\mathcal{K}_m(f, v_0)$. The application of f to a vector $f^k(v_0)$ of this basis yields the next basis vector $f^{k+1}(v_0)$, $k = 0, 1, \ldots, m - 2$, and the application of f to the last vector $f^{m-1}(v_0)$ yields a linear combination of all basis vectors, since $f^m(v_0) \in \mathcal{K}_m(f, v_0)$. Due to this special structure, the subspace $\mathcal{K}_m(f, v_0)$ is called a *cyclic f-invariant subspace*.

Definition 16.2 Let $\mathcal{V} \neq \{0\}$ be a K-vector space. An endomorphism $f \in \mathcal{L}(\mathcal{V}, \mathcal{V})$ is called *nilpotent*, if $f^m = 0$ holds for an $m \in \mathbb{N}$. If at the same time $f^{m-1} \neq 0$, then f is called *nilpotent of index m*.

The zero map $f = 0$ is the only nilpotent endomorphism of index $m = 1$. If $\mathcal{V} = \{0\}$, then the zero map is the only endomorphism on $\mathcal{V}$. This map is nilpotent of index $m = 1$, where in this case we omit the requirement $f^{m-1} = f^0 \neq 0$.

If f is nilpotent of index m and $v \neq 0$ is any vector with $f^{m-1}(v) \neq 0$, then $f(f^{m-1})(v) = f^m(v) = 0 = 0 \cdot f^{m-1}(v)$. Hence $f^{m-1}(v)$ is an eigenvector of f corresponding to the eigenvalue 0. Our construction in Sect. 16.2 will show that 0 is the only eigenvalue of a nilpotent endomorphism (also cp. Exercise 8.4).

Lemma 16.3 *If $\mathcal{V} \neq \{0\}$ is a K-vector space and if $f \in \mathcal{L}(\mathcal{V}, \mathcal{V})$ is nilpotent of index m, then $m \leq \dim(\mathcal{V})$.*

Proof If f is nilpotent of index m, then there exists a $v_0 \in \mathcal{V}$ with $f^{m-1}(v_0) \neq 0$ and $f^m(v_0) = 0$. By (3) in Lemma 16.1 the m vectors $v_0, \ldots, f^{m-1}(v_0)$ are linearly independent, which implies that $m \leq \dim(\mathcal{V})$. □

Example 16.4 In the vector space $K^{3,1}$ the endomorphism

$$f : K^{3,1} \to K^{3,1}, \quad \begin{bmatrix} v_1 \\ v_2 \\ v_3 \end{bmatrix} \mapsto \begin{bmatrix} 0 \\ v_1 \\ v_2 \end{bmatrix},$$

is nilpotent of index 3, since $f \neq 0$, $f^2 \neq 0$ and $f^3 = 0$.

If $\mathcal{U}$ is an f-invariant subspace of $\mathcal{V}$, then $f|_{\mathcal{U}} \in \mathcal{L}(\mathcal{U}, \mathcal{U})$, where

$$f|_{\mathcal{U}} : \mathcal{U} \to \mathcal{U}, \quad u \mapsto f(u),$$

is the restriction of f to the subspace $\mathcal{U}$ (cp. Definition 2.13).

Theorem 16.5 *Let V be a finite dimensional K-vector space and $f \in \mathcal{L}(V, V)$. Then there exist f-invariant subspaces $\mathcal{U}_1 \subseteq V$ and $\mathcal{U}_2 \subseteq V$ with $V = \mathcal{U}_1 \oplus \mathcal{U}_2$, such that $f|_{\mathcal{U}_1} \in \mathcal{L}(\mathcal{U}_1, \mathcal{U}_1)$ is bijective and $f|_{\mathcal{U}_2} \in \mathcal{L}(\mathcal{U}_2, \mathcal{U}_2)$ is nilpotent.*

Proof If $v \in \ker(f)$, then $f^2(v) = f(f(v)) = f(0) = 0$. Thus, $v \in \ker(f^2)$ and therefore $\ker(f) \subseteq \ker(f^2)$. Proceeding inductively we see that

$$\{0\} \subseteq \ker(f) \subseteq \ker(f^2) \subseteq \ker(f^3) \subseteq \cdots.$$

Since V is finite dimensional, there exists a smallest number $m \in \mathbb{N}_0$ with $\ker(f^m) = \ker(f^{m+j})$ for all $j \in \mathbb{N}$. For this number m let

$$\mathcal{U}_1 := \mathrm{im}(f^m), \quad \mathcal{U}_2 := \ker(f^m).$$

(If f is bijective, then $m = 0$, $\mathcal{U}_1 = V$ and $\mathcal{U}_2 = \{0\}$.) We now show that the spaces $\mathcal{U}_1$ and $\mathcal{U}_2$ satisfy the assertion.

First observe that $\mathcal{U}_1$ and $\mathcal{U}_2$ are both f-invariant: If $v \in \mathcal{U}_1$, then $v = f^m(w)$ for some $w \in V$, and therefore $f(v) = f(f^m(w)) = f^m(f(w)) \in \mathcal{U}_1$. If $v \in \mathcal{U}_2$, then $f^m(f(v)) = f(f^m(v)) = f(0) = 0$, and therefore $f(v) \in \mathcal{U}_2$.

We have $\mathcal{U}_1 + \mathcal{U}_2 \subseteq V$. An application of the dimension formula for linear maps (cp. Theorem 10.9) to f^m gives $\dim(V) = \dim(\mathcal{U}_1) + \dim(\mathcal{U}_2)$. If $v \in \mathcal{U}_1 \cap \mathcal{U}_2$, then $v = f^m(w)$ for some $w \in V$ (since $v \in \mathcal{U}_1$) and hence

$$0 = f^m(v) = f^m(f^m(w)) = f^{2m}(w).$$

The first equation holds since $v \in \mathcal{U}_2$. By the definition of m we have $\ker(f^m) = \ker(f^{2m})$, which implies $f^m(w) = 0$, and therefore $v = f^m(w) = 0$. From $\mathcal{U}_1 \cap \mathcal{U}_2 = \{0\}$ we obtain $V = \mathcal{U}_1 \oplus \mathcal{U}_2$.

Let now $v \in \ker(f|_{\mathcal{U}_1}) \subseteq \mathcal{U}_1$ be given. Since $v \in \mathcal{U}_1$, there exists a vector $w \in V$ with $v = f^m(w)$, which implies $0 = f(v) = f(f^m(w)) = f^{m+1}(w)$. By the definition of m we have $\ker(f^m) = \ker(f^{m+1})$, thus $w \in \ker(f^m)$, and therefore $v = f^m(w) = 0$. This implies that $\ker(f|_{\mathcal{U}_1}) = \{0\}$, i.e., $f|_{\mathcal{U}_1}$ is injective and thus also bijective (cp. Corollary 10.11).

If, on the other hand, $v \in \mathcal{U}_2$, then by definition $0 = f^m(v) = (f|_{\mathcal{U}_2})^m(v)$, and thus $(f|_{\mathcal{U}_2})^m$ is the zero map in $\mathcal{L}(\mathcal{U}_2, \mathcal{U}_2)$, so that $f|_{\mathcal{U}_2}$ is nilpotent. □

For the further development we recall some terms and results from Chapter 11. Let V be a finite dimensional K-vector space and let V^* be the dual space of V. If $\mathcal{U} \subseteq V$ and $\mathcal{W} \subseteq V^*$ are two subspaces and if the bilinear form

$$\beta : \mathcal{U} \times \mathcal{W} \to K, \quad (v, h) \mapsto h(v), \tag{16.1}$$

16.1 Cyclic f-invariant Subspaces and Duality

is non-degenerate, then $\mathcal{U}, \mathcal{W}$ is called a *dual pair* with respect to β. This requires that $\dim(\mathcal{U}) = \dim(\mathcal{W})$. For $f \in \mathcal{L}(\mathcal{U}, \mathcal{U})$ the *dual map* $f^* \in \mathcal{L}(\mathcal{U}^*, \mathcal{U}^*)$ is defined by

$$f^* : \mathcal{U}^* \to \mathcal{U}^*, \quad h \mapsto h \circ f.$$

For all $v \in \mathcal{U}$ and $h \in \mathcal{U}^*$ we have $(f^*(h))(v) = h(f(v))$. Furthermore, $(f^k)^* = (f^*)^k$ for all $k \in \mathbb{N}_0$. The set

$$\mathcal{U}^0 := \{ h \in \mathcal{V}^* \mid h(u) = 0 \text{ for all } u \in \mathcal{U} \}$$

is called the *annihilator* of $\mathcal{U}$. This set is a subspace of $\mathcal{V}^*$ (cp. Exercise 11.8). Analogously, the set

$$\mathcal{W}^0 := \{ v \in \mathcal{V} \mid h(v) = 0 \text{ for all } h \in \mathcal{W} \}$$

is called the *annihilator* of $\mathcal{W}$. This set is a subspace of $\mathcal{V}$.

Lemma 16.6 *Let $\mathcal{V}$ be a finite dimensional K-vector space, $f \in \mathcal{L}(\mathcal{V}, \mathcal{V})$, $\mathcal{V}^*$ the dual space of $\mathcal{V}$, $f^* \in \mathcal{L}(\mathcal{V}^*, \mathcal{V}^*)$ the dual map of f, and let $\mathcal{U} \subseteq \mathcal{V}$ and $\mathcal{W} \subseteq \mathcal{V}^*$ be two subspaces. Then the following assertions hold:*

(1) $\dim(\mathcal{V}) = \dim(\mathcal{W}) + \dim(\mathcal{W}^0) = \dim(\mathcal{U}) + \dim(\mathcal{U}^0)$.
(2) *If f is nilpotent of index m, then f^* is nilpotent of index m.*
(3) *If $\mathcal{W} \subseteq \mathcal{V}^*$ is an f^*-invariant subspace, then $\mathcal{W}^0 \subseteq \mathcal{V}$ is an f-invariant subspace.*
(4) *If $\mathcal{U}, \mathcal{W}$ are a the bilinear form defined in (16.1), then $\mathcal{V} = \mathcal{U} \oplus \mathcal{W}^0$.*

Proof

(1) Exercise.
(2) For all $v \in \mathcal{V}$ we have $f^m(v) = 0$ and hence,

$$0 = h(f^m(v)) = ((f^m)^*(h))(v) = ((f^*)^m(h))(v)$$

for every $h \in \mathcal{V}^*$ and $v \in \mathcal{V}$, so that f^* is nilpotent of index at most m. If $(f^*)^{m-1} = 0$, then $(f^*)^{m-1}(h) = 0$ for all $h \in \mathcal{V}^*$, and therefore $0 = ((f^*)^{m-1}(h))(v) = h(f^{m-1}(v))$ for all $v \in \mathcal{V}$. This implies that $f^{m-1} = 0$, in contradiction to the assumption that f is nilpotent of index m. Thus, f^* is nilpotent of index m.
(3) Let $w \in \mathcal{W}^0$. For every $h \in \mathcal{W}$, we have $f^*(h) \in \mathcal{W}$, and thus $0 = f^*(h)(w) = h(f(w))$. Hence $f(w) \in \mathcal{W}^0$.

(4) If $u \in \mathcal{U} \cap \mathcal{W}^0$, then $h(u) = 0$ for all $h \in \mathcal{W}$, since $u \in \mathcal{W}^0$. Since $\mathcal{U}, \mathcal{W}$ is a dual pair with respect to the bilinear form defined in (16.1), we have $u = 0$. Moreover, $\dim(\mathcal{U}) = \dim(\mathcal{W})$ and using (1) we obtain

$$\dim(\mathcal{V}) = \dim(\mathcal{W}) + \dim(\mathcal{W}^0) = \dim(\mathcal{U}) + \dim(\mathcal{W}^0).$$

From $\mathcal{U} \cap \mathcal{W}^0 = \{0\}$ we obtain $\mathcal{V} = \mathcal{U} \oplus \mathcal{W}^0$. □

Example 16.7 We consider the vector space $\mathcal{V} = \mathbb{R}^{2,1}$ with the canonical basis $B = \{e_1, e_2\}$. For the subspaces

$$\mathcal{U} = \text{span}\left\{\begin{bmatrix} 0 \\ 1 \end{bmatrix}\right\} \subset \mathcal{V},$$

$$\mathcal{W} = \left\{h \in \mathcal{V}^* \,\big|\, [h]_{B,\{1\}} = [\alpha, \alpha] \text{ for an } \alpha \in \mathbb{R}\right\} \subset \mathcal{V}^*,$$

we have

$$\mathcal{U}^0 = \left\{h \in \mathcal{V}^* \,\big|\, [h]_{B,\{1\}} = [\alpha, 0] \text{ for an } \alpha \in \mathbb{R}\right\} \subset \mathcal{V}^*,$$

$$\mathcal{W}^0 = \text{span}\left\{\begin{bmatrix} 1 \\ -1 \end{bmatrix}\right\} \subset \mathcal{V}.$$

In this example, we easily see that $\dim(\mathcal{V}) = \dim(\mathcal{W}) + \dim(\mathcal{W}^0) = \dim(\mathcal{U}) + \dim(\mathcal{U}^0)$, and that $\mathcal{U}, \mathcal{W}$ form a dual pair with respect to the bilinear form defined in (16.1) with $K = \mathbb{R}$. Moreover, $\mathcal{V} = \mathcal{U} \oplus \mathcal{W}^0$.

The following theorem presents, for a given nilpotent f, a decomposition of $\mathcal{V}$ into f-invariant subspaces. The idea of the decomposition is to construct a dual pair of subspaces $\mathcal{U} \subseteq \mathcal{V}$ and $\mathcal{W} \subseteq \mathcal{V}^*$, where $\mathcal{U}$ is f-invariant and $\mathcal{W}$ is f^*-invariant. By (3) in Lemma 16.6 then $\mathcal{W}^0$ is f-invariant and with (4) in Lemma 16.6 it follows that $\mathcal{V} = \mathcal{U} \oplus \mathcal{W}^0$.

Theorem 16.8 *Let $\mathcal{V}$ be a finite dimensional K-vector space and let $f \in \mathcal{L}(\mathcal{V}, \mathcal{V})$ be nilpotent of index m. Let $v_0 \in \mathcal{V}$ satisfy $f^{m-1}(v_0) \neq 0$ and let $h_0 \in \mathcal{V}^*$ satisfy $h_0(f^{m-1}(v_0)) \neq 0$.*
Then $m(f, v_0) = m(f^*, h_0) = m$, and the f- and f^*-invariant subspaces $\mathcal{K}_m(f, v_0) \subseteq \mathcal{V}$ and $\mathcal{K}_m(f^*, h_0) \subseteq \mathcal{V}^*$, respectively, are a dual pair with respect to the bilinear form defined in (16.1). Furthermore,

$$\mathcal{V} = \mathcal{K}_m(f, v_0) \oplus (\mathcal{K}_m(f^*, h_0))^0,$$

where $(\mathcal{K}_m(f^*, h_0))^0$ is an f-invariant subspace of $\mathcal{V}$.

16.1 Cyclic f-invariant Subspaces and Duality

Proof Let $v_0 \in \mathcal{V}$ be a vector with $f^{m-1}(v_0) \neq 0$. Since $f^m(v_0) = 0$, the space $\mathcal{K}_m(f, v_0)$ is an m-dimensional f-invariant subspace of $\mathcal{V}$ (cp. (3) in Lemma 16.1). Let $h_0 \in \mathcal{V}^*$ be a vector with

$$0 \neq h_0(f^{m-1}(v_0)) = ((f^*)^{m-1}(h_0))(v_0).$$

Then, in particular, $0 \neq (f^*)^{m-1}(h_0) \in \mathcal{L}(\mathcal{V}^*, \mathcal{V}^*)$. Since f is nilpotent of index m, also f^* is nilpotent of index m (cp. (2) in Lemma 16.6), so that

$$(f^*)^m(h_0) = 0 \in \mathcal{L}(\mathcal{V}^*, \mathcal{V}^*).$$

Therefore, $\mathcal{K}_m(f^*, h_0)$ is an m-dimensional f^*-invariant subspace of $\mathcal{V}^*$ (cp. (3) in Lemma 16.1).

It remains to show that $\mathcal{K}_m(f, v_0), \mathcal{K}_m(f^*, h_0)$ are a dual pair. Let

$$v_1 = \sum_{j=0}^{m-1} \gamma_j f^j(v_0) \in \mathcal{K}_m(f, v_0)$$

be a vector with $h(v_1) = \beta(v_1, h) = 0$ for all $h \in \mathcal{K}_m(f^*, h_0)$. We show inductively that then $\gamma_0 = \cdots = \gamma_{m-1} = 0$, and thus $v_1 = 0$.

Using $(f^*)^{m-1}(h_0) \in \mathcal{K}_m(f^*, h_0)$, our assumption on the vector v_1 yields

$$0 = ((f^*)^{m-1}(h_0))(v_1) = h_0(f^{m-1}(v_1)) = \sum_{j=0}^{m-1} \gamma_j h_0(f^{m-1+j}(v_0))$$

$$= \gamma_0 h_0(f^{m-1}(v_0)).$$

The last equation holds, since $f^{m-1+j}(v_0) = 0$ for $j = 1, \ldots, m-1$ (because $f^m = 0$). From $h_0(f^{m-1}(v_0)) \neq 0$ we obtain $\gamma_0 = 0$.

Suppose now that $\gamma_0 = \cdots = \gamma_{k-1} = 0$ for a k, $1 \leq k \leq m-1$. Using $(f^*)^{m-1-k}(h_0) \in \mathcal{K}_m(f^*, h_0)$, our assumption on the vector v_1 yields

$$0 = ((f^*)^{m-1-k}(h_0))(v_1) = h_0(f^{m-1-k}(v_1)) = \sum_{j=0}^{m-1} \gamma_j h_0(f^{m-1+j-k}(v_0))$$

$$= \gamma_k h_0(f^{m-1}(v_0)).$$

The last equation holds, since $\gamma_j = 0$ for $j = 0, \ldots, k-1$ and $f^{m-1+j-k}(v_0) = 0$ for $j = k+1, \ldots, m-1$.

We have $v_1 = 0$ as asserted, and therefore the bilinear form defined in (16.1) for the spaces $\mathcal{K}_m(f, v_0)$ and $\mathcal{K}_m(f^*, h_0)$ is non-degenerate in the first variable. Analogously, the bilinear form is non-degenerate in the second variable, and hence $\mathcal{K}_m(f, v_0), \mathcal{K}_m(f^*, h_0)$ are a dual pair.

Using (4) in Lemma 16.6 we now have $V = \mathcal{K}_m(f, v_0) \oplus (\mathcal{K}_m(f^*, h_0))^0$, where the space $(\mathcal{K}_m(f^*, h_0))^0$ is, by (3) in Lemma 16.6, an f-invariant subspace of V. $\square$

16.2 The Jordan Canonical Form

Let V be a finite dimensional K-vector space and $f \in \mathcal{L}(V, V)$. If there exists a basis B of V consisting of eigenvectors of f, then $[f]_{B,B}$ is a diagonal matrix, i.e., f is diagonalizable. A necessary and sufficient condition for this is that the characteristic polynomial P_f decomposes into linear factors over K and that in addition $g(f, \lambda_j) = a(f, \lambda_j)$ for every eigenvalue λ_j (cp. Theorem 14.14).

If P_f decomposes into linear factors but $g(f, \lambda_j) < a(f, \lambda_j)$ holds for at least one eigenvalue λ_j, then f is not diagonalizable but can still be triangulated, i.e., there exists a basis B of V such that $[f]_{B,B}$ is an upper triangular matrix (cp. Theorem 14.18). From this triangular matrix we can read off the algebraic, but usually not the geometric multiplicities of the eigenvalues. The goal of the following construction is to determine a basis B of V so that $[f]_{B,B}$ is upper triangular and in addition to the algebraic also reveals the geometric multiplicities of the eigenvalues.

Under the assumption that P_f decomposes into linear factors over K, we will construct a basis B of V for which $[f]_{B,B}$ is a *block diagonal matrix* of the form

$$[f]_{B,B} = \begin{bmatrix} J_{d_1}(\lambda_1) & & \\ & \ddots & \\ & & J_{d_m}(\lambda_m) \end{bmatrix}, \quad (16.2)$$

where each diagonal block has the form

$$J_{d_j}(\lambda_j) := \begin{bmatrix} \lambda_j & 1 & & \\ & \ddots & \ddots & \\ & & \ddots & 1 \\ & & & \lambda_j \end{bmatrix} \in K^{d_j, d_j} \quad (16.3)$$

for some $\lambda_j \in K$ and $d_j \in \mathbb{N}$, $j = 1, \ldots, m$. A matrix of the form (16.3) is called a *Jordan block* of size d_j corresponding to the eigenvalue λ_j.

In the following construction we first do *not* assume that P_f decomposes into linear factors. We only assume the existence of a single eigenvalue $\lambda_1 \in K$ of f. Using this eigenvalue, we define the endomorphism

$$g := f - \lambda_1 \operatorname{Id}_V \in \mathcal{L}(V, V).$$

16.2 The Jordan Canonical Form

By Theorem 16.5 there exist g-invariant subspaces $\mathcal{U} \subseteq \mathcal{V}$ and $\mathcal{W} \subseteq \mathcal{V}$ with

$$\mathcal{V} = \mathcal{U} \oplus \mathcal{W},$$

such that

$$g_1 := g|_{\mathcal{U}}$$

is nilpotent and $g|_{\mathcal{W}}$ is bijective. Then $\mathcal{U} \neq \{0\}$, since otherwise $\mathcal{W} = \mathcal{V}$ and $g|_{\mathcal{W}} = g|_{\mathcal{V}} = g$ would be bijective, which contradicts the assumption that λ_1 is an eigenvalue of f.

Let g_1 be nilpotent of index d_1. Then by construction $1 \leq d_1 \leq \dim(\mathcal{U})$. Let $w_1 \in \mathcal{U}$ be a vector with $g_1^{d_1-1}(w_1) \neq 0$. Since $g_1^{d_1}(w_1) = 0$, the vector $g_1^{d_1-1}(w_1)$ is an eigenvector of g_1 corresponding to the eigenvalue 0. By (3) in Lemma 16.1, the d_1 vectors

$$w_1, g_1(w_1), \ldots, g_1^{d_1-1}(w_1)$$

are linearly independent and $\mathcal{U}_1 := \mathcal{K}_{d_1}(g_1, w_1)$ is a d_1-dimensional g_1-invariant subspace of $\mathcal{U}$.

Consider the basis

$$B_1 := \left\{ g_1^{d_1-1}(w_1), \ldots, g_1(w_1), w_1 \right\}$$

of $\mathcal{U}_1$. Then the matrix representation $g_1|_{\mathcal{U}_1}$ with respect to the basis B_1 is given by

$$[g_1|_{\mathcal{U}_1}]_{B_1, B_1} = J_{d_1}(0) \in K^{d_1, d_1}.$$

This shows, in particular, that the characteristic polynomial of $g_1|_{\mathcal{U}_1}$ is given by the monomial t^{d_1}, and hence 0 is the only eigenvalue of $g_1|_{\mathcal{U}_1}$. Moreover, by construction $[g_1|_{\mathcal{U}_1}]_{B_1, B_1} = [g|_{\mathcal{U}_1}]_{B_1, B_1}$.

If $d_1 = \dim(\mathcal{U})$, then our construction is complete for the moment. If, on the other hand, $d_1 < \dim(\mathcal{U})$, then applying Theorem 16.8 to $g_1 \in \mathcal{L}(\mathcal{U}, \mathcal{U})$ shows that there exists a g_1-invariant subspace $\widetilde{\mathcal{U}} \neq \{0\}$ with $\mathcal{U} = \mathcal{U}_1 \oplus \widetilde{\mathcal{U}}$, and we consider

$$g_2 := g_1|_{\widetilde{\mathcal{U}}}.$$

This map is nilpotent of index d_2, where $1 \leq d_2 \leq d_1$. We now carry out the same construction as before:

We determine a vector $w_2 \in \widetilde{\mathcal{U}}$ with $g_2^{d_2-1}(w_2) \neq 0$. Then $g_2^{d_2-1}(w_2)$ is an eigenvector of g_2, $\mathcal{U}_2 := \mathcal{K}_{d_2}(g_2, w_2)$ is a d_2-dimensional g_2-invariant subspace of $\widetilde{\mathcal{U}} \subset \mathcal{U}$ and for the basis

$$B_2 := \left\{ g_2^{d_2-1}(w_2), \ldots, g_2(w_2), w_2 \right\}$$

of $\mathcal{U}_2$ we have

$$[g_2|_{\mathcal{U}_2}]_{B_2,B_2} = J_{d_2}(0) \in K^{d_2,d_2},$$

where again $[g_2|_{\mathcal{U}_2}]_{B_2,B_2} = [g|_{\mathcal{U}_2}]_{B_2,B_2}$ by construction.

After $k \leq \dim(\mathcal{U})$ steps this procedure terminates. We then have found a decomposition of $\mathcal{U}$ of the form

$$\mathcal{U} = \mathcal{K}_{d_1}(g_1, w_1) \oplus \ldots \oplus \mathcal{K}_{d_k}(g_k, w_k) = \mathcal{K}_{d_1}(g, w_1) \oplus \ldots \oplus \mathcal{K}_{d_k}(g, w_k).$$

In the second equation we have used that $\mathcal{K}_{d_j}(g_j, w_j) = \mathcal{K}_{d_j}(g, w_j)$ for $j = 1, \ldots, k$. If we combine the constructed bases $B_1, \ldots, B_k$ to a basis B of $\mathcal{U}$, then

$$[g|_{\mathcal{U}}]_{B,B} = \begin{bmatrix} [g|_{\mathcal{U}_1}]_{B_1,B_1} & & \\ & \ddots & \\ & & [g|_{\mathcal{U}_k}]_{B_k,B_k} \end{bmatrix} = \begin{bmatrix} J_{d_1}(0) & & \\ & \ddots & \\ & & J_{d_k}(0) \end{bmatrix}.$$

Thus, the nilpotent endomorphism $g_1 = g|_{\mathcal{U}}$ has the characteristic polynomial $t^{d_1+\ldots+d_k}$, and its only eigenvalue is 0.

We now transfer these results to

$$f = g + \lambda_1 \operatorname{Id}_V.$$

Every g-invariant subspace is f-invariant and one observes easily that

$$\mathcal{K}_{d_j}(f, w_j) = \mathcal{K}_{d_j}(g, w_j), \quad j = 1, \ldots, k$$

(cp. Exercise 16.4). Hence, it follows that

$$\mathcal{U} = \mathcal{K}_{d_1}(f, w_1) \oplus \ldots \oplus \mathcal{K}_{d_k}(f, w_k). \tag{16.4}$$

For every $j = 1, \ldots, k$ and $0 \leq \ell \leq d_j - 1$ we have

$$f\left(g^\ell(w_j)\right) = g\left(g^\ell(w_j)\right) + \lambda_1 g^\ell(w_j) = \lambda_1 g^\ell(w_j) + g^{\ell+1}(w_j), \tag{16.5}$$

where $g^{d_j}(w_j) = 0$. The matrix representation of $f|_{\mathcal{U}}$ with respect to the basis B of $\mathcal{U}$ is therefore given by

$$[f|_{\mathcal{U}}]_{B,B} = \begin{bmatrix} [f|_{\mathcal{U}_1}]_{B_1,B_1} & & \\ & \ddots & \\ & & [f|_{\mathcal{U}_k}]_{B_k,B_k} \end{bmatrix} = \begin{bmatrix} J_{d_1}(\lambda_1) & & \\ & \ddots & \\ & & J_{d_k}(\lambda_1) \end{bmatrix}. \tag{16.6}$$

16.2 The Jordan Canonical Form

The map $g|_\mathcal{W} = f|_\mathcal{W} - \lambda_1 \operatorname{Id}_\mathcal{W}$ is bijective by construction, i.e., λ_1 is not an eigenvalue of $f|_\mathcal{W}$. Therefore, $a(f, \lambda_1) = \dim(\mathcal{U}) = d_1 + \ldots + d_k$. In order to determine $g(f, \lambda_1)$, let $v \in \mathcal{U}$ be an arbitrary vector. Then there exist scalars $\alpha_{j,\ell} \in K$ with

$$v = \sum_{j=1}^{k} \sum_{\ell=0}^{d_j-1} \alpha_{j,\ell} g^\ell(w_j).$$

Using (16.5) we obtain

$$f(v) = \sum_{j=1}^{k} \sum_{\ell=0}^{d_j-1} \alpha_{j,\ell} f\left(g^\ell(w_j)\right) = \sum_{j=1}^{k} \sum_{\ell=0}^{d_j-1} \alpha_{j,\ell} \lambda_1 g^\ell(w_j) + \sum_{j=1}^{k} \sum_{\ell=0}^{d_j-1} \alpha_{j,\ell} g^{\ell+1}(w_j)$$

$$= \lambda_1 v + \sum_{j=1}^{k} \sum_{\ell=0}^{d_j-2} \alpha_{j,\ell} g^{\ell+1}(w_j).$$

The vectors in the last sum are linearly independent. Hence, $f(v) = \lambda_1 v$ if and only if $\alpha_{j,\ell} = 0$ for $j = 1, \ldots, k$ and $\ell = 0, 1, \ldots, d_j - 2$. This shows that every eigenvector of f corresponding to the eigenvalue λ_1 has the form

$$v = \sum_{j=1}^{k} \alpha_j g^{d_j-1}(w_j),$$

where at least one α_j is nonzero so that we have

$$\mathcal{V}_f(\lambda_1) = \operatorname{span}\{g^{d_1-1}(w_1), \ldots, g^{d_k-1}(w_k)\}.$$

Since $g^{d_1-1}(w_1), \ldots, g^{d_k-1}(w_k)$ are linearly independent, it follows that $g(f, \lambda_1) = k$. The geometric multiplicity of the eigenvalue λ_1 therefore is equal to the number of Jordan blocks corresponding to the eigenvalue λ_1 in the matrix representation (16.6). Furthermore, we observe that in every subspace $\mathcal{K}_{d_j}(f, w_j)$, the endomorphism f has exactly one (linearly independent) eigenvector corresponding to the eigenvalue λ_1.

We summarize these results in the following theorem.

Theorem 16.9 *Let $\mathcal{V}$ be a finite dimensional K-vector space and let $f \in \mathcal{L}(\mathcal{V}, \mathcal{V})$. If $\lambda_1 \in K$ is an eigenvalue of f, then the following assertions hold:*

(1) There exist f-invariant subspaces $\{0\} \neq \mathcal{U} \subseteq \mathcal{V}$ and $\mathcal{W} \subset \mathcal{V}$ with $\mathcal{V} = \mathcal{U} \oplus \mathcal{W}$. The map $f|_\mathcal{U} - \lambda_1 \operatorname{Id}_\mathcal{U}$ is nilpotent and the map $f|_\mathcal{W} - \lambda_1 \operatorname{Id}_\mathcal{W}$ is bijective. In particular, λ_1 is not an eigenvalue of $f|_\mathcal{W}$.

(2) The subspace $\mathcal{U}$ from (1) can be written as

$$\mathcal{U} = \mathcal{K}_{d_1}(f, w_1) \oplus \ldots \oplus \mathcal{K}_{d_k}(f, w_k)$$

for some vectors $w_1, \ldots, w_k \in \mathcal{U}$, where $\mathcal{K}_{d_j}(f, w_j)$ is a d_j-dimensional f-invariant subspace of $\mathcal{V}$, $j = 1, \ldots, k$. This is called a cyclic decomposition of $\mathcal{U}$.

(3) There exists a basis B of $\mathcal{U}$ with

$$[f|_{\mathcal{U}}]_{B,B} = \begin{bmatrix} J_{d_1}(\lambda_1) & & \\ & \ddots & \\ & & J_{d_k}(\lambda_1) \end{bmatrix}.$$

(4) We have $a(\lambda_1, f) = d_1 + \ldots + d_k$ and $g(\lambda_1, f) = k$.

If f has a further eigenvalue $\lambda_2 \neq \lambda_1$, then it is an eigenvalue of the restriction $f|_{\mathcal{W}} \in \mathcal{L}(\mathcal{W}, \mathcal{W})$ and we can apply Theorem 16.9 to $f|_{\mathcal{W}}$. The vector space $\mathcal{W}$ then is the direct sum of the form $\mathcal{W} = \mathcal{X} \oplus \mathcal{Y}$, where $f|_{\mathcal{X}} - \lambda_2 \operatorname{Id}_{\mathcal{X}}$ is nilpotent and $f|_{\mathcal{Y}} - \lambda_2 \operatorname{Id}_{\mathcal{Y}}$ is bijective. The space $\mathcal{X}$ has a cyclic decomposition analogously to (2) in Theorem 16.9, and there exists a matrix representation of $f|_{\mathcal{X}}$ analogously to (3).

This construction can be carried out for all eigenvalues of f. If the characteristic polynomial P_f decomposes into linear factors over K, then we finally obtain a cyclic decomposition of the entire space $\mathcal{V}$, which gives the following theorem.

Theorem 16.10 *Let $\mathcal{V}$ be a finite dimensional K-vector space and let $f \in \mathcal{L}(\mathcal{V}, \mathcal{V})$. If the characteristic polynomial P_f decomposes into linear factors over K, then there exists a basis B of $\mathcal{V}$, such that*

$$[f]_{B,B} = \begin{bmatrix} J_{d_1}(\lambda_1) & & \\ & \ddots & \\ & & J_{d_m}(\lambda_m) \end{bmatrix}, \tag{16.7}$$

where $\lambda_1, \ldots, \lambda_m \in K$ are the (not necessarily pairwise distinct) eigenvalues of f. For every eigenvalue λ_j of f then $a(\lambda_j, f)$ is equal to the sum of the sizes of all Jordan blocks corresponding to λ_j in (16.7), and $g(\lambda_j, f)$ is equal to the number of Jordan blocks corresponding to λ_j in (16.7). A matrix representation of the form (16.7) is called a Jordan canonical form[4] of f.

---

[4] Camille Jordan (1838–1922) derived this form 1870. Two years earlier, Karl Weierstraß (1815–1897) proved a result that implies the Jordan canonical form.

16.2 The Jordan Canonical Form

The assumption in Theorem 16.10, that P_f decomposes into linear factors over K, is not only sufficient for the existence of a matrix representation as in (16.7), but also necessary. Indeed, if f has a matrix representation of the form (16.7), then we immediately see that $P_f = \prod_{j=1}^{m}(t - \lambda_j)^{d_j}$.

From Theorem 14.14 we know that $f \in \mathcal{L}(\mathcal{V}, \mathcal{V})$ is diagonalizable if and only if P_f decomposes into linear factors over K and $g(\lambda_j, f) = a(\lambda_j, f)$ holds for every eigenvalue λ_j of f. If P_f decomposes into linear factors, then the Jordan canonical form (16.7) shows that $g(\lambda_j, f) = a(\lambda_j, f)$ if and only if every Jordan block corresponding to λ_j is of size 1.

The Fundamental Theorem of Algebra yields the following corollary of Theorem 16.10.

Corollary 16.11 *If $\mathcal{V}$ is a finite dimensional $\mathbb{C}$-vector space, then every $f \in \mathcal{L}(\mathcal{V}, \mathcal{V})$ has a Jordan canonical form.*

The following uniqueness result justifies the name *canonical form*.

Theorem 16.12 *Let $\mathcal{V}$ be a finite dimensional K-vector space. If $f \in \mathcal{L}(\mathcal{V}, \mathcal{V})$ has a Jordan canonical form, then it is unique up to the order of the Jordan blocks on the diagonal.*

Proof Let $\dim(\mathcal{V}) = n$ and let B_1, B_2 be two bases of $\mathcal{V}$ with

$$A_1 = [f]_{B_1, B_1} = \begin{bmatrix} J_{d_1}(\lambda_1) & & \\ & \ddots & \\ & & J_{d_m}(\lambda_m) \end{bmatrix} \in K^{n,n},$$

as well as

$$A_2 = [f]_{B_2, B_2} = \begin{bmatrix} J_{c_1}(\mu_1) & & \\ & \ddots & \\ & & J_{c_k}(\mu_k) \end{bmatrix} \in K^{n,n}.$$

For a given eigenvalue λ_j, $1 \leq j \leq m$, we define

$$r_s^{(1)}(\lambda_j) := \text{rank}\left((A_1 - \lambda_j I_n)^s\right), \quad s = 0, 1, 2, \ldots .$$

Then

$$d_s^{(1)}(\lambda_j) := r_{s-1}^{(1)}(\lambda_j) - r_s^{(1)}(\lambda_j), \quad s = 1, 2, \ldots,$$

is equal to the number of Jordan blocks $J_\ell(\lambda_j) \in K^{\ell,\ell}$ on the diagonal of A_1 with $\ell \geq s$. The number of Jordan blocks corresponding to the eigenvalue λ_j with *exact* size s therefore is given by

$$d_s^{(1)}(\lambda_j) - d_{s+1}^{(1)}(\lambda_j) = r_{s-1}^{(1)}(\lambda_j) - 2r_s^{(1)}(\lambda_j) + r_{s+1}^{(1)}(\lambda_j) \qquad (16.8)$$

(cp. Example 16.13).

The matrices A_1 and A_2 are similar and, therefore, have the same eigenvalues, i.e.,

$$\{\lambda_1, \ldots, \lambda_m\} = \{\mu_1, \ldots, \mu_k\}.$$

Furthermore,

$$\text{rank}\left((A_1 - \alpha I_n)^m\right) = \text{rank}\left((A_2 - \alpha I_n)^m\right)$$

for all $\alpha \in K$ and $m \in \mathbb{N}_0$.

In particular, for every λ_j there exists $\mu_i \in \{\mu_1, \ldots, \mu_k\}$ with $\mu_i = \lambda_j$ and for this μ_i and the matrix A_2 we get

$$r_s^{(2)}(\mu_i) := \text{rank}\left((A_2 - \mu_i I_n)^s\right) = r_s^{(1)}(\lambda_j), \quad s = 0, 1, 2, \ldots.$$

Now (16.8) shows that the matrix A_2 has, up to reordering, the same Jordan blocks on the diagonal as the matrix A_1. $\square$

Example 16.13 This example illustrates the construction in the proof of Theorem 16.12. If

$$A = \begin{bmatrix} J_2(1) & & \\ & J_1(1) & \\ & & J_2(0) \end{bmatrix} = \begin{bmatrix} 1 & 1 & & & \\ & 1 & & & \\ & & 1 & & \\ & & & 0 & 1 \\ & & & & 0 \end{bmatrix} \in \mathbb{R}^{5,5}, \qquad (16.9)$$

then $(A - 1 \cdot I_5)^0 = I_5$,

$$A - 1 \cdot I_5 = \begin{bmatrix} 0 & 1 & & & \\ & 0 & & & \\ & & 0 & & \\ & & & -1 & 1 \\ & & & & -1 \end{bmatrix}, \quad (A - 1 \cdot I_5)^2 = \begin{bmatrix} 0 & 0 & & & \\ & 0 & & & \\ & & 0 & & \\ & & & 1 & -2 \\ & & & & 1 \end{bmatrix},$$

and we get

$$r_0(1) = 5, \quad r_1(1) = 3, \quad r_s(1) = 2, \quad s \geq 2,$$

$$d_1(1) = 2, \quad d_2(1) = 1, \quad d_s(1) = 0, \quad s \geq 3,$$

$$d_1(1) - d_2(1) = 1, \quad d_2(1) - d_3(1) = 1, \quad d_s(1) - d_{s+1}(1) = 0, \quad s \geq 3.$$

16.3 The Minimal Polynomial and the Frobenius Canonical Form

Let $\mathcal{V}$ be a finite dimensional K-vector space and let $f \in \mathcal{L}(\mathcal{V}, \mathcal{V})$, where we do not assume that P_f decomposes into linear factors. From the Cayley-Hamilton theorem (Theorem 8.6) we know that $P_f(f) = 0 \in \mathcal{L}(\mathcal{V}, \mathcal{V})$, i.e., there exists a monic polynomial of degree at most $\dim(\mathcal{V})$ which annihilates the endomorphism f. Let $p_1, p_2 \in K[t]$ be two monic polynomials of smallest possible degree with $p_1(f) = p_2(f) = 0$. Then $(p_1 - p_2)(f) = 0$, and since p_1 and p_2 are monic, $p_1 - p_2 \in K[t]$ is a polynomial with $\deg(p_1 - p_2) < \deg(p_1) = \deg(p_2)$. The minimality assumption on $\deg(p_1)$ and $\deg(p_2)$ implies that $p_1 - p_2 = 0$, i.e., $p_1 = p_2$. Thus, for every $f \in \mathcal{L}(\mathcal{V}, \mathcal{V})$ there exists a uniquely determined monic polynomial of minimal degree which annihilates f. This justifies the following definition.

Definition 16.14 If $\mathcal{V}$ is finite dimensional K-vector space and $f \in \mathcal{L}(\mathcal{V}, \mathcal{V})$, then the uniquely determined monic polynomial of minimal degree that annihilates f is called the *minimal polynomial of f*. We denote this polynomial by M_f.

By construction we always have $\deg(M_f) \leq \deg(P_f) = \dim(\mathcal{V})$.

Lemma 16.15 *If $\mathcal{V}$ is a finite dimensional K-vector space and $f \in \mathcal{L}(\mathcal{V}, \mathcal{V})$, then the minimal polynomial M_f divides every polynomial that annihilates f and is, in particular, a divisor of the characteristic polynomial P_f.*

Proof For $p = 0$ we have $p(f) = 0$ and M_f divides p. If $p \in K[t] \setminus \{0\}$ is a polynomial with $p(f) = 0$, then $\deg(M_f) \leq \deg(p)$. Using division with remainder (cp. Theorem 15.4), there exist uniquely determined polynomials $q, r \in K[t]$ with $p = q \cdot M_f + r$ and $\deg(r) < \deg(M_f)$. Thus,

$$0 = p(f) = q(f)M_f(f) + r(f) = r(f).$$

The minimality of $\deg(M_f)$ implies that $r = 0$, and hence M_f divides p. □

Definition 16.14 can be formulated also for matrices $A \in K^{n,n}$, and then the analogous version of Lemma 16.15 shows that $M_A | P_A$. A matrix A is called *derogatory*, if M_A is a proper divisor of P_A, i.e., if $\deg(M_A) < \deg(P_A)$, and *non-derogatory*, if $M_A = P_A$. The same terms are used for an endomorphism f on a finite dimensional K-vector space, i.e., f is called *derogatory* if $\deg(M_f) < \deg(P_f)$, and *non-derogatory* if $M_f = P_f$.

We now want to determine the minimal polynomial of an endomorphism f on a finite dimensional real or complex vector space using the Jordan canonical form of f.

We start our investigation by considering the powers of a Jordan block $J_d(\lambda) \in K^{d,d}$, where $K = \mathbb{R}$ or $K = \mathbb{C}$. Since I_d and $J_d(0)$ commute,

$$J_d(\lambda)^k = (\lambda I_d + J_d(0))^k = \sum_{j=0}^{k} \binom{k}{j} \lambda^{k-j} J_d(0)^j = \sum_{j=0}^{k} \frac{p^{(j)}(\lambda)}{j!} J_d(0)^j,$$

for every $k \in \mathbb{N}_0$, where $p^{(j)}$ is the jth derivative of the polynomial $p = t^k$ with respect to t. Using this result for the monomials $p = t^k$ we can obtain the following result.

Lemma 16.16 *If $J_d(\lambda) \in K^{d,d}$ and $p \in K[t]$ has $\deg(p) = k \geq 0$, where $K = \mathbb{R}$ or $K = \mathbb{C}$, then*

$$p(J_d(\lambda)) = \sum_{j=0}^{k} \frac{p^{(j)}(\lambda)}{j!} J_d(0)^j. \tag{16.10}$$

Proof Exercise. □

Considered as a linear map from $K^{d,1}$ to $K^{d,1}$, the matrix $J_d(0)$ represents an "upshift", since

$$J_d(0) \begin{bmatrix} \alpha_1 \\ \alpha_2 \\ \vdots \\ \alpha_d \end{bmatrix} = \begin{bmatrix} \alpha_2 \\ \vdots \\ \alpha_d \\ 0 \end{bmatrix} \quad \text{for all} \quad \begin{bmatrix} \alpha_1 \\ \alpha_2 \\ \vdots \\ \alpha_d \end{bmatrix} \in K^{d,1}.$$

Clearly,

$$J_d(0)^\ell \neq 0, \quad \ell = 0, 1, \ldots, d-1, \quad J_d(0)^d = 0,$$

and hence the linear map $J_d(0)$ is nilpotent of index d. The sum on the right hand side of (16.10) therefore has at most d nonzero terms, even when $\deg(p) > d$.

Moreover, the right hand side of (16.10) shows that $p(J_d(\lambda))$ is an upper triangular matrix with constant entries on its diagonals. A matrix with constant

16.3 The Minimal Polynomial and the Frobenius Canonical Form

diagonals is called a *Toeplitz matrix*.[5] In particular, on the main diagonal we have the entry $p(\lambda)$. From (16.10) we see that $p(J_d(\lambda)) = 0$ holds if and only if

$$p(\lambda) = p'(\lambda) = \cdots = p^{(d-1)}(\lambda) = 0.$$

Thus we have shown the following result.

Lemma 16.17 *If $J_d(\lambda) \in K^{d,d}$ and $p \in K[t]$, where $K = \mathbb{R}$ or $K = \mathbb{C}$, then the following assertions hold:*

(1) The matrix $p(J_d(\lambda))$ is invertible if and only if λ is not a root of p.
(2) We have $p(J_d(\lambda)) = 0 \in K^{d,d}$ if and only if λ is a d-fold root of p, i.e., if the polynomial $(t-\lambda)^d$ is a divisor of p.

If P_f decomposes into linear factors, then we can explicitly construct M_f using the Jordan canonical form of f.

Lemma 16.18 *Let $\mathcal{V}$ be a finite dimensional K-vector space with $K = \mathbb{R}$ or $K = \mathbb{C}$. If $f \in \mathcal{L}(\mathcal{V}, \mathcal{V})$ has a Jordan canonical form with pairwise distinct eigenvalues $\widetilde{\lambda}_1, \ldots, \widetilde{\lambda}_k$ and if $\widetilde{d}_1, \ldots, \widetilde{d}_k$ are the respective maximal sizes of the corresponding Jordan blocks, then*

$$M_f = \prod_{j=1}^{k} (t - \widetilde{\lambda}_j)^{\widetilde{d}_j}.$$

Thus, f is derogatory if and only if it has an eigenvalue with more than one corresponding Jordan block.

Proof We know from Lemma 16.15 that M_f is a divisor of P_f. Therefore,

$$M_f = \prod_{j=1}^{k} (t - \widetilde{\lambda}_j)^{\ell_j}$$

for some exponents $\ell_1, \ldots, \ell_k$. If

$$A = \begin{bmatrix} J_{d_1}(\lambda_1) & & \\ & \ddots & \\ & & J_{d_m}(\lambda_m) \end{bmatrix}$$

---
[5] Otto Toeplitz (1881–1940).

is a Jordan canonical form of f, then $M_f(f) = 0 \in \mathcal{L}(\mathcal{V}, \mathcal{V})$ is equivalent to $M_f(A) = 0 \in K^{n,n}$, where $n = \dim(\mathcal{V})$. We have $M_f(A) = 0$ if and only if $M_f(J_{d_j}(\lambda_j)) = 0$ for $j = 1, \ldots, m$. For this it is necessary and sufficient that $M_f(J_{\widetilde{d}_j}(\widetilde{\lambda}_j)) = 0$ for $j = 1, \ldots, k$. By Lemma 16.17 this holds if and only if every of the polynomials $(t - \widetilde{\lambda}_j)^{\widetilde{d}_j}$, $j = 1, \ldots, k$, is a divisor of M_f. Therefore, M_f has the desired form.

Finally, $\deg(M_f) < \deg(P_f)$ holds if and only if $\widetilde{d}_1 + \ldots + \widetilde{d}_k < n$, i.e., f has an eigenvalue with more than one corresponding Jordan blocks. □

Example 16.19 If f is an endomorphism with the Jordan canonical form A in (16.9), then f is derogatory and we have

$$P_f = (t-1)^3 t^2, \quad M_f = (t-1)^2 t^2$$

and

$$M_f(A) = (A - 1 \cdot I_5)^2 A^2 = \begin{bmatrix} 0 & 0 & & & \\ & 0 & & & \\ \hline & & 0 & & \\ \hline & & & 1 & -2 \\ & & & & 1 \end{bmatrix} \begin{bmatrix} 1 & 2 & & & \\ & 1 & & & \\ \hline & & 1 & & \\ \hline & & & 0 & 0 \\ & & & & 0 \end{bmatrix},$$

which shows that $M_f(A) = 0 \in \mathbb{R}^{5,5}$ and $M_f(f) = 0 \in \mathcal{L}(\mathcal{V}, \mathcal{V})$.

Using duality theory, as in the proof of the Jordan canonical form, we now derive the Frobenius canonical form for endomorphisms on a finite dimensional K-vector space. As the Jordan canonical form, this is a block-diagonal matrix, where the blocks are not upper triangular but companion matrices of polynomials. Moreover, in contrast to the derivation of the Jordan canonical form, we do not need to assume the existence of eigenvalues of f.

Let $\mathcal{V}$ be a finite dimensional K-vector space, $f \in \mathcal{L}(\mathcal{V}, \mathcal{V})$ and let $v_0 \in \mathcal{V} \setminus \{0\}$ with grade $m = m(f, v_0)$ be given. Then the vectors $v_0, f(v_0), \ldots, f^{m-1}(v_0)$ are linearly independent, and

$$f^m(v_0) = -\alpha_0 v_0 - \alpha_1 f(v_0) - \ldots - \alpha_{m-1} f^{m-1}(v_0)$$

for uniquely determined $\alpha_0, \alpha_1, \ldots, \alpha_{m-1} \in K$. We can write this equation as

$$M_{f,v_0}(f)(v_0) = 0 \quad \text{with} \quad M_{f,v_0} := t^m + \alpha_{m-1} t^{m-1} + \ldots + \alpha_1 t + \alpha_0 \in K[t].$$

Form the uniqueness of the coefficients $\alpha_0, \alpha_1, \ldots, \alpha_{m-1}$ it follows immediately that the polynomial M_{f,v_0} is the uniquely determined monic polynomial of minimal degree (among all polynomials $p \in K[t]$) with the property $p(f)(v_0) = 0$. We call

16.3 The Minimal Polynomial and the Frobenius Canonical Form

M_{f,v_0} the *minimal polynomial of v_0 with respect to f*. Analogously to Lemma 16.15 we obtain the following result.

Lemma 16.20 *Let V be a finite dimensional K-vector space, $f \in \mathcal{L}(V, V)$, $v_0 \in V\setminus\{0\}$ and let M_{f,v_0} be the minimal polynomial of v_0 with respect to f. If $p \in K[t]$ is a polynomial with $p(f)(v_0) = 0$, then $M_{f,v_0} | p$ and, hence, in particular $M_{f,v_0} | M_f$.*

Proof Exercise. □

The following lemma will be important in our derivation of the Frobenius canonical form.

Lemma 16.21 *If V is a finite dimensional K-vector space and $f \in \mathcal{L}(V, V)$, then there exists a vector $v_0 \in V$ with $M_{f,v_0} = M_f$. We call such a vector v_0 maximal with respect to f.*

Proof Let two arbitrary vectors $v, w \in V \setminus \{0\}$ be given. We show first that there exists a vector $u \in V$ with $M_{f,u} = \text{lcm}(M_{f,v}, M_{f,w})$. As in (15.3) we can factorize the monic minimal polynomials $M_{f,v}$ and $M_{f,w}$ as

$$M_{f,v} = (q_1^{c_1} \cdot \ldots \cdot q_\ell^{c_\ell}) \cdot (q_{\ell+1}^{c_{\ell+1}} \cdot \ldots \cdot q_m^{c_m}) =: g_1 \cdot g_2,$$

$$M_{f,w} = (q_1^{d_1} \cdot \ldots \cdot q_\ell^{d_\ell}) \cdot (q_{\ell+1}^{d_{\ell+1}} \cdot \ldots \cdot q_m^{d_m}) =: h_1 \cdot h_2.$$

Here $q_1, \ldots, q_m \in K[t]$ are pairwise distinct monic irreducible polynomials and $c_j, d_j \in \mathbb{N}_0$, $j = 1, \ldots, m$ with

$$c_j \geq d_j, \quad j = 1, \ldots, \ell, \quad \text{and} \quad c_j < d_j, \quad j = \ell+1, \ldots, m.$$

Hence, $\text{lcm}(M_{f,v}, M_{f,w}) = g_1 \cdot h_2$ and $\gcd(g_1, h_2) = 1$. Furthermore, we have

$$g_1 = M_{f,g_2(f)(v)} \quad \text{and} \quad h_2 = M_{f,h_1(f)(w)}.$$

We now define the vector $u := g_2(f)(v) + h_1(f)(w) \in V$, for which

$$(g_1 \cdot h_2)(f)(u) = h_2(f)((g_1 \cdot g_2)(f)(v)) + g_1(f)((h_1 \cdot h_2)(f)(w)) = 0,$$

where we have used that $(g_1 \cdot g_2)(f)(v) = M_{f,v}(f)(v) = 0$ and $(h_1 \cdot h_2)(f)(w) = M_{f,w}(f)(w) = 0$.

By Lemma 16.20 with $p = g_1 \cdot h_2$ it follows that $M_{f,u} | (g_1 \cdot h_2)$. Moreover, $g_2(f)(v) = u - h_1(f)(w)$, and hence

$$((M_{f,u} \cdot h_2)(f))(g_2(f)(v)) = h_2(f)(M_{f,u}(f)(u)) - M_{f,u}(f)((g_1 \cdot g_2)(f)(w)) = 0.$$

By Lemma 16.20 with $p = M_{f,u} \cdot h_2$ it follows that $M_{f,g_2(f)(v)} | (M_{f,u} \cdot h_2)$, and thus $g_1 | (M_{f,u} \cdot h_2)$. This implies that $g_1 | M_{f,u}$, since $\gcd(g_1, h_2) = 1$. From the equation $h_1(f)(w) = u - g_2(f)(v)$ we conclude analogously that $h_2 | M_{f,u}$, and hence $(g_1 \cdot h_2) | M_{f,u}$. Together with $M_{f,u} | (g_1 \cdot h_2)$, we finally obtain $M_{f,u} = g_1 \cdot h_2 = \mathrm{lcm}(M_{f,v}, M_{f,w})$.

Let $\{v_1, \ldots, v_n\}$ be an arbitrary basis of $\mathcal{V}$. If we factorize the polynomials $M_{f,v_1}, \ldots, M_{f,v_n}$ as the polynomials $M_{f,v}$ and $M_{f,w}$ above, then we obtain

$$M_f = \mathrm{lcm}(M_{f,v_1}, \ldots, M_{f,v_n}).$$

Using repeatedly that for any $v, w \in \mathcal{V} \setminus \{0\}$ there exists $u \in \mathcal{V}$ with $M_{f,u} = \mathrm{lcm}(M_{f,v}, M_{f,w})$, shows the existence of a vector $v_0 \in \mathcal{V}$ with $M_{f,v_0} = \mathrm{lcm}(M_{f,v_1}, \ldots, M_{f,v_n}) = M_f$. □

The main goal in the first part of the proof of Lemma 16.21 is to factorize $\mathrm{lcm}(M_{f,v}, M_{f,w})$ into a product of two coprime polynomials. Assuming the existence of (prime) factorizations of $M_{f,v}$ and $M_{f,w}$ as in (15.3), we easily obtain this factorization in the form $\mathrm{lcm}(M_{f,v}, M_{f,w}) = g_1 \cdot h_2$, where $\gcd(g_1, h_2) = 1$. The proof then continues by constructing a vector u with $M_{f,u} = g_1 \cdot h_2$. We point out that instead of assuming that factorizations as in (15.3) are given, we can obtain the required factorization of $\mathrm{lcm}(M_{f,v}, M_{f,w})$ into a product of two coprime polynomials using only greatest common divisors, and hence via Algorithm 15.15 (the Euclidean algorithm).

In order to show how this can be achieved, let $g := \gcd(M_{f,v}, M_{f,w})$. Then there exist coprime polynomials $p, q \in K[t]$ such that

$$M_{f,v} = g \cdot p \quad \text{and} \quad M_{f,w} = g \cdot q,$$

and therefore $\mathrm{lcm}(M_{f,v}, M_{f,w}) = g \cdot p \cdot q$. If $\gamma := \deg(g) = 0$, then $\mathrm{lcm}(M_{f,v}, M_{f,w}) = (g \cdot p) \cdot q$ is a factorization into a product of two coprime polynomials. If $\gamma > 0$, we consider $h := \gcd(g, q^\gamma)$ so that $g = h \cdot k$ for some $k \in K[t]$. We claim that then

$$\mathrm{lcm}(M_{f,v}, M_{f,w}) = (h \cdot q) \cdot (k \cdot p),$$

where $h \cdot q$ and $k \cdot p$ are coprime. Indeed, suppose that $r \in K[t]$ is irreducible with $r | (h \cdot q)$ and $r | (k \cdot p)$. Because of the definition of h, all its irreducible factors are divisors of q, and hence $r | q$. Since p and q are coprime, and $r | (k \cdot p)$, we must have $r | k$, and hence $r | g$. Let $\rho \geq 1$ be the largest integer with $r^\rho | g$, i.e., $g = r^\rho \cdot \widetilde{g}$, where r and $\widetilde{g}$ are coprime. Then $\rho \leq \gamma$, and thus $r^\rho | q^\gamma$, which implies $r^\rho | h$, i.e., $h = r^\rho \cdot \widetilde{h}$ for some $\widetilde{h} \in K[t]$. Since $r | k$, we also have $k = r \cdot \widetilde{k}$ for some $\widetilde{k} \in K[t]$ and, therefore,

$$g = h \cdot k = (r^\rho \cdot \widetilde{h}) \cdot (r \cdot \widetilde{k}) = r^\rho \cdot (r \cdot \widetilde{h} \cdot \widetilde{k}),$$

16.3 The Minimal Polynomial and the Frobenius Canonical Form

which contradicts that $g = r^\rho \cdot \tilde{g}$, where r and $\tilde{g}$ are coprime. Thus, there exists no irreducible polynomial that divides both $h \cdot q$ and $k \cdot p$, which implies that these polynomials are coprime.

We are now ready to prove existence and uniqueness of the Frobenius canonical form. For the definition of the companion matrix of a polynomial, see Lemma 8.4.

Theorem 16.22 *Let $\mathcal{V}$ be a finite dimensional K-vector space and $f \in \mathcal{L}(\mathcal{V}, \mathcal{V})$. Then there exist a basis B of $\mathcal{V}$ and non-constant monic polynomials $p_1, \ldots, p_k \in K[t]$ with $p_1 = M_f$ and $p_{i+1} | p_i$ for $i = 1, \ldots, k-1$, such that*

$$[f]_{B,B} = \begin{bmatrix} C_{p_1} & & \\ & \ddots & \\ & & C_{p_k} \end{bmatrix}, \tag{16.11}$$

where C_{p_i} is the companion matrix of p_i, $i = 1, \ldots, k$. A matrix representation of the form (16.11) is called Frobenius canonical form. *This canonical form of f is unique.*

Proof Let $n := \dim(\mathcal{V})$ and let $v_1 \in \mathcal{V}$ be a maximal vector with respect to f, i.e., $M_{f,v_1} = M_f$ and $m = m(f, v_1) = \deg(M_{f,v_1}) = \deg(M_f) \leq n$. The vectors

$$u_1 := v_1, \ u_2 := f(v_1), \ \ldots, \ u_m := f^{m-1}(v_1) \in \mathcal{V}$$

are linearly independent and $\mathcal{U}_1 := \text{span}\{u_1, \ldots, u_m\} = \mathcal{K}_m(f, v_1)$ is a cyclic f-invariant subspace of $\mathcal{V}$. We set $p_1 := M_{f,v_1} = M_f$. If

$$p_1 = \alpha_0 + \ldots + \alpha_{m-1} t^{m-1} + t^m,$$

then $f(u_i) = u_{i+1}$ for $i = 1, \ldots, m-1$, and $p_1(f)(v_1) = 0$ yields

$$f(u_m) = f^m(v_1) = -\alpha_0 v_1 - \ldots - \alpha_{m-1} f^{m-1}(v_1) = -\alpha_0 u_1 - \ldots - \alpha_{m-1} u_m.$$

For the basis $B_{\mathcal{U}_1} := \{u_1, \ldots, u_m\}$ we then have $[f|_{\mathcal{U}_1}]_{B_{\mathcal{U}_1}, B_{\mathcal{U}_1}} = C_{p_1}$, where

$$C_{p_1} = \begin{bmatrix} 0 & & & -\alpha_0 \\ 1 & \ddots & & \vdots \\ & \ddots & 0 & -\alpha_{m-2} \\ & & 1 & -\alpha_{m-1} \end{bmatrix} \in K^{m,m}$$

is the companion matrix of p_1. (For $m = 1$ we have $C_{p_1} = [-\alpha_0]$.)

If $m = n$, and thus $\mathcal{U}_1 = \mathcal{V}$, then we set $B := B_{\mathcal{U}_1}$, so that $[f]_{B,B} = C_{p_1}$ and we are finished. (In this case $[f]_{B,B}$ is non-derogatory.)

If on the other hand $m < n$, then we extend the basis $B_{\mathcal{U}_1} := \{u_1, \ldots, u_m\}$ of $\mathcal{U}_1$ to a basis $\widetilde{B} := \{u_1, \ldots, u_m, u_{m+1}, \ldots, u_n\}$ of $\mathcal{V}$. Let $\widetilde{B}^* := \{u_1^*, \ldots, u_m^*, u_{m+1}^*, \ldots, u_n^*\} \subseteq \mathcal{V}^*$ be the dual basis of $\widetilde{B}$. Then, in particular,

$$u_m^*(u_i) = \begin{cases} 0, & 1 \le i \le m-1, \\ 1, & i = m. \end{cases} \tag{16.12}$$

Let $f^* \in \mathcal{L}(\mathcal{V}^*, \mathcal{V}^*)$ be the dual map of f. By Theorem 11.7,

$$A := [f]_{\widetilde{B}, \widetilde{B}} = ([f^*]_{\widetilde{B}^*, \widetilde{B}^*})^T.$$

For every polynomial $p \in K[t]$ we have $p(A) = (p(A^T))^T$ and, therefore, $p(A) = 0$ if and only if $p(A^T) = 0$. This immediately implies that $M_f = M_A = M_{A^T} = M_{f^*}$.

Let $\mathcal{W}_1 := \mathcal{K}_m(f^*, u_m^*) = \mathrm{span}\{u_m^*, \ldots, (f^*)^{m-1}(u_m^*)\}$. Since $m = \deg(M_f) = \deg(M_{f^*})$, we see that $\mathcal{W}_1$ is an f^*-invariant subspace of $\mathcal{V}^*$. From (3) in Lemma 16.6 it then follows that $\mathcal{W}_1^0$ is an f-invariant subspace of $\mathcal{V}$. We now show that $\mathcal{U}_1 \cap \mathcal{W}_1^0 = \{0\}$. For this let $u \in \mathcal{U}_1 \cap \mathcal{W}_1^0$. Then

$$u = \beta_0 v_1 + \beta_1 f(v_1) + \ldots + \beta_{m-1} f^{m-1}(v_1) = \beta_0 u_1 + \beta_1 u_2 + \ldots + \beta_{m-1} u_m \tag{16.13}$$

for some $\beta_0, \beta_1, \ldots, \beta_{m-1} \in K$. Since $u \in \mathcal{W}_1^0$, we have $u_m^*(u) = 0$ and, using (16.12), we obtain

$$0 = u_m^*(u) = \beta_0 u_m^*(u_1) + \beta_1 u_m^*(u_2) + \ldots + \beta_{m-1} u_m^*(u_m) = \beta_{m-1}.$$

If $m = 1$ we have shown that $u = 0$. If $m > 1$ we use $f^*(u_m^*)(u) = 0$ as well as (16.13) with $\beta_{m-1} = 0$ and obtain

$$0 = f^*(u_m^*)(u) = u_m^*(f(u)) = \beta_0 u_m^*(u_2) + \beta_1 u_m^*(u_3) + \ldots + \beta_{m-2} u_m^*(u_m) = \beta_{m-2}.$$

From $(f^*)^k(u_m^*)(u) = 0$ we obtain $\beta_{m-k-1} = 0$ for $k = 0, 1, \ldots, m-1$ and, therefore, $u = 0$.

Now $\mathcal{U}_1 \cap \mathcal{W}_1^0 = \{0\}$ shows that

$$\mathcal{V} = \mathcal{U}_1 \oplus \mathcal{W}_1^0,$$

where the subspaces $\mathcal{U}_1$ and $\mathcal{W}_1^0$ are both f-invariant, and where $\dim(\mathcal{W}_1^0) = n - m < \dim(\mathcal{V})$. If $B_{\mathcal{W}_1^0} := \{\widetilde{u}_{m+1}, \ldots, \widetilde{u}_n\}$ is an arbitrary basis of $\mathcal{W}_1^0$, then $B_1 := \{u_1, \ldots, u_m, \widetilde{u}_{m+1}, \ldots, \widetilde{u}_n\}$ is a basis of $\mathcal{V}$ and, by construction,

$$[f]_{B_1, B_1} = \begin{bmatrix} C_{p_1} & \\ & [f|_{\mathcal{W}_1^0}]_{B_{\mathcal{W}_1^0}, B_{\mathcal{W}_1^0}} \end{bmatrix}.$$

16.3 The Minimal Polynomial and the Frobenius Canonical Form

Since $M_f([f]_{B_1,B_1}) = 0$, we have $M_f([f|_{\mathcal{W}_1^0}]_{B_{\mathcal{W}_1^0},B_{\mathcal{W}_1^0}}) = 0$, and thus by Lemma 16.15 the minimal polynomial of $f|_{\mathcal{W}_1^0}$ is a divisor of M_f.

We can now apply the same construction as before to $f|_{\mathcal{W}_1^0}$. For this, we determine a maximal vector $v_2 \in \mathcal{W}_1^0$ with respect to $f|_{\mathcal{W}_1^0}$ and the decomposition $\mathcal{W}_1^0 = \mathcal{U}_2 \oplus \mathcal{W}_2^0$, where $\mathcal{U}_2$ is a cyclic f-invariant subspace of $\mathcal{W}_1^0$. Then, for the subspace $\mathcal{W}_2^0$, which is also f-invariant, we have $\dim(\mathcal{W}_2^0) < \dim(\mathcal{W}_1^0)$. This yields a basis B_2 of $\mathcal{V}$ with

$$[f]_{B_2,B_2} = \begin{bmatrix} C_{p_1} & & \\ & C_{p_2} & \\ & & [f|_{\mathcal{W}_2^0}]_{B_{\mathcal{W}_2^0},B_{\mathcal{W}_2^0}} \end{bmatrix},$$

where C_{p_2} is the companion matrix of the minimal polynomial p_2 of $f|_{\mathcal{W}_1^0}$. After finitely many steps we obtain $\mathcal{W}_k^0 = \{0\}$ and $\mathcal{V} = \mathcal{U}_1 \oplus \ldots \oplus \mathcal{U}_k$, as well as a corresponding basis B such that $[f]_{B,B}$ has the desired form.

To show the uniqueness, let

$$A_1 = \begin{bmatrix} C_{p_1} & & \\ & \ddots & \\ & & C_{p_k} \end{bmatrix} \quad \text{und} \quad A_2 = \begin{bmatrix} C_{q_1} & & \\ & \ddots & \\ & & C_{q_\ell} \end{bmatrix}$$

be two matrix representations of f in Frobenius canonical form, i.e. $p_1, \ldots, p_k$, $q_1, \ldots, q_\ell \in K[t]$ are non-constant monic polynomials with $p_1 = q_1 = M_f$ as well as $p_{i+1}|p_i$, $i = 1, \ldots, k-1$, and $q_{i+1}|q_i$, $i = 1, \ldots, \ell - 1$. If $k = 1$, then $p_1 = q_1 = M_f = P_f$ and thus also $\ell = 1$ and $A_1 = A_2$.

Let $k > 1$ and hence also $\ell > 1$. The matrices $A_1, A_2 \in K^{n,n}$ are similar since they are matrix representations of the same endomorphism f. Hence, $A_1 = SA_2S^{-1}$ for a matrix $S \in GL_n(K)$. Since $p_{i+1}|p_i$ for $i = 1, \ldots, k-1$, it follows that

$$p_2(A_1) = \begin{bmatrix} p_2(C_{p_1}) & & & \\ & p_2(C_{p_2}) & & \\ & & \ddots & \\ & & & p_2(C_{p_k}) \end{bmatrix} = \begin{bmatrix} p_2(C_{p_1}) & & & \\ & 0 & & \\ & & \ddots & \\ & & & 0 \end{bmatrix}$$

$$= p_2(SA_2S^{-1}) = Sp_2(A_2)S^{-1} = S \begin{bmatrix} p_2(C_{p_1}) & & & \\ & p_2(C_{q_2}) & & \\ & & \ddots & \\ & & & p_2(C_{q_\ell}) \end{bmatrix} S^{-1}.$$

This implies, in particular, that $p_2(C_{q_2}) = 0$, and hence $q_2 | p_2$ by Lemma 16.15. Analogously, we can show that $p_2 | q_2$ and, therefore, $p_2 = q_2$. Inductively we obtain $k = \ell$ and $p_i = q_i$ for $i = 1, \ldots, k$, which means that $A_1 = A_2$. □

The Frobenius canonical form is also called *rational canonical form*. This is motivated by the fact that this form can be computed for every field with "rational operations". Recall that the key ingredient for determining the required maximal vectors is the computation of the greatest common divisor of certain given polynomials. As indicated after the proof of Lemma 16.21, this can be done using the Euclidean algorithm, which requires only (finitely many) additions and multiplications. We do not discuss this in more detail here, but we point out again that the existence of the Frobenius canonical form does not depend on the existence of eigenvalues of f. Thus, in contrast to the Jordan canonical form, no roots of polynomials have to be determined.

The "matrix version" of Theorem 16.22 is the following:

For every matrix $A \in K^{n,n}$, there exist unique non-constant monic polynomials $p_1, \ldots, p_k \in K[t]$ with $p_1 = M_A$ and $p_{i+1} | p_i$ for $i = 1, \ldots, k-1$ as well as a matrix $S \in GL_n(K)$ with

$$S^{-1} A S = \begin{bmatrix} C_{p_1} & & \\ & \ddots & \\ & & C_{p_k} \end{bmatrix},$$

where C_{p_i} is the companion matrix of p_i, for $i = 1, \ldots, k$. The uniquely determined n monic polynomials

$$p_1, \ldots, p_k, \; p_{k+1} = \cdots = p_n = 1$$

associated with a given matrix $A \in K^{n,n}$ are called *invariant factors* of A.

If $\mathcal{V}$ is a finite dimensional $\mathbb{R}$- or $\mathbb{C}$-vector space and $f \in \mathcal{L}(\mathcal{V}, \mathcal{V})$ has a Jordan canonical form, then we can read off the uniquely determined polynomials $p_1, \ldots, p_k$ in Theorem 16.22 from the Jordan canonical form. From Lemma 16.18 we know that $p_1 = M_f$ is the product of the minimal polynomials of the respective largest Jordan blocks corresponding to every of the pairwise distinct eigenvalues of f. By the divisibility conditions $p_{i+1} | p_i$ for $i = 1, \ldots, k-1$, it then follows that p_2 is the product of the minimal polynomials of the respective second largest Jordan blocks corresponding to every of the pairwise distinct eigenvalues of f, and analogously for $p_3, \ldots, p_k$.

Example 16.23 If $\mathcal{V}$ is a 10-dimensional $\mathbb{C}$-vector space and if $f \in \mathcal{L}(\mathcal{V}, \mathcal{V})$ has the Jordan canonical form

$$A = \mathrm{diag}(J_3(0), J_2(0), J_2(1), J_2(1), J_1(1)),$$

then the 10 invariant factors of A are given by

$$p_1 = t^3(t-1)^2, \quad p_2 = t^2(t-1)^2, \quad p_3 = t-1, \quad p_4 = \cdots = p_{10} = 1,$$

and the Frobenius canonical form of f is the matrix $\mathrm{diag}(C_{p_1}, C_{p_2}, C_{p_3})$ with

$$C_{p_1} = \begin{bmatrix} 0 & & & & 0 \\ 1 & 0 & & & 0 \\ & 1 & 0 & & 0 \\ & & 1 & 0 & -1 \\ & & & 1 & 2 \end{bmatrix}, \quad C_{p_2} = \begin{bmatrix} 0 & & & 0 \\ 1 & 0 & & 0 \\ & 1 & 0 & -1 \\ & & 1 & 2 \end{bmatrix}, \quad C_{p_3} = [1].$$

16.4 Computation of the Jordan Canonical Form

The Jordan canonical form is of great importance in theoretical Linear Algebra. In practical applications, however, where usually matrices over $K = \mathbb{R}$ or $K = \mathbb{C}$ are considered, it is not so relevant, since there is no numerically stable method for computing the Jordan canonical form of a general matrix in finite precision arithmetic. The reason for the lack of such a method is that the entries of the Jordan canonical form do not depend continuously on the entries of the given matrix.

Example 16.24 Consider the matrix

$$A(\varepsilon) = \begin{bmatrix} \varepsilon & 1 \\ 0 & 0 \end{bmatrix}, \quad \varepsilon \in \mathbb{R}.$$

For every given $\varepsilon \neq 0$, the matrix $A(\varepsilon)$ has the two distinct eigenvalues ε and 0, and hence the diagonal matrix

$$J(\varepsilon) = \begin{bmatrix} \varepsilon & 0 \\ 0 & 0 \end{bmatrix}$$

is a Jordan canonical form of $A(\varepsilon)$. However, for $\varepsilon \to 0$, we obtain

$$A(\varepsilon) \to \begin{bmatrix} 0 & 1 \\ 0 & 0 \end{bmatrix} \quad \text{and} \quad J(\varepsilon) \to \begin{bmatrix} 0 & 0 \\ 0 & 0 \end{bmatrix}.$$

Thus, $J(\varepsilon)$ does *not* converge to the Jordan canonical form of $A(0)$ for $\varepsilon \to 0$.

A similar example is given by the matrices in Exercise 8.6: While $A(0)$ is a Jordan block of size n corresponding to the eigenvalue 1, for every $\varepsilon \neq 0$ we obtain a diagonalizable matrix $A(\varepsilon) \in \mathbb{C}^{n,n}$ with n pairwise distinct eigenvalues.

> **MATLAB-Minute 10**
>
> Let
>
> $$A = T^{-1} \begin{bmatrix} 1 & 0 \\ 1 & 1 \end{bmatrix} T \in \mathbb{C}^{2,2},$$
>
> where $T \in \mathbb{C}^{2,2}$ is a random matrix constructed with the command T=rand(2). Construct several such matrices and always compute the eigenvalues using the command eig(A). Display the eigenvalues in format long.
>
> One observes that the two eigenvalues are real or complex conjugates, and that they always have an error starting from the 8th digit after the decimal point, i.e., an error on the order of 10^{-8}. This does not happen by chance, but is due to the behavior of the eigenvalues under perturbations, which arise from rounding errors in the computer.

We now derive a method for the computation of the Jordan canonical form of an endomorphism f on a finite dimensional K-vector space $\mathcal{V}$. We assume that P_f decomposes into linear factors over K, and that the roots of P_f, i.e., the eigenvalues of f, are known. The construction follows the important steps in the existence proof of the Jordan canonical form in Sect. 16.2.

Suppose that λ is an eigenvalue of f and that f has a corresponding Jordan block of size s. Then there exist s linearly independent vectors $t_1, \ldots, t_s$ with $[f]_{\widehat{B},\widehat{B}} = J_s(\lambda)$ for $\widehat{B} = \{t_1, \ldots, t_s\}$. With $t_0 := 0$ and writing Id instead of $\mathrm{Id}_\mathcal{V}$ for simplicity of notation, we then have

$$(f - \lambda \,\mathrm{Id})(t_1) = t_0,$$
$$(f - \lambda \,\mathrm{Id})(t_2) = t_1,$$
$$\vdots$$
$$(f - \lambda \,\mathrm{Id})(t_s) = t_{s-1},$$

hence $t_{s-j} = (f - \lambda \,\mathrm{Id})^j (t_s)$ for $j = 0, 1, \ldots, s$.

The vectors $t_s, t_{s-1}, \ldots, t_1$ form a sequence as the one we have constructed in the context of the Krylov subspaces, and

$$\mathrm{span}\{t_s, t_{s-1}, \ldots, t_1\} = \mathcal{K}_s(f - \lambda \,\mathrm{Id}, t_s).$$

The reverse sequence

$$t_1, t_2, \ldots, t_s$$

16.4 Computation of the Jordan Canonical Form

is called a *Jordan chain* of f corresponding to the eigenvalue λ. The vector t_1 is an eigenvector of f corresponding to λ. For the vector t_2 we then have $(f - \lambda \operatorname{Id})(t_2) \neq 0$ and

$$(f - \lambda \operatorname{Id})^2(t_2) = (f - \lambda \operatorname{Id})(t_1) = 0.$$

Hence $t_2 \in \ker((f - \lambda \operatorname{Id})^2) \setminus \ker(f - \lambda \operatorname{Id})$, and in general

$$t_j \in \ker((f - \lambda \operatorname{Id})^j) \setminus \ker((f - \lambda \operatorname{Id})^{j-1}), \quad j = 1, \ldots, s.$$

This motivates the following definition.

Definition 16.25 Let $\mathcal{V}$ be a finite dimensional K-vector space, let $f \in \mathcal{L}(\mathcal{V}, \mathcal{V})$ have the eigenvalue $\lambda \in K$, and let $k \in \mathbb{N}$. A vector $v \in \mathcal{V}$ with

$$v \in \ker((f - \lambda \operatorname{Id})^k) \setminus \ker((f - \lambda \operatorname{Id})^{k-1})$$

is called a *principal vector of level k* of f corresponding to the eigenvalue λ.

Principal vectors of level one are eigenvectors. Principal vectors of higher levels can be considered generalizations of eigenvectors, and they are therefore sometimes called *generalized eigenvectors*.

For the computation of the Jordan canonical form of f, we thus need to know the number and lengths of the Jordan chains corresponding to the different eigenvalues of f. These correspond to the number and sizes of the Jordan blocks of f. If F is a matrix representation of f with respect to an arbitrary basis, then (cp. the proof of Theorem 16.12)

$$\begin{aligned} d_s(\lambda) &:= \operatorname{rank}((F - \lambda I)^{s-1}) - \operatorname{rank}((F - \lambda I)^s) \\ &= \dim(\operatorname{im}((f - \lambda \operatorname{Id})^{s-1})) - \dim(\operatorname{im}((f - \lambda \operatorname{Id})^s)) \\ &= \dim(\mathcal{V}) - \dim(\ker((f - \lambda \operatorname{Id})^{s-1})) - (\dim(\mathcal{V}) - \dim(\ker((f - \lambda \operatorname{Id})^s))) \\ &= \dim(\ker((f - \lambda \operatorname{Id})^s)) - \dim(\ker((f - \lambda \operatorname{Id})^{s-1})) \end{aligned}$$

is the number of Jordan blocks corresponding to λ of size at least s. This implies, in particular, that

$$d_s(\lambda) \geq d_{s+1}(\lambda) \geq 0, \quad s = 1, 2, \ldots,$$

and $d_s(\lambda) - d_{s+1}(\lambda)$ is the number of Jordan blocks of exact size s corresponding to λ. There exists a smallest number $m \in \mathbb{N}$ with

$$\{0\} = \ker((f - \lambda \operatorname{Id})^0) \subset \ker((f - \lambda \operatorname{Id})^1) \subset \cdots \subset \ker((f - \lambda \operatorname{Id})^m)$$
$$= \ker((f - \lambda \operatorname{Id})^{m+1}).$$

Hence $d_s(\lambda) = 0$ for all $s \geq m + 1$, so that there is no Jordan block corresponding to λ of size $m + 1$ or larger.

In order to compute the Jordan canonical form, we proceed as follows:

(1) Determine the eigenvalues of f.
(2) For every eigenvalue λ of f carry out the following steps:
   (a) Determine the smallest number $m \in \mathbb{N}$ with

$$\ker((f - \lambda \operatorname{Id})^0) \subset \ker((f - \lambda \operatorname{Id})^1) \subset \cdots \subset \ker((f - \lambda \operatorname{Id})^m)$$
$$= \ker((f - \lambda \operatorname{Id})^{m+1}).$$

   Then $\dim(\ker((f - \lambda \operatorname{Id})^m)) = a(\lambda, f)$.
   (b) For $s = 1, \ldots, m$ determine

$$d_s(\lambda) = \dim(\ker((f - \lambda \operatorname{Id})^s)) - \dim(\ker((f - \lambda \operatorname{Id})^{s-1})) > 0.$$

   If $s \geq m + 1$, then $d_s(\lambda) = 0$, and

$$d_1(\lambda) = \dim(\ker(f - \lambda \operatorname{Id})) = g(\lambda, f)$$

   is the number of Jordan blocks corresponding to λ.
   (c) To simplify notation, we write $d_s := d_s(\lambda)$ and determine the Jordan chains as follows:
      (i) Since $d_m - d_{m+1} = d_m$, there exist d_m Jordan blocks of size m. For each of these blocks we determine a Jordan chain of d_m principal vectors of level m, i.e., vectors

$$t_{1,m}, t_{2,m}, \ldots, t_{d_m,m} \in \ker((f - \lambda \operatorname{Id})^m) \setminus \ker((f - \lambda \operatorname{Id})^{m-1})$$

      with the following property:
      If $\alpha_1, \ldots, \alpha_{d_m} \in K$ with $\sum_{i=1}^{d_m} \alpha_i t_{i,m} \in \ker((f - \lambda \operatorname{Id})^{m-1})$, then $\alpha_1 = \cdots = \alpha_{d_m} = 0$. Here the first index in $t_{i,j}$ indicates the number of the chain, and the second indicates the level of the principal vector (from $\ker((f - \lambda \operatorname{Id})^j)$ and not $\ker((f - \lambda \operatorname{Id})^{j-1})$).
      (ii) For $j = m, m - 1, \ldots, 2$ we proceed as follows:

16.4 Computation of the Jordan Canonical Form

When we have determined d_j principal vectors of level j, say $t_{1,j}, t_{2,j}, \ldots, t_{d_j,j}$, we apply $f - \lambda \,\mathrm{Id}$ to each of these vectors, hence

$$t_{i,j-1} := (f - \lambda \,\mathrm{Id})(t_{i,j}), \quad 1 \leq i \leq d_j,$$

in order to determine the principal vectors of level $j - 1$.

If $\alpha_1, \ldots, \alpha_{d_j} \in K$ with $\sum_{i=1}^{d_j} \alpha_i t_{i,j-1} \in \ker((f - \lambda \,\mathrm{Id})^{j-2})$, then

$$0 = (f - \lambda \,\mathrm{Id})^{j-2}\left(\sum_{i=1}^{d_j} \alpha_i t_{i,j-1}\right) = (f - \lambda \,\mathrm{Id})^{j-1}\left(\sum_{i=1}^{d_j} \alpha_i t_{i,j}\right),$$

and thus $\sum_{i=1}^{d_j} \alpha_i t_{i,j} \in \ker((f - \lambda \,\mathrm{Id})^{j-1})$ giving $\alpha_1 = \cdots = \alpha_{d_j} = 0$.

If $d_{j-1} > d_j$, then there exist $d_{j-1} - d_j$ Jordan blocks of size $j - 1$. For these we need the Jordan chains of length $j - 1$. Thus we extend the already computed

$$t_{1,j-1}, t_{2,j-1}, \ldots, t_{d_j,j-1} \in \ker((f - \lambda \,\mathrm{Id})^{j-1}) \setminus \ker((f - \lambda \,\mathrm{Id})^{j-2})$$

to d_{j-1} principal vectors of level $(j-1)$ (but only if $d_{j-1} > d_j$) via

$$t_{1,j-1}, t_{2,j-1}, \ldots, t_{d_{j-1},j-1} \in \ker((f - \lambda \,\mathrm{Id})^{j-1}) \setminus \ker((f - \lambda \,\mathrm{Id})^{j-2}),$$

where the following must hold: If $\alpha_1, \ldots, \alpha_{d_{j-1}} \in K$ with $\sum_{i=1}^{d_{j-1}} \alpha_i t_{i,j-1} \in \ker((f - \lambda \,\mathrm{Id})^{j-2})$, then $\alpha_1 = \cdots = \alpha_{d_{j-1}} = 0$.

After completing the step for $j = 2$, we have obtained (linearly independent) vectors $t_{1,1}, t_{2,1}, \ldots, t_{d_1,1} \in \ker(f - \lambda \,\mathrm{Id})$. Since $\dim(\ker(f - \lambda \,\mathrm{Id})) = d_1$, we have found a basis of $\ker(f - \lambda \,\mathrm{Id})$. In this way we have determined d_1 different Jordan chains that we combine as follows:

$$T_\lambda := \{t_{1,1}, t_{1,2}, \ldots, t_{1,m}; \, t_{2,1}, t_{2,2}, \ldots, t_{2,*}; \ldots; t_{d_1,1}, \ldots, t_{d_1,*}\}.$$

Each chain begins with an eigenvector, followed by principal vectors of increasing levels. Here we use the convention that the chains are ordered decreasingly according to their length.

(3) Jordan chains are linearly independent, if their first vectors (the eigenvectors) are linearly independent. (Show this as an exercise.) Thus, if $\lambda_1, \ldots, \lambda_\ell$ are the pairwise distinct eigenvalues of f, then

$$T = \{T_{\lambda_1}, \ldots, T_{\lambda_\ell}\}$$

is a basis, for which $[f]_{T,T}$ is in Jordan canonical form.

Example 16.26 We interpret the matrix

$$F = \begin{bmatrix} 5 & 0 & 1 & 0 & 0 \\ 0 & 1 & 0 & 0 & 0 \\ -1 & 0 & 3 & 0 & 0 \\ 0 & 0 & 0 & 1 & 0 \\ 0 & 0 & 0 & 0 & 4 \end{bmatrix} \in \mathbb{R}^{5,5}$$

as endomorphism on $\mathbb{R}^{5,1}$.

(1) The eigenvalues of F are the roots of $P_F = (t-1)^2(t-4)^3$. In particular, P_F decomposes into linear factors and F has a Jordan canonical form.
(2) We now consider the different eigenvalues of F:
   (a) For the eigenvalue $\lambda_1 = 1$ we obtain

$$\ker(F - I) = \ker \left(\begin{bmatrix} 4 & 0 & 1 & 0 & 0 \\ 0 & 0 & 0 & 0 & 0 \\ -1 & 0 & 2 & 0 & 0 \\ 0 & 0 & 0 & 0 & 0 \\ 0 & 0 & 0 & 0 & 3 \end{bmatrix} \right) = \mathrm{span}\{e_2, e_4\}.$$

Here $\dim(\ker(F - I)) = 2 = a(1, F)$.
For the eigenvalue $\lambda_2 = 4$ we obtain

$$\ker(F - 4I) = \ker \left(\begin{bmatrix} 1 & 0 & 1 & 0 & 0 \\ 0 & -3 & 0 & 0 & 0 \\ -1 & 0 & -1 & 0 & 0 \\ 0 & 0 & 0 & -3 & 0 \\ 0 & 0 & 0 & 0 & 0 \end{bmatrix} \right) = \mathrm{span}\{e_1 - e_3, e_5\},$$

$$\ker((F - 4I)^2) = \ker \left(\begin{bmatrix} 0 & 0 & 0 & 0 & 0 \\ 0 & 9 & 0 & 0 & 0 \\ 0 & 0 & 0 & 0 & 0 \\ 0 & 0 & 0 & 9 & 0 \\ 0 & 0 & 0 & 0 & 0 \end{bmatrix} \right) = \mathrm{span}\{e_1, e_3, e_5\}.$$

Here $\dim(\ker((F - 4I)^2)) = 3 = a(4, F)$.
   (b) For $\lambda_1 = 1$ we have $d_1(1) = \dim(\ker(F - I)) = 2$.
   For $\lambda_2 = 4$ we have $d_1(4) = \dim(\ker(F - 4I)) = 2$ and $d_2(4) = \dim(\ker((F - 4I)^2)) - \dim(\ker(F - 4I)) = 3 - 2 = 1$.

(c) Computation of the Jordan chains:

- For $\lambda_1 = 1$ we have $m = 1$. As principal vectors of level one we choose $t_{1,1} = e_2$ and $t_{2,1} = e_4$. These form a basis of $\ker(F - I)$: If $\alpha_1, \alpha_2 \in \mathbb{R}$ with $\alpha_1 e_2 + \alpha_2 e_4 = 0$, then $\alpha_1 = \alpha_2 = 0$. For $\lambda_1 = 1$ we are finished.
- For $\lambda_2 = 4$ we have $m = 2$, and we choose a principal vector of level two, say $t_{1,2} = e_1$. For this vector we have: If $\alpha_1 \in \mathbb{R}$ with $\alpha_1 e_1 \in \text{span}\{e_1 - e_3, e_5\}$, then $\alpha_1 = 0$. We compute

$$t_{1,1} := (F - 4I)t_{1,2} = e_1 - e_3.$$

Since $d_1(4) = 2 > 1 = d_2(4)$, we have to add to $t_{1,1}$ another principal vector of level one, and we choose $t_{2,1} = e_5$. Since the vectors are linearly independent, $\alpha_1 t_{1,1} + \alpha_2 t_{2,1} \in \ker((F - 4I)^0) = \{0\}$ implies that $\alpha_1 = \alpha_2 = 0$.

In this way we get

$$T_{\lambda_1} = \begin{bmatrix} 0 & 0 \\ 1 & 0 \\ 0 & 0 \\ 0 & 1 \\ 0 & 0 \end{bmatrix} \quad \text{and} \quad T_{\lambda_2} = \begin{bmatrix} 1 & 1 & 0 \\ 0 & 0 & 0 \\ -1 & 0 & 0 \\ 0 & 0 & 0 \\ 0 & 0 & 1 \end{bmatrix}.$$

(3) The coordinate transformation matrix is $T = [T_{\lambda_1} \ T_{\lambda_2}]$, and the Jordan canonical form of F is

$$\begin{bmatrix} 1 & & & & \\ & 1 & & & \\ & & 4 & 1 & \\ & & & 4 & \\ & & & & 4 \end{bmatrix} = T^{-1}FT, \quad \text{where} \quad T^{-1} = \begin{bmatrix} 0 & 1 & 0 & 0 & 0 \\ 0 & 0 & 0 & 1 & 0 \\ 0 & 0 & -1 & 0 & 0 \\ 1 & 0 & 1 & 0 & 0 \\ 0 & 0 & 0 & 0 & 1 \end{bmatrix}.$$

Exercises

(In the following exercises K is an arbitrary field.)

16.1 Prove Lemma 16.1 (1).
16.2 Prove Lemma 16.6 (1).

16.3 Let

$$A = \begin{bmatrix} 0 & 1 & 1 \\ 1 & 0 & 1 \\ 1 & 1 & 0 \end{bmatrix} \in \mathbb{R}^{3,3}, \quad v_1 = \begin{bmatrix} 1 \\ 1 \\ 1 \end{bmatrix} \in \mathbb{R}^{3,1}, \quad v_2 = \begin{bmatrix} 0 \\ 1 \\ 1 \end{bmatrix} \in \mathbb{R}^{3,1}.$$

Determine the Krylov subspaces $\mathcal{K}_j(A, v_i)$, $i = 1, 2$, for all $j \in \mathbb{N}$.

16.4 Let $\mathcal{V}$ be a finite dimensional K-vector space, $f \in \mathcal{L}(\mathcal{V}, \mathcal{V})$, $v \in \mathcal{V}$ and $\lambda \in K$. Show that $\mathcal{K}_j(f, v) = \mathcal{K}_j(f - \lambda \operatorname{Id}_\mathcal{V}, v)$ for all $j \in \mathbb{N}$. Conclude that the grade of v with respect to f is equal to the grade of v with respect to $f - \lambda \operatorname{Id}_\mathcal{V}$.

16.5 Let $f \in \mathcal{L}(\mathbb{R}^{3,1}, \mathbb{R}^{3,1})$ with $f(e_1) = -e_1$, $f(e_2) = e_2 + e_3$, and $f(e_3) = -e_3$.

(a) Determine the characteristic polynomial P_f and a basis of the eigenspace $\mathcal{V}_f(-1)$.

(b) Show that $\dim(\mathcal{K}_3(f, v)) \leq 2$ holds for every $v \in \mathbb{R}^{3,1}$.

16.6 Prove Lemma 16.16.

16.7 Let $\mathcal{V}$ be a finite dimensional Euclidean or unitary vector space and let $f \in \mathcal{L}(\mathcal{V}, \mathcal{V})$ be selfadjoint and nilpotent. Show that then $f = 0$.

16.8 Let $\mathcal{V} \neq \{0\}$ be a finite dimensional K-vector space, let $f \in \mathcal{L}(\mathcal{V}, \mathcal{V})$ be nilpotent of index m and suppose that P_f decomposes into linear factors. Show the following assertions:

(a) $P_f = t^n$ with $n = \dim(\mathcal{V})$.

(b) $M_f = t^m$.

(c) There exists a vector $v \in \mathcal{V}$ of grade m with respect f.

(d) For every $\lambda \in K$ we have $M_{f - \lambda \operatorname{Id}_\mathcal{V}} = (t + \lambda)^m$.

16.9 Let $\mathcal{V}$ be a finite dimensional $\mathbb{C}$-vector space, and let $f, g \in \mathcal{L}(\mathcal{V}, \mathcal{V})$ be commuting and nilpotent. Show that then $f + g$ is also nilpotent.

16.10 Let $\mathcal{V}$ be a finite dimensional K-vector space and $f \in \mathcal{L}(\mathcal{V}, \mathcal{V})$. Show the following assertions:

(a) $\ker(f^j) \subseteq \ker(f^{j+1})$ for all $j \geq 0$ and there exists an $m \geq 0$ with $\ker(f^m) = \ker(f^{m+1})$. For this m we have $\ker(f^m) = \ker(f^{m+j})$ for all $j \geq 1$.

(b) $\operatorname{im}(f^j) \supseteq \operatorname{im}(f^{j+1})$ for all $j \geq 0$ and there exists an $\ell \geq 0$ with $\operatorname{im}(f^\ell) = \operatorname{im}(f^{\ell+1})$. For this ℓ we have $\operatorname{im}(f^\ell) = \operatorname{im}(f^{\ell+j})$ for all $j \geq 1$.

(c) If $m, \ell \geq 0$ are minimal with $\ker(f^m) = \ker(f^{m+1})$ and $\operatorname{im}(f^\ell) = \operatorname{im}(f^{\ell+1})$, then $m = \ell$.

(Theorem 16.5 now implies that $\mathcal{V} = \ker(f^m) \oplus \operatorname{im}(f^m)$ is a decomposition of $\mathcal{V}$ into f-invariant subspaces.)

16.11 Determine the Jordan canonical form of the matrices

$$A = \begin{bmatrix} 1 & -1 & 0 & 0 \\ 1 & -1 & 0 & 0 \\ 3 & 0 & 3 & -3 \\ 4 & -1 & 3 & -3 \end{bmatrix} \in \mathbb{R}^{4,4}, \quad B = \begin{bmatrix} 2 & 10 & 0 & 0 \\ -1 & 1 & 1 & 0 & 0 \\ -1 & 0 & 3 & 0 & 0 \\ -1 & -1 & 0 & 1 & 1 \\ -2 & -1 & 1 & -1 & 3 \end{bmatrix} \in \mathbb{R}^{5,5}$$

using the method presented in Sect. 16.4. Determine also the minimal polynomial.

16.12 Determine the Jordan canonical form and the minimal polynomial of the linear map

$$f : \mathbb{C}[t]_{\leq 3} \to \mathbb{C}[t]_{\leq 3}, \quad \alpha_0 + \alpha_1 t + \alpha_2 t^2 + \alpha_3 t^3 \mapsto \alpha_1 + \alpha_2 t + \alpha_3 t^3.$$

16.13 Determine (up to the order of the blocks) all matrices J in Jordan canonical form with $P_J = (t+1)^3(t-1)^3$ and $M_J = (t+1)^2(t-1)^2$.

16.14 Let $A \in \mathbb{C}^{3,3}$ with $A^2 = 2A - I_3$. Determine (up to the order of the blocks) all possible Jordan canonical forms of A.

16.15 Let $\mathcal{V} \neq \{0\}$ be a finite dimensional K-vector space, $f \in \mathcal{L}(\mathcal{V}, \mathcal{V})$, and suppose that P_f decomposes into linear factors. Show the following assertions:
   (a) $P_f = M_f$ holds if and only if $g(\lambda, f) = 1$ for all eigenvalues λ of f.
   (b) f is diagonalizable if and only if M_f has only simple roots, i.e., roots with multiplicity one.
   (c) A root of $\lambda \in K$ of M_f is simple if and only if $\ker(f - \lambda \operatorname{Id}_\mathcal{V}) = \ker((f - \lambda \operatorname{Id}_\mathcal{V})^2)$.

16.16 Let $\mathcal{V}$ be a K-vector space of dimension 2 or 3 and let $f \in \mathcal{L}(\mathcal{V}, \mathcal{V})$ with P_f decomposing into linear factors. Show that the Jordan canonical form of f is uniquely determined by P_f and M_f. Why does this not hold any longer if $\dim(\mathcal{V}) \geq 4$?

16.17 Prove Lemma 16.20.

16.18 Show that two matrices $A, B \in K^{n,n}$ are similar if and only if they have the same Frobenius canonical form, i.e., the same invariant factors.

16.19 Let $A \in K^{n,n}$ be a matrix for which the characteristic polynomial decomposes into linear factors. Show that there exists a diagonalizable matrix D and a nilpotent matrix N with $A = D + N$ and $DN = ND$.

16.20 Let

$$A = \begin{bmatrix} \alpha & \beta & 0 \\ \beta & 0 & 0 \\ 0 & 0 & \alpha \end{bmatrix} \in \mathbb{R}^{3,3}.$$

(a) Determine the eigenvalues of A in dependence of $\alpha, \beta \in \mathbb{R}$.

(b) Determine all $\alpha, \beta \in \mathbb{R}$ such that $A \notin GL_3(\mathbb{R})$.
(c) Determine the minimal polynomial M_A in dependence of $\alpha, \beta \in \mathbb{R}$.

16.21 Let $A \in K^{n,n}$ be a matrix that has a Jordan canonical form. We define

$$I_n^R := [\delta_{i,n+1-j}] = \begin{bmatrix} & & 1 \\ & \cdot^{\cdot^{\cdot}} & \\ 1 & & \end{bmatrix}, \quad J_n^R(\lambda) := \begin{bmatrix} & & & \lambda \\ & & \cdot^{\cdot^{\cdot}} & 1 \\ & \cdot^{\cdot^{\cdot}} & \cdot^{\cdot^{\cdot}} & \\ \lambda & 1 & & \end{bmatrix} \in K^{n,n}.$$

Show the following assertions:
(a) $I_n^R J_n(\lambda) I_n^R = J_n(\lambda)^T$.
(b) A and A^T are similar.
(c) $J_n(\lambda) = I_n^R J_n^R(\lambda)$.
(d) A can be written as a product of two symmetric matrices.

16.22 Let $\lambda \in K \setminus \{0\}$.
(a) Show that $J_n(\lambda)$ is invertible with

$$J_n(\lambda)^{-1} = [c_{ij}], \quad c_{ij} = \begin{cases} (-1)^{j-i} \lambda^{i-j-1}, & i \leq j, \\ 0, & i > j. \end{cases}$$

(b) Determine a Jordan canonical form of $J_n(\lambda)^{-1}$.
(c) Let $A \in GL_n(K)$ be such that P_A decomposes into linear factors over K. Show that A^{-1} has a Jordan canonical form and determine this form based on the Jordan canonical form of A.

16.23 Determine for the matrix

$$A = \begin{bmatrix} 5 & 1 & 1 \\ 0 & 5 & 1 \\ 0 & 0 & 4 \end{bmatrix} \in \mathbb{R}^{3,3}$$

two symmetric matrices $S_1, S_2 \in \mathbb{R}^{3,3}$ with $A = S_1 S_2$.

17 Matrix Functions and Systems of Differential Equations

In this chapter we give an introduction to the area of matrix functions. We first define general matrix functions and derive their most important properties. Using the examples of network analysis and chemical reactions, we illustrate how matrix functions arise naturally in applications. The network analysis example involves the exponential function of matrices, and we study the properties of this important function in detail. The analysis of chemical reaction kinetics leads to a system of ordinary differential equations, whose solution again is based on the matrix exponential function.

17.1 Matrix Functions and the Matrix Exponential Function

In the following we will study functions that yield for a given $n \times n$ matrix again an $n \times n$ matrix. A possible definition of such a function is given by the entrywise application of scalar functions to the matrix. For instance, one could define for $A = [a_{ij}] \in \mathbb{C}^{n,n}$ the function $\sin(A)$ by $\sin(A) := [\sin(a_{ij})]$. However, such a definition is not compatible with the matrix multiplication, since in general already $A^2 \neq [a_{ij}^2]$.

The following definition of the *primary matrix function*, which can be found in the authoritative book of Nick Higham[1] [5, Definition 1.1–1.2], will turn out to be consistent with the matrix multiplication. Since this definition is based on the Jordan canonical form, we assume for simplicity that $A \in \mathbb{C}^{n,n}$. Our considerations also apply to square matrices over $\mathbb{R}$, as long as they have a Jordan canonical form.

---

[1] Nicholas (Nick) John Higham (1961–2024).

Definition 17.1 Let $A \in \mathbb{C}^{n,n}$ have the Jordan canonical form

$$J = \operatorname{diag}(J_{d_1}(\lambda_1), \ldots, J_{d_m}(\lambda_m)) = S^{-1}AS,$$

and let $\Omega \subset \mathbb{C}$ be such that $\{\lambda_1, \ldots, \lambda_m\} \subseteq \Omega$. A function $f : \Omega \to \mathbb{C}$ is said to be *defined on the spectrum of* A, if the values

$$f^{(j)}(\lambda_i) \quad \text{for} \quad i = 1, \ldots, m \quad \text{and} \quad j = 0, 1 \ldots, d_i - 1 \tag{17.1}$$

exist. Here $f^{(j)}(\lambda_i)$, $j = 1, \ldots, d_i - 1$, is the jth derivative of the function $f(\lambda)$ with respect to λ evaluated at λ_i. If $\lambda_i \in \mathbb{R}$, then this is the real derivative, and for $\lambda_i \in \mathbb{C} \setminus \mathbb{R}$ it is the complex derivative. Moreover, we assume that equal eigenvalues that occur in different Jordan blocks are mapped to the same values in (17.1).

If f is defined on the spectrum of A then the *primary matrix function* $f(A)$ is defined by

$$f(A) := S f(J) S^{-1}, \text{ where } f(J) := \operatorname{diag}(f(J_{d_1}(\lambda_1)), \ldots, f(J_{d_m}(\lambda_m))) \tag{17.2}$$

and

$$f(J_{d_i}(\lambda_i)) := \begin{bmatrix} f(\lambda_i) & f'(\lambda_i) & \frac{f''(\lambda_i)}{2!} & \cdots & \frac{f^{(d_i-1)}(\lambda_i)}{(d_i-1)!} \\ & f(\lambda_i) & f'(\lambda_i) & \ddots & \vdots \\ & & \ddots & \ddots & \frac{f''(\lambda_i)}{2!} \\ & & & \ddots & f'(\lambda_i) \\ & & & & f(\lambda_i) \end{bmatrix} \quad \text{for } i = 1, \ldots, m. \tag{17.3}$$

Note that for the definition of $f(A)$ in (17.2)–(17.3) only the existence of the values in (17.1) is required. By definition, $f(A)$ and $f(J)$ are similar. The eigenvalues of $f(A)$ are given by $f(\lambda_1), \ldots, f(\lambda_m)$, where $\lambda_1, \ldots, \lambda_m$ are the eigenvalues of A. If we denote by $\Lambda(M)$ the set of eigenvalues of a matrix $M \in \mathbb{C}^{n,n}$, then we can express this observation compactly as $\Lambda(f(A)) = f(\Lambda(A))$. This result is known as the *spectral mapping theorem*.

Example 17.2 Let $A = I_2 \in \mathbb{C}^{2,2}$ and let $f(z) = \sqrt{z}$ (the square root function). If we set $f(1) = \sqrt{1} = +1$, then $f(A) = \sqrt{A} = I_2$ by Definition 17.1. If we choose the other branch of the square root function, i.e., $f(1) = \sqrt{1} = -1$, then $f(A) = \sqrt{A} = -I_2$. The matrices I_2 and $-I_2$ are *primary square roots* of $A = I_2$. Taking different branches of a function for different Jordan blocks corresponding to the same eigenvalue is incompatible with Definition 17.1. For instance, the matrices

$$X_1 = \begin{bmatrix} 1 & 0 \\ 0 & -1 \end{bmatrix} \quad \text{and} \quad X_2 = \begin{bmatrix} -1 & 0 \\ 0 & 1 \end{bmatrix}$$

are incompatible with Definition 17.1, despite the fact that $X_1^2 = I_2$ and $X_2^2 = I_2$.

17.1 Matrix Functions and the Matrix Exponential Function

All solutions $X \in \mathbb{C}^{n,n}$ of the matrix equation $X^2 = A$ are called square roots of the matrix $A \in \mathbb{C}^{n,n}$. But as Example 17.2 shows, some of these may not be primary square roots according to Definition 17.1. In the following, by $f(A)$ we will always mean a primary matrix function according to Definition 17.1, and will usually omit the term "primary".

In (16.10) we have shown that for each polynomial $p \in \mathbb{C}[t]$ of degree $k \geq 0$ we have

$$p(J_{d_i}(\lambda_i)) = \sum_{j=0}^{k} \frac{p^{(j)}(\lambda_i)}{j!} J_{d_i}(0)^j. \tag{17.4}$$

A simple comparison shows that this formula agrees with (17.3) for $f = p$. This means that the *computation* of $p(J_{d_i}(\lambda_i))$ with (17.4) leads to the same result as the *definition* of $p(J_{d_i}(\lambda_i))$ by (17.3). More generally, the following result holds.

Lemma 17.3 Let $A \in \mathbb{C}^{n,n}$ and $p = \alpha_0 + \alpha_1 t + \ldots + \alpha_k t^k \in \mathbb{C}[t]$. Then (17.2)–(17.3) with $f = p$ yields a matrix function $f(A)$ that satisfies $f(A) = \alpha_0 I_n + \alpha_1 A + \ldots + \alpha_k A^k$.

Proof Exercise. □

If we consider, in particular, the polynomial $f = t^2$ in (17.2)–(17.3), then the resulting $f(A)$ is equal to the product $A * A$. This shows that the definition of the primary matrix function $f(A)$ is consistent with the matrix multiplication.

We now will show that $f(A)$ is always a polynomial in A. For this we need the following extension of the Lagrange interpolation problem (cp. Exercise 10.18).

Lemma 17.4 Let $\lambda_1, \ldots, \lambda_k \in \mathbb{C}$ be pairwise distinct (or only $\lambda_1 \in \mathbb{C}$ if $k = 1$), $d_1, \ldots, d_k \in \mathbb{N}$, and $n = d_1 + \ldots + d_k - 1$. Then, for every choice of $n + 1$ values $f_{ij} \in \mathbb{C}$, $i = 1, \ldots, k$ and $j = 0, 1, \ldots, d_k - 1$, *the* Hermite interpolation problem[2]

$$p^{(j)}(\lambda_i) = f_{ij}, \quad i = 1, \ldots, k, \quad j = 0, 1, \ldots, d_i - 1, \tag{17.5}$$

has a uniquely determined solution $p \in \mathbb{C}[t]_{\leq n}$.

Proof Every polynomial $p \in \mathbb{C}[t]_{\leq n}$ has the form

$$p = \alpha_0 + \alpha_1 t + \ldots + \frac{\alpha_{n-1}}{(n-1)!} t^{n-1} + \frac{\alpha_n}{n!} t^n$$

---

[2] Charles Hermite (1822–1901).

for coefficients $\alpha_0, \alpha_1, \ldots, \alpha_n \in \mathbb{C}$. The derivatives of p are given by

$$p^{(j)} = \alpha_j + \ldots + \frac{\alpha_{n-1}}{(n-j-1)!} t^{n-j-1} + \frac{\alpha_n}{(n-j)!} t^{n-j}, \quad j = 0, 1, \ldots, n.$$

Thus, we can write the conditions (17.5) as

$$\alpha_j + \ldots + \frac{\lambda_i^{n-j-1}}{(n-j-1)!} \alpha_{n-1} + \frac{\lambda_i^{n-j}}{(n-j)!} \alpha_n = f_{ij}, \quad i=1,\ldots,k, \quad j=0,\ldots,d_i-1.$$

These are $n+1$ linear equations for the $n+1$ coefficients $\alpha_0, \alpha_1, \ldots, \alpha_n$ that we can write as linear system $Ax = b$ with a matrix $A \in \mathbb{C}^{n+1,n+1}$, a right hand side $b \in \mathbb{C}^{n+1,1}$, and an unknown vector $x = [\alpha_0, \alpha_1, \ldots, \alpha_n]^T$.

If $f_{ij} = 0$ for an $i \in \{1, \ldots, k\}$ and all $j = 0, 1, \ldots, d_i - 1$, then it follows from the conditions (17.5) and Corollary 15.8, that λ_i is a root of p with multiplicity d_i. If all values $f_{ij} = 0$, and thus $b = 0$, then the polynomial $p \in \mathbb{C}[z]_{\leq n}$ has the roots $\lambda_1, \ldots, \lambda_k$ with the combined multiplicities $d_1 + \ldots + d_k = n + 1 > \deg(p)$. Then we have $p = 0$, and thus $x = 0$. Since for the linear system $b = 0$ implies $x = 0$, the matrix A must be invertible. Therefore, for every right hand side b, i.e., for every choice of the values f_{ij}, there exists a unique solution x, i.e., a unique polynomial $p \in \mathbb{C}[t]_{\leq n}$ that satisfies the $n+1$ conditions (17.5). $\square$

For $d_1 = \cdots = d_m = 1$ in (17.5) we obtain the Lagrange interpolation problem from Exercise 10.18.

The following theorem is of great practical and theoretical importance.

Theorem 17.5 Let $A \in \mathbb{C}^{n,n}$ have the minimal polynomial M_A, and let $f(A)$ be as in Definition 17.1. Then there exists a uniquely determined polynomial $p \in \mathbb{C}[t]$ of degree at most $\deg(M_A) - 1$ with $f(A) = p(A)$. In particular, $Af(A) = f(A)A$, $f(A^T) = f(A)^T$ as well as $f(VAV^{-1}) = Vf(A)V^{-1}$ for all $V \in GL_n(\mathbb{C})$.

Proof If $J = \mathrm{diag}(J_{d_1}(\lambda_1), \ldots, J_{d_m}(\lambda_m)) = S^{-1}AS$ is a Jordan canonical form of A, then according to Lemma 16.18 the minimal polynomial of A has the form

$$M_A = \prod_{j=1}^{k} (t - \widetilde{\lambda}_j)^{\widetilde{d}_j},$$

where $\widetilde{\lambda}_1, \ldots, \widetilde{\lambda}_k$ are the pairwise distinct eigenvalues of A, and $\widetilde{d}_1, \ldots, \widetilde{d}_k$ are the largest sizes of a corresponding Jordan block. If $f(A)$ is given as in Definition 17.1, then Lemma 17.4 yields a uniquely determined polynomial $p \in \mathbb{C}[t]_{\leq \deg(M_A)-1}$ with

$$p^{(j)}(\widetilde{\lambda}_i) = f^{(j)}(\widetilde{\lambda}_i), \quad i = 1, \ldots, k, \quad j = 0, 1, \ldots, \widetilde{d}_i - 1.$$

17.1 Matrix Functions and the Matrix Exponential Function

This polynomial satisfies $p(J_i(\lambda_i)) = f(J_i(\lambda_i))$ for each Jordan block of A, and hence also $p(A) = f(A)$. The other assertions now can be easily seen, since p is a polynomial. $\square$

Using Theorem 17.5 we can show that the primary matrix function $f(A)$ in Definition 17.1 is independent of the choice of the Jordan canonical form of A. We already know from Theorem 16.12, that the Jordan canonical form of A is unique up to the order of the Jordan blocks. If

$$J = \mathrm{diag}(J_{d_1}(\lambda_1), \ldots, J_{d_m}(\lambda_m)) = S^{-1}AS,$$

$$\widetilde{J} = \mathrm{diag}(J_{\widetilde{d}_1}(\widetilde{\lambda}_1), \ldots, J_{\widetilde{d}_m}(\widetilde{\lambda}_m)) = \widetilde{S}^{-1}A\widetilde{S}$$

are two Jordan canonical forms of A, then $\widetilde{J} = P^T J P$ for a permutation matrix $P \in \mathbb{R}^{n,n}$, where the matrices J and $\widetilde{J}$ are the same up to the order of diagonal blocks. Hence

$$\begin{aligned}
f(J) &= \mathrm{diag}(f(J_{d_1}(\lambda_1)), \ldots, f(J_{d_m}(\lambda_m))) \\
&= P\left(P^T \mathrm{diag}(f(J_{d_1}(\lambda_1)), \ldots, f(J_{d_m}(\lambda_m)))P\right)P^T \\
&= P\left(\mathrm{diag}(f(J_{\widetilde{d}_1}(\widetilde{\lambda}_1)), \ldots, f(J_{\widetilde{d}_m}(\widetilde{\lambda}_m)))\right)P^T \\
&= Pf(\widetilde{J})P^T.
\end{aligned}$$

Theorem 17.5 applied to the matrix J yields the existence of a polynomial p with $f(J) = p(J)$. Thus, we get

$$f(A) = Sf(J)S^{-1} = Sp(J)S^{-1} = p(A) = p(\widetilde{S}\widetilde{J}\widetilde{S}^{-1}) = \widetilde{S}P^T p(J) P\widetilde{S}^{-1}$$
$$= \widetilde{S}P^T f(J) P\widetilde{S}^{-1} = \widetilde{S}f(\widetilde{J})\widetilde{S}^{-1}.$$

Let us now consider the exponential function $f(z) = e^z$ that is infinitely often complex differentiable throughout $\mathbb{C}$. In particular, e^z is defined (in the sense of Definition 17.1) on the spectrum of every given matrix

$$A = S\mathrm{diag}(J_{d_1}(\lambda_1), \ldots, J_{d_m}(\lambda_m))S^{-1} \in \mathbb{C}^{n,n}.$$

If $t \in \mathbb{C}$ is arbitrary (but fixed), then the derivatives of the function e^{tz} with respect to the variable z are given by

$$\frac{d^j}{dz^j} e^{tz} = t^j e^{tz}, \quad j = 0, 1, 2, \ldots.$$

We will use the notation $\exp(M)$ instead of e^M for the exponential function of a matrix M. For every Jordan block $J_d(\lambda)$ of A we then have, by (17.3) with $f(z) = e^{tz}$,

$$\exp(t J_d(\lambda)) = e^{t\lambda} \begin{bmatrix} 1 & t & \frac{t^2}{2!} & \cdots & \frac{t^{d-1}}{(d-1)!} \\ & 1 & t & \ddots & \vdots \\ & & \ddots & \ddots & \frac{t^2}{2!} \\ & & & \ddots & t \\ & & & & 1 \end{bmatrix} = e^{t\lambda} \sum_{k=0}^{d-1} \frac{1}{k!} (t J_d(0))^k, \quad (17.6)$$

and the matrix exponential function $\exp(tA)$ is given by

$$\exp(tA) = S \operatorname{diag}(\exp(t J_{d_1}(\lambda_1)), \ldots, \exp(t J_{d_m}(\lambda_m))) S^{-1}. \quad (17.7)$$

The parameter t will be used in the next section in the context of linear differential equations.

In Analysis it is shown that for every $z \in \mathbb{C}$ the function e^z can be represented by the absolutely convergent series

$$e^z = \sum_{j=0}^{\infty} \frac{z^j}{j!}.$$

Using this series and the equation $(J_d(0))^\ell = 0$ for all $\ell \geq d$, we obtain

$$\exp(t J_d(\lambda)) = e^{t\lambda} \sum_{\ell=0}^{d-1} \frac{1}{\ell!} (t J_d(0))^\ell = \left(\sum_{j=0}^{\infty} \frac{(t\lambda)^j}{j!} \right) \cdot \left(\sum_{\ell=0}^{\infty} \frac{1}{\ell!} (t J_d(0))^\ell \right)$$

$$= \sum_{j=0}^{\infty} \left(\sum_{\ell=0}^{j} \frac{(t\lambda)^{j-\ell}}{(j-\ell)!} \cdot \frac{1}{\ell!} (t J_d(0))^\ell \right)$$

$$= \sum_{j=0}^{\infty} \frac{t^j}{j!} \left(\sum_{\ell=0}^{j} \binom{j}{\ell} \lambda^{j-\ell} J_d(0)^\ell \right)$$

$$= \sum_{j=0}^{\infty} \frac{t^j}{j!} (\lambda I_d + J_d(0))^j$$

$$= \sum_{j=0}^{\infty} \frac{1}{j!} (t J_d(\lambda))^j. \quad (17.8)$$

17.1 Matrix Functions and the Matrix Exponential Function

In this derivation we have used the absolute convergence of the exponential series and the finiteness of the series with the matrix $J_d(0)$. This allows the application of the *Cauchy product formula*[3] for absolutely convergent series, which is also proven in Analysis.

Lemma 17.6 *If $A \in \mathbb{C}^{n,n}$, $t \in \mathbb{C}$ and $\exp(tA)$ is the matrix exponential function in (17.6)–(17.7), then*

$$\exp(tA) = \sum_{j=0}^{\infty} \frac{1}{j!}(tA)^j.$$

Proof In (17.8) we have shown this already for Jordan blocks. The assertion then follows from

$$\sum_{j=0}^{\infty} \frac{1}{j!}(tSJS^{-1})^j = S\left(\sum_{j=0}^{\infty} \frac{1}{j!}(tJ)^j\right)S^{-1}$$

and the representation (17.7) of the matrix exponential function. □

We immediately see from Lemma 17.6 that for a matrix $A \in \mathbb{R}^{n,n}$ and every real t the matrix exponential function $\exp(tA)$ is a real matrix.

The following result presents further important properties of the matrix exponential function.

Lemma 17.7 *If the two matrices $A, B \in \mathbb{C}^{n,n}$ commute, then $\exp(A + B) = \exp(A)\exp(B)$. For every matrix $A \in \mathbb{C}^{n,n}$ we have $\exp(A) \in GL_n(\mathbb{C})$ with $(\exp(A))^{-1} = \exp(-A)$.*

Proof If A and B commute, then the Cauchy product formula yields

$$\exp(A)\exp(B) = \left(\sum_{j=0}^{\infty}\frac{1}{j!}A^j\right)\left(\sum_{\ell=0}^{\infty}\frac{1}{\ell!}B^\ell\right) = \sum_{j=0}^{\infty}\left(\sum_{\ell=0}^{j}\frac{1}{\ell!}A^\ell\frac{1}{(j-\ell)!}B^{j-\ell}\right)$$

$$= \sum_{j=0}^{\infty}\left(\frac{1}{j!}\sum_{\ell=0}^{j}\binom{j}{\ell}A^\ell B^{j-\ell}\right) = \sum_{j=0}^{\infty}\frac{1}{j!}(A+B)^j$$

$$= \exp(A+B).$$

---

[3] Augustin Louis Cauchy (1789–1857).

Here we have used the binomial formula for commuting matrices (cp. Exercise 4.12).

Since A and $-A$ commute, we have

$$\exp(A)\exp(-A) = \exp(A - A) = \exp(0) = \sum_{j=0}^{\infty} \frac{1}{j!} 0^j = I_n,$$

and hence $\exp(A) \in GL_n(\mathbb{C})$ with $(\exp(A))^{-1} = \exp(-A)$. □

For non-commuting matrices the statements in Lemma 17.7 in general do not hold (cp. Exercise 17.10).

MATLAB-Minute 11

Compute the matrix exponential function $\exp(A)$ for the matrix

$$A = \begin{bmatrix} 1 & -1 & 3 & 4 & 5 \\ -1 & -2 & 4 & 3 & 5 \\ 2 & 0 & -3 & 1 & 5 \\ 3 & 0 & 0 & -2 & -3 \\ 4 & 0 & 0 & -3 & -5 \end{bmatrix} \in \mathbb{R}^{5,5}$$

using the command E1=expm(A). (Look at help expm.)
Also compute a diagonalization of A using the command [S,D]=eig(A), and form the matrix exponential function $\exp(A)$ as E2=S*expm(D)/S.
Compare the matrices E1 and E2 and compute the relative error norm(E1-E2)/norm(E2). (Look at help norm.)

Example 17.8 Let $A = [a_{ij}] \in \mathbb{R}^{n,n}$ be a symmetric matrix with $a_{ii} = 0$ and $a_{ij} \in \{0, 1\}$ for all $i, j = 1, \ldots, n$. We identify the matrix A with a *graph* $G_A = (V_A, E_A)$ consisting of a set of n *vertices* $V_A = \{1, \ldots, n\}$ and a set of *edges* $E_A \subseteq V_A \times V_A$. For $i = 1, \ldots, n$ the row i of A is identified with the vertex $i \in V_A$, and every entry $a_{ij} = 1$ is identified with an edge $(i, j) \in E_A$. Due to the symmetry of A, we have $a_{ij} = 1$ if and only if $a_{ji} = 1$. We therefore consider in the following the elements of E_A as *unordered pairs*, i.e., $(i, j) = (j, i)$. The following example illustrates this identification:

$$A = \begin{bmatrix} 0 & 1 & 1 & 1 & 0 \\ 1 & 0 & 0 & 1 & 1 \\ 1 & 0 & 0 & 0 & 1 \\ 1 & 1 & 0 & 0 & 0 \\ 0 & 1 & 1 & 0 & 0 \end{bmatrix}$$

17.1 Matrix Functions and the Matrix Exponential Function

is identified with $G_A = (V_A, E_A)$, where

$$V_A = \{1, 2, 3, 4, 5\}, \quad E_A = \{(1,2), (1,3), (1,4), (2,4), (2,5), (3,5)\},$$

and the graph G_A can be displayed as follows:

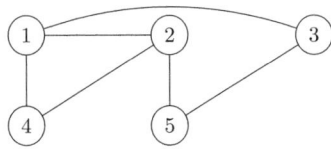

A *path* of length m from the vertex k_1 to the vertex k_{m+1} is an ordered list of vertices $k_1, k_2, \ldots, k_{m+1}$, where $(k_i, k_{i+1}) \in E_A$ for $i = 1, \ldots, m$. If $k_1 = k_{m+1}$, then this is a closed path of length m. In the above example, paths from 1 to 4 are given by 1, 2, 4 and 1, 2, 5, 3, 1, 2, 4; these have the lengths 2 and 6, respectively. In the mathematical field of Graph Theory one usually assumes that the vertices in a path are pairwise distinct. Our deviation from this convention is motivated by the following interpretation of a matrix A and its powers:

An entry $a_{ij} = 1$ in the matrix A means that there exists a path of length 1 from vertex i to vertex j, i.e., the vertices i and j are adjacent. If $a_{ij} = 0$, then no such path exists. The matrix A is therefore called the *adjacency matrix* of the graph G_A. If we square the adjacency matrix, then the entry in the (i, j) position is given by

$$(A^2)_{ij} = \sum_{\ell=1}^{n} a_{i\ell} a_{\ell j}.$$

In the sum on the right hand side, we obtain for a given ℓ a 1 if and only if $(i, \ell) \in E_A$ and $(\ell, j) \in E_A$. The sum on the right had side therefore is equal to the number of vertices that are adjacent to both i and j. Hence the (i, j) entry of A^2 is equal to the number of pairwise distinct paths from i to j ($i \neq j$), or the pairwise distinct closed paths from i to i of length 2 in G_A. More generally, one can show the following.

Theorem 17.9 *Let $A = [a_{ij}] \in \mathbb{R}^{n,n}$ be a symmetric adjacency matrix, i.e., $A = A^T$ with $a_{ii} = 0$ and $a_{ij} \in \{0, 1\}$ for all $i, j = 1, \ldots, n$, and let G_A be the graph identified with A. Then for each $m \in \mathbb{N}$ the (i, j) entry of A^m is equal to the number of pairwise distinct paths from i to j ($i \neq j$) or the pairwise distinct closed paths from i to i of length m in G_A.*

Proof Exercise. □

For the matrix A in Example 17.8 we obtain

$$A^2 = \begin{bmatrix} 3 & 1 & 0 & 1 & 2 \\ 1 & 3 & 2 & 1 & 0 \\ 0 & 2 & 2 & 1 & 0 \\ 1 & 1 & 1 & 2 & 1 \\ 2 & 0 & 0 & 1 & 2 \end{bmatrix} \quad \text{and} \quad A^3 = \begin{bmatrix} 2 & 6 & 5 & 4 & 1 \\ 6 & 2 & 1 & 4 & 5 \\ 5 & 1 & 0 & 2 & 4 \\ 4 & 4 & 2 & 2 & 2 \\ 1 & 5 & 4 & 2 & 0 \end{bmatrix}.$$

The 3 pairwise distinct closed paths of length 2 from 1 to 1 are

$$1, 2, 1, \quad 1, 3, 1, \quad 1, 4, 1$$

and the 4 pairwise distinct paths of length 3 from 1 to 4 are

$$1, 2, 1, 4, \quad 1, 3, 1, 4, \quad 1, 4, 1, 4, \quad 1, 4, 2, 4.$$

Numerous real world applications involve networks that can be modeled mathematically using graphs. Examples include social, biological, telecommunication or airline networks. The properties of such networks are studied in the interdisciplinary area of *Network Science*. An important task is to identify participants in the network that are central in the sense that their functionality has a significant impact on the entire network. If the network has been modeled by a graph, then we can study the *centrality* of the vertices. For example, a vertex can be considered central if it is connected to a large part of the graph via many short closed paths. Longer connections are usually less important, and thus paths should be scaled down according to their length. If we use the scaling factor $1/m!$ for a path of length m, then for the vertex i in the graph G_A with the adjacency matrix A we obtain a centrality measure of the form

$$\left(\frac{1}{1!} A + \frac{1}{2!} A^2 + \frac{1}{3!} A^3 + \dots \right)_{ii}.$$

The relative ordering of the vertices according to this formula is not changed when we add the constant 1. We then obtain the *centrality* of the vertex i as

$$\left(I + A + \frac{1}{2} A^2 + \frac{1}{3!} A^3 + \dots \right)_{ii} = (\exp(A))_{ii}.$$

Another important quantity is the so-called *communicability* between the vertices i and j for $i \neq j$, which is given by the weighted sum of the pairwise distinct paths from i to j, i.e., by

$$\left(I + A + \frac{1}{2} A^2 + \frac{1}{3!} A^3 + \dots \right)_{ij} = (\exp(A))_{ij}.$$

17.2 Systems of Linear Ordinary Differential Equations

For the above matrix A the MATLAB function `expm` yields

$$\exp(A) = \begin{bmatrix} 3.7630 & 3.1953 & 2.2500 & 2.7927 & 1.8176 \\ 3.1953 & 3.7630 & 1.8176 & 2.7927 & 2.2500 \\ 2.2500 & 1.8176 & 2.4881 & 1.2749 & 1.9204 \\ 2.7927 & 2.7927 & 1.2749 & 2.8907 & 1.2749 \\ 1.8176 & 2.2500 & 1.9204 & 1.2749 & 2.4881 \end{bmatrix},$$

where we have rounded the entries to four decimal places. The vertices 1 and 2 have the largest centrality, followed by 4, 3 and 5. If we would define the centrality of a vertex as the number of adjacent vertices, then in this example we could not distinguish between the vertices 3, 4 and 5. The largest communicability in this example exists between the vertices 1 and 2.

Further information concerning the analysis of networks using adjacency matrices and matrix functions can be found in the article [4].

17.2 Systems of Linear Ordinary Differential Equations

A differential equation describes a relationship between a desired function and its derivatives. Such equations are used in all areas of science and engineering for modeling physical phenomena. Ordinary differential equations involve a function of one variable and its derivatives, while partial differential equations involve functions of several variables and their partial derivatives. In this section we focus on ordinary differential equations of first order, i.e., those in which only the function and its first derivative occur.

A simple example for the modeling with ordinary differential equations of first order is the increase or decrease of a biological population, such as bacteria in a petri dish. Let $y = y(t)$ be the size of the population at time t. If there is enough food and if the external conditions (e.g. temperature or pressure) are constant, then the population grows with a (real) rate $k > 0$, that is proportional to the current number of individuals. This can be described by the equation

$$\dot{y} := \frac{d}{dt} y = ky. \qquad (17.9)$$

Clearly, one can also take $k < 0$, and then the population shrinks.

We are then looking for a function $y : D \subset \mathbb{R} \to \mathbb{R}$ that satisfies (17.9). The general solution of (17.9) is given by the exponential function

$$y(t) = ce^{tk},$$

where $c \in \mathbb{R}$ is an arbitrary constant. For a unique solution of (17.9) we need to know the size of the population at a given initial time t_0. In this way we obtain the *initial value problem*

$$\dot{y} = ky, \quad y(t_0) = y_0,$$

which, as we will show below, is solved uniquely by the function

$$y(t) = e^{(t-t_0)k} y_0.$$

Example 17.10 In a chemical reaction certain initial substances (called educts or reactants) are transformed into other substances (called products). Reactions can be distinguished concerning their order. Here we only discuss reactions of first order, where the reaction rate is determined by only one educt. In reactions of second and higher order one typically obtains *nonlinear* differential equations, which are beyond our focus in this chapter.

If, for example, the educt A_1 is transformed into the product A_2 with the rate $-k_1 < 0$, then we write this reaction symbolically as

$$A_1 \xrightarrow{k_1} A_2,$$

and we model it mathematically by the ordinary differential equation

$$\dot{y}_1 = -k_1 y_1.$$

Here the value $y_1(t)$ is the concentration of the substance A_1 at time t. For the concentration of the product A_2, which grows with the rate $k_1 > 0$, we have the corresponding equation $\dot{y}_2 = k_1 y_1$.

It may happen that a reaction of first order develops in both directions. If A_1 transforms into A_2 with the rate $-k_1$, and A_2 transforms into A_1 with the rate $-k_2$, i.e.,

$$A_1 \underset{k_2}{\overset{k_1}{\rightleftarrows}} A_2,$$

then we can model this reaction mathematically by the system of linear ordinary differential equations

$$\dot{y}_1 = -k_1 y_1 + k_2 y_2,$$
$$\dot{y}_2 = k_1 y_1 - k_2 y_2.$$

17.2 Systems of Linear Ordinary Differential Equations

Combining the functions y_1 and y_2 in a vector valued function $y = [y_1, y_2]^T$, we can write this system as

$$\dot{y} = Ay, \quad \text{where} \quad A = \begin{bmatrix} -k_1 & k_2 \\ k_1 & -k_2 \end{bmatrix}.$$

The derivative of the function y is always considered entrywise,

$$\dot{y} = \begin{bmatrix} \dot{y}_1 \\ \dot{y}_2 \end{bmatrix}.$$

Reactions can also have several steps. For example, a reaction of the form

$$A_1 \xrightarrow{k_1} A_2 \underset{k_3}{\overset{k_2}{\rightleftarrows}} A_3 \xrightarrow{k_4} A_4$$

leads to the differential equations

$$\dot{y}_1 = -k_1 y_1,$$
$$\dot{y}_2 = k_1 y_1 - k_2 y_2 + k_3 y_3,$$
$$\dot{y}_3 = k_2 y_2 - (k_3 + k_4) y_3,$$
$$\dot{y}_4 = k_4 y_3,$$

and thus to the system

$$\dot{y} = Ay, \quad \text{where} \quad A = \begin{bmatrix} -k_1 & 0 & 0 & 0 \\ k_1 & -k_2 & k_3 & 0 \\ 0 & k_2 & -(k_3 + k_4) & 0 \\ 0 & 0 & k_4 & 0 \end{bmatrix}.$$

The sum of the entries in each column of A is equal to zero, since for every decrease in a substance with a certain rate other substances increase with the same rate.

In summary, a chemical reaction of first order leads to a system of linear ordinary differential equations of first order that can be written as $\dot{y} = Ay$ with a (real) square matrix A.

We now derive the general theory for systems of linear (real or complex) ordinary differential equations of first order of the form

$$\dot{y} = Ay + g, \quad t \in [0, a]. \tag{17.10}$$

Here $A \in K^{n,n}$ is a given matrix, a is a given positive real number, $g : [0, a] \to K^{n,1}$ is a given (piecewise) continuous function, $y : [0, a] \to K^{n,1}$ is the desired solution, and we assume that $K = \mathbb{R}$ or $K = \mathbb{C}$. If $g(t) = 0 \in K^{n,1}$ for all $t \in [0, a]$, then the system (17.10) is called *homogeneous*, otherwise it is called *non-homogeneous*. For a given system of the form (17.10), the system

$$\dot{y} = Ay, \quad t \in [0, a], \tag{17.11}$$

is called the *associated homogeneous system*.

Lemma 17.11 *The solutions of the homogeneous system (17.11) form a subspace of the (infinite dimensional) K-vector space of the continuously differentiable functions from the interval $[0, a]$ to $K^{n,1}$.*

Proof We will show the required properties according to Lemma 9.5. The function $w = 0$ is continuously differentiable on $[0, a]$ and solves the homogeneous system (17.11). Thus, the solution set of this system is not empty. If

$$w_1, w_2 : [0, a] \to K^{n,1}$$

are continuously differentiable solutions and if $\alpha_1, \alpha_2 \in K$, then $w = \alpha_1 w_1 + \alpha_2 w_2$ is continuously differentiable on $[0, a]$, and

$$\dot{w} = \alpha_1 \dot{w}_1 + \alpha_2 \dot{w}_2 = \alpha_1 A w_1 + \alpha_2 A w_2 = A w,$$

i.e., the function w is a solution of the homogeneous system. $\square$

The following characterization of the solutions of the non-homogeneous system (17.10) is analogous to the characterization of the solution set of a non-homogeneous linear system of equations in Lemma 6.2 (also cp. (8) in Lemma 10.7).

Lemma 17.12 *If $w_1 : [0, a] \to K^{n,1}$ is a solution of the non-homogeneous system (17.10), then every other solution y can be written as $y = w_1 + w_2$, where w_2 is a solution of the associated homogeneous system (17.11).*

Proof If w_1 and y are solutions of (17.10), then The difference $w_2 := y - w_1$ thus is a solution of the associated homogeneous system and $y = w_1 + w_2$. $\square$

In order to describe the solutions of systems of ordinary differential equations, we consider for a given matrix $A \in K^{n,n}$ the matrix exponential function $\exp(tA)$ from Lemma 17.6 or (17.6)–(17.7), where we now consider $t \in [0, a]$ as *real variable*. The power series of the matrix exponential function in Lemma 17.6 converges, and it can be differentiated termwise with respect to the variable t, where again the

17.2 Systems of Linear Ordinary Differential Equations

derivative of a matrix with respect to the variable t is considered entrywise. This yields

$$\frac{d}{dt}\exp(tA) = \frac{d}{dt}\left(I + (tA) + \frac{1}{2}(tA)^2 + \frac{1}{6}(tA)^3 + \ldots\right)$$

$$= A + tA^2 + \frac{1}{2}t^2A^3 + \ldots$$

$$= A\exp(tA).$$

The same result is obtained by the entrywise differentiation of the matrix $\exp(tA)$ in (17.6)–(17.7) with respect to t. With

$$M(t) := \begin{bmatrix} 1 & t & \frac{t^2}{2!} & \cdots & \frac{t^{d-1}}{(d-1)!} \\ & 1 & t & \ddots & \vdots \\ & & \ddots & \ddots & \frac{t^2}{2!} \\ & & & \ddots & t \\ & & & & 1 \end{bmatrix}$$

we obtain

$$\frac{d}{dt}\exp(tJ_d(\lambda)) = \frac{d}{dt}\left(e^{t\lambda}M(t)\right)$$

$$= \lambda e^{t\lambda}M(t) + e^{t\lambda}\dot{M}(t)$$

$$= \lambda e^{t\lambda}M(t) + e^{t\lambda}J_d(0)M(t)$$

$$= (\lambda I_d + J_d(0))\, e^{t\lambda}M(t)$$

$$= J_d(\lambda)\exp(tJ_d(\lambda)),$$

which also gives $\frac{d}{dt}\exp(tA) = A\exp(tA)$.

Theorem 17.13

(1) The unique solution of the homogeneous differential equation system (17.11) for a given initial condition $y(0) = y_0 \in K^{n,1}$ is given by the function $y = \exp(tA)y_0$.
(2) The set of all solutions of the homogeneous differential equation system (17.11) forms an n-dimensional K-vector space with the basis $\{\exp(tA)e_1, \ldots, \exp(tA)e_n\}$.

Proof

(1) If $y(t) = \exp(tA)y_0$, then

$$\dot{y} = \frac{d}{dt}(\exp(tA)y_0) = \left(\frac{d}{dt}\exp(tA)\right)y_0 = (A\exp(tA))y_0$$

$$= A(\exp(tA)y_0) = Ay,$$

and $y(0) = \exp(0)y_0 = I_n y_0 = y_0$. Hence y is a solution of (17.11) that satisfies the initial condition. If w is another such solution and $u := \exp(-tA)w$, then

$$\dot{u} = \frac{d}{dt}(\exp(-tA)w) = -A\exp(-tA)w + \exp(-tA)\dot{w}$$

$$= \exp(-tA)(\dot{w} - Aw) = 0 \in K^{n,1},$$

which shows that the function u has constant entries. In particular, we then have $u = u(0) = w(0) = y_0 = y(0)$ and $w = \exp(tA)y_0$, where we have used that $\exp(-tA) = (\exp(tA))^{-1}$ (cp. Lemma 17.7).

(2) Each of the functions $\exp(tA)e_j : [0, a] \to K^{n,1}$, $j = 1, \ldots, n$, solves the homogeneous system $\dot{y} = Ay$. Since the matrix $\exp(tA) \in K^{n,n}$ is invertible for every $t \in [0, a]$ (cp. Lemma 17.7), these functions are linearly independent.

If $\widetilde{y}$ is an arbitrary solution of $\dot{y} = Ay$, then $\widetilde{y}(0) = y_0$ for some $y_0 \in K^{n,1}$. By (1) then $\widetilde{y}$ is the unique solution of the initial value problem with $y(0) = y_0$, so that $\widetilde{y}(t) = \exp(tA)y_0$. As a consequence, $\widetilde{y}$ is a linear combination of the functions $\exp(tA)e_1, \ldots, \exp(tA)e_n$. $\square$

To describe the solution of the non-homogeneous system (17.10), we need the integral of functions of the form

$$w = \begin{bmatrix} w_1 \\ \vdots \\ w_n \end{bmatrix} : [0, a] \to K^{n,1}.$$

For every fixed $t \in [0, a]$ we define

$$\int_0^t w(s)ds := \begin{bmatrix} \int_0^t w_1(s)ds \\ \vdots \\ \int_0^t w_n(s)ds \end{bmatrix} \in K^{n,1},$$

17.2 Systems of Linear Ordinary Differential Equations

i.e., we apply the integral entrywise to the function w, and we assume that the functions w_j, $j = 1, \ldots, n$, are integrable. By this definition we have

$$\frac{d}{dt}\left(\int_0^t w(s)\,ds\right) = w(t)$$

for all $t \in [0, a]$. We can now determine an explicit solution formula for systems of linear differential equations based on the so-called *Duhamel integral*.[4]

Theorem 17.14 *The unique solution of the non-homogeneous differential equation system (17.10) with the initial condition $y(0) = y_0 \in K^{n,1}$ is given by*

$$y(t) = \exp(tA)y_0 + \exp(tA)\int_0^t \exp(-sA)g(s)\,ds. \tag{17.12}$$

Proof The derivative of the function y defined in (17.12) is

$$\dot{y} = \frac{d}{dt}(\exp(tA)y_0) + \frac{d}{dt}\left(\exp(tA)\int_0^t \exp(-sA)g(s)\,ds\right)$$

$$= A\exp(tA)y_0 + A\exp(tA)\int_0^t \exp(-sA)g(s)\,ds + \exp(tA)\exp(-tA)g$$

$$= A\exp(tA)y_0 + A\exp(tA)\int_0^t \exp(-sA)g(s)\,ds + g$$

$$= Ay + g.$$

Furthermore, we have

$$y(0) = \exp(0)y_0 + \exp(0)\int_0^0 \exp(-sA)g(s)\,ds = y_0,$$

so that y also satisfies the initial condition.

Let now $\tilde{y}$ be another solution of (17.10) that satisfies the initial condition. By Lemma 17.12 we then have $\tilde{y} = y + w$, where w solves the homogeneous system (17.11). Therefore, $w(t) = \exp(tA)c$ for some $c \in K^{n,1}$ (cp. (2) in Theorem 17.13). For $t = 0$ we obtain $y_0 = y_0 + c$, where $c = 0$ and hence $\tilde{y} = y$. □

In the above theorems we have shown that for the explicit solution of systems of linear ordinary differential equations of first order we have to compute the matrix exponential function. While we have introduced this function using the Jordan canonical form of the given matrix, numerical computations based on the

---

[4] Jean-Marie Constant Duhamel (1797–1872).

Jordan canonical form are not advisable (cp. Example 16.24). Because of its significant practical relevance, numerous different algorithms for computing the matrix exponential function have been proposed. But, as shown in the article [12], no existing algorithm is completely satisfactory.

Example 17.15 The example from circuit simulation presented in Sect. 1.5 lead to the system of ordinary differential equations

$$\frac{d}{dt} I = -\frac{R}{L} I - \frac{1}{L} V_C + \frac{1}{L} V_S,$$

$$\frac{d}{dt} V_C = \frac{1}{C} I.$$

Using (17.12) and the initial values $I(0) = I^0$ and $V_C(0) = V_C^0$, we obtain the solution

$$\begin{bmatrix} I \\ V_C \end{bmatrix} = \exp\left(t \begin{bmatrix} -R/L & -1/L \\ 1/C & 0 \end{bmatrix} \right) \begin{bmatrix} I^0 \\ V_C^0 \end{bmatrix}$$

$$+ \int_0^t \exp\left((t-s) \begin{bmatrix} -R/L & -1/L \\ 1/C & 0 \end{bmatrix} \right) \begin{bmatrix} V_S(s)/L \\ 0 \end{bmatrix} ds.$$

Let us also consider an example from Mechanics.

Example 17.16 A weight with mass $m > 0$ is attached to a spring with the spring constant $\mu > 0$. Let $x_0 > 0$ be the initial distance of the weight from its equilibrium position; see the figure below. We want to determine the position $x(t)$ of the weight at time $t \geq 0$, where $x(0) = x_0$.

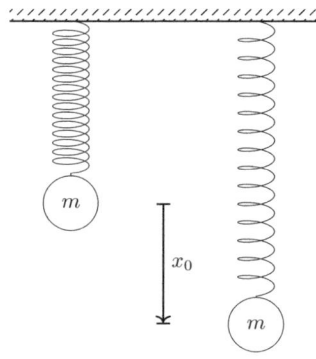

The extension of the spring is described by *Hooke's law*.[5] The corresponding ordinary differential equation of second order is

$$\ddot{x} = \frac{d^2}{dt^2}x = -\frac{\mu}{m}x,$$

with initial conditions $x(0) = x_0$ and $\dot{x}(0) = v_0$, where $v_0 > 0$ is the initial velocity of the weight. We can write this differential equation of second order for x as a system of first order by introducing the velocity v as new variable. The velocity is given by the derivative of the position with respect to time, i.e., $v = \dot{x}$, and thus for the acceleration we have $\dot{v} = \ddot{x}$, which yields the system

$$\dot{y} = Ay, \quad \text{where} \quad A = \begin{bmatrix} 0 & 1 \\ -\frac{\mu}{m} & 0 \end{bmatrix} \quad \text{and} \quad y = \begin{bmatrix} x \\ v \end{bmatrix}.$$

The initial condition then is $y(0) = y_0 = [x_0, v_0]^T$.

By Theorem 17.13, the unique solution of this homogeneous initial value problem is given by the function $y(t) = \exp(tA)y_0$. We consider A as an element of $\mathbb{C}^{2,2}$. The eigenvalues of A are the two complex (non-real) numbers $\lambda_1 = i\rho$ and $\lambda_2 = -i\rho = \overline{\lambda}_1$, where $\rho := \sqrt{\frac{\mu}{m}}$. Corresponding eigenvectors are

$$s_1 = \begin{bmatrix} 1 \\ i\rho \end{bmatrix} \in \mathbb{C}^{2,1}, \quad s_2 = \begin{bmatrix} 1 \\ -i\rho \end{bmatrix} \in \mathbb{C}^{2,1}$$

and thus

$$\exp(tA)y_0 = S \begin{bmatrix} e^{it\rho} & 0 \\ 0 & e^{-it\rho} \end{bmatrix} S^{-1} y_0, \quad S = \begin{bmatrix} 1 & 1 \\ i\rho & -i\rho \end{bmatrix} \in \mathbb{C}^{2,2}.$$

Exercises

17.1 Construct a matrix $A = [a_{ij}] \in \mathbb{C}^{2,2}$ with $A^3 \neq [(a_{ij})^3]$.

17.2 Determine all solutions $X \in \mathbb{C}^{2,2}$ of the matrix equation $X^2 = I_2$, and classify which of these solutions are primary square roots of I_2.

17.3 Determine a matrix $X \in \mathbb{C}^{2,2}$ with real entries and $X^2 = -I_2$.

17.4 Determine A^{2020} for the matrix $A = \begin{bmatrix} 1 & 1 & 0 \\ 0 & 1 & 0 \\ 0 & 0 & -1 \end{bmatrix}$.

17.5 Prove Lemma 17.3.

---

[5] Sir Robert Hooke (1635–1703).

17.6 Prove the following assertions for $A \in \mathbb{C}^{n,n}$:
 (a) $\det(\exp(A)) = \exp(\operatorname{trace}(A))$.
 (b) If $A^H = -A$, then $\exp(A)$ is unitary.
 (c) If $A^2 = I$, then $\exp(A) = \frac{1}{2}(e + \frac{1}{e})I + \frac{1}{2}(e - \frac{1}{e})A$.

17.7 Let $A = S \operatorname{diag}(J_{d_1}(\lambda_1), \ldots, J_{d_m}(\lambda_m)) S^{-1} \in \mathbb{C}^{n,n}$ with $\operatorname{rank}(A) = n$. Determine the primary matrix function $f(A)$ for $f(z) = z^{-1}$. Does this function also exist if $\operatorname{rank}(A) < n$?

17.8 Let $\log : \{z = re^{i\varphi} \mid r > 0, -\pi < \varphi < \pi\} \to \mathbb{C}, re^{i\varphi} \mapsto \ln(r) + i\varphi$, be the principle branch of the complex logarithm (where ln denotes the real natural logarithm). Show that this function is defined on the spectrum of

$$A = \begin{bmatrix} 0 & 1 \\ -1 & 0 \end{bmatrix} \in \mathbb{C}^{2,2},$$

and compute $\log(A)$ as well as $\exp(\log(A))$.

17.9 Compute

$$\exp\left(\begin{bmatrix} 0 & 1 \\ -1 & 0 \end{bmatrix}\right), \quad \exp\left(\begin{bmatrix} -1 & 1 \\ -1 & -3 \end{bmatrix}\right), \quad \sin\left(\begin{bmatrix} \pi & 1 & 1 \\ 0 & \pi & 1 \\ 0 & 0 & \pi \end{bmatrix}\right).$$

17.10 Construct two matrices $A, B \in \mathbb{C}^{2,2}$ with $\exp(A + B) \neq \exp(A)\exp(B)$.

17.11 Prove Theorem 17.9.

17.12 Let

$$A = \begin{bmatrix} 5 & 1 & 1 \\ 0 & 5 & 1 \\ 0 & 0 & 4 \end{bmatrix} \in \mathbb{R}^{3,3}.$$

Compute $\exp(tA)$ for $t \in \mathbb{R}$ and solve the homogeneous system of differential equations $\dot{y} = Ay$ with the initial condition $y(0) = [1, 1, 1]^T$.

17.13 Compute the matrix $\exp(tA)$ from Example 17.16 explicitly and thus show that $\exp(tA) \in \mathbb{R}^{2,2}$ (for $t \in \mathbb{R}$), despite the fact that the eigenvalues and eigenvectors of A are not real.

Special Classes of Endomorphisms 18

In this chapter we discuss some classes of endomorphisms (or square matrices) whose eigenvalues and eigenvectors have special properties. Such properties only exist under further assumptions, and in this chapter our assumptions concern the relationship between the given endomorphism and its adjoint endomorphism. Thus, we focus on Euclidean or unitary vector spaces. This leads to the classes of normal, orthogonal, unitary and selfadjoint endomorphisms. Each of these classes has a natural counterpart in the set of square (real or complex) matrices.

18.1 Normal Endomorphisms

We start with the definition of a normal[1] endomorphism or matrix.

Definition 18.1. Let $\mathcal{V}$ be a finite dimensional Euclidean or unitary vector space. An endomorphism $f \in \mathcal{L}(\mathcal{V}, \mathcal{V})$ is called *normal* if $f \circ f^{ad} = f^{ad} \circ f$. A matrix $A \in \mathbb{R}^{n,n}$ or $A \in \mathbb{C}^{n,n}$ is called *normal* if $A^T A = A A^T$ or $A^H A = A A^H$, respectively.

For all $z \in \mathbb{C}$ we have $\overline{z}z = |z|^2 = z\overline{z}$. The property of normality can therefore be interpreted as a generalization of this property of complex numbers.

We will first study the properties of normal endomorphisms on a finite dimensional *unitary* vector space $\mathcal{V}$. Recall the following results:

(1) If B is an orthonormal basis of $\mathcal{V}$ and if $f \in \mathcal{L}(\mathcal{V}, \mathcal{V})$, then $([f]_{B,B})^H = [f^{ad}]_{B,B}$ (cp. Theorem 13.12).
(2) Every $f \in \mathcal{L}(\mathcal{V}, \mathcal{V})$ can be unitarily triangulated (cp. Corollary 14.21, Schur's theorem). This does not hold in general in the Euclidean case, since not every real polynomial decomposes into linear factors over $\mathbb{R}$.

---

[1] Otto Toeplitz (1881–1940) introduced this term in 1918 in the context of bilinear forms.

Using these results we obtain the following characterization of normal endomorphisms on a unitary vector space.

Theorem 18.2 *If V is a finite dimensional unitary vector space, then $f \in \mathcal{L}(V, V)$ is normal if and only if there exists an orthonormal basis B of V such that $[f]_{B,B}$ is a diagonal matrix, i.e., f is unitarily diagonalizable.*

Proof Let $f \in \mathcal{L}(V, V)$ be normal and let B be an orthonormal basis of V such that $R := [f]_{B,B}$ is an upper triangular matrix. Then $R^H = [f^{ad}]_{B,B}$, and from $f \circ f^{ad} = f^{ad} \circ f$ we obtain

$$RR^H = [f \circ f^{ad}]_{B,B} = [f^{ad} \circ f]_{B,B} = R^H R,$$

i.e., R is normal.

We now show by induction on n, that every normal upper triangular matrix $R \in \mathbb{C}^{n,n}$ is diagonal. This is obvious for $n = 1$.

Let the assertion hold for an $n \geq 1$, and let $R \in \mathbb{C}^{n+1,n+1}$ be upper triangular and normal. We write R as

$$R = \begin{bmatrix} R_1 & r_1 \\ 0 & \alpha_1 \end{bmatrix},$$

where $R_1 \in \mathbb{C}^{n,n}$ is upper triangular, $r_1 \in \mathbb{C}^{n,1}$, and $\alpha_1 \in \mathbb{C}$. Then

$$\begin{bmatrix} R_1 R_1^H + r_1 r_1^H & \overline{\alpha}_1 r_1 \\ \alpha_1 r_1^H & |\alpha_1|^2 \end{bmatrix} = RR^H = R^H R = \begin{bmatrix} R_1^H R_1 & R_1^H r_1 \\ r_1^H R_1 & r_1^H r_1 + |\alpha_1|^2 \end{bmatrix}.$$

From $|\alpha_1|^2 = r_1^H r_1 + |\alpha_1|^2$ we obtain $r_1^H r_1 = 0$, hence $r_1 = 0$ and $R_1 R_1^H = R_1^H R_1$. By the induction hypothesis, $R_1 \in \mathbb{C}^{n,n}$ is diagonal, and therefore

$$R = \begin{bmatrix} R_1 & 0 \\ 0 & \alpha_1 \end{bmatrix}$$

is diagonal as well.

Conversely, suppose that there exists orthonormal basis B of V such that $[f]_{B,B}$ is diagonal. Then $[f^{ad}]_{B,B} = ([f]_{B,B})^H$ is diagonal and, since diagonal matrices commute, we have

$$[f \circ f^{ad}]_{B,B} = [f]_{B,B}[f^{ad}]_{B,B} = [f^{ad}]_{B,B}[f]_{B,B} = [f^{ad} \circ f]_{B,B},$$

which implies $f \circ f^{ad} = f^{ad} \circ f$, and hence f is normal. □

The application of this theorem to the unitary vector space $V = \mathbb{C}^{n,1}$ with the standard scalar product and a matrix $A \in \mathbb{C}^{n,n}$ viewed as element of $\mathcal{L}(V, V)$ yields the following "matrix version".

18.1 Normal Endomorphisms

Corollary 18.3 *A matrix $A \in \mathbb{C}^{n,n}$ is normal if and only if there exists a unitary matrix $Q \in \mathbb{C}^{n,n}$ and a diagonal matrix $D \in \mathbb{C}^{n,n}$ such that $A = QDQ^H$, i.e., A is unitarily diagonalizable.*

The columns of the matrix Q in Corollary 18.3 form an orthonormal basis of $\mathbb{C}^{n,1}$ (with respect to the standard scalar product) consisting of eigenvectors of A and the diagonal entries of D are the corresponding eigenvalues.

The following theorem presents another characterization of normal endomorphisms on a unitary vector space.

Theorem 18.4 *If $\mathcal{V}$ is a finite dimensional unitary vector space, then $f \in \mathcal{L}(\mathcal{V}, \mathcal{V})$ is normal if and only if there exists a polynomial $p \in \mathbb{C}[t]$ with $p(f) = f^{ad}$.*

Proof If $p(f) = f^{ad}$ for a polynomial $p \in \mathbb{C}[t]$, then

$$f \circ f^{ad} = f \circ p(f) = p(f) \circ f = f^{ad} \circ f,$$

and hence f is normal.

Conversely, if f is normal, then there exists an orthonormal basis B of $\mathcal{V}$, such that $[f]_{B,B} = \operatorname{diag}(\lambda_1, \ldots, \lambda_n)$. Furthermore,

$$[f^{ad}]_{B,B} = ([f]_{B,B})^H = \operatorname{diag}(\overline{\lambda}_1, \ldots, \overline{\lambda}_n).$$

Let $p \in \mathbb{C}[t]$ be a polynomial with $p(\lambda_j) = \overline{\lambda}_j$ for $j = 1, \ldots, n$. Such a polynomial can be explicitly constructed using the Lagrange basis of $\mathbb{C}[t]_{\leq n-1}$ (cp. Exercise 10.18). Then

$$[f^{ad}]_{B,B} = \operatorname{diag}(\overline{\lambda}_1, \ldots, \overline{\lambda}_n) = \operatorname{diag}(p(\lambda_1), \ldots, p(\lambda_n)) = p(\operatorname{diag}(\lambda_1, \ldots, \lambda_n))$$
$$= p([f]_{B,B}) = [p(f)]_{B,B},$$

and hence also $f^{ad} = p(f)$. $\square$

Several other characterizations of normal endomorphisms on a finite dimensional unitary vector space and of normal matrices $A \in \mathbb{C}^{n,n}$ can be found in the article [7] (see also Exercise 15.22).

We now consider the Euclidean case, where we focus on real square matrices. All the results can be formulated analogously for normal endomorphisms on a finite dimensional Euclidean vector space.

Let $A \in \mathbb{R}^{n,n}$ be normal, i.e., $A^T A = AA^T$. Then A also satisfies $A^H A = AA^H$ and when A is considered as an element of $\mathbb{C}^{n,n}$, it is unitarily diagonalizable, i.e., $A = QDQ^H$ holds for a unitary matrix $Q \in \mathbb{C}^{n,n}$ and a diagonal matrix $D \in \mathbb{C}^{n,n}$.

Despite the fact that A has real entries, neither Q nor D will be real in general, since A as an element of $\mathbb{R}^{n,n}$ may not be diagonalizable. For instance,

$$A = \begin{bmatrix} 1 & 2 \\ -2 & 1 \end{bmatrix} \in \mathbb{R}^{2,2}$$

is a normal matrix that is not diagonalizable (over $\mathbb{R}$). Considered as element of $\mathbb{C}^{2,2}$, it has the eigenvalues $1 + 2\mathbf{i}$ and $1 - 2\mathbf{i}$ and it is unitarily diagonalizable.

To discuss the case of real normal matrices in more detail, we first prove a "real version" of Schur's theorem.

Theorem 18.5 *For every matrix $A \in \mathbb{R}^{n,n}$ there exists an orthogonal matrix $U \in \mathbb{R}^{n,n}$ with*

$$U^T A U = R = \begin{bmatrix} R_{11} & \cdots & R_{1m} \\ & \ddots & \vdots \\ & & R_{mm} \end{bmatrix} \in \mathbb{R}^{n,n},$$

where for every $j = 1, \ldots, m$ either $R_{jj} \in \mathbb{R}^{1,1}$ or

$$R_{jj} = \begin{bmatrix} r_1^{(j)} & r_2^{(j)} \\ r_3^{(j)} & r_4^{(j)} \end{bmatrix} \in \mathbb{R}^{2,2} \quad \text{with} \quad r_3^{(j)} \neq 0.$$

In the second case R_{jj} has, considered as complex matrix, a pair of complex conjugate eigenvalues of the form $\alpha_j \pm \mathbf{i}\beta_j$ with $\alpha_j \in \mathbb{R}$ and $\beta_j \in \mathbb{R} \setminus \{0\}$. The matrix R is called a real Schur form of A.

Proof We proceed via induction on n. For $n = 1$ we have $A = [a_{11}] = R$ and $U = [1]$.

Suppose that the assertion holds for some $n \geq 1$ and let $A \in \mathbb{R}^{n+1,n+1}$ be given. We consider A as an element of $\mathbb{C}^{n+1,n+1}$. Then A has an eigenvalue $\lambda = \alpha + \mathbf{i}\beta \in \mathbb{C}$, $\alpha, \beta \in \mathbb{R}$, corresponding to the eigenvector $v = x + \mathbf{i}y \in \mathbb{C}^{n+1,1}$, $x, y \in \mathbb{R}^{n+1,1}$, and we have $Av = \lambda v$. Dividing this equation into its real and imaginary parts, we obtain the two real equations

$$Ax = \alpha x - \beta y \quad \text{and} \quad Ay = \beta x + \alpha y. \tag{18.1}$$

We have two cases:

Case 1: $\beta = 0$ Then the two equations in (18.1) are $Ax = \alpha x$ and $Ay = \alpha y$. Thus at least one of the real vectors x or y is an eigenvector corresponding to the real eigenvalue α of A. Without loss of generality we assume that this is the vector x and that $\|x\|_2 = 1$. We extend x by the vectors $w_2, \ldots, w_{n+1}$ to an

18.1 Normal Endomorphisms

orthonormal basis of $\mathbb{R}^{n+1,1}$ with respect to the standard scalar product. The matrix $U_1 := [x, w_2, \ldots, w_{n+1}] \in \mathbb{R}^{n+1,n+1}$ then is orthogonal and satisfies

$$U_1^T A U_1 = \begin{bmatrix} \alpha & \star \\ 0 & A_1 \end{bmatrix}$$

for a matrix $A_1 \in \mathbb{R}^{n,n}$. By the induction hypothesis there exists an orthogonal matrix $U_2 \in \mathbb{R}^{n,n}$ such that $R_1 := U_2^T A_1 U_2$ has the desired form. The matrix

$$U := U_1 \begin{bmatrix} 1 & 0 \\ 0 & U_2 \end{bmatrix}$$

is orthogonal and satisfies

$$U^T A U = \begin{bmatrix} 1 & 0 \\ 0 & U_2^T \end{bmatrix} U_1^T A U_1 \begin{bmatrix} 1 & 0 \\ 0 & U_2 \end{bmatrix} = \begin{bmatrix} \alpha & \star \\ 0 & R_1 \end{bmatrix} =: R,$$

where R has the desired form.

Case 2: $\beta \neq 0$ We first show that x, y are linearly independent. If $x = 0$, then using $\beta \neq 0$ in the first equation in (18.1) implies that also $y = 0$. This is not possible, since the eigenvector $v = x + \mathbf{i}y$ must be nonzero. Thus, $x \neq 0$, and using $\beta \neq 0$ in the second equation in (18.1) implies that also $y \neq 0$. If $x, y \in \mathbb{R}^{n+1,1} \setminus \{0\}$ are linearly dependent, then there exists a $\mu \in \mathbb{R} \setminus \{0\}$ with $y = \mu x$. The two equations in (18.1) then can be written as

$$Ax = (\alpha - \beta\mu)x \quad \text{and} \quad Ax = \frac{1}{\mu}(\beta + \alpha\mu)x,$$

which implies that $\beta(1 + \mu^2) = 0$. Since $1 + \mu^2 \neq 0$ for all $\mu \in \mathbb{R}$, this implies $\beta = 0$, which contradicts the assumption that $\beta \neq 0$. Consequently, x, y are linearly independent.

We can combine the two equations in (18.1) to the system

$$A[x, y] = [x, y] \begin{bmatrix} \alpha & \beta \\ -\beta & \alpha \end{bmatrix},$$

where $\operatorname{rank}([x, y]) = 2$. Applying the Gram-Schmidt method with respect to the standard scalar product of $\mathbb{R}^{n+1,1}$ to the matrix $[x, y] \in \mathbb{R}^{n+1,2}$ yields

$$[x, y] = [q_1, q_2] \begin{bmatrix} r_{11} & r_{12} \\ 0 & r_{22} \end{bmatrix} =: QR_1,$$

with $Q^T Q = I_2$ and $R_1 \in GL_2(\mathbb{R})$. It then follows that

$$AQ = A[x, y]R_1^{-1} = [x, y]\begin{bmatrix} \alpha & \beta \\ -\beta & \alpha \end{bmatrix} R_1^{-1} = QR_1 \begin{bmatrix} \alpha & \beta \\ -\beta & \alpha \end{bmatrix} R_1^{-1}.$$

The real matrix

$$R_2 := R_1 \begin{bmatrix} \alpha & \beta \\ -\beta & \alpha \end{bmatrix} R_1^{-1}$$

has, considered as element of $\mathbb{C}^{2,2}$, the pair of complex conjugate eigenvalues $\alpha \pm i\beta$ with $\beta \neq 0$. In particular, the $(2, 1)$-entry of R_2 is nonzero, since otherwise R_2 would have two real eigenvalues.

We again extend q_1, q_2 by vectors $w_3, \ldots, w_{n+1}$ to an orthonormal basis of $\mathbb{R}^{n+1,1}$ with respect to the standard scalar product. (For $n = 1$ the list $w_3, \ldots, w_{n+1}$ is empty.) Then $U_1 := [Q, w_3, \ldots, w_{n+1}] \in \mathbb{R}^{n+1,n+1}$ is orthogonal and we have

$$U_1^T A U_1 = U_1^T [AQ, A[w_3, \ldots, w_{n+1}]] = U_1^T [QR_2, A[w_3, \ldots, w_{n+1}]]$$
$$= \begin{bmatrix} R_2 & \star \\ 0 & A_1 \end{bmatrix}$$

for a matrix $A_1 \in \mathbb{R}^{n-1,n-1}$. Analogously to the first case, an application of the induction hypothesis to this matrix yields the desired matrices R and U. $\square$

Theorem 18.5 implies the following result for real normal matrices.

Corollary 18.6 *A matrix $A \in \mathbb{R}^{n,n}$ is normal if and only if there exists an orthogonal matrix $U \in \mathbb{R}^{n,n}$ with*

$$U^T A U = \mathrm{diag}(R_1, \ldots, R_m),$$

where, for every $j = 1, \ldots, m$ either $R_j \in \mathbb{R}^{1,1}$ or

$$R_j = \begin{bmatrix} \alpha_j & \beta_j \\ -\beta_j & \alpha_j \end{bmatrix} \in \mathbb{R}^{2,2} \quad \text{with} \quad \beta_j \neq 0.$$

In the second case the matrix R_j has, considered as complex matrix, a pair of complex conjugate eigenvalues of the form $\alpha_j \pm i\beta_j$.

Proof Exercise. $\square$

18.2 Orthogonal and Unitary Endomorphisms

Example 18.7 The matrix

$$A = \frac{1}{2} \begin{bmatrix} 0 & \sqrt{2} & -\sqrt{2} \\ -\sqrt{2} & 1 & 1 \\ \sqrt{2} & 1 & 1 \end{bmatrix} \in \mathbb{R}^{3,3}$$

has, considered as a complex matrix, the eigenvalues $1, \mathbf{i}, -\mathbf{i}$. It is therefore neither diagonalizable nor can it be triangulated over $\mathbb{R}$. For the orthogonal matrix

$$U = \frac{1}{2} \begin{bmatrix} 0 & 2 & 0 \\ -\sqrt{2} & 0 & \sqrt{2} \\ \sqrt{2} & 0 & \sqrt{2} \end{bmatrix} \in \mathbb{R}^{3,3}$$

the transformed matrix

$$U^T A U = \begin{bmatrix} 0 & 1 & 0 \\ -1 & 0 & 0 \\ 0 & 0 & 1 \end{bmatrix}$$

is in real Schur form.

18.2 Orthogonal and Unitary Endomorphisms

In this section we extend the concept of orthogonal and unitary matrices to endomorphisms.

Definition 18.8 Let $\mathcal{V}$ be a finite dimensional Euclidean or unitary vector space. An endomorphism $f \in \mathcal{L}(\mathcal{V}, \mathcal{V})$ is called *orthogonal* or *unitary*, respectively, if $f^{ad} \circ f = \text{Id}_\mathcal{V}$.

If $f^{ad} \circ f = \text{Id}_\mathcal{V}$, then $f^{ad} \circ f$ is bijective and hence f is injective (cp. (3) in Theorem 2.20). Corollary 10.11 implies that f is bijective. Hence f^{ad} is the unique inverse of f, and we also have $f \circ f^{ad} = \text{Id}_\mathcal{V}$ (cp. our remarks following Definition 2.22).

Note that an orthogonal or unitary endomorphism f is normal, and therefore all results from the previous section also apply to f.

Lemma 18.9 *Let $\mathcal{V}$ be a finite dimensional Euclidean or unitary vector space and let $f \in \mathcal{L}(\mathcal{V}, \mathcal{V})$ be orthogonal or unitary, respectively. If B is an orthonormal basis of $\mathcal{V}$, then $[f]_{B,B}$ is an orthogonal or unitary matrix, respectively.*

Proof Let $\dim(\mathcal{V}) = n$. For every orthonormal basis B of $\mathcal{V}$ we have

$$I_n = [\mathrm{Id}_\mathcal{V}]_{B,B} = [f^{ad} \circ f]_{B,B} = [f^{ad}]_{B,B}[f]_{B,B} = ([f]_{B,B})^H [f]_{B,B},$$

and thus $[f]_{B,B}$ is orthogonal or unitary, respectively. (In the Euclidean case $([f]_{B,B})^H = ([f]_{B,B})^T$.) □

We next show that an orthogonal or unitary endomorphism is characterized by the fact that it does not change the scalar product of arbitrary vectors.

Lemma 18.10 *Let $\mathcal{V}$ be a finite dimensional Euclidean or unitary vector space with the scalar product $\langle \cdot, \cdot \rangle$. Then $f \in \mathcal{L}(\mathcal{V}, \mathcal{V})$ is orthogonal or unitary, respectively, if and only if $\langle f(v), f(w) \rangle = \langle v, w \rangle$ for all $v, w \in \mathcal{V}$.*

Proof If f is orthogonal or unitary and if $v, w \in \mathcal{V}$, then

$$\langle v, w \rangle = \langle \mathrm{Id}_\mathcal{V}(v), w \rangle = \langle (f^{ad} \circ f)(v), w \rangle = \langle f(v), f(w) \rangle.$$

On the other hand, suppose that $\langle v, w \rangle = \langle f(v), f(w) \rangle$ for all $v, w \in \mathcal{V}$. Then

$$0 = \langle v, w \rangle - \langle f(v), f(w) \rangle = \langle v, w \rangle - \langle v, (f^{ad} \circ f)(w) \rangle$$
$$= \langle v, (\mathrm{Id}_\mathcal{V} - f^{ad} \circ f)(w) \rangle.$$

Since the scalar product is non-degenerate and v can be chosen arbitrarily, we have $(\mathrm{Id}_\mathcal{V} - f^{ad} \circ f)(w) = 0$ for all $w \in \mathcal{V}$, and hence $\mathrm{Id}_\mathcal{V} = f^{ad} \circ f$. □

We have the following corollary (cp. Lemma 12.14).

Corollary 18.11 *If $\mathcal{V}$ is a finite dimensional Euclidean or unitary vector space with the scalar product $\langle \cdot, \cdot \rangle$, $f \in \mathcal{L}(\mathcal{V}, \mathcal{V})$ is orthogonal or unitary, respectively, and $\| \cdot \| = \langle \cdot, \cdot \rangle^{1/2}$ is the norm induced by the scalar product, then $\| f(v) \| = \| v \|$ for all $v \in \mathcal{V}$.*

For the vector space $\mathcal{V} = \mathbb{C}^{n,1}$ with the standard scalar product and induced norm $\|v\|_2 = (v^H v)^{1/2}$ as well as a unitary matrix $A \in \mathbb{C}^{n,n}$, we have $\|Av\|_2 = \|v\|_2$ for all $v \in \mathbb{C}^{n,1}$. Thus,

$$\|A\|_2 = \sup_{v \in \mathbb{C}^{n,1} \setminus \{0\}} \frac{\|Av\|_2}{\|v\|_2} = 1$$

(cp. (6) in Example 12.4). This holds analogously for orthogonal matrices $A \in \mathbb{R}^{n,n}$.

We now study the eigenvalues and eigenvectors of orthogonal and unitary endomorphisms.

18.2 Orthogonal and Unitary Endomorphisms

Lemma 18.12 *Let $\mathcal{V}$ be a finite dimensional Euclidean or unitary vector space and let $f \in \mathcal{L}(\mathcal{V}, \mathcal{V})$ be orthogonal or unitary, respectively. If λ is an eigenvalue of f, then $|\lambda| = 1$.*

Proof Let $\langle \cdot, \cdot \rangle$ be the scalar product on $\mathcal{V}$. If $f(v) = \lambda v$ with $v \neq 0$, then

$$\langle v, v \rangle = \langle f(v), f(v) \rangle = \langle \lambda v, \lambda v \rangle = |\lambda|^2 \langle v, v \rangle,$$

and $\langle v, v \rangle \neq 0$ implies that $|\lambda| = 1$. $\square$

The statement of Lemma 18.12 holds, in particular, for unitary and orthogonal matrices. However, one should keep in mind that an orthogonal matrix (or an orthogonal endomorphism) may not have an eigenvalue. For example, the orthogonal matrix

$$A = \begin{bmatrix} 0 & -1 \\ 1 & 0 \end{bmatrix} \in \mathbb{R}^{2,2}$$

has the characteristic polynomial $P_A = t^2 + 1$, which has no real roots. If considered as an element of $\mathbb{C}^{2,2}$, the matrix A has the eigenvalues $\mathbf{i}$ and $-\mathbf{i}$.

Theorem 18.13

(1) If $A \in \mathbb{C}^{n,n}$ is unitary, then there exists a unitary matrix $Q \in \mathbb{C}^{n,n}$ with

$$Q^H A Q = \mathrm{diag}(\lambda_1, \ldots, \lambda_n)$$

and $|\lambda_j| = 1$ for $j = 1, \ldots, n$.
(2) If $A \in \mathbb{R}^{n,n}$ is orthogonal, then there exists an orthogonal matrix $U \in \mathbb{R}^{n,n}$ with

$$U^T A U = \mathrm{diag}(R_1, \ldots, R_m),$$

where for every $j = 1, \ldots, m$ either $R_j = [\lambda_j] \in \mathbb{R}^{1,1}$ with $\lambda_j = \pm 1$ or

$$R_j = \begin{bmatrix} c_j & s_j \\ -s_j & c_j \end{bmatrix} \in \mathbb{R}^{2,2} \quad \text{with} \quad s_j \neq 0 \quad \text{and} \quad c_j^2 + s_j^2 = 1.$$

Proof

(1) A unitary matrix $A \in \mathbb{C}^{n,n}$ is normal and hence unitarily diagonalizable (cp. Corollary 18.3). By Lemma 18.12, all eigenvalues of A have absolute value 1.

(2) An orthogonal matrix A is normal and hence by Corollary 18.6 there exists an orthogonal matrix $U \in \mathbb{R}^{n,n}$ with $U^T A U = \text{diag}(R_1, \ldots, R_m)$, where either $R_j \in \mathbb{R}^{1,1}$ or

$$R_j = \begin{bmatrix} \alpha_j & \beta_j \\ -\beta_j & \alpha_j \end{bmatrix} \in \mathbb{R}^{2,2}$$

with $\beta_j \neq 0$. In the first case then $R_j = [\lambda_j]$ with $|\lambda_j| = 1$ by Lemma 18.12. Since A and U are orthogonal, also $U^T A U$ is orthogonal, and hence every diagonal block R_j is orthogonal as well. From $R_j^T R_j = I_2$ we obtain $\alpha_j^2 + \beta_j^2 = 1$, so that R_j has the desired form. $\square$

We now study two important classes of orthogonal matrices.

Example 18.14 Let $i, j, n \in \mathbb{N}$ with $1 \leq i < j \leq n$ and let $\alpha \in \mathbb{R}$. We define

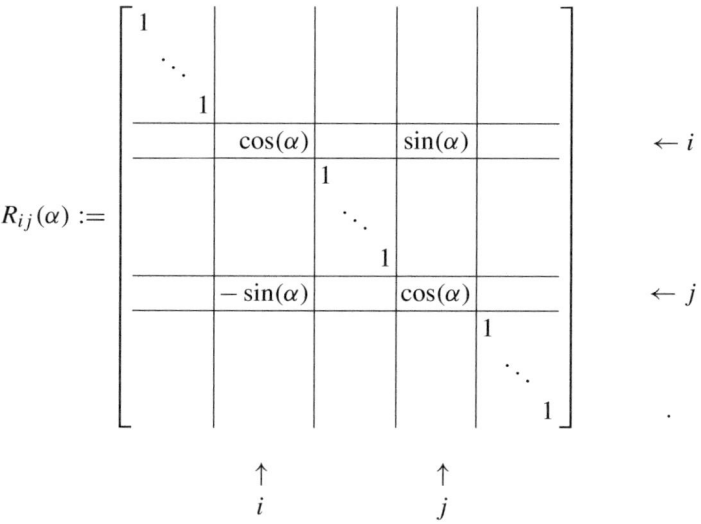

The matrix $R_{ij}(\alpha) = [r_{ij}] \in \mathbb{R}^{n,n}$ is equal to the identity matrix I_n except for its entries

$$r_{ii} = \cos(\alpha), \quad r_{ij} = -\sin(\alpha), \quad r_{ji} = \sin(\alpha), \quad r_{jj} = \cos(\alpha).$$

For $n = 2$ we have the matrix

$$R_{12}(\alpha) = \begin{bmatrix} \cos(\alpha) & \sin(\alpha) \\ -\sin(\alpha) & \cos(\alpha) \end{bmatrix},$$

18.2 Orthogonal and Unitary Endomorphisms

which satisfies

$$R_{12}(\alpha)^T R_{12}(\alpha) = \begin{bmatrix} \cos(\alpha) & -\sin(\alpha) \\ \sin(\alpha) & \cos(\alpha) \end{bmatrix} \begin{bmatrix} \cos(\alpha) & \sin(\alpha) \\ -\sin(\alpha) & \cos(\alpha) \end{bmatrix}$$

$$= \begin{bmatrix} \cos^2(\alpha) + \sin^2(\alpha) & 0 \\ 0 & \cos^2(\alpha) + \sin^2(\alpha) \end{bmatrix}$$

$$= I_2 = R_{12}(\alpha) R_{12}(\alpha)^T,$$

i.e., $R_{12}(\alpha)$ is orthogonal. We easily see that each of the matrices $R_{ij}(\alpha) \in \mathbb{R}^{n,n}$ is orthogonal.

The multiplication of a vector $v \in \mathbb{R}^{n,1}$ with the matrix $R_{ij}(\alpha)$ results in a clockwise rotation of v by the angle α in the (i, j)-coordinate plane. The multiplication with $R_{ij}(\alpha)^T = R_{ij}(-\alpha)$ results in a counterclockwise rotation. In Numerical Mathematics, the matrices $R_{ij}(\alpha)$ are called *Givens rotations*.[2]

This is illustrated in the figure below for the vector $v = [1.0, 0.75]^T \in \mathbb{R}^{2,1}$ and the matrices $R_{12}(\frac{\pi}{2})^T$ and $R_{12}(\frac{2\pi}{3})^T$, which represent counterclockwise rotations by 90° and 120°, respectively.

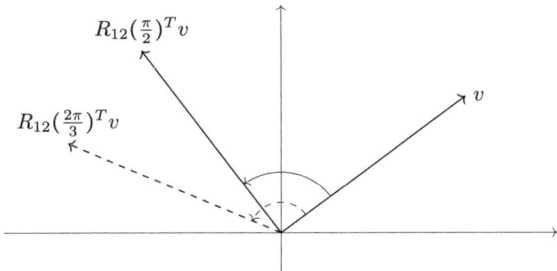

Example 18.15 For $u \in \mathbb{R}^{n,1} \setminus \{0\}$ we define the *Householder matrix*

$$H(u) := I_n - \frac{2}{u^T u} u u^T \in \mathbb{R}^{n,n}, \qquad (18.2)$$

and for $u = 0$ we set $H(0) := I_n$. For every $u \in \mathbb{R}^{n,1}$ then $H(u)$ is an orthogonal matrix (cp. Exercise 12.23). The multiplication of a vector $v \in \mathbb{R}^{n,1}$ with the matrix $H(u)$ describes a reflection of v at the hyperplane

$$(\text{span}\{u\})^\perp = \{y \in \mathbb{R}^{n,1} \mid u^T y = 0\},$$

---

[2] Wallace Givens (1910–1993), pioneer of Numerical Linear Algebra.

i.e., the hyperplane of vectors that are orthogonal to u with respect to the standard scalar product. This is illustrated in the figure below for the vector $v = [1.75, 0.5]^T \in \mathbb{R}^{2,1}$ and the Householder matrix

$$H(u) = \begin{bmatrix} 0 & 1 \\ 1 & 0 \end{bmatrix},$$

which corresponds to $u = [-1, 1]^T \in \mathbb{R}^{2,1}$.

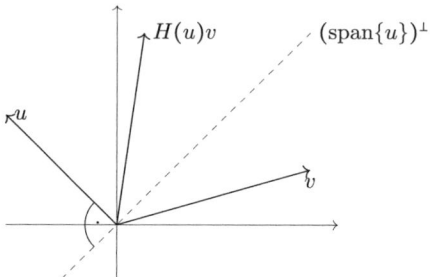

MATLAB-Minute 12

Let $u = [5, 3, 1]^T \in \mathbb{R}^{3,1}$. Apply the command norm(u) to compute the Euclidean norm of u and form the Householder matrix H=eye(3)-(2/(u'*u))*(u*u'). Check the orthogonality of H via the computation of norm(H'*H-eye(3)). Form the vector v=H*u and compare the Euclidean norms of u and v.

18.3 Selfadjoint Endomorphisms

We have already studied selfadjoint endomorphisms f on a finite dimensional Euclidean or unitary vector space. The defining property for this class of endomorphisms is $f = f^{ad}$ (cp. Definition 13.13).

Obviously, selfadjoint endomorphisms are normal and hence the results of Sect. 18.1 hold. We now strengthen some of these results.

Lemma 18.16 *For a finite dimensional Euclidean or unitary vector space $\mathcal{V}$ and $f \in \mathcal{L}(\mathcal{V}, \mathcal{V})$, the following statements are equivalent:*

(1) f is selfadjoint.
(2) For every orthonormal basis B of $\mathcal{V}$ we have $[f]_{B,B} = ([f]_{B,B})^H$.
(3) There exists an orthonormal basis B of $\mathcal{V}$ with $[f]_{B,B} = ([f]_{B,B})^H$.

(In the Euclidean case $([f]_{B,B})^H = ([f]_{B,B})^T$.)

18.3 Selfadjoint Endomorphisms

Proof In Corollary 13.14 we have already shown that (1) implies (2), and obviously (2) implies (3). If (3) holds, then $[f]_{B,B} = ([f]_{B,B})^H = [f^{ad}]_{B,B}$ (cp. Theorem 13.12), and hence $f = f^{ad}$, so that (1) holds. □

We have the following strong result on the diagonalizability of selfadjoint endomorphisms in both the Euclidean and the unitary case.

Theorem 18.17 *Let $\mathcal{V}$ be a finite dimensional Euclidean or unitary vector space. An endomorphism $f \in \mathcal{L}(\mathcal{V}, \mathcal{V})$ is selfadjoint if and only if there exists an orthonormal basis B of $\mathcal{V}$ such that $[f]_{B,B}$ is a real diagonal matrix.*

Proof Consider first the unitary case. If f is selfadjoint, then f is normal and hence unitarily diagonalizable (cp. Theorem 18.2). Let B be an orthonormal basis of $\mathcal{V}$ such that $[f]_{B,B}$ is a diagonal matrix. Then $[f]_{B,B} = [f^{ad}]_{B,B} = ([f]_{B,B})^H$ implies that the diagonal entries of $[f]_{B,B}$, which are the eigenvalues of f, are real. On the other hand, if B is an orthonormal basis of $\mathcal{V}$ such that $[f]_{B,B}$ is a real diagonal matrix, then $[f]_{B,B} = ([f]_{B,B})^H = [f^{ad}]_{B,B}$, and hence $f = f^{ad}$.

Let $\mathcal{V}$ be an n-dimensional Euclidean vector space. If $\widetilde{B} = \{v_1, \ldots, v_n\}$ is an orthonormal basis of $\mathcal{V}$, then $[f]_{\widetilde{B}, \widetilde{B}}$ is symmetric and in particular normal. By Corollary 18.6, there exists an orthogonal matrix $U = [u_{ij}] \in \mathbb{R}^{n,n}$ with

$$U^T [f]_{\widetilde{B},\widetilde{B}} U = \mathrm{diag}(R_1, \ldots, R_m),$$

where for $j = 1, \ldots, m$ either $R_j \in \mathbb{R}^{1,1}$ or

$$R_j = \begin{bmatrix} \alpha_j & \beta_j \\ -\beta_j & \alpha_j \end{bmatrix} \in \mathbb{R}^{2,2} \quad \text{with} \quad \beta_j \neq 0.$$

Since $U^T [f]_{\widetilde{B},\widetilde{B}} U$ is symmetric, a 2×2 block R_j with $\beta_j \neq 0$ cannot occur. Thus, $U^T [f]_{\widetilde{B},\widetilde{B}} U$ is a real diagonal matrix.

We define the basis $B = \{w_1, \ldots, w_n\}$ of $\mathcal{V}$ by

$$(w_1, \ldots, w_n) = (v_1, \ldots, v_n) U.$$

Then, by construction, $U = [\mathrm{Id}_\mathcal{V}]_{B, \widetilde{B}}$ and hence $U^T = U^{-1} = [\mathrm{Id}_\mathcal{V}]_{\widetilde{B}, B}$. Therefore, $U^T [f]_{\widetilde{B},\widetilde{B}} U = [f]_{B,B}$. If $\langle \cdot, \cdot \rangle$ is the scalar product on $\mathcal{V}$, then $\langle v_i, v_j \rangle = \delta_{ij}$, $i, j = 1, \ldots, n$. With $U^T U = I_n$ we get

$$\langle w_i, w_j \rangle = \left\langle \sum_{k=1}^n u_{ki} v_k, \sum_{\ell=1}^n u_{\ell j} v_\ell \right\rangle = \sum_{k=1}^n \sum_{\ell=1}^n u_{ki} u_{\ell j} \langle v_k, v_\ell \rangle = \sum_{k=1}^n u_{ki} u_{kj} = \delta_{ij}.$$

Hence B is an orthonormal basis of $\mathcal{V}$.

On the other hand, if B is an orthonormal basis of $\mathcal{V}$ such that $[f]_{B,B}$ is a real diagonal matrix, then $[f]_{B,B} = ([f]_{B,B})^T = [f^{ad}]_{B,B}$, and hence $f = f^{ad}$. □

This theorem has the following "matrix version".

Corollary 18.18

(1) A matrix $A \in \mathbb{R}^{n,n}$ is symmetric if and only if there exist an orthogonal matrix $U \in \mathbb{R}^{n,n}$ and a diagonal matrix $D \in \mathbb{R}^{n,n}$ with $A = UDU^T$.
(2) A matrix $A \in \mathbb{C}^{n,n}$ is Hermitian if and only if there exist a unitary matrix $U \in \mathbb{C}^{n,n}$ and a diagonal matrix $D \in \mathbb{R}^{n,n}$ with $A = UDU^H$.

The statement (1) in this corollary is known as the *principal axes transformation*. We will briefly discuss the background of this name from the theory of bilinear forms and their applications in geometry. A symmetric matrix $A = [a_{ij}] \in \mathbb{R}^{n,n}$ defines a symmetric bilinear form on $\mathbb{R}^{n,1}$ via

$$\beta_A : \mathbb{R}^{n,1} \times \mathbb{R}^{n,1} \to \mathbb{R}, \quad (x, y) \mapsto y^T A x = \sum_{i=1}^{n} \sum_{j=1}^{n} a_{ij} x_i y_j.$$

The map

$$q_A : \mathbb{R}^{n,1} \to \mathbb{R}, \quad x \mapsto \beta_A(x, x) = x^T A x,$$

is called the *quadratic form* associated with this symmetric bilinear form.

Since A is symmetric, there exists an orthogonal matrix $U = [u_1, \ldots, u_n]$ such that $U^T A U = D = \operatorname{diag}(\lambda_1, \ldots, \lambda_n)$ is a real diagonal matrix. If $B_1 = \{e_1, \ldots, e_n\}$, then $[\beta_A]_{B_1 \times B_1} = A$. The set $B_2 = \{u_1, \ldots, u_n\}$ forms an orthonormal basis of $\mathbb{R}^{n,1}$ with respect to the standard scalar product, and $(u_1, \ldots, u_n) = (e_1, \ldots, e_n)U$, hence $U = [\operatorname{Id}_{\mathbb{R}^{n,1}}]_{B_2, B_1}$. For the change of bases from of B_1 to B_2 we obtain

$$[\beta_A]_{B_2 \times B_2} = \left([\operatorname{Id}_{\mathbb{R}^{n,1}}]_{B_2, B_1}\right)^T [\beta_A]_{B_1 \times B_1} [\operatorname{Id}_{\mathbb{R}^{n,1}}]_{B_2, B_1} = U^T A U = D$$

(cp. Theorem 11.16). Thus, the real diagonal matrix D represents the bilinear form β_A defined by A with respect to the basis B_2.

The quadratic form q_A associated with β_A is also transformed to a simpler form by this change of bases, since analogously

$$q_A(x) = x^T A x = x^T U D U^T x = y^T D y = \sum_{i=1}^{n} \lambda_i y_i^2 = q_D(y),$$

$$y = \begin{bmatrix} y_1 \\ \vdots \\ y_n \end{bmatrix} := U^T x.$$

18.3 Selfadjoint Endomorphisms

Thus, the quadratic form q_A is turned into a "sum of squares", defined by the quadratic form q_D.

The *principal axes transformation* is given by the change of bases from the canonical basis of $\mathbb{R}^{n,1}$ to the basis given by the pairwise orthonormal eigenvectors of A in $\mathbb{R}^{n,1}$. The n pairwise orthogonal subspaces span$\{u_j\}$, $j = 1, \ldots, n$, form the n *principal axes*. The geometric interpretation of this term is illustrated in the following example.

Example 18.19 For the symmetric matrix

$$A = \begin{bmatrix} 4 & 1 \\ 1 & 2 \end{bmatrix} \in \mathbb{R}^{2,2}$$

we have

$$U^T A U = \begin{bmatrix} 3 + \sqrt{2} & 0 \\ 0 & 3 - \sqrt{2} \end{bmatrix} = D,$$

with the orthogonal matrix $U = [u_1, u_2] \in \mathbb{R}^{2,2}$ and

$$u_1 = \begin{bmatrix} c \\ s \end{bmatrix}, \quad u_2 = \begin{bmatrix} -s \\ c \end{bmatrix},$$

$$c = \frac{1 + \sqrt{2}}{\sqrt{(1 + \sqrt{2})^2 + 1}} \approx 0.9239, \quad s = \frac{1}{\sqrt{(1 + \sqrt{2})^2 + 1}} \approx 0.3827.$$

With the associated quadratic form $q_A(x) = 4x_1^2 + 2x_1 x_2 + 2x_2^2$, we define the set

$$E_A = \{x \in \mathbb{R}^{2,1} \mid q_A(x) - 1 = 0\}.$$

As described above, the principal axes transformation consists in the transformation from the canonical coordinate system to a coordinate system given by an orthonormal basis of eigenvectors of A. If we carry out this transformation and replace q_A by the quadratic form q_D, we get the set

$$E_D = \left\{ y \in \mathbb{R}^{2,1} \mid q_D(y) - 1 = 0 \right\} = \left\{ [y_1, y_2]^T \in \mathbb{R}^{2,1} \mid \frac{y_1^2}{\beta_1^2} + \frac{y_2^2}{\beta_2^2} - 1 = 0 \right\},$$

where $\beta_1 = \sqrt{\dfrac{1}{3 + \sqrt{2}}} \approx 0.4760$, $\beta_2 = \sqrt{\dfrac{1}{3 - \sqrt{2}}} \approx 0.7941$.

The set E_D forms the ellipse centered at the origin of the two dimensional Cartesian coordinate system (spanned by the canonical basis vectors e_1, e_2) with axes of lengths β_1 and β_2. The elements $x \in E_A$ are given by $x = Uy$ for $y \in E_D$. The orthogonal matrix

$$U = \begin{bmatrix} c & -s \\ s & c \end{bmatrix} = \begin{bmatrix} \cos(\alpha) & -\sin(\alpha) \\ \sin(\alpha) & \cos(\alpha) \end{bmatrix} = R_{12}(\alpha)^T$$

is a Givens rotation that rotates the ellipse E_D counterclockwise by the angle $\alpha = \cos^{-1}(c) = 0.3926$ (approximately 22.5°). Hence E_A is just a "rotated version" of E_D. The right part of the following figure shows the ellipse E_A in the Cartesian coordinate system. The dashed lines indicate the respective spans of the vectors u_1 and u_2, which are the eigenvectors of A and the *principal axes* of the ellipse E_A.

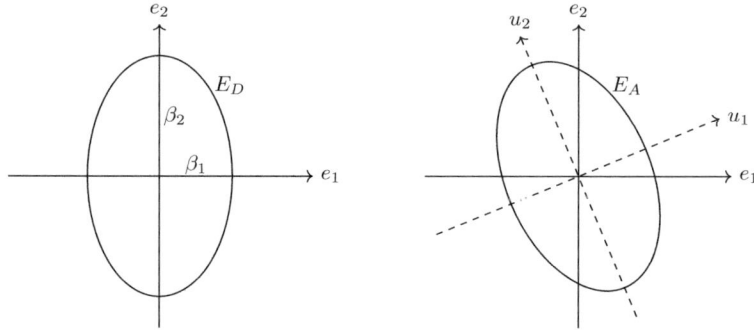

Let $A \in \mathbb{R}^{n,n}$ be symmetric. For a given vector $v \in \mathbb{R}^{n,1}$ and a scalar $\gamma \in \mathbb{R}$,

$$Q(x) = x^T A x + v^T x + \gamma, \quad x \in \mathbb{R}^{n,1}$$

is a quadratic function in n variables (the entries of the vector x). The set of zeros of this function, i.e., the set $\{x \in \mathbb{R}^{n,1} \mid Q(x) = 0\}$, is called a *hypersurface of degree 2* or a *quadric*. In Example 18.19 we have already seen quadrics in the case $n = 2$ and with $v = 0$. We next give some further examples.

Example 18.20

(1) Let $n = 3$, $A = I_3$, $v = [0, 0, 0]^T$ and $\gamma = -1$. The corresponding quadric

$$\left\{ [x_1, x_2, x_3]^T \in \mathbb{R}^{3,1} \mid x_1^2 + x_2^2 + x_3^2 - 1 = 0 \right\}$$

18.3 Selfadjoint Endomorphisms

is the surface of the ball with radius 1 around the origin:

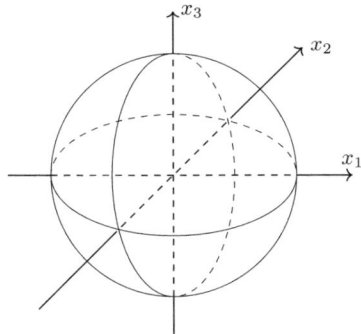

(2) Let $n = 2$, $A = \begin{bmatrix} 1 & 0 \\ 0 & 0 \end{bmatrix}$, $v = [0, 2]^T$ and $\gamma = 0$. The corresponding quadric

$$\left\{ [x_1, x_2]^T \in \mathbb{R}^{2,1} \mid x_1^2 + 2x_2 = 0 \right\}$$

is a parabola:

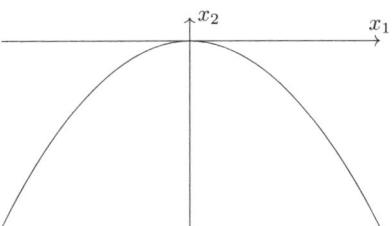

(3) Let $n = 3$, $A = \begin{bmatrix} 1 & 0 & 0 \\ 0 & 0 & 0 \\ 0 & 0 & 0 \end{bmatrix}$, $v = [0, 2, 0]^T$ and $\gamma = 0$. The corresponding quadric

$$\left\{ [x_1, x_2, x_3]^T \in \mathbb{R}^{3,1} \mid x_1^2 + 2x_2 = 0 \right\}$$

is a parabolic cylinder:

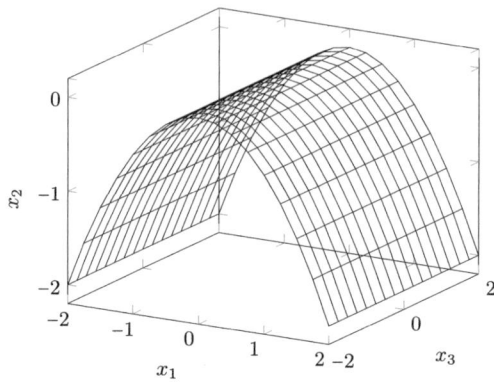

Corollary 18.18 motivates the following definition.

Definition 18.21 If $A \in \mathbb{R}^{n,n}$ is symmetric or $A \in \mathbb{C}^{n,n}$ is Hermitian with n_+ positive, n_- negative and n_0 zero eigenvalues (counted with their corresponding multiplicities), then the triple (n_+, n_-, n_0) is called the *inertia* of A.

Let us first consider, for simplicity, only the case of real symmetric matrices.

Lemma 18.22 *If* $A \in \mathbb{R}^{n,n}$ *symmetric has the inertia* (n_+, n_-, n_0), *then A and* $S_A = \mathrm{diag}(I_{n_+}, -I_{n_-}, 0_{n_0})$ *are congruent.*

Proof Let $A \in \mathbb{R}^{n,n}$ be symmetric and let $A = U \Lambda U^T$ with an orthogonal matrix $U \in \mathbb{R}^{n,n}$ and $\Lambda = \mathrm{diag}(\lambda_1, \ldots, \lambda_n) \in \mathbb{R}^{n,n}$. If A has the inertia (n_+, n_-, n_0), then we can assume without loss of generality that

$$\Lambda = \begin{bmatrix} \Lambda_{n_+} & & \\ & \Lambda_{n_-} & \\ & & 0_{n_0} \end{bmatrix} = \mathrm{diag}(\Lambda_{n_+}, \Lambda_{n_-}, 0_{n_0}),$$

where the diagonal matrices Λ_{n_+} and Λ_{n_-} contain the positive and negative eigenvalues of A, respectively, and $0_{n_0} \in \mathbb{R}^{n_0, n_0}$. We have $\Lambda = \Delta S_A \Delta$, where

$$S_A := \mathrm{diag}(I_{n_+}, -I_{n_-}, 0_{n_0}) \in \mathbb{R}^{n,n},$$
$$\Delta := \mathrm{diag}((\Lambda_{n_+})^{1/2}, (-\Lambda_{n_-})^{1/2}, I_{n_0}) \in GL_n(\mathbb{R}).$$

Here $(\mathrm{diag}(\mu_1, \ldots, \mu_m))^{1/2} = \mathrm{diag}(\sqrt{\mu_1}, \ldots, \sqrt{\mu_m})$ and thus

$$A = U \Lambda U^T = U \Delta S_A \Delta U^T = (U \Delta) S_A (U \Delta)^T. \qquad \square$$

18.3 Selfadjoint Endomorphisms

This result will be used in the proof of *Sylvester's law of inertia*.[3]

Theorem 18.23 *The inertia of a symmetric matrix $A \in \mathbb{R}^{n,n}$ is invariant under congruence, i.e., for every matrix $G \in GL_n(\mathbb{R})$ the matrices A and $G^T A G$ have the same inertia.*

Proof The assertion is trivial for $A = 0$. Let $A \neq 0$ have the inertia (n_+, n_-, n_0), then not both n_+ and n_- can be equal to zero. We assume without loss of generality that $n_+ > 0$. (If $n_+ = 0$, then the following argument can be applied for $n_- > 0$.)

By Lemma 18.22 there exist $G_1 \in GL_n(\mathbb{R})$ and $S_A = \text{diag}(I_{n_+}, -I_{n_-}, 0_{n_0})$ with $A = G_1^T S_A G_1$. Let $G_2 \in GL_n(\mathbb{R})$ be arbitrary and set $B := G_2^T A G_2$. Then B is symmetric and has an inertia $(\tilde{n}_+, \tilde{n}_-, \tilde{n}_0)$. Therefore, $B = G_3^T S_B G_3$ for $S_B = \text{diag}(I_{\tilde{n}_+}, -I_{\tilde{n}_-}, 0_{\tilde{n}_0})$ and a matrix $G_3 \in GL_n(\mathbb{R})$. If we show that $n_+ = \tilde{n}_+$ and $n_0 = \tilde{n}_0$, then also $n_- = \tilde{n}_-$.

We have

$$A = \left(G_2^{-1}\right)^T B G_2^{-1} = \left(G_2^{-1}\right)^T G_3^T S_B G_3 G_2^{-1} = G_4^T S_B G_4, \quad G_4 := G_3 G_2^{-1},$$

and $G_4 \in GL_n(\mathbb{R})$ implies that $\text{rank}(A) = \text{rank}(S_B) = \text{rank}(B)$, hence $n_0 = \tilde{n}_0$.

We set

$$G_1^{-1} = [u_1, \ldots, u_{n_+}, v_1, \ldots, v_{n_-}, w_1, \ldots, w_{n_0}] \quad \text{and}$$

$$G_4^{-1} = [\tilde{u}_1, \ldots, \tilde{u}_{\tilde{n}_+}, \tilde{v}_1, \ldots, \tilde{v}_{\tilde{n}_-}, \tilde{w}_1, \ldots, \tilde{w}_{n_0}].$$

Let $\mathcal{V}_1 := \text{span}\{u_1, \ldots, u_{n_+}\}$ and $\mathcal{V}_2 := \text{span}\{\tilde{v}_1, \ldots, \tilde{v}_{\tilde{n}_-}, \tilde{w}_1, \ldots, \tilde{w}_{n_0}\}$. Since $n_+ > 0$, we have $\dim(\mathcal{V}_1) \geq 1$. If $x \in \mathcal{V}_1 \setminus \{0\}$, then

$$x = \sum_{j=1}^{n_+} \alpha_j u_j = G_1^{-1} [\alpha_1, \ldots, \alpha_{n_+}, 0, \ldots, 0]^T$$

for some $\alpha_1, \ldots, \alpha_{n_+} \in \mathbb{R}$ that are not all zero. This implies

$$x^T A x = \sum_{j=1}^{n_+} \alpha_j^2 > 0.$$

---

[3] James Joseph Sylvester (1814–1897) proved this result for quadratic forms in 1852. He also coined the name *law of inertia* which according to him is "expressing the fact of the existence of an invariable number inseparably attached to such [bilinear] forms".

If, on the other hand, $x \in \mathcal{V}_2$, then an analogous argument shows that $x^T A x \leq 0$. Hence $\mathcal{V}_1 \cap \mathcal{V}_2 = \{0\}$, and the dimension formula for subspaces (cp. Theorem 9.34) yields

$$\underbrace{\dim(\mathcal{V}_1)}_{= n_+} + \underbrace{\dim(\mathcal{V}_2)}_{= n - \widetilde{n}_+} - \underbrace{\dim(\mathcal{V}_1 \cap \mathcal{V}_2)}_{= 0} = \dim(\mathcal{V}_1 + \mathcal{V}_2) \leq \dim(\mathbb{R}^{n,1}) = n,$$

and thus $n_+ \leq \widetilde{n}_+$. If we repeat the same construction by interchanging the roles of n_+ and $\widetilde{n}_+$, then $\widetilde{n}_+ \leq n_+$. Thus, $n_+ = \widetilde{n}_+$ and the proof is complete. □

In the following result we transfer Lemma 18.22 and Theorem 18.23 to complex Hermitian matrices.

Theorem 18.24 *Let $A \in \mathbb{C}^{n,n}$ be Hermitian with the inertia (n_+, n_-, n_0). Then there exists a matrix $G \in GL_n(\mathbb{C})$ with*

$$A = G^H \operatorname{diag}(I_{n_+}, -I_{n_-}, 0_{n_0}) G.$$

Moreover, for every matrix $G \in GL_n(\mathbb{C})$ the matrices A and $G^H A G$ have the same inertia.

Proof Exercise. □

Finally, we discuss a special class of symmetric and Hermitian matrices.

Definition 18.25 A real symmetric or complex Hermitian matrix A of size $n \times n$ is called

(1) *positive semidefinite*, if $v^H A v \geq 0$ for all $v \in \mathbb{R}^{n,1}$ resp. $v \in \mathbb{C}^{n,1}$,
(2) *positive definite*, if $v^H A v > 0$ for all $v \in \mathbb{R}^{n,1} \setminus \{0\}$ resp. $v \in \mathbb{C}^{n,1} \setminus \{0\}$.

If in (1) or (2) the reverse inequality holds, then the corresponding matrices are called *negative semidefinite* or *negative definite*, respectively.

For selfadjoint endomorphisms we define analogously: If $\mathcal{V}$ is a finite dimensional Euclidean or unitary vector space with the scalar product $\langle \cdot, \cdot \rangle$ and if $f \in \mathcal{L}(\mathcal{V}, \mathcal{V})$ is selfadjoint, then f is called *positive semidefinite* or *positive definite*, if $\langle f(v), v \rangle \geq 0$ for all $v \in \mathcal{V}$ resp. $\langle f(v), v \rangle > 0$ for all $v \in \mathcal{V} \setminus \{0\}$.

The following theorem characterizes the positive definite matrices; see Exercise 18.25 and Exercise 18.27 for variants of this result for positive semidefinite matrices resp. positive definite endomorphisms.

18.3 Selfadjoint Endomorphisms

Theorem 18.26 *If $A \in \mathbb{R}^{n,n}$ is symmetric or $A \in \mathbb{C}^{n,n}$ is Hermitian, then the following statements are equivalent:*

(1) A is positive definite.
(2) All eigenvalues of A are real and positive.
(3) There exists an invertible lower triangular matrix with $A = LL^H$ (in the real case $A = LL^T$).

Proof

(1) $\Rightarrow$ (2): The matrix A is diagonalizable with real eigenvalues (cp. Corollary 18.18). If λ is an eigenvalue with associated eigenvector v, i.e., $Av = \lambda v$, then $\lambda v^H v = v^H A v > 0$, and $v^H v > 0$ implies that $\lambda > 0$.

(2) $\Rightarrow$ (1): Let $A = U \text{diag}(\lambda_1, \ldots, \lambda_n) U^H$ be a diagonalization A with an orthogonal resp. unitary matrix U (cp. Corollary 18.18) and $\lambda_j > 0$, $j = 1, \ldots, n$. Let $v \in \mathbb{R}^{n,1} \setminus \{0\}$ resp. $v \in \mathbb{C}^{n,1} \setminus \{0\}$ be arbitrary, and let $w := U^H v$. Then $w \neq 0$ and $v = Uw$, so that

$$v^H A v = (Uw)^H U \text{diag}(\lambda_1, \ldots, \lambda_n) U^H (Uw)$$

$$= w^H \text{diag}(\lambda_1, \ldots, \lambda_n) w = \sum_{j=1}^n \lambda_j |w_j|^2 > 0.$$

(3) $\Rightarrow$ (1): If $A = LL^H$ with an invertible matrix L, then for every $v \in \mathbb{R}^{n,1} \setminus \{0\}$ resp. $v \in \mathbb{C}^{n,1} \setminus \{0\}$ we have

$$v^H A v = v^H L L^H v = \|L^H v\|_2^2 > 0.$$

(Note that here we do not need that L is lower triangular.)

(1) $\Rightarrow$ (3): Let $A = U \text{diag}(\lambda_1, \ldots, \lambda_n) U^H$ be a diagonalization of A with an orthogonal resp. unitary matrix U (cp. Corollary 18.18). Since A is positive definite, we know from (2) that $\lambda_j > 0$, $j = 1, \ldots, n$. We set

$$\Lambda^{1/2} := \text{diag}(\sqrt{\lambda_1}, \ldots, \sqrt{\lambda_n}),$$

and then have $A = (U \Lambda^{1/2})(\Lambda^{1/2} U^H) =: B^H B$. Let $B = QR$ be a QR-decomposition of the invertible matrix B, where Q is orthogonal resp. unitary, and R is invertible and upper triangular (cp. Corollary 12.13). Then $A = B^H B = (QR)^H (QR) = LL^H$, where $L := R^H$ is invertible and lower triangular. □

The factorization $A = LL^H$ in (3) is called a *Cholesky decomposition*[4] of A. Since $L = [l_{ij}]$ is invertible, the diagonal matrix $D := \operatorname{diag}(l_{11}, \ldots, l_{nn})$ is also invertible. Hence the Cholesky decomposition yields an LU-decomposition of the form

$$A = (LD^{-1})(DL^H) = \widetilde{L}\widetilde{U},$$

where $\widetilde{L} := LD^{-1}$ is lower triangular with ones on the diagonal, and $\widetilde{U} := DL^H$ is invertible and upper triangular (cp. Theorem 5.5). Thus, Theorem 18.26 shows that an LU-decomposition of a symmetric or Hermitian positive definite matrix can be computed without row permutations.

In order to compute a Cholesky decomposition of the symmetric positive definite matrix $A = [a_{ij}] \in \mathbb{R}^{n,n}$, we consider the equation

$$A = LL^T = \begin{bmatrix} l_{11} & & \\ \vdots & \ddots & \\ l_{n1} & \cdots & l_{nn} \end{bmatrix} \begin{bmatrix} l_{11} & \cdots & l_{n1} \\ & \ddots & \vdots \\ & & l_{nn} \end{bmatrix}.$$

For the first row of A we obtain

$$\begin{aligned} a_{11} = l_{11}^2 &\Rightarrow l_{11} = a_{11}^{1/2}, \\ a_{1j} = l_{11}l_{j1} &\Rightarrow l_{j1} = \frac{a_{1j}}{l_{11}}, \quad j = 2, \ldots, n. \end{aligned} \tag{18.3}$$

Analogously, for the rows $i = 2, \ldots, n$ of A we obtain

$$\begin{aligned} a_{ii} = \sum_{j=1}^{i} l_{ij}l_{ij} &\Rightarrow l_{ii} = \left(a_{ii} - \sum_{j=1}^{i-1} l_{ij}^2\right)^{1/2}, \\ a_{ij} = \sum_{k=1}^{i} l_{ik}l_{jk} &\Rightarrow l_{ji} = \frac{1}{l_{ii}}\left(a_{ij} - \sum_{k=1}^{i-1} l_{ik}l_{jk}\right), \quad j = i+1, \ldots, n. \end{aligned} \tag{18.4}$$

This algorithm can be adapted easily to a Hermitian positive definite matrix $A \in \mathbb{C}^{n,n}$.

The symmetric or Hermitian positive definite matrices are closely related to the positive definite bilinear forms on Euclidean or unitary vector spaces.

---

[4] André-Louis Cholesky (1875–1918).

Theorem 18.27 *If $\mathcal{V}$ is a finite dimensional Euclidean or unitary vector space and if β is a symmetric bilinear or Hermitian sesquilinear form on $\mathcal{V}$, respectively, then the following statements are equivalent:*

(1) *β is positive definite, i.e., $\beta(v, v) > 0$ for all $v \in \mathcal{V} \setminus \{0\}$.*
(2) *For every basis B of $\mathcal{V}$ the matrix representation $[\beta]_{B \times B}$ is (symmetric or Hermitian) positive definite.*
(3) *There exists a basis B of $\mathcal{V}$ such that the matrix representation $[\beta]_{B \times B}$ is (symmetric or Hermitian) positive definite.*

Proof Exercise. □

Exercises

18.1 Let $A \in \mathbb{R}^{n,n}$ be normal. Show that αA for every $\alpha \in \mathbb{R}$, A^k for every $k \in \mathbb{N}_0$, and $p(A)$ for every $p \in \mathbb{R}[t]$ are normal.

18.2 Let $A, B \in \mathbb{R}^{n,n}$ be normal. Are $A + B$ and AB then normal as well?

18.3 Let $A \in \mathbb{R}^{2,2}$ be normal but not symmetric. Show that then

$$A = \begin{bmatrix} \alpha & \beta \\ -\beta & \alpha \end{bmatrix}$$

for some $\alpha \in \mathbb{R}$ and $\beta \in \mathbb{R} \setminus \{0\}$.

18.4 Prove Corollary 18.6 using Theorem 18.5.

18.5 Formulate Theorem 18.5 and Corollary 18.6 for normal endomorphisms on a finite dimensional Euclidean vector space.

18.6 Show that real skew-symmetric matrices (i.e., matrices with $A = -A^T \in \mathbb{R}^{n,n}$) and complex skew-Hermitian matrices (i.e., matrices with $A = -A^H \in \mathbb{C}^{n,n}$) are normal.

18.7 Let $\mathcal{V}$ be a finite dimensional Euclidean or unitary vector space and let $f \in \mathcal{L}(\mathcal{V}, \mathcal{V})$ be *skew-adjoint*, i.e., $f = -f^{ad}$. Show the following assertions:
 (a) If $\mathcal{V}$ is unitary, then there exists an orthonormal basis B of $\mathcal{V}$ with $[f]_{B,B} = iD$, where D is a real diagonal matrix.
 (b) If $\mathcal{V}$ is Euclidean, then there exists an orthonormal basis B of $\mathcal{V}$ with

$$[f]_{B,B} = \mathrm{diag}(0, \ldots, 0, R_1, \ldots, R_m) \text{ and}$$

$$R_j = \begin{bmatrix} 0 & \beta_j \\ -\beta_j & 0 \end{bmatrix} \in \mathbb{R}^{2,2}, \ \beta_j \neq 0, \ j = 1, \ldots, m.$$

18.8 Let $\mathcal{V}$ be a Euclidean or unitary vector space with the scalar product $\langle \cdot, \cdot \rangle$ and let $f \in \mathcal{L}(\mathcal{V}, \mathcal{V})$ be skew-adjoint. Show the following assertions:
 (a) For all $v \in \mathcal{V}$ we have $\mathrm{Re}(\langle v, f(v) \rangle) = 0$.
 (b) If $\lambda \in \mathbb{R}$ is an eigenvalue of f, then $\lambda = 0$.
 (c) If $n = \dim(\mathcal{V})$ is odd, then 0 is an eigenvalue of f.

18.9 Let $\mathcal{V}$ be a finite dimensional unitary vector space and let $f \in \mathcal{L}(\mathcal{V}, \mathcal{V})$ be normal. Show the following assertions:
   (a) If $f = f^2$, then f is selfadjoint.
   (b) If $f^2 = f^3$, then $f = f^2$.
   (c) If f is nilpotent, then $f = 0$.

18.10 Let $\mathcal{V}$ be a finite dimensional real or complex vector space and let $f \in \mathcal{L}(\mathcal{V}, \mathcal{V})$ be diagonalizable. Show that there exists a scalar product on $\mathcal{V}$ such that f is normal with respect to this scalar product.

18.11 Let $A \in \mathbb{C}^{n,n}$. Show the following assertions:
   (a) A is normal if and only if there exists a normal matrix B with n distinct eigenvalues that commutes with A.
   (b) A is normal if and only if $A + aI$ is normal for every $a \in \mathbb{C}$.
   (c) Let $H(A) := \frac{1}{2}(A + A^H)$ be the Hermitian and $S(A) := \frac{1}{2}(A - A^H)$ the skew-Hermitian part of A. Show that $A = H(A) + S(A)$, $H(A)^H = H(A)$ and $S(A)^H = -S(A)$. Show, furthermore, that A is normal if and only if $H(A)$ and $S(A)$ commute.

18.12 Show that if $A \in \mathbb{C}^{n,n}$ is normal and if $f(z) = \frac{az+b}{cz+d}$ with $ad - bc \neq 0$ is defined on the spectrum of A, then $f(A) = (aA + bI)(cA + dI)^{-1}$.
   (The map $f(z)$ is called a *Möbius transformation*.[5] Such transformations play an important role in Function Theory and in many other areas of Mathematics.)

18.13 Let $\mathcal{V}$ be a finite dimensional Euclidean or unitary vector space and let $f \in \mathcal{L}(\mathcal{V}, \mathcal{V})$ be orthogonal or unitary, respectively. Show that f^{-1} exists and is again orthogonal or unitary, respectively.

18.14 Let $A \in \mathbb{C}^{n,n}$ be normal with $|\lambda| = 1$ for every eigenvalue of A. Show that A is unitary.

18.15 Let $u \in \mathbb{R}^{n,1}$ and let the Householder matrix $H(u)$ be defined as in (18.2). Show the following assertions:
   (a) For $u \neq 0$ the matrices $H(u)$ and $[-e_1, e_2, \ldots, e_n]$ are orthogonally similar, i.e., there exists an orthogonal matrix $U \in \mathbb{R}^{n,n}$ with
$$U^T H(u) U = [-e_1, e_2, \ldots, e_n].$$
   (This implies that $H(u)$ only has the eigenvalues 1 and -1 with the algebraic multiplicities $n - 1$ and 1, respectively.)
   (b) Every orthogonal matrix $U \in \mathbb{R}^{n,n}$ can be written as the product of n Householder matrices, i.e., there exist $u_1, \ldots, u_n \in \mathbb{R}^{n,1}$ with $U = H(u_1) \cdots H(u_n)$.

18.16 Let $v \in \mathbb{R}^{n,1}$ satisfy $v^T v = 1$. Show that there exists an orthogonal matrix $U \in \mathbb{R}^{n,n}$ with $Uv = e_1$.

---

[5] August Ferdinand Möbius (1790–1868).

Exercises

18.17 Let $Q \in \mathbb{C}^{n,n}$ be a unitary matrix. Show that the function $f(z) = \bar{z}$ is defined on the spectrum of Q (cp. Definition 17.1), and compute the matrix $Qf(Q)$.

18.18 Prove Theorem 18.24.

18.19 Consider the Euclidean vector space $\mathcal{V} = \mathbb{R}^{n,n}$ with the scalar product $\langle A, B \rangle = \text{trace}(B^T A)$ (cp. Example 12.2 (3)). Show that $f \in \mathcal{L}(\mathcal{V}, \mathcal{V})$, $A \mapsto A^T$, is orthogonal and selfadjoint.

18.20 Determine for the symmetric matrix

$$A = \begin{bmatrix} 10 & 6 \\ 6 & 10 \end{bmatrix} \in \mathbb{R}^{2,2}$$

an orthogonal matrix $U \in \mathbb{R}^{2,2}$ such that $U^T A U$ is diagonal. Is A positive (semi-)definite?

18.21 Let $K \in \{\mathbb{R}, \mathbb{C}\}$ and let $\{v_1, \ldots, v_n\}$ be a basis of $K^{n,1}$. Prove or disprove: A matrix $A = A^H \in K^{n,n}$ is positive definite if and only if $v_j^H A v_j > 0$ for all $j = 1, \ldots, n$.

18.22 Use Definition 18.25 to test whether the symmetric matrices

$$\begin{bmatrix} 1 & 1 \\ 1 & 1 \end{bmatrix}, \quad \begin{bmatrix} 1 & 2 \\ 2 & 1 \end{bmatrix}, \quad \begin{bmatrix} 2 & 1 \\ 1 & 2 \end{bmatrix} \in \mathbb{R}^{2,2}$$

are positive (semi-)definite. Determine in all cases the inertia.

18.23 Let

$$A = \begin{bmatrix} A_{11} & A_{12} \\ A_{12}^T & A_{22} \end{bmatrix} \in \mathbb{R}^{n,n}$$

with $A_{11} = A_{11}^T \in GL_m(\mathbb{R})$, $A_{12} \in \mathbb{R}^{m,n-m}$ and $A_{22} = A_{22}^T \in \mathbb{R}^{n-m,n-m}$. The matrix $S := A_{22} - A_{12}^T A_{11}^{-1} A_{12} \in \mathbb{R}^{m,m}$ is called the *Schur complement*[6] of A_{11} in A. Show the following *inertia additivity formula* of Haynsworth:[7]

$$(n_+(A), n_-(A), n_0(A)) = (n_+(A_{11}), n_-(A_{11}), n_0(A_{11}))$$
$$+ (n_+(S), n_-(S), n_0(S)).$$

In particular, A is positive definite if and only if A_{11} and S are both positive definite.

(For the Schur complement, see also Exercise 4.21.)

18.24 Show that $A \in \mathbb{C}^{n,n}$ is Hermitian positive definite if and only if $\langle x, y \rangle = y^H A x$ defines a scalar product on $\mathbb{C}^{n,1}$.

---

[6] Issai Schur (1875–1941).
[7] Emilie Virginia Haynsworth (1916–1985).

18.25 Prove the following version of Theorem 18.26 for positive semidefinite matrices.

If $A \in \mathbb{R}^{n,n}$ is symmetric, then the following statements are equivalent:
(1) *A is positive semidefinite.*
(2) *All eigenvalues of A are real and nonnegative.*
(3) *There exists an upper triangular matrix $L \in \mathbb{R}^{n,n}$ with $A = LL^T$.*

18.26 Show that $A \in \mathbb{C}^{n,n}$ is normal if and only if $A^H A - A A^H$ is Hermitian positive semidefinite.

18.27 Let $\mathcal{V}$ be a finite dimensional Euclidean or unitary vector space and let $f \in \mathcal{L}(\mathcal{V}, \mathcal{V})$ be selfadjoint. Show that f is positive definite if and only if all eigenvalues of f are real and positive.

18.28 Let $A \in \mathbb{R}^{n,n}$. A matrix $X \in \mathbb{R}^{n,n}$ with $X^2 = A$ is called a square root of A (cp. Sect. 17.1).
  (a) Show that a symmetric positive definite matrix $A \in \mathbb{R}^{n,n}$ has a symmetric positive definite square root.
  (b) Show that the matrix

  $$A = \begin{bmatrix} 33 & 6 & 6 \\ 6 & 24 & 12 \\ 6 & -12 & 24 \end{bmatrix} \in \mathbb{R}^{3,3}$$

  is symmetric positive definite, and compute a symmetric positive definite square root of A.
  (c) Show that the matrix $A = J_n(0)$, $n \geq 2$, does not have a square root.

18.29 Show that the matrix

$$A = \begin{bmatrix} 2 & 1 & 0 \\ 1 & 2 & 1 \\ 0 & 1 & 2 \end{bmatrix} \in \mathbb{R}^{3,3}$$

is positive definite and compute a Cholesky factorization of A using (18.3)–(18.4).

18.30 Let $A, B \in \mathbb{C}^{n,n}$ be Hermitian and let B be furthermore positive definite. Show that the polynomial $\det(tB - A) \in \mathbb{C}[t]_{\leq n}$ has exactly n real roots.

18.31 Prove Theorem 18.27.

The Singular Value Decomposition   19

The matrix decomposition introduced in this chapter is very important in many practical applications, since it yields the best possible approximation (in a certain sense) of a given matrix by a matrix of low rank. A low rank approximation can be considered a "compression" of the data represented by the given matrix. We illustrate this below with an example from image processing.

We first prove the existence of the decomposition.

Theorem 19.1 *Let $A \in \mathbb{C}^{n,m}$ with $n \geq m$ be given. Then there exist unitary matrices $V \in \mathbb{C}^{n,n}$ and $W \in \mathbb{C}^{m,m}$ such that*

$$A = V\Sigma W^H \quad \text{with} \quad \Sigma = \begin{bmatrix} \Sigma_r & 0_{r,m-r} \\ 0_{n-r,r} & 0_{n-r,m-r} \end{bmatrix} \in \mathbb{R}^{n,m}, \quad \Sigma_r = \mathrm{diag}(\sigma_1, \ldots, \sigma_r), \tag{19.1}$$

where $\sigma_1 \geq \sigma_2 \geq \cdots \geq \sigma_r > 0$ and $r = \mathrm{rank}(A)$.

Proof If $A = 0$, then we set $V = I_n$, $\Sigma = 0 \in \mathbb{C}^{n,m}$, $\Sigma_r = [\,]$, $W = I_m$, and we are finished.

Let $A \neq 0$ and $r := \mathrm{rank}(A)$. Since $n \geq m$, we have $1 \leq r \leq m$, and since $A^H A \in \mathbb{C}^{m,m}$ is Hermitian, there exists a unitary matrix $W = [w_1, \ldots, w_m] \in \mathbb{C}^{m,m}$ with

$$W^H(A^H A)W = \mathrm{diag}(\lambda_1, \ldots, \lambda_m) \in \mathbb{R}^{m,m}$$

(cp. (2) in Corollary 18.18). Without loss of generality we assume that $\lambda_1 \geq \lambda_2 \geq \cdots \geq \lambda_m$. For every $j = 1, \ldots, m$ then $A^H A w_j = \lambda_j w_j$, and hence

$$\lambda_j w_j^H w_j = w_j^H A^H A w_j = \|Aw_j\|_2^2 \geq 0,$$

i.e., $\lambda_j \geq 0$ for $j = 1, \ldots, m$. Then $\operatorname{rank}(A^H A) = \operatorname{rank}(A) = r$ (to see this, modify the proof of Lemma 10.25 for the complex case). Therefore, the matrix $A^H A$ has exactly r positive eigenvalues $\lambda_1, \ldots, \lambda_r$ and $m - r$ times the eigenvalue 0. We then define $\sigma_j := \lambda_j^{1/2}$, $j = 1, \ldots, r$, and have $\sigma_1 \geq \sigma_2 \geq \cdots \geq \sigma_r$. Let Σ_r be as in (19.1),

$$D := \begin{bmatrix} \Sigma_r & 0 \\ 0 & I_{m-r} \end{bmatrix} \in GL_m(\mathbb{R}), \quad X = [x_1, \ldots, x_m] := AWD^{-1},$$

$V_r := [x_1, \ldots, x_r]$, and $Z := [x_{r+1}, \ldots, x_m]$. Then

$$\begin{bmatrix} V_r^H V_r & V_r^H Z \\ Z^H V_r & Z^H Z \end{bmatrix} = \begin{bmatrix} V_r^H \\ Z^H \end{bmatrix} [V_r, Z] = X^H X = D^{-1} W^H A^H A W D^{-1} = \begin{bmatrix} I_r & 0 \\ 0 & 0 \end{bmatrix},$$

which implies, in particular, that $Z = 0$ and $V_r^H V_r = I_r$. We extend the vectors $x_1, \ldots, x_r$ to an orthonormal basis $\{x_1, \ldots, x_r, \widetilde{x}_{r+1}, \ldots, \widetilde{x}_n\}$ of $\mathbb{C}^{n,1}$ with respect to the standard scalar product. Then the matrix

$$V := [V_r, \widetilde{x}_{r+1}, \ldots, \widetilde{x}_n] \in \mathbb{C}^{n,n}$$

is unitary. From $X = AWD^{-1}$ and $X = [V_r, Z] = [V_r, 0]$ we finally obtain $A = [V_r, 0]DW^H$ and $A = V\Sigma W^H$ with Σ as in (19.1). □

As the proof shows, Theorem 19.1 can be formulated analogously for real matrices $A \in \mathbb{R}^{n,m}$ with $n \geq m$. In this case the two matrices V and W are orthogonal. If $n < m$, then we can apply the theorem to A^H (resp. A^T in the real case).

Definition 19.2 A decomposition of the form (19.1) is called a *singular value decomposition* or short *SVD*[1] of the matrix A. The diagonal entries of the matrix Σ_r are called *singular values* and the columns of V resp. W are called *left* resp. *right singular vectors* of A.

From (19.1) we obtain the unitary diagonalizations of the matrices $A^H A$ and AA^H,

$$A^H A = W \begin{bmatrix} \Sigma_r^2 & 0 \\ 0 & 0 \end{bmatrix} W^H \quad \text{and} \quad AA^H = V \begin{bmatrix} \Sigma_r^2 & 0 \\ 0 & 0 \end{bmatrix} V^H.$$

---

[1] Many important researchers played a role in the development of the singular value decomposition from special cases in the middle of the nineteenth century to its current general form. In the historical notes about the singular value decomposition in [6] one finds contributions of Jordan (1873), Sylvester (1889/1890) and Schmidt (1907). The current form was shown in 1939 by Carl Henry Eckart (1902–1973) and Gale J. Young (1912–1990).

19 The Singular Value Decomposition

The singular values of A are therefore uniquely determined as the positive square roots of the positive eigenvalues of $A^H A$ or $A A^H$. The unitary matrices V and W in the singular value decomposition, however, are (as the eigenvectors in general) not uniquely determined.

If we write the SVD of A in the form

$$A = V \Sigma W^H = \left(V \begin{bmatrix} I_m \\ 0 \end{bmatrix} W^H \right) \left(W \begin{bmatrix} \Sigma_r & 0 \\ 0 & 0 \end{bmatrix} W^H \right) =: UP,$$

then $U \in \mathbb{C}^{n,m}$ has orthonormal columns, i.e., $U^H U = I_m$, and $P = P^H \in \mathbb{C}^{m,m}$ is positive semidefinite with the inertia $(r, 0, m - r)$. The factorization $A = UP$ is called a *polar decomposition* of A. It can be viewed as a generalization of the polar representation of complex numbers, $z = e^{i\varphi} |z|$.

Lemma 19.3 *Suppose that the matrix $A \in \mathbb{C}^{n,m}$ with $\mathrm{rank}(A) = r$ has an SVD of the form (19.1) with $V = [v_1, \ldots, v_n]$ and $W = [w_1, \ldots, w_m]$. Then $\mathrm{im}(A) = \mathrm{span}\{v_1, \ldots, v_r\}$ and $\ker(A) = \mathrm{span}\{w_{r+1}, \ldots, w_m\}$.*

Proof For $j = 1, \ldots, r$ we have $A w_j = V \Sigma W^H w_j = V \Sigma e_j = \sigma_j v_j \neq 0$, since $\sigma_j \neq 0$. Hence these r linear independent vectors satisfy $v_1, \ldots, v_r \in \mathrm{im}(A)$. Now $r = \mathrm{rank}(A) = \dim(\mathrm{im}(A))$ implies that $\mathrm{im}(A) = \mathrm{span}\{v_1, \ldots, v_r\}$.

For $j = r+1, \ldots, m$ we have $A w_j = 0$, and hence these $m-r$ linear independent vectors satisfy $w_{r+1}, \ldots, w_m \in \ker(A)$. Then $\dim(\ker(A)) = m - \dim(\mathrm{im}(A)) = m - r$ implies that $\ker(A) = \mathrm{span}\{w_{r+1}, \ldots, w_m\}$. $\square$

An SVD of the form (19.1) can be written as

$$A = \sum_{j=1}^{r} \sigma_j v_j w_j^H.$$

Thus, A can be written as a sum of r matrices of the form $\sigma_j v_j w_j^H$, where $\mathrm{rank}\left(\sigma_j v_j w_j^H\right) = 1$. Let

$$A_k := \sum_{j=1}^{k} \sigma_j v_j w_j^H \quad \text{for some } k, \quad 1 \leq k \leq r. \tag{19.2}$$

Then $\mathrm{rank}(A_k) = k$ and, using that the matrix 2-norm is unitarily invariant (cp. Exercise 19.1), we get

$$\|A - A_k\|_2 = \|\mathrm{diag}(\sigma_{k+1}, \ldots, \sigma_r)\|_2 = \sigma_{k+1}. \tag{19.3}$$

Hence A is approximated by the matrix A_k, where the rank of the approximating matrix and the approximation error in the matrix 2-norm are explicitly known. The singular value decomposition, furthermore, yields the *best possible* approximation of A by a matrix of rank k with respect to the matrix 2-norm.

Theorem 19.4 *With A_k as in (19.2), we have $\|A - A_k\|_2 \leq \|A - B\|_2$ for every matrix $B \in \mathbb{C}^{n,m}$ with $\mathrm{rank}(B) = k$.*

Proof The assertion is clear for $k = \mathrm{rank}(A)$, since then $A_k = A$ and $\|A - A_k\|_2 = 0$.

Let $k < \mathrm{rank}(A) \leq m$. Let $B \in \mathbb{C}^{n,m}$ with $\mathrm{rank}(B) = k$ be given, then $\dim(\ker(B)) = m - k$, where we consider B as an element of $\mathcal{L}(\mathbb{C}^{m,1}, \mathbb{C}^{n,1})$. If $w_1, \ldots, w_m$ are the right singular vectors of A from (19.1), then $\mathcal{U} := \mathrm{span}\{w_1, \ldots, w_{k+1}\}$ has the dimension $k+1$. Since $\ker(B)$ and $\mathcal{U}$ are subspaces of $\mathbb{C}^{m,1}$ with $\dim(\ker(B)) + \dim(\mathcal{U}) = m + 1$, we have $\ker(B) \cap \mathcal{U} \neq \{0\}$.

Let $v \in \ker(B) \cap \mathcal{U}$ with $\|v\|_2 = 1$ be given. Then there exist $\alpha_1, \ldots, \alpha_{k+1} \in \mathbb{C}$ with $v = \sum_{j=1}^{k+1} \alpha_j w_j$ and $\sum_{j=1}^{k+1} |\alpha_j|^2 = \|v\|_2^2 = 1$. Hence

$$(A - B)v = Av - \underbrace{Bv}_{=0} = \sum_{j=1}^{k+1} \alpha_j A w_j = \sum_{j=1}^{k+1} \alpha_j \sigma_j v_j$$

and, therefore,

$$\|A - B\|_2 = \max_{\|y\|_2=1} \|(A - B)y\|_2 \geq \|(A - B)v\|_2 = \Big\|\sum_{j=1}^{k+1} \alpha_j \sigma_j v_j\Big\|_2$$

$$= \Big(\sum_{j=1}^{k+1} |\alpha_j \sigma_j|^2\Big)^{1/2} \quad \text{(since } v_1, \ldots, v_{k+1} \text{ are pairwise orthonormal)}$$

$$\geq \sigma_{k+1} \Big(\sum_{j=1}^{k+1} |\alpha_j|^2\Big)^{1/2} \quad \text{(since } \sigma_1 \geq \cdots \geq \sigma_{k+1})$$

$$= \sigma_{k+1} = \|A - A_k\|_2,$$

which completes the proof. □

MATLAB-Minute 13
The command A=magic(n) generates for $n \geq 3$ an $n \times n$ matrix A with entries from 1 to n^2, so that all row, column and diagonal sums of A are equal. The entries of A, therefore, from a "magic square".

Compute the SVD of A=magic(10) using the command [V,S,W]=svd(A). What can be said about the singular values of A and what is rank(A)? Form A_k for $k = 1, 2, \ldots, \text{rank}(A)$ as in (19.2) and verify numerically the Eq. (19.3).

The SVD is one of the most important and practical mathematical tools in almost all areas of science, engineering and social sciences, in medicine and even in psychology. Its great importance is due to the fact that the SVD allows to distinguish between "important" and "non-important" information in a given data. In practice, the latter corresponds, e.g., to measurement errors, noise in the transmission of data, or fine details in a signal or an image that do not play an important role. Often, the "important" information corresponds to the large singular values, and the "non-important" information to the small ones.

In many applications one sees, furthermore, that the singular values of a given matrix decay rapidly, so that there exist only few large and many small singular values. If this is the case, then the matrix can be approximated well by a matrix with low rank, since already for a small k the approximation error $\|A - A_k\|_2 = \sigma_{k+1}$ is small. A *low rank approximation* A_k requires little storage capacity in the computer; only k scalars and $2k$ vectors have to be stored. This makes the SVD a powerful tool in all applications where data compression is of interest.

Example 19.5 We illustrate the use of the SVD in image compression with a picture that we obtained from the research center MATHEON: Mathematics for Key Technologies.[2] The greyscale picture is shown on the left of the figure below. It consists of 286×152 pixels, where each of the pixels is given by a value between 0 and 64. These values are stored in a real 286×152 matrix A which has (full) rank 152.

---
[2] We thank Falk Ebert for his help. The original bear can be seen in front of the Mathematics building of the TU Berlin.

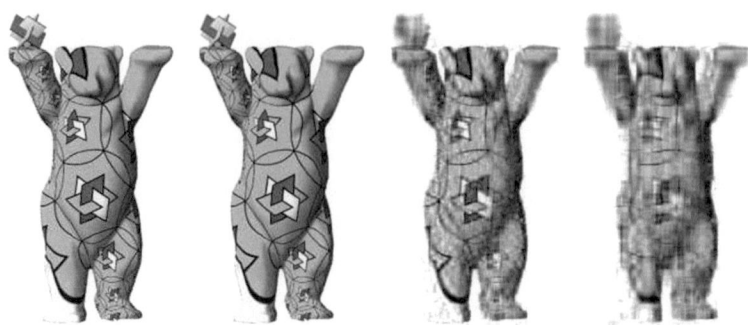

We compute an SVD $A = V\Sigma W^T$ using the command [V,S,W]=svd(A) in MATLAB. The diagonal entries of the matrix S, i.e., the singular values of A, are ordered decreasingly by MATLAB (as in Theorem 19.1). For $k = 100, 20, 10$ we now compute matrices A_k with rank k as in (19.2) using the command Ak=V(:,1:k)*S(1:k,1:k)*W(:,1:k)'. These matrices represent approximations of the original picture based on the k largest singular values and the corresponding singular vectors. The three approximations are shown next to the original picture above. The quality of the approximation decreases with decreasing k, but even the approximation for $k = 10$ shows the essential features of the "MATHEON bear".

Another important application of the SVD arises in the solution of linear systems of equations. If $A \in \mathbb{C}^{n,m}$ has an SVD of the form (19.1), we define the matrix

$$A^\dagger := W\Sigma^\dagger V^H \in \mathbb{C}^{m,n}, \quad \text{where} \quad \Sigma^\dagger := \begin{bmatrix} \Sigma_r^{-1} & 0 \\ 0 & 0 \end{bmatrix} \in \mathbb{R}^{m,n}. \tag{19.4}$$

We easily see that

$$A^\dagger A = W \begin{bmatrix} I_r & 0 \\ 0 & 0 \end{bmatrix} W^H \in \mathbb{R}^{m,m}.$$

If $r = m = n$, then A is invertible and the right hand side of the above equation is equal to the identity matrix I_n. In this case we have $A^\dagger = A^{-1}$. The matrix $A^\dagger$ can therefore be viewed as a *generalized inverse*, that in the case of an invertible matrix A is equal to the inverse of A.

Definition 19.6 The matrix $A^\dagger$ in (19.4) is called *Moore-Penrose inverse*[3] or *pseudoinverse* of A.

---

[3] Eliakim Hastings Moore (1862–1932) and Sir Roger Penrose (b. 1931), winner of the Nobel Prize in Physics in 2020.

Let $A \in \mathbb{C}^{n,m}$ and $b \in \mathbb{C}^{n,1}$ be given. If the linear system of equations $Ax = b$ has no solution, then we can try to find an $\widehat{x} \in \mathbb{C}^{m,1}$ such that $A\widehat{x}$ is "as close as possible" to b. Using the Moore-Penrose inverse we obtain the best possible approximation with respect to the Euclidean norm.

Theorem 19.7 *Let $A \in \mathbb{C}^{n,m}$ with $n \geq m$ and $b \in \mathbb{C}^{n,1}$ be given. If $A = V\Sigma W^H$ is an SVD, and $A^\dagger$ is as in (19.4), then $\widehat{x} = A^\dagger b$ satisfies*

$$\|b - A\widehat{x}\|_2 \leq \|b - Ay\|_2 \quad \text{for all} \quad y \in \mathbb{C}^{m,1},$$

and

$$\|\widehat{x}\|_2 = \left(\sum_{j=1}^{r} \left|\frac{v_j^H b}{\sigma_j}\right|^2\right)^{1/2} \leq \|y\|_2$$

for all $y \in \mathbb{C}^{m,1}$ with $\|b - A\widehat{x}\|_2 = \|b - Ay\|_2$.

Proof Let $y \in \mathbb{C}^{m,1}$ be given and let $z = [\xi_1, \ldots, \xi_m]^T := W^H y$. Then

$$\|b - Ay\|_2^2 = \|b - V\Sigma W^H y\|_2^2 = \|V(V^H b - \Sigma z)\|_2^2 = \|V^H b - \Sigma z\|_2^2$$

$$= \sum_{j=1}^{r} \left|v_j^H b - \sigma_j \xi_j\right|^2 + \sum_{j=r+1}^{n} \left|v_j^H b\right|^2$$

$$\geq \sum_{j=r+1}^{n} \left|v_j^H b\right|^2. \tag{19.5}$$

Equality holds if and only if $\xi_j = \left(v_j^H b\right)/\sigma_j$ for all $j = 1, \ldots, r$. This is satisfied if $z = W^H y = \Sigma^\dagger V^H b$. The last equation holds if and only if

$$y = W\Sigma^\dagger V^H b = A^\dagger b = \widehat{x}.$$

The vector $\widehat{x}$ therefore attains the lower bound (19.5).

The equation

$$\|\widehat{x}\|_2 = \left(\sum_{j=1}^{r} \left|\frac{v_j^H b}{\sigma_j}\right|^2\right)^{1/2}$$

is easily checked. Every vector $y \in \mathbb{C}^{m,1}$ that attains the lower bound (19.5) must have the form

$$y = W \left[\frac{v_1^H b}{\sigma_1}, \ldots, \frac{v_r^H b}{\sigma_r}, y_{r+1}, \ldots, y_m \right]^T$$

for some $y_{r+1}, \ldots, y_m \in \mathbb{C}$, which implies that $\|y\|_2 \geq \|\widehat{x}\|_2$. □

The minimization problem for the vector $\widehat{x}$ can be written as

$$\|b - A\widehat{x}\|_2 = \min_{y \in \mathbb{C}^{m,1}} \|b - Ay\|_2.$$

Let $A \in \mathbb{C}^{n,m}$ have rank $m \leq n$, and consider $\mathcal{V} = \mathbb{C}^{n,1}$ with the standard scalar product and the induced Euclidean norm $\|\cdot\|_2$. Then Theorem 12.28 with $U = \text{im}(A)$ shows that

$$\widehat{x} = A^{\dagger} b = f_{\text{im}(A)}(b),$$

where $f_{\text{im}(A)}$ is the orthogonal projection onto $\text{im}(A)$.

If

$$A = \begin{bmatrix} \tau_1 & 1 \\ \vdots & \vdots \\ \tau_m & 1 \end{bmatrix} \in \mathbb{R}^{m,2}$$

for (pairwise distinct) $\tau_1, \ldots, \tau_m \in \mathbb{R}$, then this minimization problem corresponds to the problem of linear regression and the least squares approximation in Example 12.17 that we have solved with the QR-decomposition of A. If $A = QR$ is this decomposition, then $A^{\dagger} = (A^H A)^{-1} A^H$ (cp. Exercise 19.5) and we have

$$A^{\dagger} = (R^H Q^H Q R)^{-1} R^H Q^H = R^{-1} R^{-H} R^H Q^H = R^{-1} Q^H.$$

Thus, the solution of the least-squares approximation in Example 12.17 is identical to the solution of the above minimization problem using the SVD of A.

Exercises

19.1 Show that the Frobenius norm and the matrix 2-norm are *unitarily invariant*, i.e., that $\|PAQ\|_F = \|A\|_F$ and $\|PAQ\|_2 = \|A\|_2$ for all $A \in \mathbb{C}^{n,m}$ and unitary matrices $P \in \mathbb{C}^{n,n}$, $Q \in \mathbb{C}^{m,m}$.
(*Hint:* For the Frobenius norm we can use that $\|A\|_F^2 = \text{trace}(A^H A)$.)

19.2 Use the result of Exercise 19.1 to show that $\|A\|_F = \left(\sigma_1^2 + \ldots + \sigma_r^2\right)^{1/2}$ and $\|A\|_2 = \sigma_1$, where $\sigma_1 \geq \cdots \geq \sigma_r > 0$ are the singular values of $A \in \mathbb{C}^{n,m}$. (This implies the inequality $\|A\|_2 \leq \|A\|_F$.)

19.3 Show that $\|A\|_2 = \|A^H\|_2$ and $\|A\|_2^2 = \|A^H A\|_2$ for all $A \in \mathbb{C}^{n,m}$.

19.4 Show that $\|A\|_2^2 \leq \|A\|_1 \|A\|_\infty$ for all $A \in \mathbb{C}^{n,m}$.

19.5 Let $A \in \mathbb{C}^{n,m}$ and let $A^\dagger \in \mathbb{C}^{m,n}$, be the Moore-Penrose inverse of A. Show the following assertions:
  (a) If $\text{rank}(A) = m$, then $A^\dagger = (A^H A)^{-1} A^H$.
  (b) The matrix $X = A^\dagger$ is the uniquely determined matrix that satisfies the following four matrix equations:
    (1) $AXA = A$,
    (2) $XAX = X$,
    (3) $(AX)^H = AX$,
    (4) $(XA)^H = XA$.
  (c) $(A^\dagger)^\dagger = A$.
  (d) The matrix $AA^\dagger$ is the orthogonal projection onto $\text{im}(A)$, and $A^\dagger A$ is the orthogonal projection onto $\text{im}(A^H)$ (in both cases with respect to the standard scalar product).

19.6 Let $A \in \mathbb{C}^{n,m}$ have rank $m \leq n$. Show that $A^\dagger$ is the unique solution of $\min_{X \in \mathbb{C}^{m,n}} \|I_n - AX\|_F$.

19.7 Let
$$A = \begin{bmatrix} 2 & 1 \\ 0 & 3 \\ 1 & -2 \end{bmatrix} \in \mathbb{R}^{3,2}, \quad b = \begin{bmatrix} 5 \\ 2 \\ -5 \end{bmatrix} \in \mathbb{R}^{3,1}.$$

Compute the Moore-Penrose inverse of A and a vector $\widehat{x} \in \mathbb{R}^{2,1}$ such that
  (a) $\|b - A\widehat{x}\|_2 \leq \|b - Ay\|_2$ for all $y \in \mathbb{R}^{2,1}$, and
  (b) $\|\widehat{x}\|_2 \leq \|y\|_2$ for all $y \in \mathbb{R}^{2,1}$ with $\|b - Ay\|_2 = \|b - A\widehat{x}\|_2$.

19.8 Prove the following theorem:
Let $A \in \mathbb{C}^{n,m}$ and $B \in \mathbb{C}^{\ell,m}$ with $m \leq n \leq \ell$. Then $A^H A = B^H B$ if and only if $B = UA$ for a matrix $U \in \mathbb{C}^{\ell,n}$ with $U^H U = I_n$. If A and B are real, then U can also be chosen to be real.
(*Hint:* One direction is trivial. For the other direction consider the unitary diagonalization of $A^H A = B^H B$. This yields the matrix W in the SVD of A and of B. Show the assertion using these two decompositions. This theorem and its applications can be found in the article [8].)

The Kronecker Product and Linear Matrix Equations

20

Many applications, in particular the stability analysis of differential equations, lead to linear matrix equations, such as the Sylvester equation[1] $AX + XB = C$. Here the matrices A, B, C are given and the goal is to determine a matrix X that solves the equation (we will give a formal definition below). In the description of the solutions of such equations, the Kronecker product,[2] another product of matrices, is useful. In this chapter we develop the most important properties of this product and we study its application in the context of linear matrix equations. Many more results on this topic can be found in the books [6] and [9].

Definition 20.1 If K is a field, $A = [a_{ij}] \in K^{m,m}$ and $B \in K^{n,n}$, then

$$A \otimes B := [a_{ij} B] = \begin{bmatrix} a_{11} B & \cdots & a_{1m} B \\ \vdots & & \vdots \\ a_{m1} B & \cdots & a_{mm} B \end{bmatrix} \in K^{nm,nm},$$

is called the *Kronecker product* of A and B.

The Kronecker product is sometimes called the *tensor product* of matrices. This product defines a map from $K^{m,m} \times K^{n,n}$ to $K^{mn,mn}$. The definition can be extended to non-square matrices, but for simplicity we consider here only the case of square matrices. The following lemma presents the basic computational rules of the Kronecker product.

---

[1] James Joseph Sylvester (1814–1897).
[2] Leopold Kronecker (1823–1891) is said to have used this product in his lectures in Berlin in the 1880s. It was defined formally for the first time in 1858 by Johann Georg Zehfuss (1832–1901).

© The Author(s), under exclusive license to Springer Nature Switzerland AG 2025
J. Liesen, V. Mehrmann, *Linear Algebra*, Springer Undergraduate
Mathematics Series, https://doi.org/10.1007/978-3-031-93260-1_20

Lemma 20.2 *For all square matrices A, B, C over K, the following computational rules hold:*

(1) $A \otimes (B \otimes C) = (A \otimes B) \otimes C$.
(2) $(\mu A) \otimes B = A \otimes (\mu B) = \mu (A \otimes B)$ *for all* $\mu \in K$.
(3) $(A + B) \otimes C = (A \otimes C) + (B \otimes C)$, *whenever $A + B$ is defined.*
(4) $A \otimes (B + C) = (A \otimes B) + (A \otimes C)$, *whenever $B + C$ is defined.*
(5) $(A \otimes B)^T = A^T \otimes B^T$, *and therefore the Kronecker product of two symmetric matrices is symmetric.*

Proof Exercise. □

In particular, in contrast to the standard matrix multiplication, the order of the factors in the Kronecker product does not change under transposition. The following result describes the matrix multiplication of two Kronecker products.

Lemma 20.3 *For $A, C \in K^{m,m}$ and $B, D \in K^{n,n}$ we have*

$$(A \otimes B)(C \otimes D) = (AC) \otimes (BD).$$

Hence, in particular,

(1) $A \otimes B = (A \otimes I_n)(I_m \otimes B) = (I_m \otimes B)(A \otimes I_n)$,
(2) $(A \otimes B)^{-1} = A^{-1} \otimes B^{-1}$, *if A and B are invertible.*

Proof Since $A \otimes B = [a_{ij} B]$ and $C \otimes D = [c_{ij} D]$, the block $F_{ij} \in K^{n,n}$ in the block matrix $[F_{ij}] = (A \otimes B)(C \otimes D)$ is given by

$$F_{ij} = \sum_{k=1}^{m} (a_{ik} B)(c_{kj} D) = \sum_{k=1}^{m} a_{ik} c_{kj} \, BD = \left(\sum_{k=1}^{m} a_{ik} c_{kj} \right) BD.$$

For the block matrix $[G_{ij}] = (AC) \otimes (BD)$ with $G_{ij} \in K^{n,n}$ we obtain

$$G_{ij} = g_{ij} BD, \quad \text{where} \quad g_{ij} = \sum_{k=1}^{m} a_{ik} c_{kj},$$

which shows $(A \otimes B)(C \otimes D) = (AC) \otimes (BD)$. Now (1) and (2) easily follow from this equation. □

In general the Kronecker product is non-commutative (cp. Exercise 20.2), but we have the following relationship between $A \otimes B$ and $B \otimes A$.

20 The Kronecker Product and Linear Matrix Equations

Lemma 20.4 *For $A \in K^{m,m}$ and $B \in K^{n,n}$ there exists a permutation matrix $P \in K^{mn,mn}$ with*

$$P^T(A \otimes B)P = B \otimes A.$$

Proof Exercise. □

For the computation of the determinant, trace and rank of a Kronecker product there exist simple formulas.

Theorem 20.5 *For $A \in K^{m,m}$ and $B \in K^{n,n}$ the following rules hold:*

(1) $\det(A \otimes B) = (\det A)^n (\det B)^m = \det(B \otimes A)$.
(2) $\mathrm{trace}(A \otimes B) = \mathrm{trace}(A)\,\mathrm{trace}(B) = \mathrm{trace}(B \otimes A)$.
(3) $\mathrm{rank}(A \otimes B) = \mathrm{rank}(A)\,\mathrm{rank}(B) = \mathrm{rank}(B \otimes A)$.

Proof

(1) From (1) in Lemma 20.3 and the multiplication theorem for determinants (cp. Theorem 7.15) we get

$$\det(A \otimes B) = \det((A \otimes I_n)(I_m \otimes B)) = \det(A \otimes I_n)\det(I_m \otimes B).$$

By Lemma 20.4 there exists a permutation matrix P with $A \otimes I_n = P(I_n \otimes A)P^T$. This implies that

$$\det(A \otimes I_n) = \det\left(P(I_n \otimes A)P^T\right) = \det(I_n \otimes A) = (\det A)^n.$$

Since $\det(I_m \otimes B) = (\det B)^m$, it then follows that $\det(A \otimes B) = (\det A)^n (\det B)^m$, and therefore also $\det(A \otimes B) = \det(B \otimes A)$.

(2) From $(A \otimes B) = [a_{ij}B]$ we obtain

$$\mathrm{trace}(A \otimes B) = \sum_{i=1}^{m}\sum_{j=1}^{n} a_{ii}b_{jj} = \left(\sum_{i=1}^{m} a_{ii}\right)\left(\sum_{j=1}^{n} b_{jj}\right) = \mathrm{trace}(A)\,\mathrm{trace}(B)$$

$$= \mathrm{trace}(B)\,\mathrm{trace}(A) = \mathrm{trace}(B \otimes A).$$

(3) Exercise. □

For a matrix $A = [a_1, \ldots, a_n] \in K^{m,n}$ with columns $a_j \in K^{m,1}$, $j = 1, \ldots, n$, we define

$$\mathrm{vec}(A) := \begin{bmatrix} a_1 \\ a_2 \\ \vdots \\ a_n \end{bmatrix} \in K^{mn,1}.$$

The application of vec turns the matrix A into a "column vector" and thus "vectorizes" A.

Lemma 20.6 *The map* $\mathrm{vec} : K^{m,n} \to K^{mn,1}$ *is an isomorphism. In particular, $A_1, \ldots, A_k \in K^{m,n}$ are linearly independent if and only if $\mathrm{vec}(A_1), \ldots, \mathrm{vec}(A_k) \in K^{mn,1}$ are linearly independent.*

Proof Exercise. □

We now consider the relationship between the Kronecker product and the vec map.

Theorem 20.7 *For $A \in K^{m,m}$, $B \in K^{n,n}$ and $C \in K^{m,n}$ we have*

$$\mathrm{vec}(ACB) = (B^T \otimes A)\mathrm{vec}(C).$$

Hence, in particular,

(1) $\mathrm{vec}(AC) = (I_n \otimes A)\mathrm{vec}(C)$ *and* $\mathrm{vec}(CB) = (B^T \otimes I_m)\mathrm{vec}(C)$,
(2) $\mathrm{vec}(AC + CB) = \left((I_n \otimes A) + (B^T \otimes I_m)\right)\mathrm{vec}(C)$.

Proof For $j = 1, \ldots, n$, the jth column of ACB is given by

$$(ACB)e_j = (AC)(Be_j) = \sum_{k=1}^{n} b_{kj}(AC)e_k = \sum_{k=1}^{n} (b_{kj}A)(Ce_k)$$

$$= [b_{1j}A, b_{2j}A, \ldots, b_{nj}A]\mathrm{vec}(C),$$

which implies that $\mathrm{vec}(ACB) = (B^T \otimes A)\mathrm{vec}(C)$. With $B = I_n$ resp. $A = I_m$ we obtain (1), while (1) and the linearity of vec yield (2). □

In order to study the relationship between the eigenvalues of the matrices A, B and those of the Kronecker product $A \otimes B$, we use *bivariate polynomials*, i.e., polynomials in two variables (cp. Exercise 9.16). If

$$p(t_1, t_2) = \sum_{i,j=0}^{l} \alpha_{ij} t_1^i t_2^j \in K[t_1, t_2]$$

is such a polynomial, then for $A \in K^{m,m}$ and $B \in K^{n,n}$ we define the matrix

$$p(A, B) := \sum_{i,j=0}^{l} \alpha_{ij} A^i \otimes B^j. \qquad (20.1)$$

Here we have to be careful with the order of the factors, since in general $A^i \otimes B^j \neq B^j \otimes A^i$ (cp. Exercise 20.2).

Example 20.8 For $A \in \mathbb{R}^{m,m}$, $B \in \mathbb{R}^{n,n}$ and $p(t_1, t_2) = 2t_1 + 3t_1 t_2^2 = 2t_1^1 t_2^0 + 3t_1^1 t_2^2 \in \mathbb{R}[t_1, t_2]$ we get the matrix $p(A, B) = 2A \otimes I_n + 3A \otimes B^2$.

The following result is known as *Stephanos' theorem*.[3]

Theorem 20.9 *Let $A \in K^{m,m}$ and $B \in K^{n,n}$ be two matrices that have Jordan canonical forms and the eigenvalues $\lambda_1, \ldots, \lambda_m \in K$ and $\mu_1, \ldots, \mu_n \in K$, respectively. If $p(A, B)$ is defined as in (20.1), then the following assertions hold:*

(1) The eigenvalues of $p(A, B)$ are $p(\lambda_k, \mu_\ell)$ for $k = 1, \ldots, m$ and $\ell = 1, \ldots, n$.
(2) The eigenvalues of $A \otimes B$ are $\lambda_k \cdot \mu_\ell$ for $k = 1, \ldots, m$ and $\ell = 1, \ldots, n$.
(3) The eigenvalues of $A \otimes I_n + I_m \otimes B$ are $\lambda_k + \mu_\ell$ for $k = 1, \ldots, m$ and $\ell = 1, \ldots, n$.

Proof Let $S \in GL_m(K)$ and $T \in GL_n(K)$ be such that $S^{-1}AS = J_A$ and $T^{-1}BT = J_B$ are in Jordan canonical form. The matrices J_A and J_B are upper triangular. Thus, for all $i, j \in \mathbb{N}_0$ the matrices J_A^i, J_B^j and $J_A^i \otimes J_B^j$ are upper triangular. The eigenvalues of J_A^i and J_B^j are $\lambda_1^i, \ldots, \lambda_m^i$ and $\mu_1^j, \ldots, \mu_n^j$, respectively. Thus, $p(\lambda_k, \mu_\ell)$, $k = 1, \ldots, m$, $\ell = 1, \ldots, n$, are the diagonal entries of the matrix $p(J_A, J_B)$. Using Lemma 20.3 we obtain

$$p(A, B) = \sum_{i,j=0}^{l} \alpha_{ij} (SJ_A S^{-1})^i \otimes (TJ_B T^{-1})^j = \sum_{i,j=0}^{l} \alpha_{ij} (SJ_A^i S^{-1}) \otimes (TJ_B^j T^{-1})$$

$$= \sum_{i,j=0}^{l} \alpha_{ij} \left((SJ_A^i) \otimes (TJ_B^j) \right) (S^{-1} \otimes T^{-1})$$

$$= \sum_{i,j=0}^{l} \alpha_{ij} (S \otimes T)(J_A^i \otimes J_B^j)(S \otimes T)^{-1}$$

---

[3] Named after Cyparissos Stephanos (1857–1917) who in 1900 showed besides this result also the assertion of Lemma 20.3.

$$= (S \otimes T) \left(\sum_{i,j=0}^{l} \alpha_{ij}(J_A^i \otimes J_B^j) \right) (S \otimes T)^{-1}$$

$$= (S \otimes T) p(J_A, J_B)(S \otimes T)^{-1},$$

which implies (1).

The assertions (2) and (3) follow from (1) with $p(t_1, t_2) = t_1 t_2$ and $p(t_1, t_2) = t_1 + t_2$, respectively. □

The following result on the matrix exponential function of a Kronecker product is helpful in applications that involve systems of linear differential equations.

Lemma 20.10 *For $A \in \mathbb{C}^{m,m}$, $B \in \mathbb{C}^{n,n}$ and $C := (A \otimes I_n) + (I_m \otimes B)$ we have*

$$\exp(C) = \exp(A) \otimes \exp(B).$$

Proof From Lemma 20.3 we know that the matrices $A \otimes I_n$ and $I_m \otimes B$ commute. Using Lemma 17.7 we obtain

$$\exp(C) = \exp(A \otimes I_n + I_m \otimes B) = \exp(A \otimes I_n) \exp(I_m \otimes B)$$

$$= \left(\sum_{j=0}^{\infty} \frac{1}{j!}(A \otimes I_n)^j \right) \left(\sum_{i=0}^{\infty} \frac{1}{i!}(I_m \otimes B)^i \right)$$

$$= \sum_{j=0}^{\infty} \frac{1}{j!} \sum_{i=0}^{\infty} \frac{1}{i!}(A \otimes I_n)^j (I_m \otimes B)^i$$

$$= \sum_{j=0}^{\infty} \frac{1}{j!} \sum_{i=0}^{\infty} \frac{1}{i!}(A^j \otimes B^i)$$

$$= \exp(A) \otimes \exp(B),$$

where we have used the properties of the matrix exponential series (cp. Sect. 17.1). □

For given matrices $A_j \in K^{m,m}$, $B_j \in K^{n,n}$, $j = 1, \ldots, q$, and $C \in K^{m,n}$ an equation of the form

$$A_1 X B_1 + A_2 X B_2 + \ldots + A_q X B_q = C \tag{20.2}$$

is called a *linear matrix equation* for the unknown matrix $X \in K^{m,n}$.

20 The Kronecker Product and Linear Matrix Equations

Theorem 20.11 *A matrix $\widehat{X} \in K^{m,n}$ solves (20.2) if and only if $\widehat{x} := \mathrm{vec}(\widehat{X}) \in K^{mn,1}$ solves the linear system of equations*

$$Gx = \mathrm{vec}(C), \quad \text{where} \quad G := \sum_{j=1}^{q} B_j^T \otimes A_j.$$

Proof Exercise. □

We now consider two special cases of (20.2).

Theorem 20.12 *For $A \in \mathbb{C}^{m,m}$, $B \in \mathbb{C}^{n,n}$ and $C \in \mathbb{C}^{m,n}$ the Sylvester equation*

$$AX + XB = C \tag{20.3}$$

has a unique solution if and only if A and $-B$ have no common eigenvalue. If all eigenvalues of A and B have negative real parts, then the unique solution of (20.3) is given by

$$\widehat{X} = -\int_0^\infty \exp(tA) C \exp(tB) dt.$$

(As in Sect. 17.2 the integral is defined entrywise.)

Proof Analogously to the representation in Theorem 20.11, we can write the Sylvester equation (20.3) as

$$(I_n \otimes A + B^T \otimes I_m) x = \mathrm{vec}(C).$$

If A and B have the eigenvalues $\lambda_1, \ldots, \lambda_m$ and $\mu_1, \ldots, \mu_n$, respectively, then $G = I_n \otimes A + B^T \otimes I_m$ by (3) in Theorem 20.9 has the eigenvalues $\lambda_k + \mu_\ell$, $k = 1, \ldots, m$, $\ell = 1, \ldots, n$. Thus, G is invertible, and the Sylvester equation is uniquely solvable, if and only if $\lambda_k + \mu_\ell \neq 0$ for all $k = 1, \ldots, m$ and $\ell = 1, \ldots, n$.

Let A and B be matrices with eigenvalues that have negative real parts. Then A and $-B$ have no common eigenvalues and (20.3) has a unique solution. Let $J_A = S^{-1}AS$ and $J_B = T^{-1}BT$ be Jordan canonical forms of A and B. We consider the linear differential equation

$$\frac{dZ}{dt} = AZ + ZB, \quad Z(0) = C, \tag{20.4}$$

that is solved by the function

$$Z : [0, \infty) \to \mathbb{C}^{m,n}, \quad Z(t) := \exp(tA) C \exp(tB)$$

(cp. Exercise 20.11). This function satisfies

$$\lim_{t\to\infty} Z(t) = \lim_{t\to\infty} \exp(tA) C \exp(tB)$$
$$= \lim_{t\to\infty} \underbrace{S \exp(tJ_A)}_{\to 0} \underbrace{S^{-1}CT}_{\text{constant}} \underbrace{\exp(tJ_B) T^{-1}}_{\to 0} = 0.$$

Integration of Eq. (20.4) from $t=0$ to $t=\infty$ yields

$$-C = -Z(0) = \lim_{t\to\infty}(Z(t)-Z(0)) = A\int_0^\infty Z(t)dt + \left(\int_0^\infty Z(t)dt\right) B.$$

(Here we use without proof the existence of the infinite integrals.) This implies that

$$\widehat{X} := -\int_0^\infty Z(t)dt = -\int_0^\infty \exp(tA) C \exp(tB) dt$$

is the unique solution of (20.3). □

Theorem 20.12 also gives the solution of another important matrix equation.

Corollary 20.13 *For $A, C \in \mathbb{C}^{n,n}$ the* Lyapunov equation[4]

$$AX + XA^H = -C \qquad (20.5)$$

has a unique solution $\widehat{X} \in \mathbb{C}^{n,n}$ if the eigenvalues of A have negative real parts. If, furthermore, C is Hermitian positive definite, then also $\widehat{X}$ is Hermitian positive definite.

Proof Since by assumption A and $-A^H$ have no common eigenvalues, the unique solvability of (20.5) follows from Theorem 20.12, and the solution is given by the matrix

$$\widehat{X} = -\int_0^\infty \exp(tA)(-C)\exp(tA^H) dt = \int_0^\infty \exp(tA) C \exp(tA^H) dt.$$

---

[4] Alexandr Mikhailovich Lyapunov (also Ljapunov or Liapunov; 1857–1918).

If C is Hermitian positive definite, then $\widehat{X}$ is Hermitian and for $x \in \mathbb{C}^{n,1} \setminus \{0\}$ we have

$$x^H \widehat{X} x = x^H \left(\int_0^\infty \exp(tA) C \exp\left(tA^H\right) dt \right) x$$

$$= \int_0^\infty \underbrace{x^H \exp(tA) C \exp\left(tA^H\right) x}_{>0} \, dt > 0.$$

The last inequality follows from the monotonicity of the integral and the fact that for $x \neq 0$ also $\exp(tA^H)x \neq 0$, since $\exp\left(tA^H\right)$ is invertible for every real t. □

Exercises

20.1 Prove Lemma 20.2.
20.2 Construct two square matrices A, B with $A \otimes B \neq B \otimes A$.
20.3 Prove Lemma 20.4.
20.4 Prove Theorem 20.5 (3).
20.5 Prove Lemma 20.6.
20.6 Let $A \in \mathbb{C}^{n,n}$. Show that $(I \otimes A)^k = I \otimes A^k$ and $(A \otimes I)^k = A^k \otimes I$ hold for all $k \in \mathbb{N}_0$.
20.7 Show that $A \otimes B$ is normal if $A \in \mathbb{C}^{m,m}$ and $B \in \mathbb{C}^{n,n}$ are normal. Is it true that if $A \otimes B$ is unitary, then A and B are unitary?
20.8 Use the singular value decompositions of $A = V_A \Sigma_A W_A^H \in \mathbb{C}^{m,m}$ and $B = V_B \Sigma_B W_B^H \in \mathbb{C}^{n,n}$ to derive the singular value decomposition of $A \otimes B$.
20.9 Show that for $A \in \mathbb{C}^{m,m}$ and $B \in \mathbb{C}^{n,n}$ and the matrix 2-norm, the equation $\|A \otimes B\|_2 = \|A\|_2 \|B\|_2$ holds.
20.10 Prove Theorem 20.11.
20.11 Let $A \in \mathbb{C}^{m,m}$, $B \in \mathbb{C}^{n,n}$ and $C \in \mathbb{C}^{m,n}$. Show that $Z(t) = \exp(tA) C \exp(tB)$ is the solution of the matrix differential equation $\frac{dZ}{dt} = AZ + ZB$ with the initial condition $Z(0) = C$.

A Short Introduction to MATLAB

MATLAB[1] is an interactive software system for numerical computations, simulations and visualizations. It contains a large number of predefined functions and allows users to implement their programs in so-called *m-files*.

The name MATLAB originates from *MATrix LABoratory*, which indicates the matrix orientation of the software. Indeed, matrices are the major objects in MATLAB. Due to the simple and intuitive use of matrices, we consider MATLAB well suited for teaching in the field of Linear Algebra.

In this short introduction we explain the most important ways to enter and operate with matrices in MATLAB. One can learn the essential matrix operations as well as important algorithms and concepts in the context of matrices (and Linear Algebra in general) by actively using the *MATLAB-Minutes* in this book. These only use predefined functions.

A matrix in MATLAB can be entered in form of a list of entries enclosed by square brackets. The entries in the list are ordered by rows in the natural order of the indices, i.e., from "top to bottom" and "left to right"). A new row starts after every semicolon. For example, the matrix

$$A = \begin{bmatrix} 1 & 2 & 3 \\ 4 & 5 & 6 \\ 7 & 8 & 9 \end{bmatrix} \quad \text{is entered in MATLAB by typing}$$

A = [1 2 3; 4 5 6; 7 8 9];

---

[1] MATLAB® is a registered trademark of The MathWorks, Inc.

A semicolon after the matrix A suppresses the output in MATLAB. If it is omitted then MATLAB writes out all the entered or computed quantities. For example, after entering

$$A = [1 \ 2 \ 3; 4 \ 5 \ 6; 7 \ 8 \ 9]$$

MATLAB gives the output

```
A =
    1   2   3
    4   5   6
    7   8   9
```

One can access parts of matrices by the corresponding indices. The list of indices from k to m is abbreviated by

$$k : m.$$

A colon : means all rows for given column indices, or all columns for given row indices. If A is as above, then for example

$$A(2, 1) \quad \text{is the matrix} \quad [4],$$
$$A(3, 1 : 2) \quad \text{is the matrix} \quad [7 \ 8],$$
$$A(:, 2 : 3) \quad \text{is the matrix} \quad \begin{bmatrix} 2 & 3 \\ 5 & 6 \\ 8 & 9 \end{bmatrix}.$$

There are several predefined functions that produce matrices. In particular, for given positive integers n and m,

$$\text{eye}(n) \qquad \text{the identity matrix } I_n,$$
$$\text{zeros}(n, m) \qquad \text{an } n \times m \text{ matrix with all zeros,}$$
$$\text{ones}(n, m) \qquad \text{an } n \times m \text{ matrix with all ones,}$$
$$\text{rand}(n, m) \qquad \text{an } n \times m \text{ "random matrix".}$$

Several matrices (of appropriate sizes) can be combined to a new matrix. For example, the commands

$$A = \text{eye}(2); \quad B = [4; 3]; \quad C = [2 \ -1]; \quad D = [-5]; \quad E = [A \ B; C \ D]$$

A   A Short Introduction to MATLAB

lead to

$$E = \begin{pmatrix} 1 & 0 & 4 \\ 0 & 1 & 3 \\ 2 & -1 & -5 \end{pmatrix}$$

The help function in MATLAB is started with the command `help`. In order to get information about specific functions one adds the name of the function. For example:

Input:	Information on:
`help ops`	operations and operators in MATLAB (in particular addition, multiplication, transposition)
`help matfun`	MATLAB functions that operate with matrices
`help gallery`	collection of example matrices
`help det`	determinant
`help expm`	matrix exponential function

Matrix Decompositions   B

The matrix-oriented approach in this book led us to numerous matrix decompositions that we collect in the following table:

Name/Theorem	Form	Properties of A and the factors
Echelon form (Theorem 5.2)	$A = SC$	$A \in K^{n,m}$, $S \in GL_n(K), C \in K^{n,m}$ in echelon form
LU-decomposition (Theorem 5.5)	$A = PLU$	$A \in K^{n,n}$, permutation matrix $P \in K^{n,n}$, lower triangular $L \in K^{n,n}$ with unit diagonal, upper triangular $U \in K^{n,n}$
Equivalence normal form (Theorem 5.12)	$A = Q \begin{bmatrix} I_r & 0 \\ 0 & 0 \end{bmatrix} Z$	$A \in K^{n,m}$ with $\operatorname{rank}(A) = r$, $Q \in GL_n(K), Z \in GL_m(K)$
QR-decomposition (Corollary 12.13)	$A = QR$	$A \in K^{n,m}$ with $K = \mathbb{R}$ or $K = \mathbb{C}$ and $\operatorname{rank}(A) = m$, $Q \in K^{n,m}$ with $Q^T Q = I_m$ or $Q^H Q = I_m$, upper triangular $R \in GL_m(K)$
Diagonalization (Theorem 14.14)	$A = SDS^{-1}$	$A \in K^{n,n}$ with $P_A = (t - \lambda_1) \cdot \ldots \cdot (t - \lambda_n)$ and $a(\lambda_j, A) = g(\lambda_j, A)$ for $j = 1, \ldots, n$, $S \in GL_n(K), D = \operatorname{diag}(\lambda_1, \ldots, \lambda_n)$
Triangulation (Theorem 14.18)	$A = SRS^{-1}$	$A \in K^{n,n}$ with $P_A = (t - \lambda_1) \cdot \ldots \cdot (t - \lambda_n)$, $S \in GL_n(K)$, upper triangular $R \in K^{n,n}$

Name/Theorem	Form	Properties of A and the factors
Unitary triangulation (Corollary 14.19)	$A = QRQ^H$	$A \in \mathbb{C}^{n,n}$, $Q \in \mathcal{U}(n)$, upper triangular $R \in \mathbb{C}^{n,n}$
Jordan canonical form (Theorem 16.10)	$A = SJS^{-1}$	$A \in K^{n,n}$ with $P_A = (t - \lambda_1) \cdot \ldots \cdot (t - \lambda_n)$, $S \in GL_n(K)$, $J \in K^{n,n}$ in Jordan canonical form
Frobenius canonical form (Theorem 16.22)	$A = SFS^{-1}$	$A \in K^{n,n}$, $S \in GL_n(K)$, $F \in K^{n,n}$ in Frobenius can. form
Unitary diagonalization (Corollary 18.3)	$A = QDQ^H$	$A \in \mathbb{C}^{n,n}$ normal, $Q \in \mathcal{U}(n)$, diagonal $D \in \mathbb{C}^{n,n}$
Real Schur form (Theorem 18.5)	$A = URU^T$	$A \in \mathbb{R}^{n,n}$, $U \in \mathcal{O}(n)$, block upper triangular R with 1×1 or 2×2 diagonal blocks
Principal axes transformation (Corollary 18.18)	$A = UDU^T$	$A = A^T \in \mathbb{R}^{n,n}$, $U \in \mathcal{O}(n)$, diagonal $D \in \mathbb{R}^{n,n}$
Lemma 18.22	$A = SDS^T$	$A = A^T \in \mathbb{R}^{n,n}$ with inertia (n_+, n_-, n_0), $S \in GL_n(\mathbb{R})$, $D = \mathrm{diag}(I_{n_+}, -I_{n_-}, 0_{n_0})$
Cholesky decomposition (Theorem 18.26)	$A = LL^H$	$A \in K^{n,n}$ with $K = \mathbb{R}$ or $K = \mathbb{C}$ symmetric resp. Hermitian positive definite, lower triangular $L \in GL_n(K)$
Singular value decomposition (Theorem 19.1)	$A = V\Sigma W^H$	$A \in \mathbb{C}^{n,m}$, $n \geq m$, with $\mathrm{rank}(A) = r$, $V \in \mathcal{U}(n)$, $W \in \mathcal{U}(m)$, $\Sigma = \begin{bmatrix} \Sigma_r & 0 \\ 0 & 0 \end{bmatrix}$ with $\Sigma_r = \mathrm{diag}(\sigma_1, \ldots, \sigma_r)$ and $\sigma_1 \geq \cdots \geq \sigma_r > 0$
Polar decomposition (follows from Theorem 19.1)	$A = UP$	$A \in \mathbb{C}^{n,m}$, $n \geq m$, $U \in \mathbb{C}^{n,m}$ with $U^H U = I_m$, Hermitian positive semidefinite $P \in \mathbb{C}^{m,m}$

The Greek Alphabet

	Lowercase	Uppercase
Alpha	α	A
Beta	β	B
Gamma	γ	Γ
Delta	δ	Δ
Epsilon	ϵ or ε	E
Zeta	ζ	Z
Eta	η	H
Theta	θ or ϑ	Θ
Iota	ι	I
Kappa	κ	K
Lambda	λ	Λ
My	μ	M
Ny	ν	N
Xi	ξ	Ξ
Omikron	o	O
Pi	π or ϖ	Π
Rho	ρ or ϱ	P
Sigma	σ or ς	Σ
Tau	τ	T
Ypsilon	υ	Υ
Phi	ϕ or φ	Φ
Chi	χ	X
Psi	ψ	Ψ
Omega	ω	Ω

Selected Historical Works on Linear Algebra

(We describe the content of these works using modern terms.)

- A. L. CAUCHY, *Sur l'équation à l'aide de laquelle on détermine les inégalités séculaires des mouvements des planètes*, Exercises de Mathématiques, 4 (1829).
  Proves that real symmetric matrices have real eigenvalues.
- H. GRASSMANN, *Die lineale Ausdehnungslehre, ein neuer Zweig der Mathematik*, Otto Wiegand, Leipzig, 1844.
  Contains the first development of abstract vector spaces and linear independence, including the dimension formula for subspaces.
- J. J. SYLVESTER, *Additions to the articles in the September Number of this Journal, "On a new Class of Theorems," and on Pascal's Theorem*, Philosophical Magazine, 37 (1850), pp. 363–370.
  Introduces the terms matrix and minor.
- J. J. SYLVESTER, *A demonstration of the theorem that every homogeneous quadratic polynomial is reducible by real orthogonal substitutions to the form of a sum of positive and negative squares*, Philosophical Magazine, 4 (1852), pp. 138–142.
  Proof of Sylvester's law of inertia.
- A. CAYLEY, *A memoir on the theory of matrices*, Proc. Royal Soc. of London, 148 (1858), pp. 17–37.
  First presentation of matrices as independent algebraic objects, including the basic matrix operations, the Cayley-Hamilton theorem (without a general proof) and the idea of a matrix square root.
- K. WEIERSTRASS, *Zur Theorie der bilinearen und quadratischen Formen*, Monatsber. Königl. Preußischen Akad. Wiss. Berlin, (1868), pp. 311–338.
  Proof of the Weierstrass canonical form, which implies the Jordan canonical form.
- C. JORDAN, *Traité des substitutions et des équations algébriques*, Paris, 1870.
  Contains the proof of the Jordan canonical form independent of Weierstrass' work.
- G. FROBENIUS, *Ueber lineare Substitutionen und bilineare Formen*, J. reine angew. Math., 84 (1878), pp. 1–63.
  Contains the concept of the minimal polynomial, the (arguably) first complete proof of the Cayley-Hamilton theorem, and results on equivalence, similarity and congruence of matrices (or bilinear forms).

- G. FROBENIUS, *Ueber homogene totale Differentialgleichungen*, J. reine angew. Math., 86 (1879), pp. 1–19.

  First definition of the concept of the rank.

- G. FROBENIUS, *Theorie der linearen Formen mit ganzen Coeffizienten*, J. reine angew. Math., 88 (1880), pp. 96–116.

  Contains a "rational" version of the theory of Weierstrass and a variant of the Frobenius canonical form we use today.

- G. PEANO, *Calcolo Geometrico secondo l'Ausdehnungslehre di H. Grassmann preceduto dalle operazioni della logica deduttiva*, Fratelli Bocca, Torino, 1888.

  Contains the first axiomatic definition of vector spaces, which Peano called "sistemi lineari", and studies properties of linear maps, including the (matrix) exponential function and the solution of differential equation systems.

- I. SCHUR, *Über die charakteristischen Wurzeln einer linearen Substitution mit einer Anwendung auf die Theorie der Integralgleichungen*, Math. Annalen, 66 (1909), pp. 488–510.

  Proof of the Schur form of complex matrices.

- O. TOEPLITZ, *Das algebraische Analogon zu einem Satze von Fejér*, Math. Zeitschrift, 2 (1918), pp. 187–197.

  Introduces the concept of a normal bilinear form and proves the equivalence of normality and unitary diagonalizability.

- F. D. MURNAGHAN AND A. WINTNER, *A canonical form for real matrices under orthogonal transformations*, Proc. Natl. Acad. Sci. U.S.A., 17 (1931), pp. 417–420.

  Proof of the real Schur form.

- C. ECKART AND G. YOUNG, *A principal axis transformation for non-Hermitian matrices*, Bull. Amer. Math. Soc., 45 (1939), pp. 118–121.

  Proof of the modern form of the singular value decomposition of a general complex matrix.

References

1. Bryan, K., Leise, T.: The $25,000,000,000 eigenvector: the linear algebra behind Google. SIAM Rev. **48**, 569–581 (2006)
2. Derksen, H.: The fundamental theorem of algebra and linear algebra. Am. Math. Mon. **110**, 620–623 (2003)
3. Ebbinghaus, H.-D., et al.: Numbers. Springer, New York (1991)
4. Estrada, E., Higham, D.J.: Network properties revealed through matrix functions. SIAM Rev. **52**, 696–714 (2010)
5. Higham, N.J.: Functions of Matrices: Theory and Computation. SIAM, Philadelphia (2008)
6. Horn, R.A., Johnson, C.R.: Topics in Matrix Analysis. Cambridge University Press, Cambridge (1991)
7. Horn, R.A., Johnson, C.R.: Matrix Analysis, 2nd edn. Cambridge University Press, Cambridge (2012)
8. Horn, R.A., Olkin, I.: When does $A^*A = B^*B$ and why does one want to know? Am. Math. Mon. **103**, 470–482 (1996)
9. Lancaster, P., Tismenetsky, M.: The Theory of Matrices: With Applications, 2nd edn. Academic Press, San Diego (1985)
10. Lewin, J.W.: A simple proof of Zorn's lemma. Am. Math. Mon. **98**, 353–354 (1991)
11. Loehr, N.: Advanced Linear Algebra. CRC Press, Boca Raton (2014)
12. Moler, C., Van Loan, C.: Nineteen dubious ways to compute the exponential of a matrix, twenty-five years later. SIAM Rev. **45**, 3–49 (2003)
13. Pták, V.: Eine Bemerkung zur Jordanschen Normalform von Matrizen. Acta Sci. Math. Szeged **17**, 190–194 (1956)
14. Shapiro, H.: A survey of canonical forms and invariants for unitary similarity. Linear Algebra. Appl. **147**, 101–167 (1991)

Index

A
Abuse of notation, 158
Adjacency matrix, 313
Adjoint, 222
　　Euclidean vector space, 225
　　unitary vector space, 227
Adjunct matrix, 100
Adjungate matrix, 101
Algebraic multiplicity, 238
Alternating, 98
Angle between vectors, 200
Annihilator, 189, 275
Anti-symmetric, 136, 191
Assertion, 10

B
Backward substitution, 55
Basis, 130
　　dual, 174
　　Hamel basis, 137
　　Schauder basis, 137
Basis extension theorem, 131
Bessel's identity, 208
Best approximation in a subspace, 212
Bidual space, 189
Bijective, 17
Bilinear form, 180
　　anti-symmetric, 191
　　non-degenerate, 180
　　positive definite, 347
　　skew-symmetric, 191
　　symmetric, 180
Binomial formula, 58
Bivariate polynomial, 147
Block matrix, 45
Block multiplication, 55

C
Canonical basis of $K^{n,m}$, 131
Canonical form, 22
Cartesian product, 20
Cauchy-Binet formula, 100
Cauchy-Schwarz inequality, 197
Cayley-Hamilton theorem, 115
Centralizer, 38
Characteristic polynomial
　　of an endomorphism, 238
　　of a matrix, 111
Chemical reaction, 316
Cholesky decomposition, 346
Circuit simulation, 7, 322
Codomain, 16
Column vector, 126
Commutative, 28
Commutative diagram, 163, 165, 182
Companion matrix, 113
Complex congruent matrices, 188
Complex numbers, 36
　　absolute value, 38
　　modulus, 38
Composition, 18
Congruent matrices, 185
Conjunction, 11
Contradiction, 25
Contraposition, 12
Coordinate map, 162
Coordinates, 137
Coordinate transformation matrix, 141, 163
Cosine theorem, 199
Cramer's rule, 105
Cross product, 213
Cycle, 107
Cyclic decomposition, 282
Cyclic subspace, 273

D

De Morgan laws, 25
Derivative of a polynomial, 168
Derogatory, 286
Determinant, 90
　alternating, 98
　computation via LU-decomposition, 100
　computational formulas, 96
　continuous, 91
　linear, 99
　linear form, 150
　multilinear form, 150
　multiplication theorem, 99
　normalized, 95
Diagonalizable, 240
Diagonal matrix, 53
Dimension formula
　for linear maps, 155
　for subspaces, 143
Dimension of a vector space, 134
Direct sum, 145, 250
Disjoint, 14
Disjunction, 11
Division with remainder, 254
Domain, 16
Dual basis, 174
Dual map, 177
Dual pair, 180
Dual space, 173
Duhamel integral, 321

E

Echelon form, 64
Eigenspace, 236
Eigenvalue
　algebraic multiplicity, 238
　of an endomorphism, 235
　geometric multiplicity, 236
　of a matrix, 116
Eigenvector
　of an endomorphism, 235
　of a matrix, 116
Elementary matrices, 61
Elementary row operations, 63
Empty list, 128
Empty map, 16
Empty product, 32
Empty set, 13
Empty sum, 31, 128
Endomorphism, 149
　derogatory, 286
　diagonalizable, 240
　direct sum, 250

nilpotent, 273
normal, 252, 325
orthogonal, 331
positive (semi-)definite, 344
selfadjoint, 231
simultaneous triangulation, 266
skew-adjoint, 347
triangulation, 245
unitarily diagonalizable, 252, 326
unitary, 331
unitary triangulation, 248
Equivalence, 11
Equivalence class, 22
Equivalence normal form, 78
Equivalence relation, 21
　canonical form, 22
　complex congruent matrices, 188
　congruent matrices, 185
　equivalent matrices, 77
　left equivalent matrices, 80
　normal form, 22
　similar matrices, 119
Equivalent matrices, 77, 164
Euclidean algorithm, 260
Euclidean theorem, 258
Evaluation homomorphism, 168
Exchange lemma, 132
Exchange theorem, 133
Extended coefficient matrix, 84

F

Fibonacci numbers, 244
Field, 35
Finite dimensional, 134
Fourier expansion, 208
Fréchet-Riesz isomorphism, 234
Frobenius canonical form, 291
Fundamental Theorem of Algebra, 261

G

Gaussian elimination algorithm, 63
Generalized eigenvector, 297
Generating system
　minimal, 147
Geometric multiplicity, 236
Givens rotation, 335
$GL_n(R)$, 53
Golden ratio, 244
Grade of a vector, 272
Gram-Schmidt method, 202
Graph, 312
Greatest common divisor, 259

Group, 27
    additive, 29
    homomorphism, 29
    multiplicative, 29
Group of units, 39

H

Hermite interpolation problem, 307
Hermitian, 185, 186
Hilbert matrix, 72, 80, 109
Homogeneous, 81, 318
Homomorphism, 149
Hooke's law, 323
Householder matrix, 218, 335

I

Idempotent, 39
Identity, 16
Identity matrix, 44
Image, 16, 152
Implication, 11
Index set, 14
Induction, 23
Inertia, 342
Initial value problem, 316
Injective, 17
Inner product, 194
Insurance premiums, 3, 49
Integral domain, 40
Interval, 14
Invariant factor, 294
Invariant subspace, 236
Inverse, 19
Inverse map, 19
Invertible, 19, 32, 52
Isometry, 218
Isomorphism, 149

J

Jordan block, 278
Jordan canonical form, 282
    algorithm for computing, 298
Jordan chain, 297

K

Kernel, 152
Klein, Felix (1849-1925), 35
Kronecker delta-function, 44
Kronecker product, 361
Krylov subspace, 272

L

Lagrange basis, 170
Lagrange interpolation problem, 171
Laplace expansion, 105
Least common multiple, 259
Least squares approximation, 7, 207, 358
Left adjoint, 222
Left ideal, 59
Left shift operator, 151
Linear, 3, 149
Linear matrix equation, 366
Linear factor, 238
Linear form, 173
Linear functional, 173
Linearly independent, 129
    maximal, 136
Linear map, 149
    change of bases, 164
    dual, 177
    matrix representation, 160
    rank, 166
    transpose, 179
Linear optimization problem, 6
Linear regression, 7, 205, 358
Linear span, 127
Linear system, 81
    homogeneous, 81
    non-homogeneous, 81
    solution algorithm, 86
    solution set, 81, 154
Logical values, 11
Low rank approximation, 355
Low-rank factorization, 76
LU-decomposition, 69
Lyapunov equation, 368

M

Map, 16
MATLAB-Minute, 49, 56, 68, 72, 100, 119,
    207, 249, 266, 295, 312, 336, 355
Matrix, 43
    block, 45
    column-stochastic, 120
    complex symmetric, 232
    derogatory, 286
    diagonal, 53
    diagonal entries, 44
    diagonalizable, 240
    diagonally dominant, 104
    empty, 44
    Hermitian, 186
    Hermitian part, 348
    Hermitian transpose, 186

invertibility criteria, 103, 119
invertible, 52, 71, 79, 102
left inverse, 110
negative (semi-)definite, 344
nilpotent, 123
(non-)singular, 52
normal, 325
orthogonal, 204
positive, 121
positive (semi-)definite, 344
right inverse, 110
row-stochastic, 4
skew-Hermitian, 347
skew-Hermitian part, 348
skew-symmetric, 48
square, 44
symmetric, 48
transpose, 48
triangular, 53
triangulation, 247
unimodular, 110
unitarily diagonalizable, 327
unitary, 204
unitary triangulation, 248
zero divisor, 51, 78
Matrix exponential function, 311
Matrix function, 306
Matrix operations, 45
Matrix representation
  adjoint map, 231
  bilinear form, 181
  dual map, 178
  linear map, 160
  sesquilinear form, 187
Maximal vector, 289
Metric, 217
Minimal polynomial, 285
Minimal polynomial of a vector, 289
Minor, 100
Möbius transformation, 348
Moivre-Binet formula, 245
Monic, 113
Monoid, 143
Moore-Penrose inverse, 356
Multiplication theorem for determinants, 99

N

Negative (semi-)definite, 344
Network, 314
  centrality, 314
  communicability, 314
Neutral element, 27
Nilpotency index, 273

Nilpotent, 39, 123, 273
Non-homogeneous, 81, 318
Norm, 195
  p-, 196
  L^2-, 196
  ∞-, 196
  Euclidean, 195
  Frobenius, 195
  induced by a scalar product, 198
  maximum column sum, 197
  maximum row sum, 197
  unitarily invariant, 359
Normal, 252, 325
Normal form, 22
Normed space, 195
n-tuple, 20
Nullity, 155
Null space, 152
Null vector, 126

O

One-form, 173
Ordered pair, 20
Ordinary differential equation, 315
Orthogonal basis, 200
Orthogonal complement, 209
Orthogonal endomorphism, 331
Orthogonal matrix, 204
Orthogonal projection, 211, 233
Orthogonal vectors, 200
Orthonormal basis, 200

P

PageRank algorithm, 1, 120
Parallelogram identity, 216
Parseval's identity, 208
Partial order, 136
People
  Abel, Niels Henrik (1802–1829), 28
  Bessel, Friedrich Wilhelm (1784–1846), 208
  Bézout, Étienne (1730–1783), 257
  Binet, Jacques Philippe Marie (1786–1856), 100, 245
  Boole, George (1815–1864), 39
  Cantor, Georg (1845–1918), 9, 14, 20
  Cauchy, Augustin Louis (1789–1857), 100, 108, 198, 311
  Cayley, Arthur (1821–1895), 43, 115
  Cholesky, André-Louis (1875–1918), 346
  Collatz, Lothar (1910–1990), 1

Index

Cramer, Gabriel (1704–1752), 105
de Moivre, Abraham (1667–1754), 245
De Morgan, Augustus (1806–1871), 25
Dedekind, Julius Wilhelm Richard (1831–1916), 35
Descartes, René (1596–1650), 20
Dodgson, Charles Lutwidge (1832–1898), 43
Duhamel, Jean-Marie Constant (1797–1872), 321
Eckart, Carl Henry (1902–1973), 352
Euclid of Alexandria (approx. 300 BC), 193
Fibonacci, Leonardo (ca. 1170–1240), 244
Fourier, Jean Baptiste Joseph (1768–1830), 208
Frobenius, Ferdinand Georg (1849–1917), 74, 115, 195, 271
Fréchet, Maurice René (1878–1973), 234
Gauß, Carl Friedrich (1777–1855), 24, 63, 261
Givens, Wallace (1910–1993), 335
Gram, Jørgen Pedersen (1850–1916), 202
Graßmann, Hermann Günther (1809–1877), 27, 133
Hamel, Georg (1877–1954), 137
Hamilton, Sir William Rowan (1805–1865), 48, 115, 126
Haynsworth, Emilie Virginia (1916–1985), 349
Hermite, Charles (1822–1901), 64, 185, 307
Higham, Nick (1961–2024), 305
Hilbert, David (1862–1943), ix, 27, 72
Hooke, Sir Robert (1635–1703), 323
Householder, Alston Scott (1904–1993), 218
Jordan, Marie Ennemond Camille (1838–1922), 271, 282, 352
Kirchhoff, Gustav Robert (1824–1887), x, 7
Klein, Felix (1849–1925), 35
Kronecker, Leopold (1823–1891), 35, 44, 361
Krylov, Aleksey Nikolaevich (1863–1945), 272
Lagrange, Joseph-Louis (1736–1813), 170
Laplace, Pierre-Simon (1749–1827), 104
Leibniz, Gottfried Wilhelm (1646–1716), 90
Lyapunov, Alexandr Mikhailovich (1857–1918), 368
Möbius, August Ferdinand (1790–1868), 348
Moore, Eliakim Hastings (1862–1932), 35, 356
Parseval, Marc-Antoine (1755–1836), 208
Peano, Giuseppe (1858–1932), 14, 23
Penrose, Sir Roger (b. 1931), 356
Perron, Oskar (1880–1975), 122
Pták, Vlastimil (1925–1999), 271
Pythagoras of Samos (approx. 570-495 BC), 199
Riesz, Frigyes (1880–1956), 234
Ruffini, Paolo (1765–1822), 255
Sarrus, Pierre Frédéric (1798–1861), 91
Schauder, Juliusz (1899–1943), 137
Schmidt, Erhard (1876–1959), 202, 352
Schur, Issai (1875–1941), 248, 349
Schwarz, Hermann Amandus (1843–1921), 198
Steinitz, Ernst (1871–1928), 133
Sylvester, James Joseph (1814–1897), 43, 100, 115, 155, 343, 352, 361
Toeplitz, Otto (1881–1940), 287, 325
Vandermonde, Alexandre-Théophile (1735–1796), 109
Weber, Heinrich Martin Georg (1842–1913), 35
Weierstraß, Karl Theodor Wilhelm (1815–1897), 32, 282
Wilkinson, James Hardy (1919–1986), 108
Young, Gale J. (1912–1990), 352
Zehfuss, Johann Georg (1832–1901), 361
Zorn, Max (1906–1993), 136
Permutation, 89
   associated permutation matrix, 95
   inversion, 90
Permutation matrix, 56
Perron eigenvector, 122
Pivot positions, 72
Polar decomposition, 353
Polarization identity, 191
Polynomial, 34
   common divisor, 259
   common multiple, 259
   common root, 268
   constant, 253
   coprime, 254
   degree, 34, 113, 253
   divisor, 254
   greatest common divisor, 259
   irreducible, 254
   least common multiple, 259
   monic, 113
   multiplicity of a root, 256
Positive definite, 193, 344, 347
Positive semidefinite, 344

Power set, 15
Predicate, 10
Pre-image, 16, 152
Principal axes transformation, 338
Principal vector, 297
Projection, 210, 233, 251
Proof by contraposition, 12
Pseudoinverse, 356
Pythagorean theorem, 199
    generalized, 201

Q
QR-decomposition, 203
Quadratic form, 190, 338
Quadric, 340
Quantifiers, 12
Quotient field, 41
Quotient set, 22

R
Rank, 74, 166
Rank factorization, 76
Rank-nullity theorem, 155
Rational canonical form, 294
Rational functions, 41
Reflection matrix, 335
Reflexive, 21
Relation, 20
Residue class, 23
Restriction, 16
Right adjoint, 222
Right-hand rule, 214
Right ideal, 59
Right shift operator, 151
Ring, 30
    Boolean, 39
    cancellation law, 40
    of matrices, 51
    multiplicative inverse, 32
    of polynomials, 34
Ring homomorphism, 58
Root of a polynomial, 117
    simple, 303
Rotation matrix, 335
Row vector, 126

S
Sarrus rule, 91
Scalar product, 193
Schur complement, 59, 349

Schur form
    of an endomorphism, 248
    of a matrix, 248
    real, 328
Schur's theorem, 248
Selfadjoint, 231
Sesquilinear form, 185
    non-degenerate, 185
Set, 9
    cardinality, 15
    difference, 14
    intersection, 14
    symmetric difference, 39
    union, 14
Sherman-Morrison-Woodbury formula, 59
Sign, 90
Signature, 90
Signature formula of Leibniz, 90
Similar matrices, 119, 165
Singular value decomposition (SVD), 352
Skew-adjoint, 347
Skew-Hermitian, 347
Skew-symmetric, 48, 191
Spectral mapping theorem, 306
Standard basis of $K^{n,m}$, 131
Standard scalar product of $\mathbb{C}^{n,1}$, 194
Standard scalar product of $\mathbb{R}^{n,1}$, 194
Stephanos' theorem, 365
Subfield, 36
Subgroup, 29
Subring, 32
Subset, 12
Subspace, 127
    complement, 148
    invariant, 236
Superset, 12
Surjective, 17
Sylvester equation, 367
Sylvester's law of inertia, 343
Symmetric, 21, 48, 180
Symmetric group, 90
System of linear differential equations, 317

T
Tensor product, 361
Toeplitz matrix, 287
Trace, 113, 124, 194
Transitive, 21
Transposition, 48, 92
Triangle inequality, 195
Triangular matrix, 53
Triangulation, 245

U

Unimodular, 110
Unitarily diagonalizable, 252, 326
Unitary endomorphism, 331
Unitary matrix, 204
Unitary triangulation, 248
Unit circle, 196
Unit vectors, 131

V

Vandermonde matrix, 109
Vec map, 364
Vector product, 213
Vector space, 125
    of bilinear forms, 189
    of continuous functions, 126
    Euclidean, 193
    of homomorphisms, 151
    of matrices, 126
    of polynomials, 126
    unitary, 193

W

Wilkinson matrix, 100, 108

Z

Zero divisor, 32, 51, 78
Zero matrix, 44
Zero ring, 31
Zero vector space, 128
Zorn's lemma, 136

MIX
Papier aus verantwortungsvollen Quellen
Paper from responsible sources
FSC® C105338

If you have any concerns about our products,
you can contact us on
ProductSafety@springernature.com

In case Publisher is established outside the EU,
the EU authorized representative is:
**Springer Nature Customer Service Center GmbH
Europaplatz 3, 69115 Heidelberg, Germany**

Printed by Libri Plureos GmbH
in Hamburg, Germany